BARRON'S

AP STATISTICS

TENTH EDITION

Martin Sternstein, Ph.D.
Professor Emeritus
Department of Mathematics
Ithaca College
Ithaca, New York

BARRON'S

About the Author

Dr. Martin Sternstein is a long-time College Board consultant and has been a Reader and Table Leader for the AP Statistics exam for many years. He has strong interests in national educational and social issues concerning equal access to math education for all. For two years, he was a Fulbright Professor in Liberia, West Africa, after which he developed a popular "Math in Africa" course, and he is the only mathematician to have given a presentation at the annual Conference on African Linguistics. He also taught the first U.S. course for college credit in chess theory. Dr. Sternstein and his wife live in a house they designed and built with their own hands with the help of many of the math faculty.

Author's Note

Thanks to Linda Turner, senior editor at Barron's, for her guidance. Thanks to my brother, Allan, my sister-in-law, Marilyn, my sons, Jonathan and Jeremy, my daughters-in-law, Cheryl and Asia, and my grandchildren, Jaiden, Jordan, Josiah, Luna, and Jayme, for their heartfelt love and support. Most thanks of all are due to my wife, Faith, whose love, warm encouragement, and always calm and optimistic perspective on life provide a home environment in which deadlines can be met and goals easily achieved. My sincere appreciation goes to the participants who have attended my AP Statistics workshops for teaching me just as much as I have taught them, and special thanks for their many useful suggestions are due to the following exceptional teachers:

David Bock	Ithaca High School, Ithaca, NY (retired)
Paul Buckley	Gonzaga College High School, Washington, DC
Dawn Burgess	Mount Desert Island High School, Bar Harbor, ME
Jim Conklin	Ithaca College, Ithaca, NY
Dawn Dentato	Somers High School, Lincolndale, NY
Jared Derksen	Rancho Cucamonga High School, Rancho Cucamonga, CA
Sarah Johnson	Grand Blanc High School, Grand Blanc, MI
Lee Kucera	University of California, Irvine, CA
Elizabeth Milligan	Cape Cod Community College, West Barnstable, MA
Brendan Murphy	John Bapst Memorial High School, Bangor, ME
Kenn Pendleton	Montgomery College, Germantown, MD
Roy St. Laurent	Northern Arizona University, Flagstaff, AZ
Daren Starnes	The Lawrenceville School, Lawrenceville, NJ
Josh Tabor	Canyon del Oro High School, Oro Valley, Arizona
Jane Viau	Frederick Douglass Academy, New York, NY
Dawn White	Silver Lake Regional High School, Kingston, MA
Melody Young	Sherman High School, Sherman, TX

Ithaca College
Spring 2019

Martin Sternstein

© Copyright 2019, 2017, 2015, 2013, 2012, 2010, 2007 by Kaplan, Inc., d/b/a Barron's Educational Series
© Copyright 2004, 2000, 1998 by Kaplan, Inc., d/b/a Barron's Educational Series, under the title *How to Prepare for the AP Advanced Placement Exam in Statistics*.

Published by Kaplan, Inc.,
d/b/a Barron's Educational Series
750 Third Avenue
New York, NY 10017
www.barronseduc.com

ISBN: 978-1-4380-1169-1

10 9 8 7 6 5 4 3 2 1

Kaplan, Inc., d/b/a Barron's Educational Series print books are available at special quantity discounts to use for sales promotions, employee premiums, or educational purposes. For more information or to purchase books, please call the Simon & Schuster special sales department at 866-506-1949.

Contents

THEME ONE: EXPLORATORY ANALYSIS

THEME FOUR: STATISTICAL INFERENCE

THEME FIVE: MORE PRACTICE AND REVIEW

> **This book comes with online access to three additional full-length practice tests. These online tests can be accessed at the following:**
>
> *http://bit.ly/Barrons-AP-Statistics*
>
> **Be sure to have your copy of *AP Statistics*, 10th edition on hand to complete the registration process,**

As you review the content in this book and work toward earning that **5** on your AP STATISTICS exam, here are five things that you **MUST** know:

Barron's Essential **5**

1 **Graders want to give you credit—help them!** Make them understand *what* you are doing, *why* you are doing it, and *how* you are doing it. Don't make the reader guess at what you are doing.
- **Communication** is just as important as statistical knowledge!
- Be sure you understand **exactly what you are being asked to do or find or explain**.
- *Naked* or *bald answers* will receive little or **no** credit! You must show where answers come from.
- On the other hand, don't give more than one solution to the same problem—you will receive credit only for the weaker one.

2 **Random sampling and random assignment are different ideas!**
- Random sampling is the use of chance in selecting a sample from a population and is critical in being able to generalize from a sample to a population.
  - A *simple random sample* (SRS) is when every possible sample of a given size has the same chance of being selected.
  - A *stratified random sample* is when the population is divided into homogeneous units called strata, and random samples are chosen from each strata.
  - A *cluster sample* is when the population is divided into heterogeneous units called clusters, and a random sample of the clusters is chosen.
- Random assignment in experiments is when subjects are randomly assigned to treatments.
  - This randomization evens out effects over which we have no control and allows for a valid comparison of treatments.
  - *Randomized block design* refers to when the randomization occurs only within groups of similar experimental units called blocks.

3 **Distributions describe variability!** Understand the difference between:
- a *population distribution* (variability in an entire population),
- a *sample distribution* (variability within a particular sample), and
- a *sampling distribution* (variability between samples).
- The larger the sample size, the more the **sample distribution** looks like the population distribution.
- Central Limit Theorem: the larger the sample size, the more the **sampling distribution** (probability distribution of the sample means) looks like a normal distribution.

4 **Check assumptions!**
- Be sure the assumptions to be checked are stated correctly, but **don't just state them!**
- Verifying assumptions and conditions means more than simply listing them with little check marks—you must show work or give some reason to confirm verification.
- If you refer to a graph, whether it is a histogram, boxplot, stemplot, scatterplot, residuals plot, normal probability plot, or some other kind of graph, you should **roughly draw it.** It is not enough to simply say, "I did a normal probability plot of the residuals on my calculator and it looked linear."

5 **Calculating the *P*-value is not the final step of a hypothesis test!**
- There must be a *decision* to reject or fail to reject the null hypothesis.
- You must indicate how you interpret the *P*-value, that is, you need *linkage*. So, "Given that $P = 0.007$, I reject ..." isn't enough. You need something like, "Because $P = 0.007$ is less than 0.05, there is sufficient evidence to reject ..."
- Finally, you need a conclusion *in the context* of the problem.

Introduction

In 1997, 7,667 students took the AP Statistics exam. As enrollment in AP Statistics classes increased at a higher rate than in any other AP class, more than 222,000 students took the exam in 2018. The number of students required to take statistics in college has surpassed the number of students required to take calculus. High schools across the country have recognized this trend and are developing and expanding their statistics offerings. The new Common Core mathematics standards feature statistics and probability in a primary role throughout the high school curriculum. This Barron's book is intended both as a topical review during the year and for final review in the weeks before the AP exam. Step-by-step solutions with detailed explanations are provided for the many illustrative examples and practice problems as well as for the six practice tests, including a diagnostic test.

The contents of this book cover the topics recommended by the AP Statistics Development Committee. A review of each of 13 topics is followed by multiple-choice and free-response questions on that topic. Detailed explanations are provided for all answers. It should be noted that some of the topic questions are not typical AP exam questions but rather are intended to help review the topic. There is a diagnostic exam, and there are five full-length practice exams, totaling 276 questions, all with instructive, complete answers. In addition, three full-length practice exams are available online.

Several points with regard to particular answers should be noted. First, step-by-step calculations using the given tables sometimes give minor differences from calculator answers due to round-off error. Second, calculator packages easily handle degrees of freedom that are not whole numbers, also resulting in minor answer differences. In the above cases, multiple-choice answers in this book have only one reasonable correct answer, and written explanations are necessary when answering free-response questions.

Students taking the AP Statistics exam will be furnished with a list of formulas (from descriptive statistics, probability, and inferential statistics) and tables (including standard normal probabilities, t-distribution critical values, and χ^2 critical values. While students will be expected to bring a graphing calculator with statistics capabilities to the exam, answers should not be in terms of calculator syntax. Furthermore, many students have commented that calculator usage was less than they had anticipated. However, even though the calculator is simply a tool, to be used sparingly, as needed, students should be proficient with this technology.

The exam will consist of two parts: a 90-minute section with 40 multiple-choice problems and a 90-minute free-response section with five open-ended questions and an investigative task to complete. In grading, the two sections of the exam will be given equal weight. Students have remarked that the first section involves "lots of reading," while the second section involves "lots of writing." The percentage of questions from each content area is approximately

25% data analysis, 15% experimental design, 25% probability, and 35% inference. Questions in both sections may involve reading generic computer output.

Note that in the multiple-choice section the questions are much more conceptual than computational, and thus use of the calculator is minimal. The score on the multiple-choice section is based on the number of correct answers, with no points deducted for incorrect answers. Blank answers are ignored.

In the free-response section, students must show all their work, and communication skills go hand in hand with statistical knowledge. Methods must be clearly indicated, as the problems will be graded on the correctness of the methods as well as on the accuracy of the results and explanation. That is, the free-response answers should address *why* a particular test was chosen, not just *how* the test is performed. Even if a statistical test is performed on a calculator such as the TI-84, Casio Prizm, or HP Prime, formulas need to be understood. Choice of test, in inference, must include confirmation of underlying assumptions, and answers must be stated in context, not just as numbers.

Free-response questions are scored on a 0 to 4 scale with 1 point for a *minimal* response, 2 points for a *developing* response, 3 points for a *substantial* response, and 4 points for a *complete* response. Individual parts of these questions are scored as E for *essentially* correct, P for *partially* correct, and I for *incorrect*. Note that *essentially* correct does not mean *perfect*. Work is graded *holistically*, that is, a student's complete response is considered as a whole whenever scores do not fall precisely on an integral value on the 0 to 4 scale.

Each open-ended question counts 15% of the total free-response score and the investigative task counts 25% of the free-response score. The first open-ended question is typically the most straightforward, and after doing this one to build confidence, students might consider looking at the investigative task since it counts more. Each completed AP exam will receive a grade based on a 5-point scale, with 5 the highest score and 1 the lowest score. Most colleges and universities accept a grade of 3 or better for credit or advanced placement or both.

While a review book such as this can be extremely useful in helping prepare students for the AP exam (practice problems, practice more problems, and practice even more problems are the three strongest pieces of advice), nothing can substitute for a good high school teacher and a good textbook. This author personally recommends the following texts from among the many excellent books on the market: *Stats: Modeling the World* by Bock, Bullard, Velleman, and DeVeaux; *The Practice of Statistics* by Starnes, Tabor, Yates, and Moore; *Workshop Statistics: Discovery with Data* by Rossman and Chance; *Statistics: Learning from Data* (AP Edition) by Peck and Olsen; and *Statistics: The Art and Science of Learning from Data* by Agresti and Franklin.

Other wonderful sources of information are the College Board's websites: *www.collegeboard.org* for students and parents, and *www.apcentral.collegeboard.com* for teachers.

A good piece of advice is for the student from day one to develop critical practices (like checking assumptions and conditions), to acquire strong technical skills, and to always write clear and thorough, yet to the point, interpretations in context. Final answers to most problems should be not numbers, but rather sentences explaining and analyzing numerical results. To help develop skills and insights to tackle AP free response questions (which often choose contexts students haven't seen before), pick up newspapers and magazines and figure out how to apply what you are learning to better understand articles in print that reference numbers, graphs, and statistical studies.

The student who uses this Barron's review book should study the text and illustrative examples carefully and try to complete the practice quiz problems before referring to the solution keys. Simply reading the detailed explanations to the answers without first striving to work through the problems on one's own is not the best approach. There is an old adage: *Mathematics is not a spectator sport!* Teachers clearly may use this book with a class in many profitable ways. Ideally, each individual topic review, together with practice quiz problems, should be assigned after the topic has been covered in class. The full-length practice tests should be reserved for final review shortly before the AP exam.

Although high-stakes assessments heighten anxiety and can actually hinder learning, recent research has shown that frequent testing, done right, can be an effective way to learn. With this in mind, the 10th edition has been reorganized to include two practice quizzes at the end of each of the first 13 topics. To test your understanding and *to produce a better recall of concepts*, at the appropriate times, take the first quiz on a topic, use the given answers to carefully check it over, and when ready, take the next quiz on the same topic. You will develop confidence and a deeper understanding of the material!

Topic 14 helps prepare you for free-response Question 6, the last question on the exam, also called the *Investigative Task*. The purpose of this question is to test your ability to take what you've learned throughout the course and apply that knowledge in a novel way. Topic 15 gives a series of comprehensive review questions covering the whole AP Statistics curriculum and set in a context that aims to give an appreciation of the power of statistics to shed insights into some of society's most pressing issues.

The AP Statistics course is absolutely the most important course offered in the high school curriculum. If you work with your teacher and study hard, you will find this to be an enjoyable course, you will do well on the AP exam, and you will develop into a more thoughtful citizen of this world! The 10th edition of *Barron's AP Statistics* has been updated to adapt to the ongoing changes in topic emphasis and scoring guidelines of the AP exam.

> Remember, this book comes with online access to three additional full-length practice tests. These online tests can be accessed at the following:
>
> *http://bit.ly/Barrons-AP-Statistics*
>
> Be sure to have your copy of *AP Statistics*, 10th edition on hand to complete the registration process,

ANSWER SHEET
Diagnostic Test

1. Ⓐ Ⓑ Ⓒ Ⓓ Ⓔ
2. Ⓐ Ⓑ Ⓒ Ⓓ Ⓔ
3. Ⓐ Ⓑ Ⓒ Ⓓ Ⓔ
4. Ⓐ Ⓑ Ⓒ Ⓓ Ⓔ
5. Ⓐ Ⓑ Ⓒ Ⓓ Ⓔ
6. Ⓐ Ⓑ Ⓒ Ⓓ Ⓔ
7. Ⓐ Ⓑ Ⓒ Ⓓ Ⓔ
8. Ⓐ Ⓑ Ⓒ Ⓓ Ⓔ
9. Ⓐ Ⓑ Ⓒ Ⓓ Ⓔ
10. Ⓐ Ⓑ Ⓒ Ⓓ Ⓔ

11. Ⓐ Ⓑ Ⓒ Ⓓ Ⓔ
12. Ⓐ Ⓑ Ⓒ Ⓓ Ⓔ
13. Ⓐ Ⓑ Ⓒ Ⓓ Ⓔ
14. Ⓐ Ⓑ Ⓒ Ⓓ Ⓔ
15. Ⓐ Ⓑ Ⓒ Ⓓ Ⓔ
16. Ⓐ Ⓑ Ⓒ Ⓓ Ⓔ
17. Ⓐ Ⓑ Ⓒ Ⓓ Ⓔ
18. Ⓐ Ⓑ Ⓒ Ⓓ Ⓔ
19. Ⓐ Ⓑ Ⓒ Ⓓ Ⓔ
20. Ⓐ Ⓑ Ⓒ Ⓓ Ⓔ

21. Ⓐ Ⓑ Ⓒ Ⓓ Ⓔ
22. Ⓐ Ⓑ Ⓒ Ⓓ Ⓔ
23. Ⓐ Ⓑ Ⓒ Ⓓ Ⓔ
24. Ⓐ Ⓑ Ⓒ Ⓓ Ⓔ
25. Ⓐ Ⓑ Ⓒ Ⓓ Ⓔ
26. Ⓐ Ⓑ Ⓒ Ⓓ Ⓔ
27. Ⓐ Ⓑ Ⓒ Ⓓ Ⓔ
28. Ⓐ Ⓑ Ⓒ Ⓓ Ⓔ
29. Ⓐ Ⓑ Ⓒ Ⓓ Ⓔ
30. Ⓐ Ⓑ Ⓒ Ⓓ Ⓔ

31. Ⓐ Ⓑ Ⓒ Ⓓ Ⓔ
32. Ⓐ Ⓑ Ⓒ Ⓓ Ⓔ
33. Ⓐ Ⓑ Ⓒ Ⓓ Ⓔ
34. Ⓐ Ⓑ Ⓒ Ⓓ Ⓔ
35. Ⓐ Ⓑ Ⓒ Ⓓ Ⓔ
36. Ⓐ Ⓑ Ⓒ Ⓓ Ⓔ
37. Ⓐ Ⓑ Ⓒ Ⓓ Ⓔ
38. Ⓐ Ⓑ Ⓒ Ⓓ Ⓔ
39. Ⓐ Ⓑ Ⓒ Ⓓ Ⓔ
40. Ⓐ Ⓑ Ⓒ Ⓓ Ⓔ

Diagnostic Test

SECTION I

Questions 1–40

Spend 90 minutes on this part of the exam.

> **Directions:** The questions or incomplete statements that follow are each followed by five suggested answers or completions. Choose the response that best answers the question or completes the statement.

1. The statistician for a professional basketball team calculates the percentages of points scored through 3-point shots, 2-point shots, and foul shots over two seasons. They are summarized in the following segmented bar chart.

Which of the following is an *incorrect* conclusion?

(A) More points were scored through foul shots in the second season than in the first.

(B) In the first season, twice as many points were scored through 2-point shots than through foul shots.

(C) In the second season, the same number of points were scored through 3-point shots and through foul shots.

(D) In both seasons, the same proportion of total points were scored through 2-point shots.

(E) In the first season, a greater proportion of the points were scored through 3-point shots than in the second season.

2. Is there a linear relationship between calories and sodium content in beef hot dogs? A study of 20 beef hot dogs gives the following regression output:

```
Dependent variable is: Sodium

Predictor        Coef    SE Coef        T       P
Constant      -228.33      77.97    -2.93   0.009
Calories        4.0133     0.4922    8.15   0.000

S = 48.5799    R-Sq = 78.7%    R-Sq(adj) = 77.5%
```

Which of the following gives a 99% confidence interval for the slope of the regression line?

(A) $4.0133 \pm 2.861 \left(\dfrac{0.4922}{\sqrt{20}} \right)$

(B) $4.0133 \pm (2.861)(0.4922)$

(C) $4.0133 \pm (2.878)(0.4922)$

(D) $4.0133 \pm 2.861 \left(\dfrac{48.5799}{\sqrt{20}} \right)$

(E) $4.0133 \pm 2.878 \left(\dfrac{48.5799}{\sqrt{20}} \right)$

3. In tossing a fair coin, which of the following sequences is more likely to appear?

(A) HHHHH
(B) HTHTHT
(C) HTHHTTH
(D) TTHTHHTH
(E) All are equally likely.

4. An entomologist hypothesizes that the mean life expectancy of a particular species of insect is 12.5 days. Researchers believing that the true mean is less than 12.5 days plan a hypothesis test at the 5% significance level on a random sample of 50 of these insects. If the alternative hypothesis is correct, for which of the following values of μ will the power of the test be greatest?

(A) 9
(B) 11
(C) 12.5
(D) 14
(E) 17

5. A simple random sample is defined by

   (A) the method of selection
   (B) how representative the sample is of the population
   (C) whether or not a random number generator is used
   (D) the assignment of different numbers associated with the outcomes of some chance situation
   (E) examination of the outcome

6. Can shoe size be predicted from height? In a random sample of 50 adults, the standard deviation in heights was 8.7 cm, while the standard deviation in shoe size was 2.3. The least squares regression equation was:
   Predicted shoe size = –33.6 + 0.25 (*Height in cm*). What was the correlation coefficient?

   (A) $\dfrac{(0.25)(8.7)}{2.3}$

   (B) $\dfrac{(0.25)(2.3)}{8.7}$

   (C) $\dfrac{2.3}{8.7/\sqrt{50}}$

   (D) $\dfrac{8.7}{2.3/\sqrt{50}}$

   (E) There is not enough information to calculate the correlation coefficient.

Questions 7–9 refer to the following situation:

   A researcher would like to show that a new oral diabetes medication he developed helps control blood sugar level better than insulin injection. He plans to run a hypothesis test at the 5% significance level.

7. What would be a Type I error?

   (A) The researcher concludes he has sufficient evidence that his new medication helps more than insulin injection, and his medication really is better than insulin injection.
   (B) The researcher concludes he has sufficient evidence that his new medication helps more than insulin injection, when in reality his medication is not better than insulin injection.
   (C) The researcher concludes he does not have sufficient evidence that his new medication helps more than insulin injection, and his medication really is not better than insulin injection.
   (D) The researcher concludes he does not have sufficient evidence that his new medication helps more than insulin injection, when in reality his medication is better than insulin injection.
   (E) The researcher concludes he has sufficient evidence that his new medication controls blood sugar level the same as insulin injection, and in reality there is a difference.

8. What would be a Type II error?

   (A) The researcher concludes he has sufficient evidence that his new medication helps more than insulin injection, and his medication really is better than insulin injection.
   (B) The researcher concludes he has sufficient evidence that his new medication helps more than insulin injection, when in reality his medication is not better than insulin injection.
   (C) The researcher concludes he does not have sufficient evidence that his new medication helps more than insulin injection, and his medication really is not better than insulin injection.
   (D) The researcher concludes he does not have sufficient evidence that his new medication helps more than insulin injection, when in reality his medication is better than insulin injection.
   (E) The researcher concludes he has sufficient evidence that his new medication controls blood sugar level the same as insulin injection, and in reality there is a difference.

9. The researcher thinks he can improve his chances by running five such identical hypotheses tests, each using a different group of diabetic volunteers, hoping that at least one of the tests will show that his new oral diabetes medication helps control blood sugar level better than insulin injection. What is the probability of committing at least one Type I error?

   (A) 0.05
   (B) $5(0.05)(0.95)^4$
   (C) $1 - (0.95)^5$
   (D) $(0.95)^5$
   (E) 0.95

_____

10. A financial analyst determines the yearly research and development investments for 50 blue chip companies. She notes that the distribution is distinctly not bell-shaped. If the 50 dollar amounts are converted to z-scores, what can be said about the standard deviation of the 50 z-scores?

    (A) It depends on the distribution of the raw scores.
    (B) It is less than the standard deviation of the raw scores.
    (C) It is greater than the standard deviation of the raw scores.
    (D) It is equal to the standard deviation of the raw scores.
    (E) It equals 1.

11. Tossing a fair die has outcomes {1,2,3,4,5,6} with mean 3.5 and standard deviation 1.708. If a fair die is thrown three times, and the mean of the resulting triplet is calculated, the mean and standard deviation of the set of means from all possible such triplets is

   (A) $\bar{x} = 3.5$, $\sigma_{\bar{x}} = 0.569$
   (B) $\bar{x} = 3.5$, $\sigma_{\bar{x}} = 0.986$
   (C) $\bar{x} = 6.062$, $\sigma_{\bar{x}} = 1.208$
   (D) $\bar{x} = 10.5$, $\sigma_{\bar{x}} = 0.569$
   (E) $\bar{x} = 10.5$, $\sigma_{\bar{x}} = 0.986$

12. A 40-question multiple-choice statistics exam is graded as number correct minus ¼ number incorrect, so scores can take values from –10 to +40. Suppose the standard deviation for one class's results is reported to be –3.14. What is the proper conclusion?

   (A) More students received negative scores than positive scores.
   (B) At least half the class received negative scores.
   (C) Some students must have received negative scores.
   (D) Some students must have received positive scores.
   (E) An error was made in calculating the standard deviation.

13. Of the 423 seniors graduating this year from a city high school, 322 plan to go on to college. When the principal asks an AP student to calculate a 95% confidence interval for the proportion of this year's graduates who plan to go to college, the student says that this would be inappropriate. Why?

   (A) The independence assumption may have been violated (students tend to do what their friends do).
   (B) There is no evidence that the data come from a normal or nearly normal population (GPAs help determine college admission and may be skewed).
   (C) Randomization was not used.
   (D) There is a difference between a confidence interval and a hypothesis test with regard to the proportion of graduates planning on college.
   (E) Some other reason.

14. An AP Statistics student in a large high school plans to survey his fellow students with regard to their preference between using a laptop or using an iPad. Which of the following survey methods is unbiased?

   (A) The student comes to school early and surveys the first 50 students who arrive.
   (B) The student passes a survey card to every student with instructions to fill it out at home and drop the filled out card in a box by the school entrance the next day.
   (C) The student posts the survey on his Facebook page, asking everyone to respond.
   (D) The student goes to all of the high school sports events for a week, hands out the survey, and waits for each student to fill it out and hand it back.
   (E) All the above sampling methods are biased.

15. The mean combined SAT score for students in one state is 1758 with a standard deviation of 213, while for a second state the mean is 1725 with a standard deviation of 228. Assuming both distributions are approximately normal, what is the probability that a randomly selected student in the first state scores higher than a randomly selected student in the second state?

    (A) 0.458
    (B) 0.500
    (C) 0.524
    (D) 0.542
    (E) 0.559

16. In a random sample of 10 insects of a newly discovered species, an entomologist measures an average life expectancy of 17.3 days with a standard deviation of 2.3 days. Assuming all conditions for inference are met, what is a 95% confidence interval for the mean life expectancy for insects of this species?

    (A) $17.3 \pm 1.96\left(\dfrac{2.3}{\sqrt{9}}\right)$

    (B) $17.3 \pm 1.96\left(\dfrac{2.3}{\sqrt{10}}\right)$

    (C) $17.3 \pm 2.228\left(\dfrac{2.3}{\sqrt{9}}\right)$

    (D) $17.3 \pm 2.228\left(\dfrac{2.3}{\sqrt{10}}\right)$

    (E) $17.3 \pm 2.262\left(\dfrac{2.3}{\sqrt{10}}\right)$

17. A coin is weighted so that heads is twice as likely to occur as tails. The coin is flipped repeatedly until a tail occurs. Let X be the number of flips made. What is the most probable value for X?

    (A) 1
    (B) 2
    (C) 3
    (D) 4
    (E) 5

18. Suppose we are interested in determining whether or not a student's score on the AP Statistics exam is a reasonable predictor of the student's GPA in the first year of college. Which of the following is the best statistical test?

(A) Two-sample *t*-test of population means
(B) Linear regression *t*-test
(C) Chi-square test of independence
(D) Chi-square test of homogeneity
(E) Chi-square test of goodness of fit

19. Following are the graphs of three normal curves and three cumulative distribution graphs:

Which normal curve corresponds to which cumulative curve?

(A) *X*-1, *Y*-2, *Z*-3
(B) *X*-1, *Y*-3, *Z*-2
(C) *X*-2, *Y*-1, *Z*-3
(D) *X*-3, *Y*-1, *Z*-2
(E) *X*-3, *Y*-2, *Z*-1

20. A campus has 55% male students. Suppose 30% of the male students pick basketball as their favorite sport compared to 20% of the females. If a randomly chosen student picks basketball as the student's favorite sport, what is the probability the student is male?

(A) $\dfrac{0.30}{0.30+0.20}$

(B) $\dfrac{0.55}{0.30+0.20}$

(C) $\dfrac{0.30}{(0.55)(0.30)+(0.45)(0.20)}$

(D) $\dfrac{0.55}{(0.55)(0.30)+(0.45)(0.20)}$

(E) $\dfrac{(0.55)(0.30)}{(0.55)(0.30)+(0.45)(0.20)}$

21. The kelvin is a unit of measurement for temperature; 0 K is absolute zero, the temperature at which all thermal motion ceases. Conversion from Fahrenheit to Kelvin is given by $K = 5/9 \times (F - 32) + 273$. The average daily temperature in Monrovia, Liberia, is 78.35°F with a standard deviation of 6.3°F. If a scientist converts Monrovia daily temperatures to the Kelvin scale, what will be the new mean and standard deviation?

(A) Mean, 25.75 K; standard deviation, 3.5 K
(B) Mean, 231.75 K; standard deviation, 3.5 K
(C) Mean, 298.75 K; standard deviation, 3.5 K
(D) Mean, 298.75 K; standard deviation, 258.72 K
(E) Mean, 298.75 K; standard deviation, 276.5 K

22. A cattle veterinarian is considering two experimental designs to compare two sources of bovine growth hormone, or BVH, to spur increased milk production in Guernsey cattle. Design 1 involves flipping a coin as each cow enters the stockade, and if *heads*, giving it BVH from bovine cadavers, and if *tails*, giving it BVH from engineered *E. coli*. Design 2 involves flipping a coin as each cow enters the stockade, and if *heads*, giving it BVH from bovine cadavers for a specified period of time and then switching to BVH from engineered *E. coli* for the same period of time, and if *tails*, the order is reversed. With both designs, daily milk production is noted. Which of the following is accurate?

(A) Neither design uses randomization since there is no indication that cows will be randomly picked from the population of all Guernsey cattle.
(B) Design 1 is a *completely randomized design*, while Design 2 is a *block design*.
(C) Both designs use *double-blinding*, but neither uses a *placebo*.
(D) In the second design, BVH from bovine cadavers and BVH from engineered *E. coli* are *confounded*.
(E) One of the two designs is actually an *observational study*, while the other is an *experiment*.

23. The purpose of the linear regression *t*-test is

    (A) to determine if there is a linear association between two numerical variables
    (B) to find a confidence interval for the slope of a regression line
    (C) to find the *y*-intercept of a regression line
    (D) to be able to calculate residuals
    (E) to be able to determine the consequences of Type I and Type II errors

24. A fair die is tossed 12 times, and the number of 3's is noted. This is repeated 200 times. Which of the following distributions is the most likely to occur?

    (A)

    (B)

    (C)

    (D)

    (E)

25. Which of the following is a true statement about sampling?

    (A) If the sample is random, the size of the sample usually doesn't matter.
    (B) If the sample is random, the size of the population usually doesn't matter.
    (C) A sample of less than 1% of the population is too small for statistical inference.
    (D) A sample of more than 10% of the population is too large for statistical inference.
    (E) All of the above are true statements.

26. Suppose, in a study of mated pairs of soldier beetles, it is found that the measure of the elytron (hardened forewing) length is always 0.5 millimeters longer in the female. What is the correlation between elytron lengths of mated females and males?

    (A) −1
    (B) −0.5
    (C) 0
    (D) 0.5
    (E) 1

27. A random sample of 100 individuals who were singled out at an international airport security checkpoint is reviewed, and the individuals are classified according to region of origin:

Region of origin	United States	Europe	Arabic	Asia, non-Arabic	Other
Number singled out	41	19	15	13	12

The proportion of travelers in each category who use this airport follows:

Region of origin	United States	Europe	Arabic	Asia, non-Arabic	Other
Proportion	0.64	0.12	0.08	0.09	0.07

We wish to test whether the distribution of people singled out is the same as the distribution of people who use the airport with regard to region of origin. What is the appropriate χ^2 statistic?

(A) $\dfrac{(41-64)^2}{64} + \dfrac{(19-12)^2}{12} + \dfrac{(15-8)^2}{8} + \dfrac{(13-9)^2}{9} + \dfrac{(12-7)^2}{7}$

(B) $\dfrac{(41-64)^2}{41} + \dfrac{(19-12)^2}{19} + \dfrac{(15-8)^2}{15} + \dfrac{(13-9)^2}{13} + \dfrac{(12-7)^2}{12}$

(C) $\dfrac{(0.41-0.64)^2}{0.64} + \dfrac{(0.19-0.12)^2}{0.12} + \dfrac{(0.15-0.08)^2}{0.08} + \dfrac{(0.13-0.09)^2}{0.09} + \dfrac{(0.12-0.07)^2}{0.07}$

(D) $\dfrac{(0.41-0.64)^2}{0.41} + \dfrac{(0.19-0.12)^2}{0.19} + \dfrac{(0.15-0.08)^2}{0.15} + \dfrac{(0.13-0.09)^2}{0.13} + \dfrac{(0.12-0.07)^2}{0.12}$

(E) $\dfrac{(41-64)^2}{20} + \dfrac{(19-12)^2}{20} + \dfrac{(15-8)^2}{20} + \dfrac{(13-9)^2}{20} + \dfrac{(12-7)^2}{20}$

28. The age distribution for a particular debilitating disease has a mean greater than the median. Which of the following graphs most likely illustrates this distribution?

(A)

(B)

(C)

(D)

(E)

29. Suppose we have a random variable X where the probability associated with the value k is $\binom{10}{k}(0.38)^k(0.62)^{10-k}$ for $k = 0, \ldots, 10$.

What is the mean of X?

(A) 0.38
(B) 0.62
(C) 3.8
(D) 5.0
(E) 6.2

30. It is hypothesized that high school varsity pitchers throw fastballs at an average of 80 mph. A random sample of varsity pitchers are timed with radar guns resulting in a 95% confidence interval of (74.5, 80.5). Which of the following is a correct statement?

(A) There is a 95% chance that the mean fastball speed of all varsity pitchers is 80 mph.

(B) There is a 95% chance that the mean fastball speed of all varsity pitchers is 77.5 mph.

(C) Most of the interval is below 80, so there is evidence at the 5% significance level that the mean of all varsity pitchers is something different than 80 mph.

(D) The test $H_0: \mu = 80$, $H_a: \mu \neq 80$ is not significant at the 5% significance level, but it would be at the 1% level.

(E) It is likely that the true mean fastball speed of all varsity pitchers is within 3 mph of the sample mean fastball speed.

31. A recent study noted prices and battery lives of 10 top-selling tablet computers. The data follow:

	1	2	3	4	5	6	7	8	9	10
Cost	303	450	260	480	540	390	350	400	600	450
Battery life (hr)	8.5	10	7	11	10	9	8	9.5	11	9.5

The residual plot of the least squares model is

What is the model's predicted battery life for the tablet computer costing $480?

(A) 10 hr
(B) 10.5 hr
(C) 11 hr
(D) 11.5 hr
(E) 12 hr

32. Should college athletes be required to give their coaches their Facebook IDs and passwords? A survey of student-athletes is to be taken. The statistician believes that Division I, II, and III players may differ in their views, so she selects a random sample of athletes from each Division to survey. This is a

    (A) simple random sample
    (B) stratified sample
    (C) cluster sample
    (D) systematic sample
    (E) convenience sample

33. Which of the following use of a random number table would be appropriate to simulate tossing 3 fair coins and noting the number of heads?

    (A) Assign "0,1" to 0 heads, "2,3" to 1 head, "4,5" to 2 heads, "6,7" to 3 heads, and ignore "8,9."
    (B) Assign "0,1" to 0 heads, "2,3,4" to 1 head, "5,6,7" to 2 heads, and "8,9" to 3 heads.
    (C) Assign "0" to 0 heads, "1,2" to 1 head, "3,4" to 2 heads, "5" to 3 heads, and ignore "6,7,8,9."
    (D) Assign "0" to 0 heads, "1,2,3" to 1 head, "4,5,6" to 2 heads, "7" to 3 heads, and ignore "8,9."
    (E) Assign "0" to 0 heads, "1,2,3,4" to 1 head, "5,6,7,8" to 2 heads, and "9" to 3 heads.

34. The population of the Greater Tokyo area is 34,400,000 and of Karachi is 17,200,000. A random sample of citizens is to be taken in each city, and confidence intervals for the mean age in each city will be calculated. Assuming roughly equal sample standard deviations, to obtain the same margin of error for each confidence interval,

    (A) the sample sizes should be the same
    (B) the sample in Greater Tokyo should be twice the size of the sample in Karachi
    (C) the sample in Karachi should be twice the size of the sample in Greater Tokyo
    (D) the sample in Greater Tokyo should be four times the size of the sample in Karachi
    (E) the sample in Karachi should be four times the size of the sample in Greater Tokyo

35. The *midhinge* is defined to be the average of the first and third quartiles. If the midhinge is 20 and the interquartile range is also 20, what is the first quartile?

    (A) 0
    (B) 10
    (C) 20
    (D) 30
    (E) Impossible to determine from the given information

36. Which of the following is an *incorrect* statement?

(A) Statistics are random variables with their own probability distributions.
(B) The standard error does not depend on the size of the population.
(C) Bias means that, on average, our estimate of a parameter is different from the true value of the parameter.
(D) There are some statistics for which the sampling distribution is not approximately normal, no matter how large the sample size.
(E) The larger the sample size, the closer the sample distribution is to a normal distribution.

37. For male Air Force cadets, the recommended fitness level with regard to the number of push-ups is 34. In a test whether or not current classes of recruits can meet this standard, a t-test of H_0: $\mu = 34$ against H_a: $\mu < 34$ gives a P-value of 0.068. Using this data, among the following, which is the largest level of confidence for a two-sided confidence interval that does not contain 34?

(A) 85%
(B) 90%
(C) 92%
(D) 95%
(E) 96%

38. A particular car is tested for stopping distance in feet on wet pavement at 30 mph using tires with one tread design and then tires with another tread design. For each set of tires, the test is repeated 30 times, and the following parallel boxplots give a comparison of the resulting five-number summaries.

Which of the following is a reasonable conclusion?

(A) Distribution I is skewed right, while distribution II is bell-shaped.
(B) Distribution I is skewed left, while distribution II is a normal distribution.
(C) The mean of distribution I is greater than the mean of distribution II.
(D) The range of distribution I is approximately 46 – 33 = 13.
(E) The upper 50% of the values in distribution I are all greater than the lower 50% of the values in distribution II.

39. In American roulette there are 18 red pockets, 18 black pockets, and two green pockets (labeled 0 and 00). The ball is equally likely to land in any of the 38 pockets. What is the probability that a player ends up with a positive outcome, that is, makes money, after 50 equal bets on "red" (that is, for each of 50 spins of the wheel, the player wins or loses the specified identical dollar bet depending on whether or not the ball lands in a red or non-red pocket, respectively)?

(A) 0.105
(B) 0.212
(C) 0.303
(D) 0.408
(E) 0.500

40. Do middle school and high school students have different views on what makes someone popular? Random samples of 100 middle school and 100 high school students yield the following counts with regard to three choices: lots of money, good at sports, and handsome or pretty:

	Money	Sports	Looks
Middle school	22	48	30
High school	36	24	40

A chi-square test of homogeneity yields which of the following test statistics?

(A) $\dfrac{(22-29)^2}{22}+\dfrac{(48-36)^2}{48}+\dfrac{(30-35)^2}{30}+\dfrac{(36-29)^2}{36}+\dfrac{(24-36)^2}{24}+\dfrac{(40-35)^2}{40}$

(B) $\dfrac{(22-29)^2}{29}+\dfrac{(48-36)^2}{36}+\dfrac{(30-35)^2}{35}+\dfrac{(36-29)^2}{29}+\dfrac{(24-36)^2}{36}+\dfrac{(40-35)^2}{35}$

(C) $\left(\dfrac{22}{29}\right)^2+\left(\dfrac{48}{36}\right)^2+\left(\dfrac{30}{35}\right)^2+\left(\dfrac{36}{29}\right)^2+\left(\dfrac{24}{36}\right)^2+\left(\dfrac{40}{35}\right)^2$

(D) $\dfrac{(22)(29)}{58}+\dfrac{(48)(36)}{72}+\dfrac{(30)(35)}{70}+\dfrac{(36)(29)}{58}+\dfrac{(24)(36)}{72}+\dfrac{(40)(35)}{70}$

(E) $\sqrt{(22-29)^2+(48-36)^2+(30-35)^2+(36-29)^2+(24-36)^2+(40-35)^2}$

STOP

If there is still time remaining, you may review your answers.

Part A

QUESTIONS 1–5

Spend about 65 minutes on this part of the exam.
Percentage of Section II grade—75

> **Directions:** You must show all work and indicate the methods you use. You will be graded on the correctness of your methods and on the accuracy of your results and explanations.

1. A horticulturist plans a study on the use of compost tea for plant disease management. She obtains 16 identical beds, each containing a random selection of five mini-pink rose plants. She plans to use two different composting times (two and five days), two different compost preparations (aerobic and anaerobic), and two different spraying techniques (with and without adjuvants). Midway into the growing season she will check all plants for rose powdery mildew disease.

    (a) List the complete set of treatments.

    (b) Describe a completely randomized design for the treatments above.

    (c) Explain the advantage of using only mini-pink roses in this experiment.

    (d) Explain a disadvantage of using only mini-pink roses in this experiment.

2. A top-100, 7.0-rated tennis pro wishes to compare a new Wilson N1 racquet against his current model. He strings the new racquet with the same Luxilon strings at 60 pounds tension that he uses on his old racquet. From past testing he knows that the average forehand cross court volley with his old racquet is 82 miles per hour (mph). On an indoor court, using a ball machine set at 70 mph, the same speed he had his old racquet tested against, he takes 47 swings with the new racquet. An associate with a speed gun records an average of 83.5 mph with a standard deviation of 3.4 mph. Assuming that the 47 swings represents a random sample of his swings, is there statistical evidence that his speed with the new racquet is an improvement over the old? Justify your answer.

3. In October 2016, a comprehensive residential college in upstate New York reported undergraduate enrollment by ethnic/racial categories as follows: 2.7% non-Hispanic Black, 3.7% Asian or Pacific Islander, 4.0% Hispanic, 80.0% non-Hispanic White, and the rest other/unknown. While racial/ethnic status is not considered in the admissions process, an admissions counselor is interested in whether or not the makeup of the new freshman class will change, and plans to do a statistical analysis on an appropriately drawn simple random sample.

   (a) What statistical test/procedure should be used?

   (b) State the null and alternative hypotheses. Is the test appropriate for an intended sample size of 200?

   (c) If the admissions counselor performs the indicated test on the following data, is there statistical evidence of a change in ethnic/racial composition? Explain.

	Non-Hispanic Black	Asian or Pacific Islander	Hispanic	Non-Hispanic White	Other/ Unknown
Number of Students	3	4	14	150	29

   (d) Suppose the data was obtained by noting the racial/ethnic status of a simple random sample of 200 potential new students visiting the campus during fall 2016. Did the test/procedure target the intended population? Explain.

4. Concrete is made by mixing sand and pebbles with water and cement and then hardening through hydration. Different densities result from different proportions of the aggregates. Assume that concrete densities are normally distributed with mean 2317 kilograms per cubic meter and standard deviation 128 kilograms per cubic meter.

   (a) What is the probability that a given concrete density is over 2400 kg/m^3?

   (b) In a random sample of five independent concrete densities, what is the probability that a majority have densities over 2400 kg/m^3?

   (c) What is the probability that the mean of the five independent concrete densities is over 2400 kg/m^3?

5. A small art gallery in Laguna Beach has the choice of stocking either oil paintings or finger paintings for a given tourist season. The oil paintings require a substantial investment, but the potential returns are also greater. The return (profit or loss) depends on whether or not the tourists that season are primarily serious art collectors or more casual buyers. A sales analysis gives the following expectations.

Season return ($1000)	Type of tourists	
Stock decision	Art collectors	Casual buyers
Oil paintings	135	−35
Finger paintings	−5	25

Let p be the probability that the type of tourist is primarily art collectors, so $(1 - p)$ is the probability of primarily casual buyers.

(a) As a function of p, what is the expected return for stocking oil paintings?

(b) As a function of p, what is the expected return for stocking finger paintings?

(c) For what value of p are the two expected returns the same, and what does it mean in context for p to be greater or less than this value?

(d) In a random sample of similar establishments in similar tourist regions, 33 out of 150 reported seasons with tourists who were primarily art collectors. Construct a 95% confidence interval for the proportion of all similar establishments with tourists who are primarily art collectors.

(e) Use the above results to justify a decision to stock finger paintings.

SECTION II

Part B

QUESTION 6

Spend about 25 minutes on this part of the exam.
Percentage of Section II grade—25

6. A national retail chain classifies cashiers as "entry level" for the first ten years. The series of boxplots below shows the relationship between yearly wages (in $1000) and years of experience (YOE) for a random sample of these employees. Below the boxplots are computer regression outputs.

```
Dependent variable is: WAGES
Var       Coef      s.e. Coef t       p
Constant  11.1113   0.2031     54.7   ≤ 0.0001
YOE       0.910485  0.03273    27.8   ≤ 0.0001
R-sq = 88.8%   R-sq(adj) = 88.6%
s = 0.9402 with 100 − 2 = 98 degrees of freedom
```

(a) Discuss the relationship between salary and experience based on the boxplots.

(b) Discuss how conditions for regression inference are met.

(c) Determine a 95% confidence interval for the regression slope, and interpret in context.

(d) Using only the given information, give a rough estimate of the probability that a salary is at least $1000 over what is predicted by the regression line.

If there is still time remaining, you may review your answers.

Answer Key

1. **A**	9. **C**	17. **A**	25. **B**	33. **D**
2. **C**	10. **E**	18. **B**	26. **E**	34. **A**
3. **A**	11. **B**	19. **A**	27. **A**	35. **B**
4. **A**	12. **E**	20. **E**	28. **D**	36. **E**
5. **A**	13. **E**	21. **C**	29. **C**	37. **A**
6. **A**	14. **E**	22. **B**	30. **E**	38. **C**
7. **B**	15. **D**	23. **A**	31. **A**	39. **C**
8. **D**	16. **E**	24. **A**	32. **B**	40. **B**

Answers Explained

SECTION I

1. **(A)** Without knowing the actual number of points scored each season, only proportions, not numbers of points, can be compared between seasons.

2. **(C)** The critical t-values with $df = 20 - 2 = 18$ are ± 2.878. Thus, we have $b_1 \pm t^* \times SE(b_1) = 4.0133 \pm (2.878)(0.4922)$.

3. **(A)** The shortest sequence has a greater probability than any longer sequence.

4. **(A)** Power, the probability of rejecting a false null hypothesis, will be the greatest for parameter values farthest from the hypothesized value, in the direction of the alternative hypothesis.

5. **(A)** A simple random sample may or may not be representative of the population. It is a method of selection in which every possible sample of the desired size has an equal chance of being selected.

6. **(A)** A formula relating the given statistics is $b = r\dfrac{s_y}{s_x}$ which in this case gives $0.25 = r\dfrac{2.3}{8.7}$ and thus $r = \dfrac{(0.25)(8.7)}{2.3}$.

7. **(B)** The null hypothesis is that the new medication is no better than insulin injection, while the alternative hypothesis is that the new medication is better. A Type I error means a mistaken rejection of a true null hypothesis.

8. **(D)** A Type II error means a mistaken failure to reject a false null hypothesis.

9. **(C)** Running a hypothesis test at the 5% significance level means that the probability of committing a Type I error is 0.05. Then the probability of not committing a Type I error is 0.95. Assuming the tests are independent, the probability of not committing a Type I error on any of the five tests is $(0.95)^5$, and the probability of at least one Type I error is $1 - (0.95)^5$.

10. **(E)** No matter what the distribution of raw scores, the set of z-scores always has mean 0 and standard deviation 1.

11. **(B)** In the sampling distribution of $\bar{x}$, the mean is equal to the population mean, and the standard deviation is equal to the population standard deviation divided by the square root of the sample size, in this case, $\sigma_{\bar{x}} = \dfrac{1.708}{\sqrt{3}} = 0.986$.

12. **(E)** The standard deviation can never be negative.

13. **(E)** When a complete census is taken (all 423 seniors were in the study), the population proportion is known and a confidence interval has no meaning.

14. **(E)** The method described in (A) is a convenience sample, (B) and (C) are voluntary response surveys, and (D) suffers from undercoverage bias.

15. **(D)** Since each set is normally distributed so is the set of differences, $X_1 - X_2$. We calculate $\mu_{x_1 - x_2} = 1758 - 1725 = 33$ and $\sigma_{x_1 - x_2} = \sqrt{213^2 + 228^2} = 312.0$.

    So $P(X_1 > X_2) = P(X_1 - X_2 > 0) = P\left(z > \dfrac{0-33}{312.0}\right) = P(z > -0.1058) = 0.542$.

16. **(E)** With $df = n - 1 = 10 - 1 = 9$ and 95% confidence, the critical t-values are ± 2.262.
    Also $SE(\bar{x}) = \dfrac{s}{\sqrt{n}} = \dfrac{2.3}{\sqrt{10}}$.

17. **(A)** $P(X = 1) = \dfrac{1}{3}$, $P(X = 2) = \left(\dfrac{2}{3}\right)\left(\dfrac{1}{3}\right)$, $P(X = 3) = \left(\dfrac{2}{3}\right)^2\left(\dfrac{1}{3}\right)$, $P(X = 4) = \left(\dfrac{2}{3}\right)^3\left(\dfrac{1}{3}\right)$,

    $P(X = 5) = \left(\dfrac{2}{3}\right)^4\left(\dfrac{1}{3}\right), \ldots$, and we see that the distribution has its maximum value at $X = 1$.

18. **(B)** The chi-square tests all involve counts, and comparing means doesn't make sense in this context.

19. **(A)** The median of X is -2, and this is also true of distribution 1 (note that a horizontal line from 0.5 strikes curve 1 above -2 on the x-axis). Y has a smaller standard deviation than Z (tighter clustering around the mean), so Y must correspond to distribution 2, which shows almost all values are between -1 and 1.

20. **(E)**

$P(\text{B-ball}) = (0.55)(0.30) + (0.45)(0.20)$

$P(\text{male}|\text{B-ball}) = \dfrac{(0.55)(0.30)}{(0.55)(0.30) + (0.45)(0.20)}$

21. **(C)** Adding the same constant to every value in a set adds the same constant to the mean but leaves the standard deviation unchanged. Multiplying every value in a set by the same constant multiplies the mean and standard deviation by that constant. So the new mean is $5/9 \times (78.35 - 32) + 273 = 298.75$, and the new standard deviation is $5/9 \times 6.3 = 3.5$.

22. **(B)** Design 2 is an example of a matched pairs design, a special case of a block design; here, each subject is compared to itself with respect to the two treatments. Both designs definitely use randomization with regard to assignment of treatments, but since they do not use randomization in selecting subjects from the general population, care must be taken in generalizing any conclusions. It's not clear whether or not the researchers who do the observations and measurements know which treatment individual cows are receiving, so there is no way to conclude if there is or is not blinding. The two sources of BVH are different treatments, and so they are not being confounded. In both designs treatments are randomly applied, so neither is an observational study.

23. **(A)** The linear regression t-test has null hypothesis H_0: $\beta = 0$ that there is no linear relationship; if the P-value is small enough, then there is evidence of a linear association, that is, there is evidence that $\beta \neq 0$.

24. **(A)** We have a binomial distribution with mean $= np = 12\left(\frac{1}{6}\right) = 2$. Answer (A) is the only reasonable choice.

25. **(B)** The size of the sample always matters; the larger the sample, the greater the power of statistical tests. One percent of a large population is large. Larger samples are better, but if the sample is greater than 10% of the population, the best statistical techniques are not those covered in the AP curriculum. While not in the AP curriculum, if the population is small, its size may matter, as shown by the finite population correction factor.

26. **(E)** The points on the scatterplot all fall on the straight line:

$$Female\ length = Male\ length + 0.5$$

27. **(A)** $\chi^2 = \sum \dfrac{(observed - expected)^2}{expected}$ and expected are found by multiplying the proportions times the sample size of 100.

28. **(D)** The distributions in (A), (B), and (C) appear roughly symmetric, so the mean and median will be roughly the same. The distribution in (D) is skewed to the right, so the mean will most likely be greater than the median, while the distribution in (E) is skewed to the left, so the mean will most likely be less than the median.

29. **(C)** This is a binomial with $n = 10$ and $p = 0.38$, so the mean is $np = 10(0.38) = 3.8$.

30. **(E)** Answers (A), (B), and (C) are common misconceptions. Since the 95% confidence interval contains 80, a two-sided test would not be significant at the 5% significance level or lower. The interval can be expressed as 77.5 ± 3, that is, we are 95% confident that the true mean fastball speed is within 3 mph of 77.5 mph.

31. **(A)** *Residual = Observed – Predicted*, so 1.0 = 11 – *Predicted* and *Predicted* = 10.

32. **(B)** Stratified sampling is when the population is divided into homogeneous groups (the three Divisions in this example), and a random sample of individuals is chosen from each group.

33. **(D)** There are 8 outcomes {TTT, TTH, THT, THH, HTT, HTH, HHT, HHH}, so P(0 heads) = 1/8, P(1 head) = 3/8, P(2 heads) = 3/8, and P(3 heads) = 1/8. Thus, we assign one digit to the results of 0 heads and 3 heads and 3 digits to the results of 1 head and 2 heads and ignore the other 2 digits (of the 10 available digits).

34. **(A)** The margin of error, $\pm t^* \dfrac{s}{\sqrt{n}}$, depends on n, the sample size, not the population size.

35. **(B)** We have $(Q_1 + Q_3)/2 = 20$ or $Q_3 + Q_1 = 40$, and $Q_3 - Q_1 = 20$, which algebraically gives $Q_1 = 10$ and $Q_3 = 30$ [add the equations to obtain $2Q_3 = 60$ so $Q_3 = 30$; then plugging into either equation and solving for Q_1 gives $Q_1 = 10$].

36. **(E)** The larger the sample size, the closer the sample distribution is to the population distribution. The central limit theorem roughly says that if multiple samples of size n are drawn randomly and independently from a population, then the histogram of the means of those samples will be approximately normal. Statistics have probability distributions called sampling distributions. The standard error is based on the spread of the sample and on the sample size. The central limit theorem does not apply to all statistics as it does to sample means. Many sampling distributions are not normal; for example, the sampling distribution of the sample max is not a normal distribution. An estimator of a parameter is unbiased if we have a method that, through repeated samples, is on average the same value as the parameter.

37. **(A)** With 0.068 in a tail, the confidence interval with 34 at one end would have a confidence level of 1 – 2(0.068) = 0.864, so anything higher than 86.4% confidence will contain 34.

38. **(C)** From a boxplot there is no way of telling if a distribution is bell-shaped (very different distributions can have the same five-number summary). Distribution I appears strongly skewed right, and so its mean is probably much greater than its median, while distribution II appears roughly symmetric, and so its mean is probably close to its median. The interquartile range, not the range, in I is 13.

39. **(C)** To make money, there must be more wins than losses, so with 50 plays, we need to calculate $P(X > 25)$. We have a binomial distribution with $n = 50$ and probability of success $p = 18/38$. On a calculator such as the TI-84 we find $P(X > 25) = 1 - P(X \le 25) = $ `1-binomcdf(50,18/38,25)` = 0.303 [or use Bcd on the Casio Prizm or BINOMIAL_CDF on the HP Prime].

40. **(B)** $\chi^2 = \sum \dfrac{(observed - expected)^2}{expected}$ and cell calculations [expected value of a cell equals (row total)(column total)/(table total)] or χ^2-test on a calculator such as the TI-84, Casio Prizm, or HP Prime will yield expected cells of 29, 36, 35, 29, 36, 35.

SECTION II

1. (a) There are $2 \times 2 \times 2 = 8$ different treatments:

Two-day, aerobic, with adjuvant

Two-day, aerobic, without adjuvant

Two-day, anaerobic, with adjuvant

Two-day, anaerobic, without adjuvant

Five-day, aerobic, with adjuvant

Five-day, aerobic, without adjuvant

Five-day, anaerobic, with adjuvant

Five-day, anaerobic, without adjuvant

(b) We must randomly assign the treatment combinations to the beds. (Roses have already been randomly assigned to the beds.) With 8 treatments and 16 beds, each treatment should be assigned to 2 beds. For example, give each bed a random number between 1 and 16 (use a computer to generate 16 random integers between 1 and 16 *without replacement*), and then assign the first treatment in the above list to the beds with the numbers 1 and 2, assign the second treatment in the above list to the beds with the numbers 3 and 4, and so on.

(c) Using only mini-pink roses in this experiment gives reduced variability and increases the likelihood of determining differences among the treatments.

(d) Using only mini-pink roses in this experiment limits the scope and makes it difficult to generalize the results to other species of roses.

SCORING

Part (a) is essentially correct for correctly listing all eight treatment combinations and is incorrect otherwise.

Part (b) is essentially correct if each treatment combination is randomly assigned to two beds of roses. Part (b) is partially correct if each treatment is randomly assigned to two beds but the method is unclear, or if a method of randomization is correctly described but the method may not assure that each treatment is assigned to *two* beds.

Part (c) is essentially correct for noting *reduced variability* and for explaining that this increases likelihood of determining differences among the treatments. Part (c) is partially correct for only one of these two components.

Part (d) is essentially correct noting *limited scope* and for explaining that this makes generalization to other species difficult. Part (d) is partially correct for only one of these two components.

Count partially correct answers as one-half an essentially correct answer.

4 Complete Answer   Four essentially correct answers.

3 Substantial Answer   Three essentially correct answers.

2 Developing Answer   Two essentially correct answers.

1 Minimal Answer   One essentially correct answer.

Use a holistic approach to decide a score totaling between two numbers.

2. Part 1: State the correct hypotheses.

$H_0: \mu = 82$ and $H_a: \mu > 82$

Part 2: Identify the correct test and check appropriate conditions.

One-sample t-test for a population mean

Conditions: Random sample (given), $n = 47$ is less than 10% of all possible swings with the new racquet, and $n = 47 \geq 30$ is sufficiently large for the CLT to apply.

Part 3: Calculate the test statistic t and the P-value.

Calculator software (such as T-Test on the TI-84 or 1-Sample t-Test on the Casio Prizm) gives $t = 3.0246$ and $P = 0.00203$.

Part 4: Give a conclusion in context with linkage to the P-value.

With this small a P-value, $0.00203 < 0.05$, there is sufficient evidence to reject H_0, that is, there is sufficient evidence that his mean speed with the new racquet is an improvement over the old.

SCORING

Part 1 either is essentially correct or is incorrect.

Part 2 is essentially correct if the test is correctly identified and the randomization and sample size assumptions are checked. Part 2 is partially correct if only one of these two elements is correct.

Part 3 is essentially correct if the t-value and P are stated. Part 3 is partially correct if only one of these two elements is correct.

Part 4 is essentially correct if the correct conclusion is given in context and the conclusion is linked to the P-value. Part 4 is partially correct if the correct conclusion is given in context but there is no linkage to the P-value.

Count partially correct answers as one-half an essentially correct answer.

4	**Complete Answer**	Four essentially correct answers.
3	**Substantial Answer**	Three essentially correct answers.
2	**Developing Answer**	Two essentially correct answers.
1	**Minimal Answer**	One essentially correct answer.

Use a holistic approach to decide a score totaling between two numbers.

3. (a) Chi-square goodness-of-fit test

(b) H_0: The new freshman class is distributed 2.7% non-Hispanic Black, 3.7% Asian or Pacific Islander, 4.0% Hispanic, 80.0% non-Hispanic White, and 9.6% other/unknown.

H_a: The new freshman class has a distribution different from that in October 2018.

Randomization is given and the expected cell frequencies—2.7% $\times$ 200 = 5.4, 3.7% $\times$ 200 = 7.4, 4.0% $\times$ 200 = 8.0, 80.0% $\times$ 200 = 160.0, and 9.6% $\times$ 200 = 19.2—are all at least 5.

(c)

$$\chi^2 = \sum \frac{(obs-exp)^2}{exp} = \frac{(3-5.4)^2}{5.4} + \frac{(4-7.4)^2}{7.4} + \frac{(14-8)^2}{8} + \frac{(150-160)^2}{160} + \frac{(29-19.2)^2}{19.2} = 12.76$$

with $df = 5 - 1 = 4$, and a P-value of 0.013. (Or use χ^2GOF-Test on a calculator to obtain χ^2 and P.) With such a small P-value, $0.013 < 0.05$, there is sufficient evidence to reject H_0 and conclude that there is statistical evidence of a change in ethnic/racial composition.

(d) No, this test/procedure targeted students visiting the campus, and such students might be different from the targeted population of students making up the new freshman class. For example, some students who eventually make up the freshman class might not have the funds or the time to visit the campus. Or it can be argued that even if all potential students do visit the campus, there is no reason to conclude that the distribution of visiting students is the same as the distribution of students who both are accepted and decide to attend this college.

SCORING

Part (a–b) is essentially correct if the correct test is named, the hypotheses are correctly stated, and the assumption of all expected cell frequencies being at least five is checked. Part (a–b) is partially correct if only two of these three elements are correct. Part (a–b) is incorrect if only one of these three elements is correct.

Part (c1) is essentially correct if the chi-square value and the P-value are given correctly, and is partially correct if only one of these values is given correctly.

Part (c2) is essentially correct if the correct conclusion is given in context and the conclusion is linked to the P-value. Part (c2) is partially correct if the correct conclusion is given in context but there is no linkage to the P-value.

Part (d) is essentially correct or incorrect. It is essentially correct for both stating that the intended population is not targeted and giving a clear argument for this answer.

Count partially correct answers as one-half an essentially correct answer.

4	**Complete Answer**	Four essentially correct answers.
3	**Substantial Answer**	Three essentially correct answers.
2	**Developing Answer**	Two essentially correct answers.
1	**Minimal Answer**	One essentially correct answer.

Use a holistic approach to decide a score totaling between two numbers.

4. (a) Normal distribution with $\mu = 2317$ and $\sigma = 128$.

$$P(x > 2400) = P\left(z > \frac{2400 - 2317}{128}\right) = P(z > 0.6484) = 0.258$$

(b) $P(at\ least\ 3\ out\ of\ 5\ are > 2400) = 10(0.258)^3(0.742)^2 + 5(0.258)^4(0.742) + (0.258)^5 = 0.112$

TIP

Full credit will be given in 4(b) for: Binomial, $n = 5$, $p = 0.258$ $P(x \geq 3) = 0.112$

[On the TI-84, $1 - \text{binomcdf}(5, 0.258, 2) = 0.112$ or use Bcd on the Casio Prizm or BINOMIAL_CDF on the HP Prime]

(c) The distribution of $\bar{x}$ is normal with mean $\mu_{\bar{x}} = 2317$ and standard deviation

$$\sigma_{\bar{x}} = \frac{128}{\sqrt{5}} = 57.24.$$

So $P(\bar{x} > 2400) = P\left(z > \frac{2400 - 2317}{57.24}\right) = P(z > 1.450) = 0.0735$

SCORING

Part (a) is essentially correct if the distribution is named (normal), the parameters are given (mean and standard deviation), the cutoff with direction is indicated (> 2400), and the correct probability is calculated. Simply writing normalcdf($2400, \infty, 2317, 128$) = 0.112 is a partially correct response; however, writing normalcdf(lower bound = 2400, upper bound = ∞, mean = 2317, standard deviation = 128) = 0.112 is essentially correct.

Part (b) is essentially correct if the correct probability is calculated and the derivation is clear. Part (b) is partially correct for indicating a binomial with $n = 5$ and $p =$ answer from (a) but calculating incorrectly. Simply writing 1-binomcdf(5,.258,2) = 0.112 is also a partially correct response.

Part (c) is essentially correct for specifying a normal distribution and both $\mu_{\bar{x}}$ and $\sigma_{\bar{x}}$ and correctly calculating the probability. Part (c) is partially correct for specifying a normal distribution and both $\mu_{\bar{x}}$ and $\sigma_{\bar{x}}$ but incorrectly calculating the probability, or for failing to specify a normal distribution and both $\mu_{\bar{x}}$ and $\sigma_{\bar{x}}$ but correctly calculating the probability.

4 Complete Answer   All three parts essentially correct.

3 Substantial Answer   Two parts essentially correct and one part partially correct.

2 Developing Answer   Two parts essentially correct OR one part essentially correct and one or two parts partially correct OR all three parts partially correct.

1 Minimal Answer   One part essentially correct OR two parts partially correct.

5. (a) $\sum xP(x) = 135p + (-35)(1 - p) = -35 + 170p$

   (b) $\sum xP(x) = (-5)p + 25(1 - p) = 25 - 30p$

   (c) $-35 + 170p = 25 - 30p$ gives $p = 0.3$. When $p > 0.3$, the expected return for oil paintings is greater than that for finger paintings, and when $p < 0.3$, the expected return for finger paintings is greater than for oil paintings. (These statements follow from the positive slope of $R = -35 + 170p$ and the negative slope of $R = 25 - 30p$.)

   (d) First, name the confidence interval: a 95% confidence interval for the proportion p of all similar establishments with tourists who are primarily art collectors.

   Second, check conditions: We are given that this is a random sample, it is reasonable to assume that the sample is less than 10 percent of all similar establishments, and the sample size is large enough ($n\hat{p} = 33$ and $n(1 - \hat{p}) = 117$ are both greater than 10).

   Third, correct mechanics: calculator software (such as `1-PropZInt` on the TI-84 or INTR,Z-1-PROP on the Casio Prizm) gives (0.154, 0.286).

   Fourth, interpret in context: We are 95% confident that the true proportion of all similar establishments with tourists who are primarily art collectors is between 0.154 and 0.286.

   (e) Stock finger paintings. The entire interval in (d) is below $p = 0.3$, so based on (c), the expected return for finger paintings is greater than for oil paintings for all p in this interval.

SCORING

Parts (a) and (b) are scored together. They are essentially correct if both answers are correct and partially correct if one answer is correct.

Part (c) is essentially correct for a correct calculation of the intersection together with a correct conclusion of what it means when p is greater or less than 0.3. Part (c) is partially correct if the intersection is not correctly calculated, but the conclusions are correct based on the incorrect intersection value.

Part (d) is essentially correct if steps 2, 3, and 4 are correct. (Step 1 is only a restatement from the question.) Part 3 is partially correct if only two of these three steps are correct.

Part (e) is essentially correct if the correct conclusion is given with clear linkage to the results from both (c) and (d). Part (e) is partially correct if the explanation of the linkage is present but weak.

Give 1 point for each essentially correct part and one-half point for each partially correct part.

4	**Complete Answer**	4 points
3	**Substantial Answer**	3 points
2	**Developing Answer**	2 points
1	**Minimal Answer**	1 point

Use a holistic approach to decide a score totaling between two numbers.

TIP

Read carefully and recognize that sometimes very different tests are required in different parts of the same problem.

6. (a) Salary and years of experience exhibit a positive, approximately linear relationship. As years of experience increase, so does the median salary. The third-quartile, Q_3, salary also increases with years, and so does the Q_1 salary with the exception of one year. As years of experience increase, there are only minor differences in measures of variability in the salaries, with roughly the same range (except for the last two) and fairly consistent interquartile ranges.

(b) First, the boxplots indicate that an overall scatterplot pattern would be roughly linear. Second, the residual plot shows no pattern. Third, the histogram of residuals appears roughly normal (unimodal, symmetric, and without clear skewness or outliers).

(c) With $df = 98$ and 0.025 in each tail, the critical t-values are ±1.984. $b_1 \pm ts_{b_1} = 0.910 \pm 1.984(0.03273) = 0.910 \pm 0.065$. We are 95% confident that for each additional year of experience, the average increase in predicted salary is between $845 and $975.

(d) The sum and thus the mean of the residuals is always 0. The standard deviation of the residuals is σ, which can be estimated with $s = 0.9402$. With a roughly normal distribution, we have $P(X \geq 1) = 0.14$.

SCORING

Part (a) is essentially correct for correctly noting a linear relationship, noting that as years of experience increase so does the median salary (or Q_3 or generally Q_1), and noting that measures of variability (range or IQR) stay roughly the same. Part (a) is partially correct for correctly noting two of the three features.

Part (b) is essentially correct for noting the three conditions (roughly linear scatterplot, no pattern in the residual plot, and roughly normal histogram of residuals). Part (b) is partially correct for correctly noting two of the three conditions.

Part (c) is essentially correct for both a correct calculation of the confidence interval and a correct interpretation in context. Part (c) is partially correct for a correct calculation without the interpretation in context, or for a correct interpretation based on an incorrect calculation.

Part (d) is essentially correct noting that the distribution of residuals is roughly normal with mean 0 and standard deviation 0.9402, and then using this to correctly calculate the probability. Part (d) is partially correct for correctly noting the distribution of residuals but incorrectly calculating the probability, or for making a calculation based on a normal distribution with mean 0 but using an incorrect standard deviation.

Count partially correct answers as one-half an essentially correct answer.

4	**Complete Answer**	Four essentially correct answers.
3	**Substantial Answer**	Three essentially correct answers.
2	**Developing Answer**	Two essentially correct answers.
1	**Minimal Answer**	One essentially correct answer.

Use a holistic approach to decide a score totaling between two numbers.

AP SCORE FOR THE DIAGNOSTIC TEST

Multiple-Choice section (40 questions)

Number correct $\times$ 1.25 = _____

Free-Response section (5 open-ended questions plus an investigative task)

Question 1 _____ $\times$ 1.875 = _____
out of 4

Question 2 _____ $\times$ 1.875 = _____
out of 4

Question 3 _____ $\times$ 1.875 = _____
out of 4

Question 4 _____ $\times$ 1.875 = _____
out of 4

Question 5 _____ $\times$ 1.875 = _____
out of 4

Question 6 _____ $\times$ 3.125 = _____
out of 4

Total points from Multiple-Choice and Free-Response sections = _____

Conversion chart based on a recent AP exam:

Total points	AP Score
70–100	5
57–69	4
44–56	3
33–43	2
0–32	1

STUDY GUIDE FOR THE DIAGNOSTIC TEST
MULTIPLE-CHOICE QUESTIONS

Note in which Themes your missed questions fall. Then give special note to the Topics corresponding to the missed questions. Additionally, whenever you wish to test yourself on a particular Theme, go back to the designated questions.

Theme One: Exploratory Analysis

Question 1	Topic Four: Exploring Categorical Data
Question 6	Topic Three: Exploring Bivariate Data
Question 10	Topic Two: Summarizing Distributions
Question 12	Topic Two: Summarizing Distributions
Question 19	Topic One: Graphical Displays
Question 21	Topic Two: Summarizing Distributions
Question 26	Topic Three: Exploring Bivariate Data
Question 28	Topic One: Graphical Displays
	Topic Two: Summarizing Distributions
Question 31	Topic Three: Exploring Bivariate Data
Question 35	Topic Two: Summarizing Distributions
Question 38	Topic One: Graphical Displays
	Topic Two: Summarizing Distributions

Theme Two: Collecting and Producing Data

Question 5	Topic Five: Planning and Conducting Surveys
Question 14	Topic Five: Planning and Conducting Surveys
Question 22	Topic Six: Planning and Conducting Experiments
Question 25	Topic Five: Planning and Conducting Surveys
Question 32	Topic Five: Planning and Conducting Surveys

Theme Three: Probability

Question 3	Topic Seven: Probability as Relative Frequency
Question 9	Topic Seven: Probability as Relative Frequency
	Topic Twelve: Tests of Significance—Proportions and Means
Question 11	Topic Ten: Sampling Distributions
Question 15	Topic Eight: Combining and Transforming Random Variables
	Topic Nine: The Normal Distribution
Question 17	Topic Seven: Probability as Relative Frequency
Question 20	Topic Seven: Probability as Relative Frequency
Question 24	Topic Seven: Probability as Relative Frequency
Question 29	Topic Seven: Probability as Relative Frequency
Question 33	Topic Seven: Probability as Relative Frequency
Question 36	Topic Ten: Sampling Distributions
Question 39	Topic Seven: Probability as Relative Frequency

Theme Four: Statistical Inference

Theme One: Exploratory Analysis

Graphical Displays

1

→ **BAR CHARTS**

→ **DOTPLOTS**

→ **HISTOGRAMS**

→ **STEMPLOTS**

→ **CENTER AND SPREAD**

→ **CLUSTERS AND GAPS**

→ **OUTLIERS**

→ **MODES**

→ **SHAPE**

→ **CUMULATIVE RELATIVE FREQUENCY PLOTS**

→ **SKEWNESS**

There are a variety of ways to organize and arrange data. Much information can be put into tables, but these arrays of bare figures tend to be spiritless and sometimes even forbidding. Some form of graphical display is often best for seeing patterns and shapes and for presenting an immediate impression of everything about the data. Among the most common visual representations of data are dotplots, bar charts, histograms, and stemplots. It is important to remember that all graphical displays should be clearly labeled, leaving no doubt what the picture represents—**AP Statistics scoring guides harshly penalize the lack of titles, labels, and scales!**

The first thing to do with data is to draw a picture—always.

TIP

Just because a variable has numerical values doesn't necessarily mean that it's quantitative.

BAR CHARTS

Bar charts are useful with regard to *categorical* (*or qualitative*) *variables*, that is, variables that note the category to which each individual belongs. This is in contrast to *quantitative variables*, which take on numerical values. Sizes can be measured as frequencies or percents.

➡ **EXAMPLE 1.1** _____

In a survey taken during the first week of January 2017, 1100 parents wanted to keep the school year to the current 180 days, 300 wanted to shorten it to 160 days, 500 wanted to extend it to 200 days, and 100 expressed no opinion. (Or noting that there were 2000 parents surveyed, percentages can be calculated.)

TIP

Graphs must have appropriate labeling and scaling, or they will lose credit!

DOTPLOTS

Dotplots can be used with categorical or quantitative variables.

➡ **EXAMPLE 1.2** _____

When asked to choose their favorite hip-hop artist, 8 students chose Eminem, 5 picked 2Pac, 6 picked Notorius B.I.G., 3 picked Nas, 5 picked Dr. Dre, and 3 picked Jay-Z. These data can be displayed in the following dotplot.

The dotplot below shows the lengths of stay (in days) for all patients admitted to a rural hospital during the first week in January 2017.

HISTOGRAMS

Histograms, useful for large data sets involving quantitative variables, show counts or percents falling either at certain values or between certain values. While the AP Statistics exam does not stress construction of histograms, there are often questions on interpreting given histograms.

To construct a histogram using the TI-84, go to STAT → EDIT and put the data in a list, then turn a STAT PLOT on, choose the histogram icon under Type, specify the list where the data is, and use ZoomStat and/or adjust the WINDOW. Note that X_{SCL} determines the width of the bin or class. Using the Casio Prizm, put the data in a List, under Graph Type pick Hist, and then go back to Select and Draw.

➥ **EXAMPLE 1.4** _____

Suppose there are 2200 seniors in a city's 6 high schools. Four hundred of the seniors are taking no AP classes, 500 are taking one, 900 are taking two, 300 are taking three, and 100 are taking four. These data can be displaced in the following histogram:

(continued)

Sometimes, instead of labeling the vertical axis with frequencies, it is more convenient or more meaningful to use *relative frequencies*, that is, frequencies divided by the total number in the population.

Number of AP classes	Frequency	Relative frequency
0	400	400/2200 = 0.18
1	500	500/2200 = 0.23
2	900	900/2200 = 0.41
3	300	300/2200 = 0.14
4	100	100/2200 = 0.05

Note that the shape of the histogram is the same whether the vertical axis is labeled with frequencies or with relative frequencies. Sometimes we show both frequencies and relative frequencies on the same graph.

➥ EXAMPLE 1.5 _____

Consider the following histogram of the numbers of pairs of shoes owned by 2000 women.

What can we learn from this histogram? For example, none of the women had fewer than 5 or more than 60 pairs of shoes. One hundred sixty of the women had 18 pairs of shoes. Twenty women had 5 pairs of shoes. Half the total area is less than or equal to 19, so half the women have 19 or fewer pairs of shoes. Fifteen percent of the area is more than 30, so 15 percent of the women have more than 30 pairs of shoes. Five percent of the area is more than 50, so 5 percent of the women have more than 50 pairs of shoes.

Consider the following histogram of exam scores, where the vertical axis has not been labeled.

What can we learn from this histogram?

Answer: It is impossible to determine the actual frequencies, that is, we have no idea if there were 25 students, 100 students, or any particular number of students who took the exam. However, we can determine the relative frequencies by noting the fraction of the total area that is over any interval.

We can divide the area into ten equal portions, and then note that $\frac{1}{10}$ or 10% of the area is between 60 and 70, so 10% of the students scored between 60 and 70. Similarly, 40% scored between 70 and 80, 30% scored between 80 and 90, and 20% scored between 90 and 100.

Although it is usually not possible to divide histograms so nicely into ten equal areas, the principle of relative frequencies corresponding to relative areas still applies. Also note how this example shows the number of exam scores falling *between* certain values, whereas the previous two examples showed the number of AP classes taken and number of shoes owned for *each* value.

TIP

Relative frequencies are the usual choice when comparing distributions of different size populations.

STEMPLOTS

Although a histogram may show how many scores fall into each grouping or interval, the exact values of individual scores are lost. An alternative pictorial display, called a *stemplot* (also called a stem-and-leaf display) retains this individual information and is useful for giving a quick overview of a distribution, displaying the relative density and shape of the data. A stemplot contains two columns separated by a vertical line. The left column contains the stems, and the right column contains the leaves.

➡ EXAMPLE 1.7 _____

Bisphenol A (BPA) is an industrial chemical that is found in many hard plastic bottles. Recent studies have shown a possible link between BPA exposure and childhood obesity. In one study of 27 elementary school children, urinary BPA levels in nanograms/milliliter (ng/mL) were as follows: {0.2, 0.4, 0.7, 0.7, 0.8, 0.8, 0.9, 1.0, 1.0, 1.3, 1.4, 1.4, 1.4, 1.7, 1.9, 2.1, 2.4, 2.5, 2.8, 2.8, 3.0, 3.3, 3.3, 3.8, 4.2, 4.5, 5.2}

```
0 | 2 4 7 7 8 8 9
1 | 0 0 3 4 4 4 7 9
2 | 1 4 5 8 8
3 | 0 3 3 8
4 | 2 5
5 | 2
```

Urinary BPA level
(0|2 means 0.2 ng/mL)

Note: Those with urine BPA level of 2 ng/mL or higher had more than twice the risk of being overweight.

➡ EXAMPLE 1.8 _____

How many nonstop pushups can a 15–18-year-old teenager do? In one study in a mixed gender high school gym class, the numbers of pushups were {2, 5, 7, 10, 12, 12, 14, 16, 16, 18, 19, 20, 21, 29, 32, 34, 35, 37, 37, 38, 39, 39, 42, 44, 50}

```
0 | 2 5 7
1 | 0 2 2 4 6 6 8 9
2 | 0 1 9
3 | 2 4 5 7 7 8 9 9
4 | 2 4
5 | 0
```

Number of pushups
(5|0 means 50 pushups)

CENTER AND SPREAD

Looking at a graphical display, we see that two important aspects of the overall pattern are

1. the *center*, which separates the values (or area under the curve in the case of a histogram) roughly in half, and
2. the *spread*, that is, the scope of the values from smallest to largest.

In the histogram of Example 1.4, the center is 2 AP classes while the spread is from 0 to 4 AP classes.

In the histogram of Example 1.5 the center is about 19, and the spread is from 5 to 60; in the histogram of Example 1.6, the center is about 80, and the spread is from about 60 to about 100.

In the stemplot of Example 1.7, the center is 1.7 (middle of the 27 values), and the spread is from 0.2 to 5.2; in the stemplot of Example 1.8, the center is 21 (middle of the 25 values), and the spread is from 2 to 50.

CLUSTERS AND GAPS

Other important aspects of the overall pattern are

1. *clusters*, which show natural subgroups into which the values fall (for example, the salaries of teachers in Ithaca, NY, fall into three overlapping clusters, one for public school teachers, a higher one for Ithaca College professors, and an even higher one for Cornell University professors), and
2. *gaps*, which show holes where no values fall (for example, the Office of the Dean sends letters to students being put on the honor roll and to those being put on academic warning for low grades; thus the GPA distribution of students receiving letters from the Dean has a huge middle gap).

➡ EXAMPLE 1.9

Hodgkin's lymphoma is a cancer of the lymphatic system, the system that drains excess fluid from the blood and protects against infection. Consider the following histogram:

Simply saying that the average age at diagnosis for female cases is around 50 clearly misses something. The distribution of ages at diagnosis for female cases of Hodgkin's lymphoma is bimodal with two distinct clusters, centered at 25 and 75.

Pay attention to outliers! They deserve special consideration.

There is a formal mathematical definition of an outlier as a value at least a certain distance away from Q_1 or Q_3. You will learn how to calculate this distance in Topic 2.

TIP

Some distributions have many little ups (and downs), which should not be confused with modes.

OUTLIERS

Extreme values, called outliers, are found in many distributions. Sometimes they are the result of errors in measurements and deserve scrutiny; however, outliers can also be the result of natural chance variation. Outliers may occur on one side or both sides of a distribution.

MODES

Some distributions have one or more major peaks, called modes. (The values with the peaks above them are the modes.) With exactly one or two such peaks, the distribution is said to be unimodal or bimodal, respectively. But every little bump in the data is not a mode! You should always look at the big picture and decide whether or not two (or more) phenomena are affecting the histogram.

➡ **EXAMPLE 1.10** _____

The histogram below shows employee computer usage (number accessing the Internet) at given times at a company main office.

Note that this is a *bimodal* distribution. Computer usage at this company appears heaviest at midmorning and midafternoon, with a dip in usage during the noon lunch hour. There is an evening outlier possibly indicating employees returning after dinner (or perhaps custodial cleanup crews taking an Internet break!).

Note that, as illustrated above, it is usually instructive to look for reasons behind outliers and modes.

SHAPE

Distributions come in an endless variety of shapes; however, certain common patterns are worth special mention:

> **IMPORTANT**
>
> When describing a distribution, always comment on Shape, Outliers, Center, and Spread (SOCS). Or, alternatively, Center, Unusual values, Shape, and Spread (CUSS). And always describe in context.

1. A *symmetric* distribution is one in which the two halves are mirror images of each other. For example, the weights of all people in some organizations fall into symmetric distributions with two mirror-image bumps, one for men's weights and one for women's weights.

2. A distribution is *skewed to the right* if it spreads far and thinly toward the higher values. For example, ages of nonagenarians (people in their 90s) is a distribution with sharply decreasing numbers as one moves from 90-year-olds to 99-year-olds.

3. A distribution is *skewed to the left* if it spreads far and thinly toward the lower values. For example, scores on an easy exam show a distribution bunched at the higher end with few low values.

4. A *bell-shaped* distribution is symmetric with a center mound and two sloping tails. For example, the distribution of IQ scores across the general population is roughly symmetric with a center mound at 100 and two sloping tails.

5. A distribution is *uniform* if its histogram is a horizontal line. For example, tossing a fair die and noting how many spots (pips) appear on top yields a uniform distribution with 1 through 6 all equally likely.

TIP

In the real world, distributions are rarely perfectly symmetric, perfectly bell-shaped, or perfectly uniform, so we usually say "roughly" or "approximately" symmetric, bell-shaped, or uniform.

Symmetric   Skewed to the right   Skewed to the left

Bell-shaped   Uniform

Even when a basic shape is noted, it is important also to note if some of the data deviate from this shape.

CUMULATIVE RELATIVE FREQUENCY PLOTS

Sometimes we sum frequencies and show the result visually in a cumulative relative frequency plot (also known as an *ogive*).

➡ **EXAMPLE 1.11** _____

The following graph shows 2019 school enrollment in the United States by age.

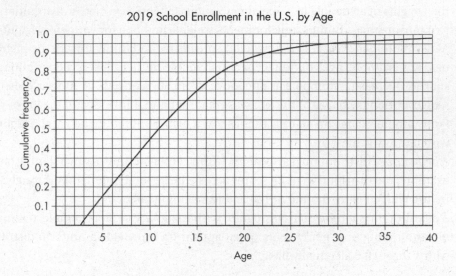

What can we learn from this cumulative relative frequency plot? For example, going up to the graph from age 5, we see that 0.15 or 15% of school enrollment is below age 5. Going over to the graph from 0.5 on the vertical axis, we see that 50% of the school enrollment is below and 50% is above a middle age of 11. Going up from age 30, we see that 0.95 or 95% of the enrollment is below age 30, and thus 5% is above age 30. Going over from 0.25 and 0.75 on the vertical axis, we see that the middle 50% of school enrollment is between ages 6 and 7 at the lower end and age 16 at the upper end.

CUMULATIVE RELATIVE FREQUENCY AND SKEWNESS

A distribution skewed to the left has a cumulative frequency plot that rises slowly at first and then steeply later, while a distribution skewed to the right has a cumulative frequency plot that rises steeply at first and then slowly later.

➡ EXAMPLE 1.12

Consider the essay grading policies of three teachers, Abrams, who gives very high scores, Brown, who gives equal numbers of low and high scores, and Connors, who gives very low scores. Histograms of the grades (with 1 the highest score and 4 the lowest score) are as follows:

These translate into the following cumulative frequency plots:

SUMMARY

- The four keys to describing a distribution are shape, center, spread, and unusual features.
- Note clusters, gaps, modes, and outliers.
- Always provide context.
- Look for reasons behind any unusual features.
- A few common shapes arise from symmetric, skewed to the right, skewed to the left, bell-shaped, and uniform distributions.
- For categorical (qualitative) data, dotplots and bar charts give useful displays.
- For quantitative data, dotplots, histograms, cumulative relative frequency plots (ogives), and stemplots give useful displays.
- In a histogram, relative area corresponds to relative frequency.

QUIZ 1

Multiple-Choice Questions

Directions: The questions that follow are each followed by five suggested answers. Choose the response that best answers the question.

1. The times, in minutes, taken by 155 high school students to complete an Angry Birds level are summarized in the histogram below.

Based on the histogram, which of the following must be true?

(A) The minimum time taken by any of these students was 5 minutes.
(B) The maximum time taken by any of these students was 35 minutes.
(C) If the times are arranged in order, the middle time would be between 10 and 15 minutes.
(D) If the times are arranged in order, the middle time would be between 15 and 20 minutes.
(E) The same number of students took less than 15 minutes as took over 20 minutes.

2. Which of the following is a true statement?

(A) Stemplots are useful both for quantitative and categorical data sets.
(B) Stemplots are equally useful for small and very large data sets.
(C) Stemplots can show symmetry, gaps, clusters, and outliers.
(D) Stemplots may or may not show individual values.
(E) Stems may be skipped if there is no data value for a particular stem.

Questions 3–6 refer to the following four histograms:

(A)   (B)

(C)   (D)

3. To which of the above four histograms does the following cumulative relative frequency plot correspond?

(A) A
(B) B
(C) C
(D) D

4. To which of the above four histograms does the following cumulative relative frequency plot correspond?

1.0

(A) A
(B) B
(C) C
(D) D

5. To which of the above four histograms does the following cumulative relative frequency plot correspond?

(A) A
(B) B
(C) C
(D) D

6. To which of the above four histograms does the following cumulative relative frequency plot correspond?

(A) A
(B) B
(C) C
(D) D

7. An auto manufacturer claims that the autos it sells have an average fuel efficiency of 35 mpg. A consumer group believes that the true figure is lower. The consumer group obtains 5 of the company's autos and determines an average of 32 mpg. Is this significant? The company runs a simulation by randomly picking 5 cars 160 times from a fleet of cars of known 35 mpg efficiency and calculating the resulting averages to show that 32 was possible. The company makes the following dotplot.

Fuel Economy (mpg)

Does the company's argument seem reasonable given this dotplot?

(A) Yes, because the dotplot shows that anything is possible in the real world.

(B) Yes, because the dotplot shows that every average was not 35.

(C) Yes, because the dotplot shows that 32 mpg is a possible average of 5 cars from a fleet of cars of known 35 mpg efficiency.

(D) No, because the dotplot shows that anything is possible in the real world.

(E) No, because the possibility of picking 5 cars with an average of 32 mpg or lower from a fleet of cars of known 35 mpg efficiency is very small, only 4 out of 160, or 0.025.

Free-Response Questions

1. The miles run by 25 randomly selected runners during a given week are given below.

Miles run	5–10	10–15	15–20	25–30
Frequency	3	11	7	4

   (a) Construct a histogram.

   (b) What percent of the runners ran under 15 miles?

   (c) Construct a cumulative relative frequency plot.

   (d) If each of the four intervals is divided in half and the corresponding numbers given, what percent would be under 15 miles?

2. The dotplot below shows the numbers of goals scored by the 20 teams playing in a city's high school soccer games on a particular day.

Goals scored by each team

   (a)   Describe the distribution.

   (b)   One superstar scored six goals, but his team still lost. What are all possible final scores for that game? Explain.

   (c)   Is it possible that all the teams scoring exactly two goals won their games? Explain.

3. The winning percentages for a major league baseball team over the past 22 years are shown in the following stemplot:

```
46 | 08
47 | 1479
48 | 5889
49 | 347
50 |
51 |
52 | 58
53 | 256
54 | 489
55 | 6
```

(55 | 6 means 55.6%)

(a)   Interpret the lowest value.
(b)   Describe the distribution.
(c)   Give a reason that one might argue that the team is more likely to lose a given game than win it.
(d)   Give a reason that one might argue that the team is more likely to win a given game than lose it.

QUIZ 2

Multiple-Choice Questions

> **Directions:** The questions that follow are each followed by five suggested answers. Choose the response that best answers the question.

1. The stemplot below shows ages of CEOs of a select group of corporations.

$$
\begin{array}{r|l}
2 & 28 \\
3 & 36 \\
4 & 27 \\
5 & 0258 \\
6 & 3379 \\
7 & 48 \\
8 & 3 \\
9 & 0 \\
\end{array}
$$

(9 | 0 means an age of 90)

Which of the following is *not* a correct statement about this distribution?

(A) The distribution is roughly bell-shaped.
(B) The distribution is skewed left and right.
(C) The center is around 60.
(D) The spread is from 22 to 90.
(E) There are no outliers.

2. Which of the following is an *incorrect* statement?

(A) In histograms, relative areas correspond to relative frequencies.
(B) In histograms, frequencies can be determined from relative heights.
(C) Symmetric histograms may have multiple peaks.
(D) Two students working with the same set of data may come up with histograms that look different.
(E) Displaying outliers may be more problematic when using histograms than when using stemplots.

Questions 3–7 refer to the following five cumulative relative frequency plots:

3. To which of the above cumulative relative frequency plots does the following histogram correspond?

(A) A
(B) B
(C) C
(D) D
(E) E

4. To which of the above cumulative relative frequency plots does the following histogram correspond?

(A) A
(B) B
(C) C
(D) D
(E) E

5. To which of the above cumulative relative frequency plots does the following histogram correspond?

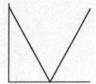

(A) A
(B) B
(C) C
(D) D
(E) E

6. To which of the above cumulative relative frequency plots does the following histogram correspond?

(A) A
(B) B
(C) C
(D) D
(E) E

7. To which of the above cumulative relative frequency plots does the following histogram correspond?

(A) A
(B) B
(C) C
(D) D
(E) E

Free-Response Questions

1. A college basketball team keeps records of career average points per game of players playing at least 75% of team games during their college careers. The cumulative relative frequency plot below summarizes statistics of players graduating over the past 10 years.

(a) Interpret the point (20, 0.4) in context.

(b) Interpret the intersection of the plot with the horizontal axis in context.

(c) Interpret the horizontal section of plot from 5 to 7 points per game in context.

(d) The players with the top 10% of the career average points per game achievements will be listed on a plaque. What is the cutoff score for being included on the plaque?

(e) What proportion of the players averaged between 10 and 20 points per game?

2. A company engineer creates a diagnostic measurement, $W = \dfrac{Max + Min}{2}$, which should be at least 24.10 in a sample of size 12 if certain machinery is operating correctly. To explore this diagnostic measurement, the machine is perfectly calibrated. Then 100 random samples of size 12 of the product are taken from the assembly line. For each of these 100 samples, the diagnostic measurement W is calculated and shown plotted below.

Each day, one sample of size 12 is taken from the assembly line and the diagnostic measurement W is calculated. If W drops too low, a decision to recalibrate the machinery is made.

(a) From the dotplot above, estimate a measure of center and a measure of variability for the distribution.

(b) For the dotplot above, do there appear to be any outliers (no calculations required)? Justify your answer.

One day the random sample is {24.2, 24.84, 25.05, 23.43, 23.9, 25.01, 23.01, 24.5, 24.23, 23.76, 24.69, 23.21}.

(c) Based on the dotplot above, does the engineer have sufficient evidence to conclude that recalibration is necessary? Justify your answer.

3. In a random sample of 32 newlyweds, the numbers of months they dated before becoming engaged are displayed in the following stemplot.

```
12 | 0 1 1 3 9
13 | 0 1 3 4 4 7 7 9
14 | 2 2 5 6 8 8 8 9
15 | 3 3 7 7
16 | 5 6
17 | 0 7
18 | 4
19 | 2
20 |
21 |
22 |
23 | 1
```

Leaf unit = 1.0 months

(a) Describe the distribution of the data in this stemplot.

(b) Describe the shape of the cumulative frequency plot of this data. (You are not asked to draw this plot.)

ANSWERS EXPLAINED

Quiz 1

MULTIPLE-CHOICE QUESTIONS

1. **(D)** The minimum time is somewhere between 5 and 10 minutes, but might not be exactly 5 minutes. Similarly, the maximum time is somewhere between 30 and 35 minutes. With 155 times, the middle time will be the 78th time if the times are arranged in order. There are 70 times less than 15 minutes and 40 times between 15 and 20 minutes, so the 78th time must be between 15 and 20 minutes.

2. **(C)** Stemplots are not used for categorical data sets, are too unwieldy to be used for very large data sets, and show every individual value. Stems should never be skipped over—gaps are important to see.

3. **(C)** A cumulative relative frequency plot that rises at a constant rate to start and then slowly at the end corresponds to a histogram which is horizontal to start and then has little area under the curve at the end.

4. **(A)** A cumulative relative frequency plot that rises slowly at first, then quickly in the middle, and finally slowly again at the end corresponds to a histogram with little area under the curve on the ends and much greater area in the middle.

5. **(D)** A cumulative relative frequency plot that rises slowly at first and then rises at a constant rate corresponds to a histogram with little area under the curve early on and then a horizontal section at the end.

6. **(B)** A cumulative relative frequency plot that rises quickly at first, then levels off, and finally rises quickly at the end corresponds to a histogram with little area under the curve in the middle and much greater area on both ends.

7. **(E)** The consumer group randomly picked 5 of the cars and found an average efficiency of only 32 mpg. The chance of picking 5 cars with an average of 32 mpg or lower from a fleet of cars of real 35 mpg efficiency is so small, only 4 out of 160, or 0.025, that it seems reasonable to conclude that the consumer group's set of 5 cars was not picked from a fleet of cars with 35 mpg efficiency and that the company's claim is wrong.

FREE-RESPONSE QUESTIONS

1. (a)

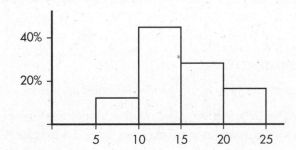

   (b) $\dfrac{3+11}{25} = \dfrac{14}{25} = 56\%$

(c)

Miles

(d) The answer would still be 56%.

2. (a) A complete answer considers shape, center, and spread; unusual features like outliers; and mentions context.
Shape: unimodal, skewed right, outlier at 10
Center: around 2 or 3
Spread: from 0 to 10 goals were scored by each team

(b) If the player scored six goals, his/her team must have scored either 7 or 10, but they lost, so they scored 7, and the only possible final score is that they lost by a score of 10 to 7.

(c) No, there were six teams that scored exactly two goals, but there were only five teams that scored less than two goals, so not all the two-goal teams could have won.

3. (a) The lowest winning percentage over the past 22 years was 46.0%.

(b) A complete answer considers shape, center, and spread; unusual features like clusters and gaps; and mentions context.
Shape: two clusters, each somewhat bell-shaped
Center: around 50%
Spread: the winning percentages were between 46.0% and 55.6%

(c) The team had more losing seasons (13) than winning seasons (9).

(d) The cluster of winning percentages is further above 50% than the cluster of losing percentages is below 50%.

Quiz 2

MULTIPLE-CHOICE QUESTIONS

1. **(B)** There is no such thing as being skewed both left and right.

2. **(B)** Histograms give information about relative frequencies (relative areas correspond to relative frequencies) and may or may not have an axis with actual frequencies. Symmetric histograms can have any number of peaks. Choice of width and number of classes changes the appearance of a histogram. Stemplots clearly show outliers; however, in histograms outliers may be hidden in large class widths.

3. **(B)** A histogram with little area under the curve early and much greater area later results in a cumulative relative frequency plot that rises slowly at first and then at a much faster rate later.

4. **(C)** A histogram with large area under the curve early and much less area later results in a cumulative relative frequency plot that rises quickly at first and then at a much slower rate later.

5. **(E)** A histogram with little area under the curve in the middle and much greater area on both ends results in a cumulative relative frequency plot that rises quickly at first, then almost levels off, and finally rises quickly at the end.

6. **(D)** A histogram with little area under the curve on the ends and much greater area in the middle results in a cumulative relative frequency plot that rises slowly at first, then quickly in the middle, and finally slowly again at the end.

7. **(A)** Uniform distributions result in cumulative relative frequency plots that rise at constant rates, thus linear.

FREE-RESPONSE QUESTIONS

1. (a) 40% of the players averaged fewer than 20 points per game.

   (b) All the players averaged at least 3 points per game.

   (c) No players averaged between 5 and 7 points per game because the cumulative relative frequency was 10% for both 5 and 7 points.

   (d) Go over to the plot from 0.9 on the vertical axis, and then down to the horizontal axis to result in 28 points per game.

   (e) Reading up to the plot and then over from 10 and from 20 shows that 0.25 of the players averaged under 10 points per game and 0.4 of the players averaged under 20 points per game. Thus, $0.4 - 0.25 = 0.15$ gives the proportion of players who averaged between 10 and 20 points per game.

2. (a) The center is roughly between 24.10 and 24.11, and the data are spread from 24.01 to 24.20.

   (b) There seems to be two "low" data points, 24.01 and 24.02, and one "high" data point, 24.20. These three data points are distinctly separated from the other points.

   (c) For this day's sample, $W = \frac{25.05 + 23.01}{2} = 24.03$. A value of 24.03 or less occurred only twice in the 100 samples. Thus, if the machinery was operating properly, a W measurement of 24.03 would be very unusual. The conclusion should be to recalibrate the machine.

3. (a) A complete answer considers shape, center, spread, and unusual features, and mentions context.

   Shape: the distribution of months of dating before becoming engaged is skewed right.

   Center: around 14.5 months

   Spread: from 12.0 to 23.1 months of dating before becoming engaged.

   Unusual feature: there appears to be an outlier at 23.1 months.

   (b) The cumulative frequency plot of this data rises rapidly between 12 and 15 months and then rises at a much slower pace. There will be a horizontal piece between 20 and 23 months because with no data between these values, the proportion of the data less than 20 is the same as that less than 23.

Summarizing Distributions

2

- → **MEASURING THE CENTER**
- → **MEASURING SPREAD**
- → **MEASURING POSITION**
- → **EMPIRICAL RULE**
- → **HISTOGRAMS**
- → **BOXPLOTS**
- → **COMPARING DISTRIBUTIONS**
- → **CHANGING UNITS**

Given a raw set of data, often we can detect no overall pattern. Perhaps some values occur more frequently, a few extreme values may stand out, and the range of values is usually apparent. The presentation of data, including summarizations and descriptions, and involving such concepts as representative or average values, measures of dispersion, positions of various values, and the shape of a distribution, falls under the broad topic of *descriptive statistics*. This aspect of statistics is in contrast to *statistical analysis*, the process of drawing inferences from limited data, a subject discussed in later topics.

MEASURING THE CENTER: MEDIAN AND MEAN

The word *average* is used in phrases common to everyday conversation. People speak of bowling and batting averages or the average life expectancy of a battery or a human being. Actually the word *average* is derived from the French *avarie*, which refers to the money that shippers contributed to help compensate for losses suffered by other shippers whose cargo did not arrive safely (i.e., the losses were shared, with everyone contributing an average amount). In common usage *average* has come to mean a representative score or a typical value or the center of a distribution. Mathematically, there are a variety of ways to define the average of a set of data. In practice, we use whichever method is most appropriate for the particular case under consideration. However, beware of a headline with the word *average;* the writer has probably chosen the method that emphasizes the point he or she wishes to make.

In the following paragraphs we consider the two primary ways of denoting an average:

1. The *median*, which is the middle number of a set of numbers arranged in numerical order.
2. The *mean*, which is found by summing items in a set and dividing by the number of items.

➡ EXAMPLE 2.1 _____

Consider the following set of home run distances (in feet) to center field in 13 ballparks: {387, 400, 400, 410, 410, 410, 414, 415, 420, 420, 421, 457, 461}. What is the average?

Answer: The median is 414 (there are six values below 414 and six values above), while the mean is

$$\frac{387 + 400 + 400 + 410 + 410 + \cdots + 457 + 461}{13} = 417.3 \text{ feet}$$

Median

REMEMBER

Don't forget to put the data in order before finding the median.

The word *median* is derived from the Latin *medius* which means "middle." The values under consideration are arranged in ascending or descending order. If there is an odd number of values, the median is the middle one. If there is an even number, the median is found by adding the two middle values and dividing by 2. Thus the median of a set has the same number of elements above it as below it.

The median is not affected by exactly how large the larger values are or by exactly how small the smaller values are. Thus it is a particularly useful measurement when the extreme values, *outliers*, are in some way suspicious or when we want to diminish their effect. For example, if ten mice try to solve a maze, and nine succeed in less than 15 minutes while one is still trying after 24 hours, the most representative value is the median (not the mean, which is over 2 hours). Similarly, if the salaries of four executives are each between $240,000 and $245,000 while a fifth is paid less than $20,000, again the most representative value is the median (the mean is under $200,000). It is often said that the median is "resistant" to extreme values.

In certain situations the median offers the most economical and quickest way to calculate an average. For example, suppose 10,000 lightbulbs of a particular brand are installed in a factory. An average life expectancy for the bulbs can most easily be found by noting how much time passes before exactly one-half of them have to be replaced. The median is also useful in certain kinds of medical research. For example, to compare the relative strengths of different poisons, a scientist notes what dosage of each poison will result in the death of exactly one-half the test animals. If one of the animals proves especially susceptible to a particular poison, the median lethal dose is not affected.

Mean

While the median is often useful in descriptive statistics, the *mean*, or more accurately, the *arithmetic mean*, is most important for statistical inference and analysis. Also, for the layperson, the average is usually understood to be the mean.

The mean of a *whole population* (the complete set of items of interest) is often denoted by the Greek letter μ (mu), while the mean of a *sample* (a part of a population) is often denoted by $\bar{x}$. For example, the mean value of the set of all houses in the United States might be $\mu = \$56,400$, while the mean value of 100 randomly chosen houses might be $\bar{x} = \$52,100$ or perhaps $\bar{x} = \$63,800$ or even $\bar{x} = \$124,000$.

NOTE

Methods of
collecting data,
including the
SRS, will be fully
discussed in
Topic 5.

In statistics we learn how to estimate a population mean from a sample mean. Throughout this book, the word *sample* often implies a *simple random sample* (SRS), that is, a sample selected in such a way that every possible sample of the desired size has an equal chance of being included. (It is also true that each element of the population will have an equal chance of being included.) In the real world, this process of random selection is often very difficult to achieve, and so we proceed, with caution, as long as we have good reason to believe that our sample is representative of the population.

Mathematically, the mean $= \dfrac{\sum x}{n}$, where $\sum x$ represents the sum of all the elements of the set under consideration and n is the actual number of elements. Σ is the uppercase Greek letter sigma.

➡ **EXAMPLE 2.2**

Suppose that the numbers of unnecessary procedures recommended by five doctors in a 1-month period are given by the set {2, 2, 8, 20, 33}. Note that the median is 8 and the mean is $\dfrac{2+2+8+20+33}{5} = 13$. If it is discovered that the fifth doctor also recommended an additional 25 unnecessary procedures, how will the median and mean be affected?

Answer: The set is now {2, 2, 8, 20, 58}. The median is still 8; however, the mean changes to $\dfrac{2+2+8+20+58}{5} = 18$.

Example 2.2 illustrates how the mean, unlike the median, is sensitive to a change in any value.

➡ **EXAMPLE 2.3**

Suppose the salaries of six employees are $3000, $7000, $15,000, $22,000, $23,000, and $38,000, respectively.

a. What is the mean salary?

Answer:

$$\frac{3000 + 7000 + 15,000 + 22,000 + 23,000 + 38,000}{6} = \$18,000$$

b. What will the new mean salary be if everyone receives a $3000 increase?

Answer:

$$\frac{6000 + 10,000 + 18,000 + 25,000 + 26,000 + 41,000}{6} = \$21,000$$

Note that $18,000 + $3000 = $21,000.

c. What if everyone receives a 10% raise?

Answer:

$$\frac{3300 + 7700 + 16,500 + 24,200 + 25,300 + 41,800}{6} = \$19,800$$

Note that 110% of $18,000 is $19,800.

Example 2.3 illustrates how adding the same constant to each value increases the mean (and median) by a like amount. Similarly, multiplying each value by the same constant multiplies the mean (and median) by a like amount.

MEASURING SPREAD: RANGE, INTERQUARTILE RANGE, VARIANCE, AND STANDARD DEVIATION

In describing a set of numbers, not only is it useful to designate an average value but it is also important to be able to indicate the *variability* or the *dispersion* of the measurements. An explosion engineer in mining operations aims for small variability—it would not be good for his 30-minute fuses actually to have a range of 10–50 minutes before detonation. On the other hand, a teacher interested in distinguishing better students from poorer students aims to design exams with large variability in results—it would not be helpful if all her students scored exactly the same. The players on two basketball teams may have the same average height, but this observation doesn't tell the whole story. If the dispersions are quite different, one team may have a 7-foot player, whereas the other has no one over 6 feet tall. Two Mediterranean holiday cruises may advertise the same average age for their passengers. One, however, may have only passengers between 20 and 25 years old, while the other has only middle-aged parents in their forties together with their children under age 10.

There are four primary ways of describing variability or dispersion:

1. The *range*, which is the difference between the largest and smallest values
2. The *interquartile range*, IQR, which is the difference between the largest and smallest values after removing the lower and upper quarters (i.e., IQR is the range of the middle 50%); that is, IQR = $Q_3 - Q_1$ = 75th percentile minus 25th percentile
3. The *variance*, which is determined by averaging the squared differences of all the values from the mean
4. The *standard deviation*, which is the square root of the variance

➡ EXAMPLE 2.4

The monthly rainfall in Monrovia, Liberia, where May through October is the rainy season and November through April the dry season, is as follows:

Month:	Jan	Feb	Mar	Apr	May	June	July	Aug	Sept	Oct	Nov	Dec
Rain (in.):	1	2	4	6	18	37	31	16	28	24	9	4

The mean is

$$\frac{1 + 2 + 4 + 6 + 18 + 37 + 31 + 16 + 28 + 24 + 9 + 4}{12} = 15 \text{ inches}$$

What are the measures of variability?

Answer: **Range:** The maximum is 37 inches (June), and the minimum is 1 inch (January). Thus the range is 37 – 1 = 36 inches of rain.

Interquartile range: Removing the lower and upper quarters leaves 4, 6, 9, 16, 18, and 24. Thus the interquartile range is 24 – 4 = 20. [The interquartile range is sometimes calculated as follows: The median of the lower half is $Q_1 = \frac{4+4}{2} = 4$, the median of

(continued)

the upper half is $Q_3 = \frac{24+28}{2} = 26$, and the interquartile range is $Q_3 - Q_1 = 22$. When there is a large number of values in the set, the two methods give the same answer.]

Variance:

$$\frac{14^2 + 13^2 + 11^2 + 9^2 + 3^2 + 22^2 + 16^2 + 1^2 + 13^2 + 9^2 + 6^2 + 11^2}{12} = 143.7$$

Standard deviation: $\sqrt{143.7} = 12.0$ inches

Range

The simplest, most easily calculated measure of variability is the *range*. The difference between the largest and smallest values can be noted quickly, and the range gives some impression of the dispersion. However, it is entirely dependent on the two extreme values and is insensitive to the ones in the middle.

One use of the range is to evaluate samples with very few items. For example, some quality control techniques involve taking periodic small samples and basing further action on the range found in several such samples.

Interquartile Range and Outliers

Finding the *interquartile range* is one method of removing the influence of extreme values on the range. It is calculated by arranging the data in numerical order, removing the upper and lower quarters of the values, and noting the range of the remaining values. That is, it is the range of the middle 50% of the values.

The numerical rule for designating outliers is to calculate 1.5 times the interquartile range (IQR) and then call a value an outlier if it is more than $1.5 \times$ IQR below the first quartile or $1.5 \times$ IQR above the third quartile.

NOTE

The description to the left is the most commonly accepted mathematical definition of *outliers* and should be used whenever you are asked to determine if a value qualifies as an outlier.

➡ **EXAMPLE 2.5**

Suppose that the starting salaries (in $1000) for college graduates who took AP Statistics in high school and at least one additional statistics class in college have the following characteristics: the smallest value is 18.8, 10% of the values are below 25.6, 25% are below 41.1, the median is 59.3, 60% are below 84.3, 75% are below 101.9, 90% are below 118.0, and the top value is 201.7.

a. What is the range?

Answer: The range is 201.7 – 18.8 = 182.9 (thousand dollars) = $182,900.

b. What is the interquartile range?

Answer: The interquartile range, that is, the range of the middle 50% of the values, is $Q_3 - Q_1 = 101.9 - 41.1 = 60.8$ (thousand dollars) = $60,800.

c. When the numerical rule is used for outliers, should either the smallest or largest value be called an outlier?

Answer: $1.5 \times$ IQR = $1.5 \times 60.8 = 91.2$. If a value is more than 91.2 below the first quartile, 41.1, or more than 91.2 above the third quartile, 101.9, then it will be called an outlier. Since the largest value, 201.7, is greater than 101.9 + 91.2 = 193.1, it is considered an outlier by the given numerical rule.

Variance

Dispersion is often the result of various chance happenings. For example, consider the motion of microscopic particles suspended in a liquid. The unpredictable motion of any particle is the result of many small movements in various directions caused by random bumps from other particles. If we average the total displacements of all the particles from their starting points, the result will not increase in direct proportion to time. If, however, we average the *squares* of the total displacements of all the particles, this result will increase in direct proportion to time.

The same holds true for the movement of paramecia. Their seemingly random motions as seen under a microscope can be described by the observation that the average of the squares of the displacements from their starting points is directly proportional to time. Also, consider ping-pong balls dropped straight down from a high tower and subjected to chance buffeting in the air. We can measure the deviations from a center spot on the ground to the spots where the balls actually strike. As the height of the tower is increased, the average of the squared deviations increases proportionately.

In a wide variety of cases we are in effect trying to measure dispersion from the mean due to a multitude of chance effects. The proper tool in these cases is the average of the squared deviations from the mean; it is called the *variance* and is denoted by σ^2 (σ is the lowercase Greek letter sigma):

$$\sigma^2 = \frac{\Sigma(x - \mu)^2}{n}$$

For circumstances specified later, the variance of a sample, denoted by s^2, is calculated as

$$s^2 = \frac{\Sigma(x - \bar{x})^2}{n - 1}$$

➡ **EXAMPLE 2.6**

TIP

Most calculators give the standard deviation, and this must be squared to find the variance.

The Points Per Game (PPG) during the 2014–2015 season of the Golden State Warriors players were {23.8, 21.7, 11.7, 10.1, 7.8, 6.3, 5.9, 7.9, 10.4, 7.1, 4.3, 4.4, 4.1, 0.9, 1.3}. What was the variance?

Answer: The variance can be quickly found on any calculator that has a simple statistical package, or it can be found as follows:

$$\mu = \frac{23.8 + 21.7 + 11.7 + 10.1 + 7.8 + 6.3 + 5.9 + 7.9 + 10.4 + 7.1 + 4.3 + 4.4 + 4.1 + 0.9 + 1.3}{15} = 8.51$$

$$\sigma^2 = \frac{(23.8 - 8.51)^2 + (21.7 - 8.51)^2 + (11.7 - 8.51)^2 + \cdots + (1.3 - 8.51)^2}{15} = 40.17$$

Standard Deviation

Suppose we wish to pick a representative value for the variability of a certain population. The preceding discussions indicate that a natural choice is the value whose square is the average of the squared deviations from the mean. Thus we are led to consider the square root of the variance. This value is called the *standard deviation*, is denoted by σ, and is calculated on your calculator or as follows:

$$\sigma = \sqrt{\frac{\Sigma(x-\mu)^2}{n}}$$

Similarly, the standard deviation of a sample is denoted by s and is calculated on your calculator or as follows:

$$s = \sqrt{\frac{\Sigma(x-\bar{x})^2}{n-1}}$$

While variance is measured in square units, standard deviation is measured in the same units as are the data.

For the various x-values, the deviations $x - \bar{x}$ are called *residuals*, and s is a "typical value" for the residuals. While s is not the average of the residuals (the average of the residuals is always 0), s does give a measure of the spread of the x-values around the sample mean.

NOTE

Residuals in the context of linear regression will be discussed in Topic 3.

➡ EXAMPLE 2.7

The number of calories in 12-ounce servings of five popular beers are {95, 152, 188, 205, 131}. Using the TI-84, 1-Var Stats gives

> 1-Var Stats
>
> $\bar{x} = 154.2$
>
> $\sum x = 771$
>
> $\sum x^2 = 126659$
>
> $Sx = 44.076$
>
> $\sigma x = 39.423$

Since these data represent a sample of beers, the standard deviation is 44.076.

NOTE

Or 1-Variable on the Casio Prizm or Statistics 1Var App on the HP Prime.

MEASURING POSITION: SIMPLE RANKING, PERCENTILE RANKING, AND *Z*-SCORE

We have seen several ways of choosing a value to represent the center of a distribution. We also need to be able to talk about the *position* of any other values. In some situations, such as wine tasting, simple rankings are of interest. Other cases, for example, evaluating college applications, may involve positioning according to percentile rankings. There are also situations in which position can be specified by making use of measurements of both central tendency and variability.

There are three important, recognized procedures for designating position:

1. *Simple ranking*, which involves arranging the elements in some order and noting where in that order a particular value falls
2. *Percentile ranking*, which indicates what percentage of all values fall below the value under consideration
3. The *z-score*, which states very specifically by how many standard deviations a particular value varies from the mean.

➡ **EXAMPLE 2.8** _____

It is recommended that the "good cholesterol," high-density lipoprotein (HDL), be present in the blood at levels of at least 40 mg/dl. Suppose a 50-member high school football team are all tested with resulting HDL levels of {53, 26, 45, 33, 64, 29, 73, 29, 21, 58, 70, 41, 48, 55, 55, 39, 57, 48, 9, 59, 56, 39, 68, 50, 65, 30, 38, 54, 49, 35, 56, 70, 43, 86, 52, 40, 28, 40, 67, 50, 47, 54, 59, 29, 29, 42, 45, 37, 51, 40}. What is the position of the HDL score of 41?

Answer: Since there are 31 higher HDL levels on the list, the 41 has a simple ranking of 32 out of 50. Eighteen HDL levels are lower, so the percentile ranking is $18/50 = 36\%$. The above list has a mean of 47.22 with a standard deviation of 15.05, so the HDL score of 41 has a z-score of $(41 - 47.22)/15.05 = -0.413$.

Simple Ranking

Simple ranking is easily calculated and easily understood. We know what it means for someone to graduate second in a class of 435, or for a player from a team of size 30 to have the seventh-best batting average. Simple ranking is useful even when no numerical values are associated with the elements. For example, detergents can be ranked according to relative cleansing ability without any numerical measurements of strength.

Percentile Ranking

Percentile ranking, another readily understood measurement of position, is helpful in comparing positions with different bases. We can more easily compare a rank of 176 out of 704 with a rank of 187 out of 935 by noting that the first has a rank of 75%, and the second, a rank of 80%. Percentile rank is also useful when the exact population size is not known or is irrelevant. For example, it is more meaningful to say that Jennifer scored in the 90th percentile on a national exam rather than trying to determine her exact ranking among some large number of test takers.

The *quartiles*, Q_1 and Q_3, lie one-quarter and three-quarters of the way up a list, respectively. Their percentile ranks are 25% and 75%, respectively. The interquartile range defined earlier can also be defined to be $Q_3 - Q_1$. The *deciles* lie one-tenth and nine-tenths of the way up a list, respectively, and have percentile ranks of 10% and 90%.

z-Score

The *z-score* is a measure of position that takes into account both the center and the dispersion of the distribution. More specifically, the z-score of a value tells how many standard deviations the value is from the mean. Mathematically, $x - \mu$ gives the raw distance from μ to x; dividing by σ converts this to number of standard deviations. Thus $z = \dfrac{x - \mu}{\sigma}$, where x is the raw score, μ is the mean, and σ is the standard deviation. If the score x is greater than the mean μ, then z is positive; if x is less than μ, then z is negative.

Given a z-score, we can reverse the procedure and find the corresponding raw score. Solving for x gives $x = \mu + z\sigma$.

➡ EXAMPLE 2.9

In 2018, because of hurricanes, refinery outages, and Middle East tensions, the average (mean) price of gasoline in one NE city hit \$3.80 per gallon with a standard deviation of \$0.05. Then \$3.90 has a z-score of $\dfrac{3.90 - 3.80}{0.05} = +2$, while \$3.65 has a z-score of $\dfrac{3.65 - 3.80}{0.05} = -3$. Alternatively, a z-score of +2.2 corresponds to a raw score of $3.80 + 2.2(0.05) = 3.80 + 0.11 = 3.91$, while a z-score of -1.6 corresponds to $3.80 - 1.6(0.05) = 3.72$.

It is often useful to portray integer z-scores and the corresponding raw scores as follows:

➡ EXAMPLE 2.10

The world's largest cinema complex (in Madrid, Spain) has one screen with attendance averaging 780 with a standard deviation of 40. Adding multiples of 40 to and subtracting multiples of 40 from the mean 780 gives

A theater attendance of 835 is converted to a z-score as follows: $\dfrac{835 - 780}{40} = \dfrac{55}{40} = 1.375$.

A z-score of -2.15 is converted to a theater attendance as follows: $780 - 2.15(40) = 694$.

EMPIRICAL RULE

The *empirical rule* (also called the *68-95-99.7 rule*) applies specifically to symmetric bell-shaped data (not to skewed data!). In this case, about 68% of the values lie within 1 standard deviation of the mean, about 95% of the values lie within 2 standard deviations of the mean, and more than 99% of the values lie within 3 standard deviations of the mean.

In the following figure the horizontal axis shows *z*-scores:

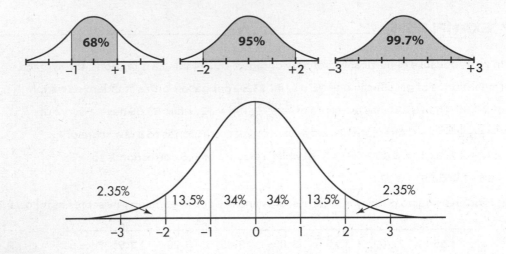

➡ **EXAMPLE 2.11** _____

Suppose that taxicabs in New York City are driven an average of 75,000 miles per year with a standard deviation of 12,000 miles. What information does the empirical rule give us?

Answer: Assuming that the distribution is bell-shaped, we can conclude that approximately 68% of the taxis are driven between 63,000 and 87,000 miles per year, approximately 95% are driven between 51,000 and 99,000 miles, and virtually all are driven between 39,000 and 111,000 miles.

The empirical rule also gives a useful quick estimate of the standard deviation in terms of the range. We can see in the figure above that 95% of the data fall within a span of 4 standard deviations (from –2 to +2 on the *z*-score line) and 99.7% of the data fall within 6 standard deviations (from –3 to +3 on the *z*-score line). It is therefore reasonable to conclude that for these data the standard deviation is roughly between one-fourth and one-sixth of the range of a finite set. Since we can find the range of a finite set almost immediately, the empirical rule technique for estimating the standard deviation is often helpful in pointing out gross arithmetic errors.

➡ **EXAMPLE 2.12** _____

If the range of a finite bell-shaped data set is 60, what is an estimate for the standard deviation?

Answer: By the empirical rule, the standard deviation is expected to be between $\left(\frac{1}{6}\right)60 = 10$ and $\left(\frac{1}{4}\right)60 = 15$. If the standard deviation is calculated to be 0.32 or 87, there is probably an arithmetic error; a calculation of 12, however, is reasonable.

However, it must be stressed that the above use of the range is not intended to provide an accurate value for the standard deviation. It is simply a tool for pointing out unreasonable answers rather than, for example, blindly accepting computer outputs.

HISTOGRAMS AND MEASURES OF CENTRAL TENDENCY

Suppose we have a detailed histogram such as

Our measures of central tendency fit naturally into such a diagram.

The *median* divides a distribution in half, so it is represented by a line that divides the area of the histogram in half.

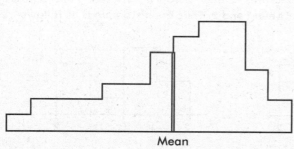

The *mean* is affected by the spacing of all the values. Therefore, if the histogram is considered to be a solid region, the mean corresponds to a line passing through the center of gravity, or balance point.

The above distribution, spread thinly far to the low side, is said to be *skewed to the left*. Note that in this case the mean is usually less than the median. Similarly, a distribution spread far to the high side is *skewed to the right*, and its mean is usually greater than its median.

➡ **EXAMPLE 2.13** _____

Suppose that the faculty salaries at a college have a median of $82,500 and a mean of $88,700. What does this indicate about the shape of the distribution of the salaries?

Answer: The median is less than the mean, and so the salaries are probably skewed to the right. There are a few highly paid professors, with the bulk of the faculty at the lower end of the pay scale.

It should be noted that the above principle is a useful, but not hard-and-fast, rule.

The set given by the dotplot below is skewed to the right; however, its median (3) is greater than its mean (2.97).

HISTOGRAMS, Z-SCORES, AND PERCENTILE RANKINGS

We have seen that relative frequencies are represented by relative areas, and so labeling the vertical axis is not crucial. If we know the standard deviation, the horizontal axis can be labeled in terms of z-scores. In fact, if we are given the percentile rankings of various z-scores, we can construct a histogram.

→ EXAMPLE 2.15

Suppose we are asked to construct a histogram from these data:

z-score:	-2	-1	0	1	2
Percentile ranking:	0	20	60	70	100

We note that the entire area is less than z-score +2 and greater than z-score -2. Also, 20% of the area is between z-scores -2 and -1, 40% is between -1 and 0, 10% is between 0 and 1, and 30% is between 1 and 2. Thus the histogram is as follows:

or

Now suppose we are given four in-between z-scores as well:

z-score	Percentile Ranking
2.0	100
1.5	80
1.0	70
0.5	65
0.0	60
-0.5	30
-1.0	20
-1.5	5
-2.0	0

Then we have:

With 1000 *z*-scores perhaps the histogram would look like

    The height at any point is meaningless; what is important is relative areas. For example, in the final diagram above, what percentage of the area is between *z*-scores of +1 and +2?

Answer: Still 30%.

What percent is to the left of 0?

Answer: Still 60%.

BOXPLOTS

A *boxplot* (also called a *box and whisker display*) is a visual representation of dispersion that shows the smallest value, the largest value, the middle (median), the middle of the bottom half of the set (Q_1), and the middle of the top half of the set (Q_3).

TIP

The IQR is the *length* of the box, not the box itself. So the median is in the box, or is between Q_1 and Q_3, but is not *in* the IQR.

➡ **EXAMPLE 2.16** _____

After an AP Statistics teacher hears that every one of her 27 students received a 3 or higher on the AP exam, she treats the class to a game of bowling. The individual student scores are 210, 130, 150, 140, 150, 210, 150, 125, 85, 200, 70, 150, 75, 90, 150, 115, 120, 125, 160, 140, 100, 95, 100, 215, 130, 160, and 200. Their students note that the greatest score is 215, the smallest is 70, the middle is 140, the middle of the top half is 160, and the middle of the bottom half is 100. A boxplot of these five numbers is

Bowling scores

TIP

Be careful about describing the shape of a distribution when all that one has is a boxplot. For example, "approximately normal" is never a possible conclusion.

    Note that the display consists of two "boxes" together with two "whiskers"—hence the alternative name. The boxes show the spread of the two middle quarters; the whiskers show the spread of the two outer quarters. This relatively simple display conveys information not immediately available from histograms or stem and leaf displays.
    Putting the above data into a list, for example, L1, on the TI-84, not only gives the five-number summary

```
1-Var Stats
minX=70
Q1=100
Med=140
Q3=160
MaxX=215
```

TIP

Note that a boxplot gives one measure of center (the median) and two measures of variability (the range and the IQR).

but also gives the boxplot itself using STAT PLOT, choosing the boxplot from among the six type choices, and then using ZoomStat or in WINDOW letting Xmin=0 and Xmax=225.

On the TI-Nspire the data can be put in a list (here called *index*), and then a simultaneous multiple view is possible.

When a distribution is strongly skewed, or when it has pronounced outliers, drawing a box-plot with its five-number summary including median, quartiles, and extremes, gives a more useful description than calculating a mean and a standard deviation.

Values more than $1.5 \times IQR$ (1.5 times the interquartile range) outside the two boxes are plotted separately as outliers. (The TI-84, Casio Prizm, and HP Prime have this boxplot option.)

➡ **EXAMPLE 2.17**

Inputting the lengths of words in a selection of Shakespeare's plays results in a calculator output of

```
1-Var Stats
minX=1
Q₁=3
Med=4
Q₃=5
maxX=12
```

Outliers consist of any word lengths less than $Q_1 - 1.5(IQR) = 3 - 1.5(5 - 3) = 0$ or greater than $Q_3 + 1.5(IQR) = 5 + 1.5(5 - 3) = 8$. A boxplot indicating outliers, together with a histogram (on the TI-84 up to three different graphs can be shown simultaneously) is

NOTE

When outliers are present, the whiskers are drawn to the smallest and largest values that are not outliers; they are *not* drawn to $Q_1 - 1.5$ (*IQR*) and $Q_3 + 1.5$ (*IQR*) unless these are actual data values.

Note: Some computer output shows *two* levels of outliers—mild (between 1.5 IQR and 3 IQR) and extreme (more than 3 IQR). In this example, the word length of 12 would be considered an extreme outlier since it is greater than 5 + 3(5 − 3) = 11.

It should be noted that two sets can have the same five-number summary and thus the same boxplots but have dramatically different distributions.

➡ EXAMPLE 2.18

Let A = {0, 5, 10, 15, 25, 30, 35, 40, 45, 50, 71, 72, 73, 74, 75, 76, 77, 78, 100} and B = {0, 22, 23, 24, 25, 26, 27, 28, 29, 50, 55, 60, 65, 70, 75, 85, 90, 95, 100}. Simple inspection indicates very different distributions, however the TI-84 gives identical boxplots with Min = 0, Q_1 = 25, Med = 50, Q_3 = 75, and Max = 100 for each.

COMPARING DISTRIBUTIONS

➡ EXAMPLE 2.19

The numbers of wins for the 30 NBA teams at the end of the 2014–2015 season is shown in the following *back-to-back stemplot*.

NOTE

The leaves radiate out from a common stem.

```
        Eastern          Western
       Conference       Conference
          8 7     1    6
            5     2    1 9
      8 8 7 3 2   3    0 8 9
        9 6 1 0   4    5 5
          3 0     5    0 1 5 5 6 6
            0     6    7
```

7 | 1 | 6 represents Eastern and Western Conference teams with 17 and 16 wins, respectively.

NOTE

Although both sets in this example are of the same size, in general you do not have to have the same number of data points on each side of the plot.

When comparing shape, center, and spread, we have:

Shape: The distribution of wins in the Eastern Conference (EC) is roughly bell-shaped, while the distribution of wins in the Western Conference (WC) is roughly skewed toward the lower values.

Center: Counting values (8th out of 15) gives medians of $m_{EC} = 38$ and $m_{WC} = 45$. Thus the WC distribution of wins has the greater center.

Spread: The range of the EC distribution of wins is 60 – 17 = 43, while the range of the WC distribution of wins is 67 – 16 = 51. Thus the WC distribution of wins has the greater spread.

➡ **EXAMPLE 2.20** _____

Following are parallel boxplots showing the daily price fluctuations of a certain common stock over the course of 5 years. What trends do the boxplots show?

The parallel boxplots show that from year to year the median daily stock price has steadily risen 20 points from about $58 to about $78, the third quartile value has been roughly stable at about $84, the yearly low has never decreased from that of the previous year, and the interquartile range has never increased from one year to the next.

➡ **EXAMPLE 2.21** _____

The graph below compares cumulative frequency plotted against age for the U.S. population in 1860 and in 1980.

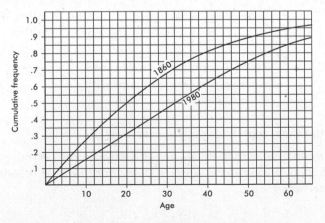

How do the medians and interquartile ranges compare?

Answer: Looking across from .5 on the vertical axis, we see that in 1860 half the population was under the age of 20, while in 1980 all the way up to age 32 must be included to encompass half the population. Looking across from .25 and .75 on the vertical axis, we see that for 1860, Q_1 = 9 and Q_3 = 35 and so the interquartile range is 35 − 9 = 26 years, while for 1980, Q_1 = 16 and Q_3 = 50 and so the interquartile range is 50 − 16 = 34 years. Thus, both the median and the interquartile range were greater in 1980 than in 1860.

EFFECT OF CHANGING UNITS

Changing units, for example, from dollars to rubles or from miles to kilometers, is common in a world that seems to become smaller all the time. It is instructive to note how measures of center and spread are affected by such changes.

Adding the same constant to every value increases the mean and median by that same constant; however, the distances between the increased values stay the same, and so the range and standard deviation are unchanged. Multiplying every value by the same constant multiplies the mean, median, range, and standard deviation by that same constant.

NOTE

One consequence is that no matter what the distribution of raw scores, the set of *z*-scores always has mean 0 and standard deviation 1.

➡ **EXAMPLE 2.22**

A set of experimental measurements of the freezing point of an unknown liquid yield a mean of 25.32 degrees Celsius with a standard deviation of 1.47 degrees Celsius. If all the measurements are converted to the Kelvin scale, what are the new mean and standard deviation?

Answer: Kelvins are equivalent to degrees Celsius plus 273.16. The new mean is thus 25.32 + 273.16 = 298.48 kelvins. However, the standard deviation remains numerically the same, 1.47 kelvins. Graphically, you should picture the whole distribution moving over by the constant 273.16; the mean moves, but the standard deviation (which measures spread) doesn't change.

➡ **EXAMPLE 2.23**

Measurements of the sizes of farms in an upstate New York county yield a mean of 59.2 hectares with a standard deviation of 11.2 hectares. If all the measurements are converted from hectares (metric system) to acres (one acre was originally the area a yoke of oxen could plow in one day), what are the new mean and standard deviation?

Answer: One hectare is equivalent to 2.471 acres. The new mean is thus 2.471 × 59.2 = 146.3 acres with a standard deviation of 2.471 × 11.2 = 27.7 acres. Graphically, multiplying each value by the constant 2.471 both moves and spreads out the distribution.

SUMMARY

- The two principal measurements of the center of a distribution are the mean and the median.
- The principal measurements of the spread of a distribution are the range (maximum value minus minimum value), the interquartile range (IQR = $Q_3 - Q_1$), the variance, and the standard deviation.
- Adding the same constant to every value in a set adds the same constant to the mean and median but leaves all the above measures of spread unchanged.
- Multiplying every value in a set by the same constant multiplies the mean, median, range, IQR, and standard deviation by that constant.
- The mean, range, variance, and standard deviation are sensitive to extreme values, while the median and interquartile range are not.
- The principal measurements of position are simple ranking, percentile ranking, and the z-score (which measures the number of standard deviations from the mean).
- No matter what the distribution of raw scores, the set of z-scores always has mean 0 and standard deviation 1.
- The empirical rule (the 68-95-99.7 rule) applies specifically to symmetric bell-shaped data.
- In skewed left data, the mean is usually less than the median, while in skewed right data, the mean is usually greater than the median.
- Boxplots visually show the five-number summary: the minimum value, the first quartile (Q_1), the median, the third quartile (Q_3), and the maximum value; and usually indicate outliers as distinct points.
- Note that two sets can have the same five-number summary and thus the same box-plots but have dramatically different distributions.
- Outliers are any values less than $Q_1 - 1.5(IQR)$ or greater than $Q_3 + 1.5(IQR)$.
- When asked for a comparison of two data sets, in addition to comparing shape and unusual features, you must state which center and which spread is greater (two separate calculations are not a comparison).

QUIZ 1

Multiple-Choice Questions

> **Directions:** The questions that follow are each followed by five suggested answers. Choose the response that best answers the question.

1. The graph below shows household income in Laguna Woods, California.

Household income distribution ($1000)

What can be said about the ratio $\dfrac{\text{Mean family income}}{\text{Median family income}}$?

(A) Approximately zero

(B) Less than one, but definitely above zero

(C) Approximately one

(D) Greater than one

(E) Cannot be answered without knowing the standard deviation

2. Students in AP Statistics class were timed in seconds while solving a series of mathematical brainteasers. One student's time had a standardized score of $z = 2.40$. If the times are all changed to minutes, what will then be the student's standardized score?

(A) $z = 0.04$

(B) $z = 0.4$

(C) $z = 1.80$

(D) $z = 2.40$

(E) The new standardized score cannot be determined without knowing the class mean.

3. Dieticians are concerned about sugar consumption in teenagers' diets (a 12-ounce can of soft drink typically has 10 teaspoons of sugar). In a random sample of 55 students, the number of teaspoons of sugar consumed for each student on a randomly selected day is tabulated. Summary statistics are noted below:

Min = 10   Max = 60   First quartile = 25   Third quartile = 38
Median = 31   Mean = 31.4 $n = 55$ $S = 11.6$

Which of the following is a true statement?

(A) None of the values are outliers.
(B) The value 10 is an outlier, and there can be no others.
(C) The value 60 is an outlier, and there can be no others.
(D) Both 10 and 60 are outliers, and there may be others.
(E) The value 60 is an outlier, and there may be others at the high end of the data set.

Questions 4–6 refer to the following five boxplots.

4. To which of the above boxplots does the following histogram correspond?

(A) A
(B) B
(C) C
(D) D
(E) E

5. To which of the above boxplots does the following histogram correspond?

(A) A
(B) B
(C) C
(D) D
(E) E

6. To which of the above boxplots does the following histogram correspond?

(A) A
(B) B
(C) C
(D) D
(E) E

7. Suppose the average score on a national test is 500 with a standard deviation of 100. If each score is increased by 25, what are the new mean and standard deviation?

(A) 500, 100
(B) 500, 125
(C) 525, 100
(D) 525, 105
(E) 525, 125

8. If quartiles $Q_1 = 20$ and $Q_3 = 30$, which of the following must be true?

  I. The median is 25.
  II. The mean is between 20 and 30.
  III. The standard deviation is at most 10.

(A) I only
(B) II only
(C) III only
(D) All must be true.
(E) None must be true.

For Questions 9 and 10 consider the following: The graph below shows cumulative proportions plotted against grade point averages for a large public high school.

9. What is the median grade point average?

    (A) 0.8
    (B) 2.0
    (C) 2.4
    (D) 2.5
    (E) 2.6

10. What is the interquartile range?

    (A) 1.0
    (B) 1.8
    (C) 2.4
    (D) 2.8
    (E) 4.0

Free-Reponse Questions

Directions: You must show all work and indicate the methods you use. You will be graded on the correctness of your methods and on the accuracy of your final answers.

1. Victims spend from 5 to 5840 hours repairing the damage caused by identity theft with a mean of 330 hours and a standard deviation of 245 hours.

   (a) What would be the mean, range, standard deviation, and variance for hours spent repairing the damage caused by identity theft if each of the victims spent an additional 10 hours?

   (b) What would be the mean, range, standard deviation, and variance for hours spent repairing the damage caused by identity theft if each of the victims' hours spent increased by 10%?

2. Are women better than men at multitasking? Suppose in one study of multitasking a random sample of 200 female and 200 male high school students were assigned several tasks at the same time, such as solving simple mathematics problems, reading maps, and answering simple questions while talking on a telephone. Total times taken to complete all the tasks are given in the histograms below.

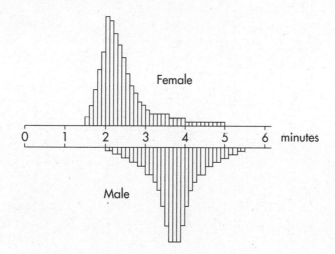

Write a few sentences comparing the distributions of times to complete all tasks by females and by males.

3. In independent random samples of 20 men and 20 women, the number of minutes spent on grooming on a given day were:

   Men: 27, 32, 82, 36, 43, 75, 45, 16, 23, 48, 51, 57, 60, 64, 39, 40, 69, 72, 54, 57

   Women: 49, 50, 35, 69, 75, 35, 49, 54, 98, 58, 22, 34, 60, 38, 47, 65, 79, 38, 42, 87

   Using back-to-back stemplots. compare the two distributions.

4. To analyze the social media behavior differences between boys and girls, Mrs. V's FDA high school AP Statistics class was asked to count the number of text messages that they sent over a long three-day weekend. The following table summarizes the data:

	Values under Q_1	Q_1	Median	Q_3	Values over Q_3
Females	15, 43, 100	130	175	358	450, 573, 1098
Males	3, 59	72	183	273	293, 337

(a) Construct parallel boxplots of this set of data.

(b) Do the data indicate that females or males had the greater mean number of texts? Explain.

QUIZ 2

Multiple-Choice Questions

> **Directions:** The questions that follow are each followed by five suggested answers. Choose the response that best answers the question.

Questions 1–3 refer to the following five histograms:

(A)   (B)   (C)

(D)   (E)

1. To which of the above histograms does the following boxplot correspond?

    (A) A
    (B) B
    (C) C
    (D) D
    (E) E

2. To which of the above histograms does the following boxplot correspond?

    (A) A
    (B) B
    (C) C
    (D) D
    (E) E

3. To which of the above histograms does the following boxplot correspond?

(A) A
(B) B
(C) C
(D) D
(E) E

___

4. Below is a boxplot of CO_2 levels (in grams per kilometer) for a sampling of year 2018 vehicles.

Suppose follow-up testing determines that the low outlier should be 10 grams per kilometer less and the two high outliers should each be 5 grams per kilometer greater. What effect, if any, will these changes have on the mean and median CO_2 levels?

(A) Both the mean and median will be unchanged.
(B) The median will be unchanged, but the mean will increase.
(C) The median will be unchanged, but the mean will decrease.
(D) The mean will be unchanged, but the median will increase.
(E) Both the mean and median will change.

5. Below is a boxplot of yearly tuition and fees of all four year colleges and universities in a Western state. The low outlier is from a private university that gives full scholarships to all accepted students, while the high outlier is from a private college catering to the very rich.

Removing both outliers will effect what changes, if any, on the mean and median costs for this state's four year institutions of higher learning?

(A) Both the mean and the median will be unchanged.
(B) The median will be unchanged, but the mean will increase.
(C) The median will be unchanged, but the mean will decrease.
(D) The mean will be unchanged, but the median will increase.
(E) Both the mean and median will change.

6. Suppose the average score on a national test is 500 with a standard deviation of 100. If each score is increased by 25%, what are the new mean and standard deviation?

   (A) 500, 100
   (B) 525, 100
   (C) 625, 100
   (D) 625, 105
   (E) 625, 125

7. A teacher is teaching two AP Statistics classes. On the final exam, the 20 students in the first class averaged 92 while the 25 students in the second class averaged only 83. If the teacher combines the classes, what will the average final exam score be?

   (A) 87
   (B) 87.5
   (C) 88
   (D) None of the above
   (E) More information is needed to make this calculation.

8. Suppose 10% of a data set lie between 40 and 60. If 5 is first added to each value in the set and then each result is doubled, which of the following is true?

   (A) 10% of the resulting data will lie between 85 and 125.
   (B) 10% of the resulting data will lie between 90 and 130.
   (C) 15% of the resulting data will lie between 80 and 120.
   (D) 20% of the resulting data will lie between 45 and 65.
   (E) 30% of the resulting data will lie between 85 and 125.

9. The following dotplot shows the speeds (in mph) of 100 fastballs thrown by a major league pitcher.

Which of the following is the best estimate of the standard deviation of these speeds?

   (A) 0.5 mph
   (B) 1.1 mph
   (C) 1.6 mph
   (D) 2.2 mph
   (E) 6.0 mph

10. The following boxplots were constructed from SAT math scores of boys and girls at a high school:

Which of the following is a possible boxplot for the combined scores of all the students?

Free-Response Questions

Directions: You must show all work and indicate the methods you use. You will be graded on the correctness of your methods and on the accuracy of your final answers.

1. Suppose a distribution has mean 300 and standard deviation 25. If the z-score of Q_1 is -0.7 and the z-score of Q_3 is 0.7, what values would be considered to be outliers?

2. (a) Draw the boxplot for the data set $\{0, 1, 2, 3, 4, 5, 6, 7, 8, 9, 10\}$.

   Very different sets can have the *same boxplot*.

   (b) Find the data set with sample size $n = 11$, with each value an integer between 0 and 10, and with the *same boxplot* as in (a), which has the smallest possible mean.

   (c) Find the data set with sample size $n = 11$, with each value an integer between 0 and 10, and with the *same boxplot* as in (a), which has the largest possible standard deviation.

3. Below are two population pyramids from the U.S. Census Bureau.

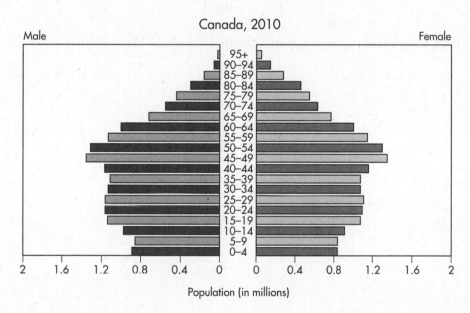

(a) The approximate median age of the Liberian population falls in which of these intervals: 0–4, 15–19, 30–34, 40–44? Explain.

(b) Which country has more children younger than 10 years of age? Explain.

(c) Does the population pyramid indicate that Canadian men or women live longer? Explain.

(d) In 2010, Liberia had recently come out of a civil war with the extensive use of child soldiers. How is this visible in the population pyramid?

4. Cumulative frequency graphs of the ages of people on three different Caribbean cruises (A, B, and C) are given below:

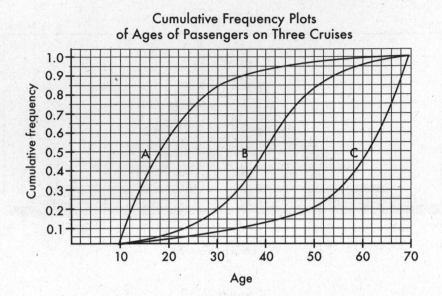

Write a few sentences comparing the distributions of ages of people on the three cruises.

ANSWERS EXPLAINED

Quiz 1

MULTIPLE-CHOICE QUESTIONS

1. **(D)** The distribution is clearly skewed right, so the mean is greater than the median, and the ratio is greater than one.

2. **(D)** Multiplying every value in a set by the same constant (in this case, by $\frac{1}{60}$) multiplies both the mean and the standard deviation by the same constant. Standardized scores (the number of standard deviations from the mean) are unchanged.

3. **(E)** Outliers are any values below $Q_1 - 1.5(IQR) = 5.5$ or above $Q_3 + 1.5(IQR) = 57.5$.

4. **(A)** The value 50 seems to split the area under the histogram in two, so the median is about 50. Furthermore, the histogram is skewed to the left with a tail from 0 to 30.

5. **(B)** Looking at areas under the curve, Q_1 appears to be around 20, the median is around 30, and Q_3 is about 40.

6. **(C)** Looking at areas under the curve, Q_1 appears to be around 10, the median is around 30, and Q_3 is about 50.

7. **(C)** Adding the same constant to every value increases the mean by that same constant; however, the distances between the increased values and the increased mean stay the same, and so the standard deviation is unchanged. Graphically, you should picture the whole distribution as moving over by a constant; the mean moves, but the standard deviation (which measures spread) doesn't change.

8. **(E)** The median is somewhere between 20 and 30, but not necessarily at 25. Even a single very large score can result in a mean over 30 and a standard deviation over 10.

9. **(C)** The median corresponds to the 0.5 cumulative proportion.

10. **(A)** The 0.25 and 0.75 cumulative proportions correspond to $Q_1 = 1.8$ and $Q_3 = 2.8$, respectively, and so the interquartile range is $2.8 - 1.8 = 1.0$.

FREE-RESPONSE QUESTIONS

1. (a) Adding 10 to each value increases the mean by 10, but leaves measures of variability unchanged, so the new mean is 340 hours while the range stays at 5835 hours, the standard deviation remains at 245 hours, and the variance remains at $245^2 = 60,025 \text{ hr}^2$.

   (b) Increasing each value by 10% (multiplying by 1.10) will increase the mean to $1.1(330) = 363$ hours, the range to $1.1(5835) = 6418.5$ hours, the standard deviation to $1.1(245) = 269.5$ hours, and the variance to $(269.5)^2 = 72,630.25 \text{ hr}^2$. (Note that the variance increases by a multiple of $(1.1)^2$ not by a multiple of 1.1.)

2. A complete answer compares shape, center, and spread and mentions context in at least one of the responses.

    Shape: The distribution of times to complete all tasks by females is skewed right (toward the higher values), whereas the distributions of times to complete all tasks by males is roughly bell-shaped.

    Center: The center of the distribution of female times (at around $2\frac{1}{4}$ minutes) is less than the center of the distribution of male times (at around $3\frac{3}{4}$ minutes).

    Spread: The spreads of the two distributions are roughly the same; the range of the female times ($5 - 1\frac{1}{2} = 3\frac{1}{2}$ minutes) equals the range of the male times ($5\frac{1}{2} - 2 = 3\frac{1}{2}$ minutes).

3.

Men		Women
6	1	
7 3	2	2
9 6 2	3	4 5 5 8 8
8 5 3 0	4	2 7 9 9
7 7 4 1	5	0 4 8
9 4 0	6	0 5 9
5 2	7	5 9
2	8	7
	9	8

(9|8 mean 98 min)

A complete answer compares shape, center, and spread and mentions context in at least one of the responses.

Shape: The men's distribution of hours grooming is roughly symmetric, whereas the women's distribution of hours grooming is skewed right (toward higher values).

Center: The center of the men's distribution is about the same as the center of the women's distribution, both about 50 min.

Spread: The spread of the men's distribution (with a range of 82 – 16 = 66 min) is less than the spread of the women's distribution (with a range of 98 – 22 = 76 min).

4. (a) For females, $Q_1 - 1.5(IQR) = 130 - 1.5(358 - 130) = -212$ and $Q_3 + 1.5(IQR) = 700$, so 1098 is an outlier. For males, $Q_1 - 1.5(IQR) = 72 - 1.5(273 - 72) = -229.5$ and $Q_3 + 1.5(IQR) = 574.5$, so there are no outliers.

Number of text messages

(b) The medians are roughly equal. The male distribution appears roughly symmetric, so the mean is close to the median; however, the female distribution shows extreme right skewness, so the mean is much greater than the median. Thus, the females had a greater mean number of text messages than did the males.

Quiz 2

MULTIPLE-CHOICE QUESTIONS

1. **(C)** The boxplot indicates that 25% of the data lie in each of the intervals 10–20, 20–35, 35–40, and 40–50. Counting boxes, only histogram C has this distribution.

2. **(D)** The boxplot indicates that 25% of the data lie in each of the intervals 10–15, 15–25, 25–35, and 35–50. Counting boxes, only histogram D has this distribution.

3. **(E)** The boxplot indicates that 25% of the data lie in each of the intervals 10–20, 20–30, 30–40, and 40–50. Counting boxes, only histogram E has this distribution.

4. **(A)** Subtracting 10 from one value and adding 5 to two values leaves the sum of the values unchanged, so the mean will be unchanged. Exactly what values the outliers take will not change what value is in the middle, so the median will be unchanged.

5. **(C)** The high outlier is further from the mean than is the low outlier, so removing both will decrease the mean. However, removing the lowest and highest values will not change what value is in the middle, so the median will be unchanged.

6. **(E)** Multiplying every value by the same constant multiplies both the mean and the standard deviation by that constant. Graphically, increasing each value by 25% (multiplying by 1.25) both moves and spreads out the distribution.

7. **(A)** The sum of the scores in one class is $20 \times 92 = 1840$, while the sum in the other is $25 \times 83 = 2075$. The total sum is $1840 + 2075 = 3915$. There are $20 + 25 = 45$ students, and so the average score is $\frac{3915}{45} = 87$.

8. **(B)** Increasing every value by 5 gives 10% between 45 and 65, and then doubling gives 10% between 90 and 130.

9. **(B)** With bell-shaped data the empirical rule applies, giving that the spread from 92 to 98 is roughly 6 standard deviations, and so one SD is about 1.

10. **(E)** The minimum of the combined set of scores must be the min of the boys since it is lower; the maximum of the combined set of scores must be the max of the girls since it is higher; the first quartile must be the same as the identical first quartiles of the two original distributions. There are no outliers (scores more than 1.5(IQR) from the first and third quartiles).

FREE-RESPONSE QUESTIONS

1. Z-scores give the number of standard deviations from the mean, so
   $Q_1 = 300 - 0.7(25) = 282.5$ and $Q_3 = 300 + 0.7(25) = 317.5$.
   The interquartile range is IQR $= 317.5 - 282.5 = 35$, and $1.5(IQR) = 1.5(35) = 52.5$.
   The standard definition of outliers encompasses all values less than $Q_1 - 52.5 = 230$ and all values greater than $Q_3 + 52.5 = 370$.

2. (a)

(b) Note that must keep Min = 0, $Q_1 = 2$, Med = 5, $Q_3 = 8$, and Max = 10, with the same totals of in-between values, so move in-between values to the left, and the answer is

$$\{0, 0, 2, 2, 2, 5, 5, 5, 8, 8, 10\}$$

(c) Note that must keep Min = 0, $Q_1 = 2$, Med = 5, $Q_3 = 8$, and Max = 10, with the same totals of in-between values, so move in-between values outward, and the answer is

$$\{0, 0, 2, 2, 2, 5, 8, 8, 8, 10, 10\}$$

3. (a) The Liberian median age lies in the 15–19 age interval because roughly 50% of the total bar lengths is above and below the 15–19 interval.

(b) Canada has more children younger than 10 years of age. There are about 1.2 million children younger than age 10 in Liberia (boys and girls) and roughly 3.5 million in Canada. (Note the difference in scales!)

(c) The population pyramid indicates that Canadian women live longer than men because all the higher age intervals show greater numbers of women than men.

(d) In the Liberian graph, the smaller 15–19 age group shows a definite break with the overall pattern. A plausible explanation is the great number of child soldiers who died in the fighting.

4. A complete answer compares shape, center, and spread and mentions context in at least one of the responses.

Shape: Cruise A, for which the cumulative frequency plot rises steeply at first, has more younger passengers, and thus a distribution skewed to the right (towards the higher ages). Cruise C, for which the cumulative frequency plot rises slowly at first and then steeply towards the end, has more older passengers, and thus a distribution skewed to the left (towards the younger ages). Cruise B, for which the cumulative frequency plot rises slowly at each end and steeply in the middle, has a more bell-shaped distribution.

Center: Considering the center to be a value separating the area under the histogram roughly in half, the centers will correspond to a cumulative frequency of 0.5. Reading across from 0.5 to the intersection of each graph, and then down to the *x*-axis, shows centers of approximately 18, 40, and 61 years, respectively. Thus, the center of distribution A is the least, and the center of distribution C is the greatest.

Spread: The spreads of the age distributions of all three cruises are the same: from 10 to 70 years.

Exploring Bivariate Data

3

→ **SCATTERPLOTS**

→ **CORRELATION AND LINEARITY**

→ **LEAST SQUARES REGRESSION LINE**

→ **RESIDUAL PLOTS**

→ **OUTLIERS AND INFLUENTIAL POINTS**

→ **TRANSFORMATIONS TO ACHIEVE LINEARITY**

Our studies so far have been concerned with measurements of a single variable. However, many important applications of statistics involve examining whether two or more variables are related to one another. For example, is there a relationship between the smoking histories of pregnant women and the birth weights of their children? Between SAT scores and success in college? Between amount of fertilizer used and amount of crop harvested?

Two questions immediately arise. First, how can the strength of an apparent relationship be measured? Second, how can an observed relationship be put into functional terms? For example, a real estate broker might not only wish to determine whether a relationship exists between the prime rate and the number of new homes sold in a month but might also find useful an expression with which to predict the number of home sales given a particular value of the prime rate.

A graphical display, called a *scatterplot*, gives an immediate visual impression of a possible relationship between two variables, while a numerical measurement, called the *correlation coefficient*, is often used as a quantitative value of the strength of a linear relationship. In either case, evidence of a relationship is *not* evidence of causation.

SCATTERPLOTS

Suppose a relationship is perceived between two quantitative variables, X and Y. We are interested in the strength and direction of this relationship and in any deviation from the basic pattern of this relationship. We graph the pairs (x, y). In this topic, we examine whether the relationship, as illustrated by the scatterplot, can be reasonably explained in terms of a linear function, that is, one whose graph is a straight line.

Recognize explanatory (*x*) and response (*y*) variables in context.

For example, we might be looking at a plot such as

NOTE

There are various so-called *lines of best fit*. We will be working with one called the *least squares regression line*.

We need to know what the term *best-fitting straight line* means and how we can find this line. Furthermore, we want to be able to gauge whether the relationship between the variables is strong enough so that finding and making use of this straight line is meaningful.

Patterns in Scatterplots

When larger values of one variable are associated with larger values of a second variable, the variables are called *positively associated*. When larger values of one are associated with smaller values of the other, the variables are called *negatively associated*.

NOTE

No matter whether the variables are positively or negatively associated, for any two points, anything is possible, and either point could be higher or lower than the other.

➡ **EXAMPLE 3.1**

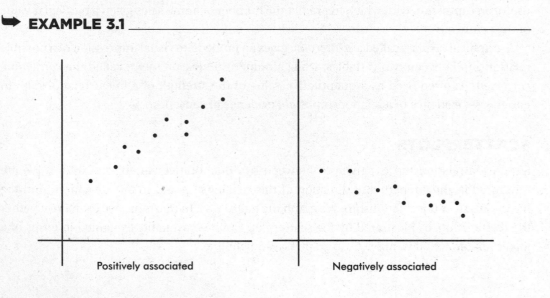

Positively associated   Negatively associated

The strength of the association is gauged by how close the plotted points are to a straight line.

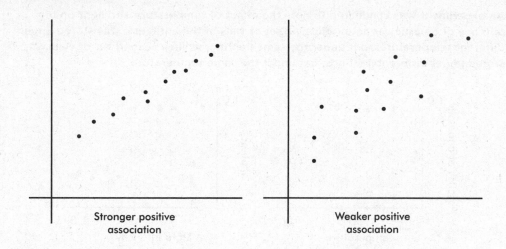

Stronger positive
association

Weaker positive
association

Sometimes different dots in a scatterplot are labeled with different symbols or different colors to show a categorical variable. The resulting labeled scatterplot might distinguish, for example, between men and women, or between stocks and bonds, and so on.

➥ EXAMPLE 3.3

The above diagram is a labeled scatterplot distinguishing men with plus signs and women with square dots to show a categorical variable.

When analyzing the overall pattern in a scatterplot, it is also important to note *clusters* and *outliers*.

➡ **EXAMPLE 3.4** _____

An experiment was conducted to note the effect of temperature and light on the potency of a particular antibiotic. One set of vials of the antibiotic was stored under different temperatures, but under the same lighting, while a second set of vials was stored under different lightings, but under the same temperature.

> **TIP**
>
> Note when the data falls in distinct groups.

In the first scatterplot note the roughly linear pattern with one outlier far outside this pattern. A possible explanation is that the antibiotic is more potent at lower temperatures, but only down to a certain temperature at which it drastically loses potency.

In the second scatterplot note the two clusters. It appears that below a certain light intensity the potency is one value, while above that intensity it is another value. In each cluster there seems to be no association between intensity and potency.

_____

> **IMPORTANT**
>
> Correlation does not imply causation!

CORRELATION AND LINEARITY

Although a scatter diagram usually gives an intuitive visual indication when a linear relationship is strong, in most cases it is quite difficult to visually judge the specific strength of a relationship. For this reason there is a mathematical measure called *correlation* (or the *correlation coefficient*). Important as correlation is, we always need to keep in mind that significant correlation does not necessarily indicate *causation* and that correlation measures the strength only of a *linear* relationship.

Correlation, designated by *r*, has the formula

$$r = \frac{1}{n-1} \sum \left(\frac{x_i - \bar{x}}{s_x} \right) \left(\frac{y_i - \bar{y}}{s_y} \right)$$

> **TIP**
>
> With standardized data (*z*-scores), *r* is the slope of the regression line.

in terms of the means and standard deviations of the two sets. We note that the formula is actually the sum of the products of the corresponding *z*-scores divided by 1 less than the sample size. However, you should be able to quickly calculate correlation using the statistical package on your calculator. (Examining the formula helps you understand where correlation is coming from, but you will NOT have to use the formula to calculate *r*.)

Note from the formula that correlation does not distinguish between which variable is called *x* and which is called *y*. The formula is also based on standardized scores (*z*-scores), and so changing units does not change the correlation. Finally, since means and standard deviations can be strongly influenced by outliers, correlation is also strongly affected by extreme values.

The value of r always falls between –1 and +1, with –1 indicating perfect negative correlation and +1 indicating perfect positive correlation. It should be stressed that a correlation at or near zero doesn't mean there isn't a relationship between the variables; there may still be a strong *nonlinear* relationship.

➡ EXAMPLE 3.5

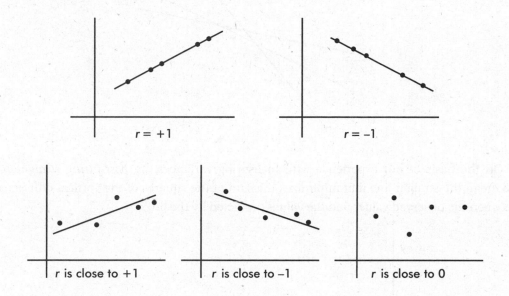

It can be shown that r^2, called the *coefficient of determination*, is the ratio of the variance of the predicted values $\hat{y}$ to the variance of the observed values y. That is, there is a partition of the y-variance, and r^2 is the proportion of this variance that is predictable from a knowledge of x. Alternatively, we can say that r^2 gives the percentage of variation in y that is explained by the variation in x. In either case, always interpret r^2 in context of the problem. Remember when calculating r from r^2 that r may be positive or negative.

While the correlation r is given as a decimal between –1.0 and 1.0, the coefficient of determination r^2 is usually given as a percentage. An r^2 of 100% is a perfect fit, with all the variation in y explained by variation in x. How large a value of r^2 is desirable depends on the application under consideration. While scientific experiments often aim for an r^2 in the 90% or above range, observational studies with r^2 of 10% to 20% might be considered informative. Note that while a correlation of 0.6 is twice a correlation of 0.3, the corresponding r^2 of 36% is *four* times the corresponding r^2 of 9%.

NOTE

We can also say that r^2 gives the percentage of variation in y that can be explained by the regression line, with x as the explanatory variable.

➡ EXAMPLE 3.6

The correlation between Total Points Scored and Total Yards Gained for the 2016 season among a set of college football teams is $r = 0.84$. What information is given by the coefficient of determination?

Answer: $r^2 = (0.84)^2 = 0.7056$. Thus, 70.56% of the variation in Total Points Scored can be explained by (or predicted by) the linear relationship between Total Points Scored and Total Yards Gained. The other 29.44% of the variation in Total Points Scored remains unexplained.

NOTE

Alternatively, you can say that 70.56% of the variation in Total Points Scored is accounted for by the variation in Total Yards Gained.

LEAST SQUARES REGRESSION LINE

What is the best-fitting straight line that can be drawn through a set of points?

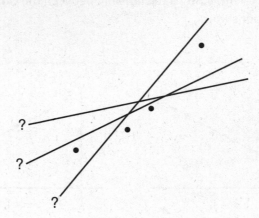

TIP

If a scatterplot indicates a non-linear relationship, don't try to force a straight-line fit.

On the basis of our experience with measuring variances, by *best-fitting straight line* we mean the straight line that minimizes the sum of the squares of the vertical differences between the observed values and the values predicted by the line.

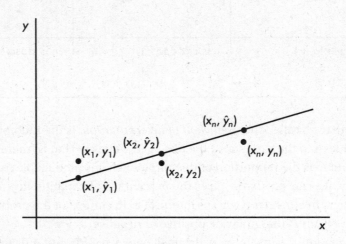

TIP

A hat over a variable means it is a predicted version of the variable.

That is, in the above figure, we wish to minimize

$$(y_1 - \hat{y}_1)^2 + (y_2 - \hat{y}_2)^2 + \cdots + (y_n - \hat{y}_n)^2$$

It is reasonable, intuitive, and correct that the best-fitting line will pass through $(\bar{x}, \bar{y})$ where $\bar{x}$ and $\bar{y}$ are the means of the variables X and Y. Then, from the basic expression for a line with a given slope through a given point, we have

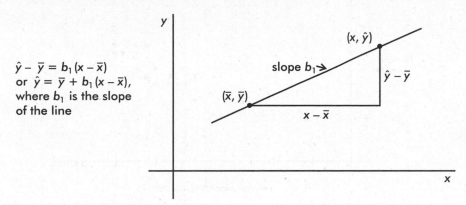

$$\hat{y} - \bar{y} = b_1(x - \bar{x})$$
or $\hat{y} = \bar{y} + b_1(x - \bar{x})$,
where b_1 is the slope
of the line

The slope b_1 can be determined from the formula

$$b_1 = r\frac{s_y}{s_x}$$

where r is the correlation and s_x and s_y are the standard deviations of the two sets. That is, each standard deviation change in x results in a change of r standard deviations in y. If you graph z-scores for the y-variable against z-scores for the x-variable, the slope of the regression line is precisely r, and, in fact, the linear equation becomes $z_y = r z_x$.

This best-fitting straight line, that is, the line that minimizes the sum of the squares of the differences between the observed values and the values predicted by the line, is called the *least squares regression line* or simply the *regression line*. It can be calculated directly by entering the two data sets and using the statistics package on your calculator.

TIP

Just because we can calculate a regression line doesn't mean it is useful.

➡ **EXAMPLE 3.7** _____

An insurance company conducts a survey of 15 of its life insurance agents. The average number of minutes spent with each potential customer and the number of policies sold in a week are noted for each agent. Letting X and Y represent the average number of minutes and the number of sales, respectively, we have

X: 25 23 30 25 20 33 18 21 22 30 26 26 27 29 20
Y: 10 11 14 12 8 18 9 10 10 15 11 15 12 14 11

Find the equation of the best-fitting straight line for the data.

(continued)

Answer: Plotting the 15 points (25, 10), (23, 11), . . . , (20, 11) gives an intuitive visual impression of the relationship:

This scatterplot indicates the existence of a relationship that appears to be *linear*; that is, the points lie roughly on a straight line. Furthermore, the linear relationship is *positive*; that is, as one variable increases, so does the other (the straight line slopes upward).

Using a calculator, we find the correlation to be $r = 0.8836$, the coefficient of determination to be $r^2 = 0.78$ (indicating that 78% of the variation in the number of policies is accounted for by the variation in the number of minutes spent), and the regression line to be

$$\hat{y} = \bar{y} + b_1(x - \bar{x})$$
$$= 12 + 0.5492(x - 25)$$
$$= -1.73 + 0.5492x$$

We also write: Policies = –1.73 + 0.5492 Minutes

Adding this to our scatterplot yields

Thus, for example, we might predict that agents who average 24 minutes per customer will average 0.5492(24) – 1.73 = 11.45 sales per week. We also note that each additional minute spent seems to produce an average 0.5492 extra sale.

EXAMPLE 3.8

Following are advertising expenditures and total sales for six detergent products:

Advertising ($1000) (*x*): 2.3   5.7   4.8   7.3   5.9   6.2

Total sales ($1000) (*y*):   77   105   96   118   102   95

Predict the total sales if $5000 is spent on advertising and interpret the slope of the regression line. What if $100,000 is spent on advertising?

Answer: With your calculator, the equation of the regression line is found to be

$$\hat{y} = 59.683 + 7.295x$$

It is also worthwhile to replace the *x* and *y* with more appropriately named variables, resulting, for example, in

$$\widehat{\text{Sales}} = 59.683 + 7.295 \text{ Advcost}$$

TIP

Use meaningful variable names.

The regression line predicts that if $5000 is spent on advertising, the resulting total sales will be 59.683 + 7.295(5) = 96.158 thousands of dollars ($96,158).

The slope of the regression line indicates that every extra $1000 spent on advertising will result in an average of $7295 in added sales.

If $100,000 is spent on advertising, we calculate 59.683 + 7.295(100) = 789.183 thousands of dollars (≈$789,000). How much confidence should we have in this answer? Not much! We are trying to use the regression line to predict a value far outside the other data values. This procedure is called *extrapolation* and must be used with great care.

TIP

Be careful about extrapolation beyond the observed *x*-values.

It should be noted that when we use the regression line to predict a *y*-value for a given *x*-value, we are actually predicting the *mean y*-value for that given *x*-value. For any given *x*-value, there are many possible *y*-values, and we are predicting their mean. So if $5000 is spent many times on advertising, various resulting total sales figures may result, but their predicted average is $96,158.

The TI-Nspire gives

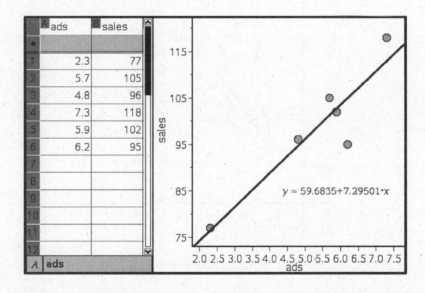

→ **EXAMPLE 3.9** _____

A random sample of 30 U.S. farm regions surveyed during the summer of 2018 produced the following statistics:

Average temperature (°F) during growing season: $\bar{x} = 81$, $s_x = 3$

Average corn yield per acre (bu.): $\bar{y} = 131$, $s_y = 5$

Correlation $r = 0.32$

Based on this study, what is the mean predicted corn yield for a region where the average growing season temperature is 76.5°F?

Answer: 76.5 is $\frac{76.5 - 81}{3} = -1.5$ standard deviations below the average temperature reported in the study, so with a correlation of $r = 0.32$, the predicted corn yield is 0.32(–1.5) = –0.48 standard deviations below the average corn yield, or 131 – 0.48(5) = 128.6 bushels per acre.

Alternatively, we could have found the linear regression equation relating these variables: $slope = r\frac{s_y}{s_x} = 0.32\left(\frac{5}{3}\right) \approx 0.533$, intercept $\approx 131 - 81(0.533) \approx 87.8$, and thus $\widehat{Yield} = 0.533\ Temp + 87.8$. Then 0.533(76.5) + 87.8 $\approx$ 128.6.

RESIDUAL PLOTS

CAREFUL!

Order is important; residual equals observed – predicted.

The difference between an observed and predicted value is called the *residual*. When the regression line is graphed on the scatterplot, the residual of a point is the vertical distance the point is from the regression line.

NOTE

When the data point is above the regression line, the residual is positive; a data point below the line gives a negative residual.

The regression line is the line that minimizes the sum of the squares of the residuals.

We calculate the predicted values from the regression line in Example 3.8 and subtract from the observed values to obtain the residuals:

x	2.3	5.7	4.8	7.3	5.9	6.2
y	77	105	96	118	102	95
$\hat{y}$	76.5	101.3	94.7	112.9	102.7	104.9
$y - \hat{y}$	0.5	3.7	1.3	5.1	−0.7	−9.9

Note that the sum of the residuals is

$$0.5 + 3.7 + 1.3 + 5.1 - 0.7 - 9.9 = 0.0$$

The above equation is true in general; that is, *the sum and thus the mean of the residuals is always zero.*

The TI-Nspire easily shows the "squares of the residuals," the sum of which is minimized by the regression line.

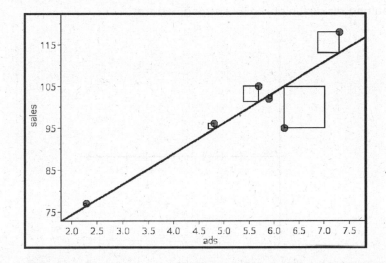

The notation for residuals is $\hat{e}_i = y_i - \hat{y}_i$ and so $\sum_{i=1}^{n} \hat{e}_i = 0$. The standard deviation of the residuals is calculated as follows:

$$s_e = \sqrt{\frac{\sum e_i^2}{n-2}} = \sqrt{\frac{\sum(y_i - \hat{y}_i)^2}{n-2}}$$

s_e gives a measure of how the points are spread around the regression line.

Plotting the residuals gives further information. In particular, a residual plot with a definite pattern is an indication that a nonlinear model will show a better fit to the data than the straight regression line. In addition to whether or not the residuals are randomly distributed, one should look at the balance between positive and negative residuals and also the size of the residuals in comparison to the associated *y*-values.

The residuals can be plotted against either the *x*-values or the $\hat{y}$-values (since $\hat{y}$ is a linear transformation of *x*, the plots are identical except for scale and a left-right reversal when the slope is negative).

It is also important to understand that a linear model may be appropriate, but weak, with a low correlation. And, alternatively, a linear model may not be the best model (as evidenced by the residual plot), but it still might be a very good model with high r^2.

➡ **EXAMPLE 3.11** _____

Suppose the drying time of a paint product varies depending on the amount of a certain additive it contains.

Additive (oz), x:	1	2	3	4	5	6	7	8	9	10
Drying time (hr), y:	4	2.1	1.5	1	1.2	1.7	2.5	3.6	4.9	6.1

Using a calculator, we find the regression line $\hat{y} = 1.062 + 0.327x$, and we find the residuals:

x:	1	2	3	4	5	6	7	8	9	10
$y - \hat{y}$:	2.61	0.39	−0.54	−1.37	−1.5	−1.32	−0.85	−0.08	0.89	1.77

The resulting residual plot shows a strong pattern:

This pattern indicates that a nonlinear model will be a better fit than a straight-line model. A scatterplot of the original data shows clearly what is happening:

The ability to interpret computer output is important not only to do well on the AP Statistics exam, but also to understand statistical reports in the business and scientific world.

➡ EXAMPLE 3.12 _____

Miles per gallon versus speed for a new model automobile is fitted with a least squares regression line. The graph of the residuals and some computer output for the regression are as follows:

Regression Analysis: MPG Versus Speed

The regression equation is
MPG = 38.9 − 0.218 Speed

Predictor	Coef	SE Coef	T	P
Constant	38.929	5.651	6.89	0.000
Speed	−0.2179	0.1119	−1.95	0.099

S = 7.252 R−Sq = 38.7% R−Sq(adj) = 28.5%

Analysis of Variance

Source	DF	SS	MS	F	P
Regression	1	199.34	199.34	3.79	0.099
Residual Error	6	315.54	52.59		
Total	7	514.88			

a. Interpret the slope of the regression line in context.

Answer: The slope of the regression line is −0.2179, indicating that, on average, the MPG drops by 0.2179 for every increase of one mile per hour in speed.

b. What is the mean predicted MPG at a speed of 30 mph?

Answer: At 30 mph, the mean predicted MPG is −0.2179(30) + 38.929, or about 32.4 MPG.

c. What was the actual MPG at a speed of 30 mph?

Answer: The residual for 30 mph is about +3.5, and since residual = actual − predicted, we estimate the actual MPG to be 32.4 + 3.5, or about 36 MPG.

d. Is a line the most appropriate model? Explain.

Answer: The fact that the residuals show such a strong curved pattern indicates that a nonlinear model would be more appropriate.

e. What does "S = 7.252" refer to?

Answer: The standard deviation of the residuals, s_e = 7.252, is a "typical value" of the residuals and gives a measure of how the points are spread around the regression line.

➡ EXAMPLE 3.13 _____

The number of youngsters playing Little League baseball in Ithaca, New York, during the years 2005–2013 is fitted with a least squares regression line. The graph of the residuals and some computer output for their regression are as follows:

Regression Analysis: Number of Players Versus Years Since 2005

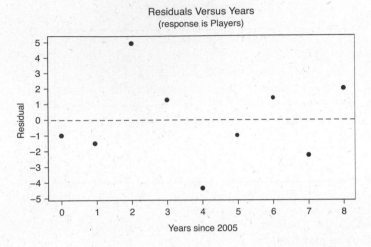

Predictor	Coef	SE Coef	T	P
Constant	123.800	1.798	68.84	0.000
Years	12.6333	0.3778	33.44	0.000

S = 2.926   R—Sq = 99.4%   R—Sq(adj) = 99.3%

Analysis of Variance

Source	DF	SS	MS	F	P
Regression	1	9576.1	9576.1	1118.45	0.000
Residual Error	7	59.9	8.6		
Total	8	9636.0			

TIP

Simply using a calculator to find a regression line is not enough; you must understand it (for example, be able to interpret the slope and intercepts in context).

a. Does it appear that a line is an appropriate model for the data? Explain.

   Answer: Yes. R-Sq = 99.4% is large, and the residual plot shows no pattern. Thus a linear model is appropriate.

b. What is the equation of the regression line (in context)?

   Answer: Predicted # of players = 123.8 + 12.6 (years since 2005)

c. Interpret the slope of the regression line in the context of the problem.

   Answer: The slope of the regression line is 12.6, indicating that, on average, the predicted number of children playing Little League baseball in Ithaca increased by an average of 12.6 players per year during the 2005–2013 time period.

d. Interpret the *y*-intercept of the regression line in the context of the problem.

   Answer: The *y*-intercept, 123.8, refers to the year 2005. Thus the number of players in Little League in Ithaca in 2005 was predicted to be around 124.

e. What is the predicted number of players in 2007?

   Answer: For 2007, *x* = 2, so the predicted number of players is 12.6(2) + 123.8 = 149.

(continued)

f. What was the actual number of players in 2007?

Answer: The residual for 2007 ($x = 2$) from the residual plot is +5, so actual – predicted = 5, and thus the actual number of players in 2007 must have been 5 + 149 = 154.

g. What years, if any, did the number of players decrease from the previous year? Explain.

Answer: The number would decrease if one residual were more than 12.6 greater than the next residual. This never happens, so the number of players never decreased.

OUTLIERS AND INFLUENTIAL POINTS

In a scatterplot, regression outliers are indicated by points falling far away from the overall pattern. That is, a point is an outlier if its residual is an outlier in the set of residuals.

➡ EXAMPLE 3.14

A scatterplot of grade point average (GPA) versus weekly television time for a group of high school seniors is as follows:

By direct observation of the scatterplot, we note that there are two outliers: one person who watches 5 hours of television weekly yet has only a 1.5 GPA, and another person who watches 25 hours weekly yet has a 3.0 GPA. Note also that while the value of 30 weekly hours of television may be considered an outlier for the television hours variable and the 0.5 GPA may be considered an outlier for the GPA variable, the point (30, 0.5) is *not* an outlier in the regression context because it does not fall off the straight-line pattern.

Scores whose removal would sharply change the regression line are called *influential scores.* Sometimes this description is restricted to points with extreme *x*-values. An influential score may have a small residual but still have a greater effect on the regression line than scores with possibly larger residuals but average *x*-values.

➡ EXAMPLE 3.15 _____

Consider the following scatterplot of six points and the regression line:

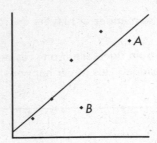

The heavy line in the scatterplot on the left below shows what happens when point *A* is removed, and the heavy line in the scatterplot on the right below shows what happens when point *B* is removed.

Note that the regression line is greatly affected by the removal of point *A* but not by the removal of point *B*. Thus point *A* is an *influential score*, while point *B* is not. This is true in spite of the fact that point *A* is closer to the original regression line than point *B*.

_____

➡ EXAMPLE 3.16 _____

Consider the effect of outliers in the *x*, the *y*, and both *x* and *y* directions on *b*, the slope of the regression line, and on *r*, the correlation:

outlier in *x* direction   outlier in *y* direction   outlier in both *x* and *y* directions

$b = 0.450, r = 0.866$ $b = 0.179, r = 0.524$ $b = 0.450, r = 0.596$ $b = 0.452, r = 0.939$

Note how an outlier in the *x* direction can dramatically affect both the slope and the correlation. Also note how an outlier in both the *x* and *y* directions, if maintaining the same pattern, will have roughly the same slope but a much increased correlation (in absolute value).

_____

MORE ON REGRESSION

The biologist, Galton, found that sons whose fathers are taller than average tend to be taller than average themselves, but on average not as tall as their fathers; their heights *regressed* towards the mean.

The regression equation, $\hat{y} = a + bx = a + \left(r\dfrac{s_y}{s_x} \right)x$ has important implications.

1. If the correlation $r = +1$, then $\hat{y} = a + \left(\dfrac{s_y}{s_x} \right)x$, and for each standard deviation s_x increase in x, the *predicted y* value increases by s_y.

2. If, for example, $r = +0.4$, then $\hat{y} = a + 0.4\left(\dfrac{s_y}{s_x} \right)x = a + \left(\dfrac{0.4s_y}{s_x} \right)x$, and for each standard deviation s_x increase in x, the *predicted y* value increases by $0.4\,s_y$.

➡️ **EXAMPLE 3.17**

Suppose x = attendance at a movie theater, y = number of boxes of popcorn sold, and we are given that there is a roughly linear relationship between *x* and *y*. Suppose further we are given the summary statistics: $\bar{x} = 250$, $s_x = 30$, $\bar{y} = 160$, $s_y = 20$, and $r = 0.8$.

a. When attendance is 250, what is the predicted number of boxes of popcorn sold?

   Answer: The least squares regression line passes through $(\bar{x}, \bar{y})$, so the predicted number of boxes of popcorn sold is 160.

b. When attendance is 295, what is the predicted number of boxes of popcorn sold?

   Answer: 295 is $(295 - 250)/30 = 1.5$ standard deviations above $\bar{x} = 250$, so the predicted number of boxes of popcorn sold is $(0.8)1.5 = 1.2$ standard deviations above $\bar{y}$, or $160 + (1.2)20 = 184$ boxes.

---

3. The regression equation for predicting x from y has the slope $r\dfrac{s_x}{s_y}$.

➡️ **EXAMPLE 3.18**

Use the same attendance and popcorn summary statistics from above.

a. When 160 boxes of popcorn are sold, what is the predicted attendance?

   Answer: The least squares regression line passes through $(\bar{y}, \bar{x})$, so the predicted attendance is 250.

b. When 184 boxes of popcorn are sold, what is the predicted attendance?

   Answer: 184 is $(184-160)/20 = 1.2$ standard deviations above $\bar{y} = 160$, so the predicted attendance is $(0.8)1.2 = 0.96$ standard deviations above $\bar{x}$ or $250 + (0.96)30 = 278.8$ people.

---

4. It is important to understand that while y versus x and x versus y have the same correlation r, the line we get from predicting y from x is *not* the equivalent line we get from predicting x from y, just solved for the other variable. For example, the slopes are not reciprocals as they were in algebra when we "solved for the other variable." We note that $r\dfrac{s_y}{s_x}$ is not the reciprocal of $r\dfrac{s_x}{s_y}$. And we see above that while an attendance of 295 gave a prediction of 184 boxes of popcorn, starting with 184 boxes of popcorn gave a predicted attendance of 278.8, not 295.

TRANSFORMATIONS TO ACHIEVE LINEARITY

Often a straight-line pattern is not the best model for depicting a relationship between two variables. A clear indication of this problem is when the scatterplot shows a distinctive curved pattern. Another indication is when the residuals show a distinctive pattern rather than a random scattering. In such a case, the nonlinear model can sometimes be revealed by transforming one or both of the variables and then noting a linear relationship. Useful transformations often result from using the *log* or *ln* buttons on your calculator to create new variables.

➡ **EXAMPLE 3.19** _____

Consider the following years and corresponding populations:

Year, x:	1950	1960	1970	1980	1990
Population (1000s), y:	50	67	91	122	165

The scatterplot and residual plot indicate that a nonlinear relationship would be an even stronger model.

Letting log y be a new variable, we obtain

x:	1950	1960	1970	1980	1990
log y:	1.70	1.83	1.96	2.09	2.22

The scatterplot and residual plot now indicate a stronger linear relationship.

(continued)

A regression analysis yields $\widehat{\log y}$ = 0.013x – 23.65. In context we have: $\widehat{\log(\text{Pop})}$ = –23.65 + 0.013(Year). So, for example, the population predicted for the year 2000 would be calculated $\widehat{\log(\text{Pop})}$ = 0.013(2000) – 23.65 = 2.35, and so Pop = $10^{2.35}$ ≈ 224 thousand, or 224,000. The "linear" equation $\widehat{\log y}$ = 0.013x – 23.65 can be re-expressed as $\hat{y} = 10^{0.013x-23.65}$ [= $10^{-23.65} \times (10^{0.013})^x$ = 2.2387E–24 × 1.0304^x].

There are many useful transformations. For example:

Log y as a linear function of x, $\log y = a + bx$, re-expresses as an *exponential*:

$$y = 10^{a+bx}, \text{ or } y = c\, 10^{bx}, \text{ where } c = 10^a$$

Log y as a linear function of log x, $\log y = a + b \log x$, re-expresses as a *power*:

$$y = 10^{a+b\log x} \text{ or } y = cx^b, \text{ where } c = 10^a$$

$\sqrt{y}$ as a linear function of x, $\sqrt{y} = a + bx$, re-expresses as a *quadratic*:

$$y = (a + bx)^2$$

$\dfrac{1}{y}$ as a linear function of x, $\dfrac{1}{y} = a + bx$, re-expresses as a *reciprocal*:

$$y = \frac{1}{a+bx}$$

y as a linear function of log x, $y = a + b \log x$, is a *logarithmic* function.

Note: Although you need to be able to recognize the need for a transformation, justify its appropriateness (residuals plot), use whichever is appropriate from above to create a linear model, and use the model to make predictions, you do not have to be able to re-express the linear equation in the manner shown above.

➡ **EXAMPLE 3.20** _____

What are possible models for the following data?

x	1	2	3	4	5
y	20	60	120	190	280

Answer: A linear fit to x and y gives $\hat{y}$ = –61 + 65x with r = 0.99.

A linear fit to x and log y gives log y = 1.139 + 0.279x with r = 0.98. This results in an *exponential* relationship:

$$\hat{y} = 10^{1.139 + 0.279x} = 13.77(10^{0.279x}) = 13.77(1.901^x)$$

A linear fit to log x and log y gives log y = 1.295 + 1.639 log x, also with r = 0.99. This results in a *power* relationship:

$$\hat{y} = 10^{1.295 + 1.639 \log x} = 19.72(x^{1.639})$$

(continued)

All three models give high correlation and are reasonable fits. Further analysis can be done by examining the residual plots:

The first two residual plots have distinct curved patterns. The third residual plot illustrates both a more random pattern and smaller residuals. (One can also create residual plots using the derived curved models, but we choose to restrict our attention to residual plots of the linear models.) Among the above three models, the power model $\hat{y} = 19.72(x^{1.639})$, appears to be best.

SUMMARY

- A scatterplot gives an immediate indication of the shape (linear or not), strength, and direction (positive or negative) of a possible relationship between two variables.
- If the relationship appears roughly linear, then the correlation coefficient, r, is a useful measurement.
- The value of r is always between –1 and +1, with positive values indicating positive association and negative values indicating negative association; and values close to –1 or +1 indicating a stronger linear association than values close to 0, which indicate a weaker linear association.
- Evidence of an association is not evidence of a cause-and-effect relationship!
- Correlation is not affected by which variable is called x and which y or by changing units.
- Correlation can be strongly affected by extreme values.
- The differences between the observed and predicted values are called residuals.
- The best-fitting straight line, called the regression line, minimizes the sum of the squares of the residuals.
- For the linear regression model, the mean of the residuals is always 0.
- A definite pattern in the residual plot indicates that a nonlinear model may fit the data better than the straight regression line.
- The coefficient of determination, r^2, gives the percentage of variation in y that is accounted for by the least squares regression line.
- Influential scores are scores whose removal would sharply change the regression line.
- Nonlinear models can sometimes be studied by transforming one or both variables and then noting a linear relationship.
- It is very important to be able to interpret generic computer output.

QUIZ 1

Multiple-Choice Questions

> **Directions:** The questions that follow are each followed by five suggested answers. Choose the response that best answers the question.

1. A rural college is considering constructing a windmill to generate electricity but is concerned over noise levels. A study is performed measuring noise levels (in decibels) at various distances (in feet) from the campus library, and a least squares regression line is calculated with a correlation of 0.74. Which of the following is a proper and most informative conclusion for an observation with a negative residual?

    (A) The measured noise level is 0.74 times the predicted noise level.
    (B) The predicted noise level is 0.74 times the measured noise level.
    (C) The measured noise level is greater than the predicted noise level.
    (D) The predicted noise level is greater than the measured noise level.
    (E) The slope of the regression line at that point must also be negative.

2. A study of department chairperson ratings and student ratings of the performance of high school statistics teachers reports a correlation of $r = 1.15$ between the two ratings. From this information we can conclude that

    (A) chairpersons and students tend to agree on who is a good teacher.
    (B) chairpersons and students tend to disagree on who is a good teacher.
    (C) there is little relationship between chairperson and student ratings of teachers.
    (D) there is strong association between chairperson and student ratings of teachers, but it would be incorrect to infer causation.
    (E) a mistake in arithmetic has been made.

Questions 3–5 refer to the following:

The relationship between winning game proportions when facing the sun and when the sun is on one's back is analyzed for a random sample of 10 professional tennis players. The computer printout for regression is below:

```
Predictor      Coef      SE Coef        T         P
Constant     0.05590     0.02368      2.36      0.046
Facing       0.92003     0.03902     23.58      0.000


S = 0.0242922    R-Sq = 98.6%    R-Sq(adj) = 98.4%


Analysis of Variance

Source           DF        SS        MS         F         P
Regression        1     0.32812   0.32812    556.03     0.000
Residual Error    8     0.00472   0.00059
Total             9     0.33284
```

3. What is the equation of the regression line, where *facing* and *back* are the winning game proportions when facing the sun and with back to the sun, respectively?

(A) $\widehat{facing} = 0.056 + 0.920\ back$

(B) $\widehat{back} = 0.056 + 0.920\ facing$

(C) $\widehat{facing} = 0.920 + 0.056\ back$

(D) $\widehat{back} = 0.920 + 0.056\ facing$

(E) $\widehat{facing} = 0.024 + 0.039\ back$

4. What is the *correlation*?

(A) −0.984
(B) −0.986
(C) 0.984
(D) 0.986
(E) 0.993

5. For one player, the winning game proportions were 0.55 and 0.59 for *facing* and *back*, respectively. What was the associated *residual*?

(A) −0.028
(B) 0.028
(C) −0.0488
(D) 0.0488
(E) 0.3608

---

6. Data are obtained for a group of college freshmen examining their SAT scores (Math + Evidence-Based Reading and Writing)) from their senior year of high school and their GPAs during their first year of college. The resulting regression equation is

$$\widehat{GPA} = 0.55 + 0.00161\ (SAT\ total) \quad \text{with} \quad r = 0.632$$

What percentage of the variation in GPAs can be explained by looking at the linear relationship between GPAs and SAT scores?

(A) 0.161%
(B) 16.1%
(C) 39.9%
(D) 63.2%
(E) This value cannot be computed from the information given.

7. Suppose the correlation between two variables is $r = 0.23$. What will the new correlation be if 0.14 is added to all values of the x-variable, every value of the y-variable is doubled, and the two variables are interchanged?

(A) 0.23

(B) 0.37

(C) 0.74

(D) –0.23

(E) –0.74

8. Consider the following scatterplot and regression analysis of the 15 data points:

Slope $b = 0.377$

Coefficient of determination $r^2 = 95.0\%$

Standard deviation of the residuals $s = 0.640$

With the addition of a data point at (35,14), which one of the following choices gives the most likely new regression statistics?

(A) Slope $b = 0.377$
Coefficient of determination $r^2 = 97.0\%$
Standard deviation of the residuals $s = 0.663$

(B) Slope $b = 0.377$
Coefficient of determination $r^2 = 97.0\%$
Standard deviation of the residuals $s = 0.617$

(C) Slope $b = 0.377$
Coefficient of determination $r^2 = 93.0\%$
Standard deviation of the residuals $s = 0.640$

(D) Slope $b = 0.377$
Coefficient of determination $r^2 = 93.0\%$
Standard deviation of the residuals $s = 0.663$

(E) Slope $b = 0.397$
Coefficient of determination $r^2 = 95.0\%$
Standard deviation of the residuals $s = 0.640$

9. A study of the number of highway deaths due to failure to wear seat belts in a year versus percentage seat belt usage in that year in 300 national regions shows a strong negative linear association. The least squares regression equation is: Predicted number of deaths = 11,100 – 305.1 (Belt usage). What does "negative" mean in this context?

   (A) If no one wore seat belts in a given region, then it is predicted that there will be 11,100 highway deaths due to failure to wear seat belts.
   (B) The correlation, r, between the number of highway deaths due to failure to wear seat belts and the percentage of seat belt usage is negative.
   (C) If a given region has a lower percentage of seat belt usage than a second region, then it will have a higher number of highway deaths due to failure to wear seat belts than the second region.
   (D) Regions with a higher percentage of seat belt usage tend to have lower numbers of highway deaths due to failure to wear seat belts.
   (E) If a region has a one percent gain in seat belt usage then it will have a reduction of 305 highway deaths due to failure to wear seat belts, on average.

10. A simple random sample of 35 world-ranked chess players provides the following statistics:

   Number of hours of study per day: $\bar{x} = 6.2$, $s_x = 1.3$
   Yearly winnings: $\bar{y} = \$208{,}000$, $s_y = \$42{,}000$
   Correlation $r = 0.15$

   Based on this data, what is the resulting linear regression equation?

   (A) $\widehat{\text{Winnings}} = 178{,}000 + 4850 \text{ Hours}$

   (B) $\widehat{\text{Winnings}} = 169{,}000 + 6300 \text{ Hours}$

   (C) $\widehat{\text{Winnings}} = 14{,}550 + 31{,}200 \text{ Hours}$

   (D) $\widehat{\text{Winnings}} = 7750 + 32{,}300 \text{ Hours}$

   (E) $\widehat{\text{Winnings}} = -52{,}400 + 42{,}000 \text{ Hours}$

11. Consider the set of points {(2, 5), (3, 7), (4, 9), (5, 12), (10, n)}. What should n be so that the correlation between the x- and y-values is 1?

   (A) 21
   (B) 24
   (C) 25
   (D) A value different from any of the above.
   (E) No value for n can make $r = 1$.

Free-Response Questions

1. The following scatterplot shows the pulse rate drop (in beats per minute) plotted against the amount of medication (in grams) of an experimental drug being field-tested in several hospitals.

A computer printout showing the results of fitting a straight line to the data by the method of least squares gives

```
PulseRateDrop = -1.68 + 8.5 Grams
R-sq = 81.9%
```

(a) Find the correlation coefficient for the relationship between pulse rate drop and grams of medication.

(b) What is the slope of the regression line and what does it signify?

(c) Predict the pulse rate drop for a patient given 2.25 grams of medication.

(d) A patient given 5 grams of medication has his pulse rate drop to zero. Does this invalidate the regression equation? Explain.

(e) How will the size of the correlation coefficient change if the 3-gram result is removed from the data set? Explain.

(f) How will the size of the slope of the least squares regression line change if the 3-gram result is removed from the data set? Explain.

2. Facebook showed a remarkable rate of growth during its early years. A scatterplot of MAUs (monthly active users in millions) during its first nine years is as follows:

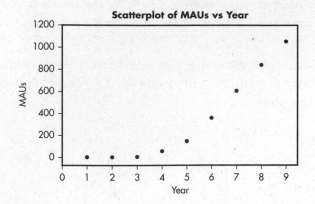

(a) Describe the relationship between MAUs and Year.

Transforming using logarithms and then fitting a linear regression model gives the following computer output and residual plot:

Regression Analysis: log(MAUs) versus log(Year)

```
The regression equation is
log(MAUs) = - 0.196 + 3.39 log(Year)

Predictor        Coef    SE Coef        T        P
Constant      -0.1963     0.1292    -1.52    0.172
log(Year)      3.3886     0.1888    17.95    0.000
S = 0.166815    R-Sq = 97.9%    R-Sq(adj) = 97.6%
```

(b) Interpret the value of r^2.

(c) Is it appropriate to use a linear regression model on the transformed variables?

3. An outlier can have a striking effect on the correlation r. Consider the following three scatterplots:

      I   II   III

(a) How do the outliers in II and III affect the correlation from I?

(b) Insert a fourth point in I such that the correlation doesn't change.

4. A continuing political debate concerns how to address the large student debt carried by college graduates. Graph (I) is a scatterplot showing the amount of student debt of 25 randomly selected college graduates and their stress level as self-reported on a 1–100 scale.

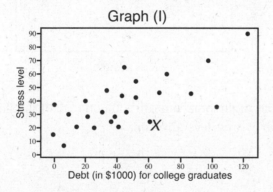

Graph (I)

Computer output from linear regression gives:

```
Predictor      Coef    SE Coef
Constant     19.376     4.782
Debt         0.42640   0.08906
```

(a) Describe the association between stress level and debt for this sample of 25 college graduates.

(b) The point labeled X represents a student with $60,000 debt and a stress level of 25. Calculate and interpret the residual of that point.

Two other possible explanatory variables with regard to stress level are Income and Health as measured on a 3 to 8 HRQoL (health related quality of life) Index. Graph (II) is a scatterplot of health plotted with the corresponding residuals from the regression of stress on debt. Graph (III) is a scatterplot of income plotted with the corresponding residuals from the regression of stress on debt.

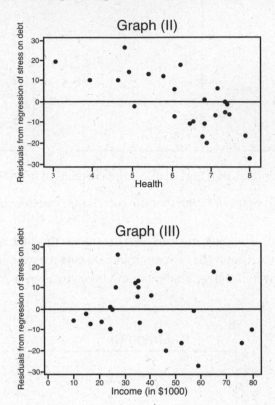

(c) Which variable, health or income, in addition to debt should be used to improve the prediction of stress level? Explain.

QUIZ 2

Multiple-Choice Questions

Directions: The questions that follow are each followed by five suggested answers. Choose the response that best answers the question.

1. A study collects data on average combined SAT scores (math, critical reading, and writing) and percentage of students who took the exam at 100 randomly selected high schools. Following is part of the computer printout for regression:

    ```
    Variable    Coefficient    s.e. of coeff    t-ratio    prob
    Constant    1576.32        12.65            124.6      ≤ 0.0001
    SAT         -2.84276       0.2461           -11.55     ≤ 0.0001
    R-squared = 76.5%     R-squared (adj) = 76.1%
    ```

   Which of the following is a correct conclusion?

   (A) SAT in the variable column indicates that SAT is the dependent (response) variable.
   (B) The correlation is 0.875.
   (C) The y-intercept indicates the mean combined SAT score if percent of students taking the exam has no effect on combined SAT scores.
   (D) The R^2 value indicates that the residual plot does not show a strong pattern.
   (E) Schools with lower percentages of students taking the exam tend to have higher average combined SAT scores.

2. Suppose the correlation is negative. Given two points from the scatterplot, which of the following is possible?

   I. The first point has a larger x-value and a smaller y-value than the second point.
   II. The first point has a larger x-value and a larger y-value than the second point.
   III. The first point has a smaller x-value and a larger y-value than the second point.

   (A) I only
   (B) II only
   (C) III only
   (D) I and III
   (E) I, II, and III

3. Suppose the regression line for a set of data, $\hat{y} = 3x + b$, passes through the point (2, 5). If $\bar{x}$ and $\bar{y}$ are the sample means of the x- and y-values, respectively, then $\bar{y} =$

   (A) $\bar{x}$.
   (B) $\bar{x} - 2$.
   (C) $\bar{x} + 5$.
   (D) $3\bar{x}$.
   (E) $3\bar{x} - 1$.

4. Which of the following statements about correlation r is true?

   (A) A correlation of 0.2 means that 20% of the points are highly correlated.
   (B) Perfect correlation, that is, when the points lie exactly on a straight line, results in $r = 0$.
   (C) Correlation is not affected by which variable is called x and which is called y.
   (D) Correlation is not affected by extreme values.
   (E) A correlation of 0.75 indicates a relationship that is 3 times as linear as one for which the correlation is only 0.25.

5. Which of the following statements about residuals from the least squares line are true?

   I. The mean of the residuals is always zero.
   II. The regression line for a residual plot is a horizontal line.
   III. A definite pattern in the residual plot is an indication that a nonlinear model will show a better fit to the data than the straight regression line.

   (A) I and II only
   (B) I and III only
   (C) II and III only
   (D) I, II, and III
   (E) None of the above gives the complete set of true responses.

6. In a study of winning percentage in home games versus average home attendance for professional baseball teams, the resulting regression line is:

   $$\widehat{\text{Winning percentage}} = 44 + 0.0003 \text{ Attendance}$$

   What is the residual if a team has a winning percentage of 55% with an average attendance of 34,000?

   (A) −11.0
   (B) −0.8
   (C) 0.8
   (D) 11.0
   (E) 23.0

7. Suppose the correlation between two variables is −0.57. If each of the y-scores is multiplied by −1, which of the following is true about the new scatterplot?

   (A) It slopes up to the right, and the correlation is −0.57.
   (B) It slopes up to the right, and the correlation is +0.57.
   (C) It slopes down to the right, and the correlation is −0.57.
   (D) It slopes down to the right, and the correlation is +0.57.
   (E) None of the above is true.

8. A study of selling prices of homes in a southern California community (in $1000) versus size of the homes (in 1000s of square feet) shows a moderate positive linear association. The least squares regression equation is: Predicted selling price = 35.3 + 214.1 (Size). What does "linear" mean in this context?

   (A) The points in the scatterplot line up in a straight line.
   (B) There is no distinct pattern in the residual plot.
   (C) The coefficient of determination, r^2, is large (close to 1).
   (D) As home size increases by 1000 square feet, the selling price tends to change by a constant amount, on average.
   (E) Each increase of 1000 square feet in home size gives an increase of 214.1 ($1000) in selling price.

9. Does lower sun exposure, as measured by greater distance from the equator, result in lower incidence of skin cancer? A study of skin cancer mortality versus latitude of 100 northern hemisphere locales shows a strong negative linear association. What does "strong" mean in this context?

   (A) More sun exposure causes greater numbers of skin cancer deaths.
   (B) More sun exposure is associated with greater numbers of skin cancer deaths.
   (C) A locale's incidence of skin cancer deaths has a linear association with its latitude.
   (D) A least squares model predicts that the greater the latitude, the lower the average incidence of skin cancer.
   (E) The actual incidence of skin cancer at a given latitude will be very close to what is predicted by a least squares model.

10. Consider n pairs of numbers. Suppose $\bar{x} = 2$, $s_x = 3$, $\bar{y} = 4$, and $s_y = 5$. Of the following, which could be the least squares line?

    (A) $\hat{y} = -2 + x$
    (B) $\hat{y} = 2x$
    (C) $\hat{y} = -2 + 3x$
    (D) $\hat{y} = \dfrac{5}{3} - x$
    (E) $\hat{y} = 6 - x$

11. Consider the three points (2, 11), (3, 17), and (4, 29). Given any straight line, we can calculate the sum of the squares of the three vertical distances from these points to the line. What is the smallest possible value this sum can be?

    (A) 6
    (B) 9
    (C) 29
    (D) 57
    (E) None of these values

Free-Response Questions

1. The following scatterplot shows the grades for research papers for a sociology professor's class plotted against the lengths of the papers (in pages).

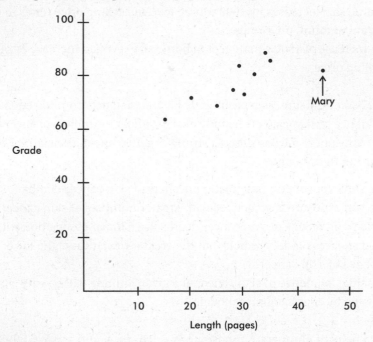

Mary turned in her paper late and was told by the professor that her grade would have been higher if she had turned it in on time. A computer printout fitting a straight line to the data (not including Mary's score) by the method of least squares gives

```
Grade = 46.51 + 1.106 Length
R–sq = 74.6%
```

(a) Find the correlation coefficient for the relationship between grade and length of paper based on these data (excluding Mary's paper).

(b) What is the slope of the regression line and what does it signify?

(c) How will the correlation coefficient change if Mary's paper is included? Explain your answer.

(d) How will the slope of the regression line change if Mary's paper is included? Explain your answer.

2. Data show a trend in winning long jump distances for an international competition over the years 1972–1992. With jumps recorded in inches and dates in years since 1900, a least squares regression line is fit to the data. The computer output and a graph of the residuals are as follows:

```
R squared = 92.1%
    Variable        Coefficient       SE of Coeff       t-ratio        Prob
    Constant          256.576            11.59            22.1         0.0001
    Year              0.95893            0.141            6.81         0.0024
```

(a) Does a line appear to be an appropriate model? Explain.
(b) What is the slope of the least squares line? Give an interpretation of the slope.
(c) What is the correlation?
(d) What is the predicted winning distance for the 1980 competition?
(e) What was the actual winning distance in 1980?

3. A scatterplot of the number of accidents per day on a particular interstate highway during a 30-day month is as follows:

(a) Draw a histogram of the frequencies of the number of accidents.
(b) Draw a boxplot of the number of accidents.
(c) Name a feature apparent in the scatterplot but not in the histogram or boxplot.
(d) Name a feature clearly shown by the histogram and boxplot but not as obvious in the scatterplot.

4. In a study of how exam scores are influenced by hours of sleep the night before the exam, a random sample of 13 students was outfitted with wristbands that tracked sleep time. The scatterplot of the resulting data follows.

(a) Describe the association between exam score and hours of sleep for these 13 students.

(b) Describe the association between exam score and hours of sleep for the 8 students who received at least 6 hours of sleep the night before the exam.

(c) Explain how the appropriateness of using linear regression is critically influenced by one student's data.

ANSWERS EXPLAINED

Quiz 1

MULTIPLE-CHOICE QUESTIONS

1. **(D)** Residual = Measured – Predicted, so if the residual is negative, the predicted must be greater than the measured (observed).

2. **(E)** The correlation r cannot take a value greater than 1.

3. **(B)** The "Predictor" column indicates the independent variable with its coefficient to the right.

4. **(E)** $r = \sqrt{0.986} = 0.993$ (The sign is the same as the sign of the slope, which in this case is positive.)

5. **(B)** $\widehat{back} = 0.056 + 0.920\,(0.55) = 0.562$, and so the $residual = 0.59 - 0.562 = 0.028$.

6. **(C)** The coefficient of determination r^2 gives the proportion of the y-variance that is predictable from a knowledge of x. In this case $r^2 = (0.632)^2 = 0.399$ or 39.9%.

7. **(A)** The correlation is not changed by adding the same number to every value of one of the variables, by multiplying every value of one of the variables by the same positive number, or by interchanging the x- and y-variables.

8. **(B)** (35,14) appears to follow the linear trend of the rest of the data. The slope wouldn't change, but with another point added to the trend, r^2 will increase, and s, the measure of typical deviation from the line, will decrease.

9. **(D)** Choice (A) is an interpretation of the y-intercept. Correlation is a measure of the strength of a linear relationship but does not by itself explain the meaning of "positive" or "negative." While the overall pattern is of a negative association, anything can be true about two points on the scatterplot. Choice (E) is an interpretation of the slope.

10. **(A)** Slope $= 0.15\left(\dfrac{42{,}000}{1.3}\right) \approx 4850$ and intercept $= 208{,}000 - 4850(6.2) \approx 178{,}000$.

11. **(E)** A scatterplot readily shows that while the first three points lie on a straight line, the fourth point does not lie on this line. Thus no matter what the fifth point is, all the points cannot lie on a straight line, and so r cannot be 1.

FREE-RESPONSE QUESTIONS

1. (a) The correlation coefficient $r = \sqrt{0.819} = 0.905$. It is positive because the slope of the regression line is positive.

   (b) The slope is 8.5, signifying that each gram of medication lowers the pulse rate by 8.5 beats per minute, on average.

   (c) $\hat{y} = -1.68 + 8.5(2.25) = 17.4$ beats per minute.

   (d) There is always danger in using a regression line to extrapolate beyond the values of x contained in the data. In this case, the 5 grams was an overdose, the patient died, and the regression line cannot be used for such values beyond the data set.

(e) Removing the 3-gram result from the data set will increase the correlation coefficient because the 3-gram result appears to be far off a regression line through the remaining points.

(f) Removing the 3-gram result from the data set will swing the regression line upward so that the slope will increase.

2. (a) There is a strong, positive, nonlinear relationship between MAUs and Year.

(b) 97.9% of the variation in log(MAUs) can be explained by the linear relationship between log(MAUs) and log(Year).

(c) The residual plot shows a definite pattern, indicating that while the linear model may be good, there is a better nonlinear model out there to describe the relationship between log(MAUs) and log(Year).

3. (a) In the first scatterplot, the points fall exactly on a downward sloping straight line, so $r = -1$. In the second scatterplot, the isolated point is an influential point, and r is close to $+1$. In the third scatterplot, the isolated point is also influential, and r is close to 0.

(b) Inserting a point that lies directly on a straight line drawn through the three points will not change the correlation of $r = -1$.

4. (a) The association between stress level and debt for this sample of 25 college graduates is linear, moderate, and positive.

(b) $19.376 + 0.42640(60) = 44.96$ and *Observed − Predicted* $= 25 - 44.96 = -19.96$. The residual value means that the stress level is 19.96 points less than what would be predicted for a student with a \$60,000 debt.

(c) Health is a better choice than Income for including with debt in a regression model for predicting stress. The relatively strong association between health and residuals from regression of stress on debt shows that health will help explain the unexplained variation (the residuals) between stress and debt.

Quiz 2

MULTIPLE-CHOICE QUESTIONS

1. **(E)** The variable column indicates the independent (explanatory) variable. The sign of the correlation is the same as the sign of the slope (negative here). In this example, the y-intercept is meaningless (predicted SAT result if no students take the exam). There can be a strong linear relation, with high R^2 value, but still a distinct pattern in the residual plot indicating that a non-linear fit may be even stronger. The negative value of the slope (-2.84276) gives that, on average, the predicted combined SAT score of a school is 2.84 points lower for each one unit higher in the percentage of students taking the exam.

2. **(E)** A negative correlation shows a tendency for higher values of one variable to be associated with lower values of the other; however, given any two points, anything is possible.

3. **(E)** Since $(2, 5)$ is on the line $y = 3x + b$, we have $5 = 6 + b$ and $b = -1$. Thus the regression line is $y = 3x - 1$. The point $(\bar{x}, \bar{y})$ is always on the regression line, and so we have $\bar{y} = 3\bar{x} - 1$.

4. **(C)** If the points lie on a straight line, $r = \pm 1$. Correlation has the formula $r = \frac{1}{n-1}\sum\left(\frac{x_i - \bar{x}}{s_x}\right)\left(\frac{y_i - \bar{y}}{s_y}\right)$ so x and y are interchangeable, and r does not depend on which variable is called x or y. However, since means and standard deviations can be strongly influenced by outliers, r too can be strongly affected by extreme values. While $r = 0.75$ indicates a better fit with a linear model than $r = 0.25$ does, we cannot say that the linearity is threefold.

5. **(D)** The sum and thus the mean of the residuals are always zero. In a good straight-line fit, the residuals show a random pattern.

6. **(C)** Predicted winning percentage $= 44 + 0.0003(34{,}000) = 54.2$, and

    Residual = Observed − Predicted = $55 - 54.2 = 0.8$.

7. **(B)** The slope and the correlation coefficient have the same sign. Multiplying every y-value by -1 changes this sign.

8. **(D)** A linear association means that as the explanatory variable (home size here) changes by a constant amount, the response variable (selling price here) also changes by a constant amount, on average. Unless there was perfect linear correlation, the points will not line up on a straight line. No distinct pattern in the residual plot just means that there is no obvious better model out there, but it doesn't necessarily say that the data is linear. The coefficient of determination indicates something about the strength of the relationship, but does not define linearity. Choice (E) is an interpretation of the slope, but again, not a definition of linearity.

9. **(E)** Association does not imply causation. While (C) refers to the existence of an association, and (B) and (D) further refer to the direction of the association, none of these statements refer to the strength of the association. "Strong" means that the points in the related scatterplot fall close to the least squares regression line. In context this means that the actual incidence of skin cancer at a given latitude will be very close to what is predicted by the least squares model.

10. **(E)** The least squares line passes through $(\bar{x}, \bar{y}) = (2, 4)$, and the slope b satisfies $b = r\frac{s_y}{s_x} = \frac{5r}{3}$. Since $-1 \le r \le 1$, we have $-\frac{5}{3} \le b \le \frac{5}{3}$.

11. **(A)** Using your calculator, find the regression line to be $\hat{y} = -8 + 9x$. The regression line, also called the least squares regression line, minimizes the sum of the squares of the vertical distances between the points and the line. In this case $(2, 10)$, $(3, 19)$, and $(4, 28)$ are on the line, and so the minimum sum is $(10 - 11)^2 + (19 - 17)^2 + (28 - 29)^2 = 6$.

FREE-RESPONSE QUESTIONS

1. (a) The correlation coefficient is $r = \sqrt{0.746} = 0.864$. It is positive because the slope of the regression line is positive.

    (b) The slope is 1.106, signifying that each additional page raises a grade by 1.106, on average.

(c) Including Mary's paper will lower the correlation coefficient because her result seems far off the regression line through the other points.

(d) Including Mary's paper will swing the regression line down and lower the value of the slope.

2. (a) Yes. The residual graph is not curved, does not show fanning, and appears to be random or scattered.

(b) The slope is 0.95893, indicating that the winning jump improves 0.95893 inches per year on average or about 3.8 inches every four years on average.

(c) With $r^2 = 0.921$, and a positive slope, the correlation r is 0.96.

(d) $0.95893(80) + 256.576 \approx 333.3$ inches

(e) The residual for 1980 is +2, and so the actual winning distance must have been $333.3 + 2 = 335.3$ inches.

3. (a)

(b)

(c) There is a roughly linear trend with daily accidents increasing during the month.

(d) The daily number of accidents is strongly skewed to the right.

TIP

In (a), a complete answer addresses form, strength, direction, and context.

4. (a) The association between exam score and hours of sleep is *nonlinear* (or *curved*), weak (or moderate), and positive.

(b) There appears to be very little, if any, association between exam score and hours of sleep for the 8 students who received at least 6 hours of sleep the night before the exam. (Because of the one very low score that also corresponds to hours of sleep just over 6, one could say that there is a very weak *positive* association.)

(c) The student who scored 75 and slept for just over 2 hours the night before the exam has such a strong influence that that particular scatterplot point taken together with all the other points suggests a nonlinear form.

Exploring Categorical Data: Two-Way Tables

4

→ **MARGINAL FREQUENCIES AND DISTRIBUTIONS**

→ **CONDITIONAL FREQUENCIES AND DISTRIBUTIONS**

While many variables such as age, income, and years of education are quantitative or numerical in nature, others such as gender, race, brand preference, mode of transportation, and type of occupation are qualitative or categorical. Quantitative variables, too, are sometimes grouped into categorical classes.

MARGINAL FREQUENCIES AND DISTRIBUTIONS

Qualitative data often encompass two categorical variables that may or may not have a dependent relationship. These data can be displayed in a *two-way table*.

➡ EXAMPLE 4.1

The Cuteness Factor: A Japanese study had volunteers look at pictures of cute baby animals, adult animals, or tasty-looking foods, after which they tested their focus in solving puzzles.

		Level of focus		
		Low	Medium	High
Pictures viewed	Baby animals	5	20	40
	Adult animals	30	40	15
	Tasty foods	55	35	10

Pictures viewed is the *row variable*, whereas level of focus is the *column variable*. One method of analyzing these data involves calculating the totals for each row and each column.

		Level of focus			
		Low	Medium	High	*Total*
Pictures viewed	Baby animals	5	20	40	*65*
	Adult animals	30	40	15	*85*
	Tasty foods	55	35	10	*100*
	Total	*90*	*95*	*65*	*250*

These totals are placed in the right and bottom margins of the table and thus are called *marginal frequencies* (or *marginal totals*). These marginal frequencies can then be put in the form of proportions or percentages. The *marginal distribution* of the level of focus is

Low: $\dfrac{90}{250} = 0.36 = 36\%$

Medium: $\dfrac{95}{250} = 0.38 = 38\%$

High: $\dfrac{65}{250} = 0.26 = 26\%$

This distribution can be displayed in a bar graph as follows;

Similarly, we can determine the marginal distribution for the pictures viewed:

Baby animals: $\dfrac{65}{250} = 0.26 = 26\%$

Adult animals: $\dfrac{85}{250} = 0.34 = 34\%$

Tasty foods: $\dfrac{100}{250} = 0.40 = 40\%$

The representative bar graph is

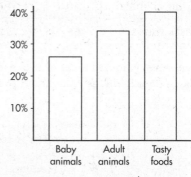

CONDITIONAL FREQUENCIES AND DISTRIBUTIONS

The marginal distributions described and calculated above do not describe or measure the relationship between the two categorical variables. For this we must consider the information in the body of the table, not just the sums in the margins.

➡ EXAMPLE 4.2

We are interested in predicting the level of focus from the pictures viewed, and so we look at *conditional frequencies* for each row separately. For example, in Example 4.1 what proportion or percentage of the participants who viewed baby animals then had each of the levels of focus?

$$\text{Low:} \qquad \frac{5}{65} = 0.077 = 7.7\%$$

$$\text{Medium:} \qquad \frac{20}{65} = 0.308 = 30.8\%$$

$$\text{High:} \qquad \frac{40}{65} = 0.615 = 61.5\%$$

This *conditional distribution* can be displayed either with groupings of bars or by a *segmented bar chart* where each segment has a length corresponding to its relative frequency:

Levels of focus after viewing baby animals

Similarly, the conditional distribution for the participants who viewed adult animals are

$$\text{Low:} \qquad \frac{30}{85} = 0.353 = 35.3\%$$

$$\text{Medium:} \qquad \frac{40}{85} = 0.471 = 47.1\%$$

$$\text{High:} \qquad \frac{15}{85} = 0.176 = 17.6\%$$

For the participants who viewed the tasty foods, we have

$$\text{Low:} \qquad \frac{55}{100} = 0.55 = 55\%$$

$$\text{Medium:} \qquad \frac{35}{100} = 0.35 = 35\%$$

$$\text{High:} \qquad \frac{10}{100} = 0.10 = 10\%$$

NOTE

Segmented bar charts are like pie charts, except that the data set is represented by a rectangular bar rather than a circle.

Both of the following bar charts give good visual pictures:

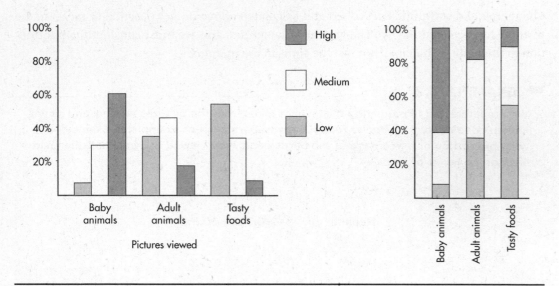

→ **EXAMPLE 4.3**_____

A study was made to compare year in high school with preference for vanilla or chocolate ice cream with the following results:

	Vanilla	Chocolate
Freshman	20	10
Sophomore	24	12
Junior	18	9
Senior	22	11

What are the conditional frequencies for each class?

Freshmen: $\frac{20}{30}$ = 0.667 prefer vanilla and $\frac{10}{30}$ = 0.333 prefer chocolate.

Sophomores: $\frac{24}{36}$ = 0.667 prefer vanilla and $\frac{12}{36}$ = 0.333 prefer chocolate.

Juniors: $\frac{18}{27}$ = 0.667 prefer vanilla and $\frac{9}{27}$ = 0.333 prefer chocolate.

Seniors: $\frac{22}{33}$ = 0.667 prefer vanilla and $\frac{11}{33}$ = 0.333 prefer chocolate.

In such a case, where all the conditional frequency distributions are identical, we say that the two variables show *perfect independence.* (However, it should be noted that even if the two variables are completely independent, the chance is very slim that a resulting two-way table will show perfect independence.)

EXAMPLE 4.4

Suppose you need heart surgery and are trying to decide between two surgeons, Dr. Fixit and Dr. Patch. You find out that each operated 250 times last year with the following results:

	Dr. F	Dr. P
Died	60	50
Survived	190	200

Whom should you go to? Among Dr. Fixit's 250 patients 190 survived, for a survival rate of $\frac{190}{250}$ = 0.76 or 76%, while among Dr. Patch's 250 patients 200 survived, for a survival rate of $\frac{200}{250}$ = 0.80 or 80%. Your choice seems clear.

However, everything may not be so clear-cut. Suppose that on further investigation you determine that the surgeons operated on patients who were in either good or poor condition with the following results:

Good condition

	Dr. F	Dr. P
Died	8	17
Survived	60	120

Poor condition

	Dr. F	Dr. P
Died	52	33
Survived	130	80

Note that adding corresponding boxes from these two tables gives the original table above.

How do the surgeons compare when operating on patients in good health? Dr. Fixit's 68 patients in good condition have a survival rate of $\frac{60}{68}$ = 0.882 or 88.2%, while Dr. Patch's 137 patients in good condition have a survival rate of $\frac{120}{137}$ = 0.876 or 87.6%. Similarly, we note that Dr. Fixit's 182 patients in poor condition have a survival rate of $\frac{130}{182}$ = 0.714 or 71.4%, while Dr. Patch's 113 patients in poor condition have a survival rate of $\frac{80}{113}$ = 0.708 or 70.8%.

Thus Dr. Fixit does better with patients in good condition (88.2% versus Dr. Patch's 87.6%) and also does better with patients in poor condition (71.4% versus Dr. P's 70.8%). However, Dr. Fixit has a lower overall patient survival rate (76% versus Dr. Patch's 80%)! How can this be?

This problem is an example *of Simpson's paradox*, where a conjecture can be reversed when more than one group is combined to form a single group. The effect of another variable is masked when the groups are combined. In this particular example, closer scrutiny reveals that Dr. Fixit operates on many more patients in poor condition than Dr. Patch, and these patients in poor condition are precisely the ones with lower survival rates. Thus even though Dr. Fixit does better with all patients, his overall rating is lower. Our original table hid the effect of the variable related to the condition of the patients.

- Two-way tables are useful in showing relationships between two categorical variables.
- The row and column totals lead to calculations of the marginal distributions.
- Focusing on single rows or columns leads to calculations of conditional distributions.
- Segmented bar charts are a useful visual tool to show conditional distributions.
- Simpson's paradox occurs when the results from a combined grouping seem to contradict the results from the individual groups.

QUIZZES ON TOPIC FOUR: EXPLORING CATEGORICAL DATA

QUIZ 1

Multiple-Choice Questions

> **Directions:** The questions that follow are each followed by five suggested answers. Choose the response that best answers the question.

Questions 1–5 are based on the following: To study the relationship between party affiliation and support for a balanced budget amendment, 500 registered voters were surveyed with the following results:

	For	Against	No opinion
Democrat	50	150	50
Republican	125	50	25
Independent	15	10	25

1. What percentage of those surveyed were Democrats?

   (A) 10%
   (B) 20%
   (C) 30%
   (D) 40%
   (E) 50%

2. What percentage of those surveyed were for the amendment and were Republicans?

   (A) 25%
   (B) 38%
   (C) 40%
   (D) 62.5%
   (E) 65.8%

3. What percentage of Independents had no opinion?

   (A) 5%
   (B) 10%
   (C) 20%
   (D) 25%
   (E) 50%

4. What percentage of those against the amendment were Democrats?

   (A) 30%
   (B) 42%
   (C) 50%
   (D) 60%
   (E) 71.4%

5. Voters of which affiliation were most likely to have no opinion about the amendment?

   (A) Democrat
   (B) Republican
   (C) Independent
   (D) Republican and Independent, equally
   (E) Democrat, Republican, and Independent, equally

   _____

6. In the following table, what value for n results in a table showing perfect independence?

20	50
30	n

   (A) 10
   (B) 40
   (C) 60
   (D) 75
   (E) 100

Free-Response Questions

1. The following table gives the numbers (in thousands) of officers and enlisted personnel by military branch in the U.S. armed forces.

	Army	Navy	Marine Corps	Air Force
Officers	88	52	20	65
Enlisted	452	276	178	258

   (a) Calculate the percentage

       i. of military men and women who are enlisted.
       ii. of military men and women who are not Marine Corps officers.
       iii. of officers who are in the Navy.

   (b) Construct a graphical display showing the association between career path (officer vs. enlisted) and military branch.

   (c) Summarize what the graphical display illustrates about the association between career path (officer vs. enlisted) and military branch.

2. In 2005, the racial composition and mean incomes of a company's employees were 150 White, 50 Black, and 50 Hispanic, with mean incomes of $37,000, $22,000, and $23,000, respectively. Ten years later in 2015, the data were 90 White, 60 Black, and 100 Hispanic, with mean incomes of $40,000, $23,000, and $25,000, respectively.

   (a) What was the mean income trend for each racial group between 2005 and 2010?

   (b) What was the mean income trend for the whole company between 2005 and 2010?

   (c) If you worked in the company's HR office, how would you explain the company's income trend to a prospective employee?

QUIZ 2

Multiple-Choice Questions

> **Directions:** The questions that follow are each followed by five suggested answers. Choose the response that best answers the question.

Questions 1–5 are based on the following: A study of music preferences in three geographic locations resulted in the following segmented bar chart:

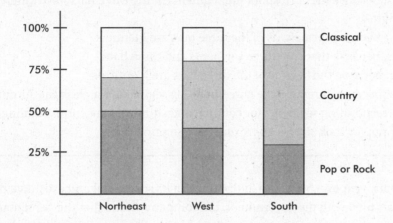

1. What percentage of those surveyed from the Northeast prefer country music?

   (A) 20%
   (B) 30%
   (C) 40%
   (D) 50%
   (E) 70%

2. Which of the following is greatest?

   (A) The percentage of those from the Northeast who prefer classical.
   (B) The percentage of those from the West who prefer country.
   (C) The percentage of those from the South who prefer pop or rock.
   (D) The above are all equal.
   (E) It is impossible to determine the answer without knowing the actual numbers of people involved.

3. Which of the following is greatest?

   (A) The number of people in the Northeast who prefer pop or rock.
   (B) The number of people in the West who prefer classical.
   (C) The number of people in the South who prefer country.
   (D) The above are all equal.
   (E) It is impossible to determine the answer without knowing the actual numbers of people involved.

4. All three bars have a height of 100%. Which of the following is true?

   (A) This is a coincidence.
   (B) This happened because each bar shows a complete distribution.
   (C) This happened because there are three bars each divided into three segments.
   (D) This happened because of the nature of musical patterns.
   (E) None of the above is true.

5. Based on the given segmented bar chart, does there seem to be a relationship between geographic location and music preference?

   (A) Yes, because the corresponding segments of the three bars have different lengths.
   (B) Yes, because the heights of the three bars are identical.
   (C) Yes, because there are three segments and three bars.
   (D) No, because the heights of the three bars are identical.
   (E) No, because summing the corresponding segments for classical, summing the corresponding segments for country, and summing the corresponding segments for pop or rock all give approximately the same total.

---

6. A company employs both men and women in its secretarial and executive positions. In reports filed with the government, the company shows that the percentage of female employees who receive raises is higher than the percentage of male employees who receive raises. A government investigator claims that the percentage of male secretaries who receive raises is higher than the percentage of female secretaries who receive raises, and that the percentage of male executives who receive raises is higher than the percentage of female executives who receive raises. Is this possible?

   (A) No, either the company report is wrong or the investigator's claim is wrong.
   (B) No, if the company report is correct, then either a greater percentage of female secretaries than of male secretaries receive raises or a greater percentage of female executives than of male executives receive raises.
   (C) No, if the investigator is correct, then by summation of the corresponding numbers, the total percentage of male employees who receive raises would have to be greater than the total percentage of female employees who receive raises.
   (D) All of the above are true.
   (E) It is possible for both the company report to be true and the investigator's claim to be correct.

Free-Response Questions

> **Directions:** You must show all work and indicate the methods you use. You will be graded on the correctness of your methods and on the accuracy of your final answers.

1. The graduate school at the University of California at Berkeley reported that in 1973 they accepted 44% of 8442 male applicants and 35% of 4321 female applicants. Concerned that one of their programs was guilty of gender bias, the graduate school analyzed admissions to the six largest graduate programs and obtained the following results:

Program	Men Accepted	Men Rejected	Women Accepted	Women Rejected
A	511	314	89	19
B	352	208	17	8
C	120	205	202	391
D	137	270	132	243
E	53	138	95	298
F	22	351	24	317

   (a) Find the percentage of men and the percentage of women accepted by each program. Comment on any pattern or bias you see.

   (b) Find the percentage of men and the percentage of women accepted overall by these six programs. Does this appear to contradict the results from part (a)?

   (c) If you worked in the Graduate Admissions Office, what would you say to an inquiring reporter who is investigating gender bias in graduate admissions?

2. Three tennis players were comparatively ranked by tournament directors voting on their play on Hardcourt, Clay, and Grass:

	Hardcourt	Clay	Grass
Player A	3	2	1
Player B	2	1	2
Player C	1	3	3

   Are these differences significant, or could the players have equal ability on all surfaces and these partciular rankings be due to chance? If the players have equal ability on all surfaces, then the expected average ranking for each would be 2.

   Consider the following test statistic: $Q = 3[(\bar{R}_A - 2)^2 + (\bar{R}_B - 2)^2 + (\bar{R}_C - 2)^2]$

   where $\bar{R}_A$, $\bar{R}_B$, and $\bar{R}_C$ are the average rankings for the three players.

(a) Calculate the test statistic Q for the above data.

One hundred simulated values of the test statistic were calculated assuming the players have equal ability on all surfaces. The results are shown in the dotplot below.

Simulated values of Q

(b) Use these simulated values and the test statistic to determine if directors' rankings provide evidence of a significant difference in the players' abilities.

ANSWERS EXPLAINED

Quiz 1

MULTIPLE-CHOICE QUESTIONS

1. **(E)** Of the 500 people surveyed, $50 + 150 + 50 = 250$ were Democrats, and $\frac{250}{500} = 0.5$ or 50%.

2. **(A)** Of the 500 people surveyed, 125 were both for the amendment and Republicans, and $\frac{125}{500} = 0.25$ or 25%.

3. **(E)** There were $15 + 10 + 25 = 50$ Independents; 25 of them had no opinion, and $\frac{25}{50} = 0.5$ or 50%.

4. **(E)** There were $150 + 50 + 10 = 210$ people against the amendment; 150 of them were Democrats, and $\frac{150}{210} = 0.714$ or 71.4%.

5. **(C)** The percentages of Democrats, Republicans, and Independents with no opinion are 20%, 12.5%, and 50%, respectively.

6. **(D)** Relative frequencies must be equal. Either looking at rows gives $\frac{20}{70} = \frac{30}{30+n}$ or looking at columns gives $\frac{20}{50} = \frac{50}{50+n}$. We could also set up a proportion $\frac{n}{30} = \frac{50}{20}$ or $\frac{n}{50} = \frac{30}{20}$. Solving any of these equations gives $n = 75$.

FREE-RESPONSE QUESTIONS

1. (a) i. $\frac{1164}{1389} = 83.80\%$

   ii. $1 - \frac{20}{1389} = 98.56\%$

   iii. $\frac{52}{225} = 23.11\%$

(b) Calculate row or column totals, and then show either a side-by-side bar graph or a segmented bar graph, showing percentages, and conditioned on either career path (officer vs. enlisted) or military branch:

(c) The Army and the Navy have about the same percentage of officers (16%), while the Air Force has a higher percentage of officers (20%), and the Marine Corps has a lower percentage of officers (10%).

OR

Among the officers and the enlisted career paths there are about the same percentage Army (39%), and about the same percentage Navy (23%), while the officers have a lower percentage Marine Corps than the enlisted (9% vs. 16%) and the officers have a higher percentage Air Force than the enlisted (29% vs. 22%).

2. (a) Each employee racial group experienced an increase in mean salary between 2005 and 2010. White employees' mean salary rose from $37,000 to $40,000, Black employees' mean salary rose from $22,000 to $23,000, and Hispanic employees' mean salary rose from $23,000 to $25,000.

(b) The mean company salary in 2005 was $\dfrac{150(37,000)+50(22,000)+50(23,000)}{250} =$ 31,200, while in 2015 the mean was $\dfrac{90(40,000)+60(23,000)+100(25,000)}{250} = 29,920$. The company's mean employee salary dropped from $31,200 to $29,920 between 2005 and 2015.

(c) You should tell the prospective employee that while it is true that the company's mean employee salary has dropped between 2005 to 2015, the mean salary paid to each racial group of employees has risen. The reason behind this apparent paradox is related to the changing employee racial composition. (You might want to address if and how the company is addressing the apparent unequal salary scales based on race!)

Quiz 2

MULTIPLE-CHOICE QUESTIONS

1. **(A)** In the bar corresponding to the Northeast, the segment corresponding to country music stretches from the 50% level to the 70% level, indicating a length of 20%.

2. **(B)** Based on lengths of indicated segments, the percentage from the West who prefer country is the greatest.

3. **(E)** The given bar chart shows percentages, not actual numbers.

4. **(B)** In a complete distribution, the probabilities sum to 1, and the relative frequencies total 100%.

5. **(A)** The different lengths of corresponding segments show that in different geographic regions different percentages of people prefer each of the music categories.

6. **(E)** It is possible for both to be correct, for example, if there were 11 secretaries (10 women, 3 of whom receive raises, and 1 man who receives a raise) and 11 executives (10 men, 1 of whom receives a raise, and 1 woman who does not receive a raise). Then 100% of the male secretaries receive raises while only 30% of the female secretaries do; and 10% of the male executives receive raises while 0% of the female executives do. However, overall 3 out of 11 women receive raises, while only 2 out of 11 men receive raises. This is an example of Simpson's paradox.

FREE-RESPONSE QUESTIONS

1. (a)

Program	Percentage of Men Accepted (%)	Percentage of Women Accepted (%)
A	62	82
B	63	68
C	37	34
D	34	35
E	28	24
F	6	7

There doesn't appear to be any real pattern; however, women seem to be favored in four of the programs, while men seem to be slightly favored in the other two programs.

(b) Overall, 1195 out of 2681 male applicants were accepted, for a 45% acceptance rate, while 559 out of 1835 female applicants were accepted, for a 30% acceptance rate. This appears to contradict the results from part *a*.

(c) You should tell the reporter that while it is true that the overall acceptance rate for women is 30% compared to the 45% acceptance rate for men, program by program women have either higher acceptance rates or only slightly lower acceptance rates than men. The reason behind this apparent paradox is that most men applied to programs A and B, which are easy to get into and have high acceptance rates. However, most women applied to programs C, D, E, and F, which are much harder to get into and have low acceptance rates.

2. (a) $\overline{R}_A = \frac{(3+2+1)}{3} = 2, \overline{R}_B = \frac{(2+1+2)}{3} = \frac{5}{3}, \overline{R}_C = \frac{(1+3+3)}{3} = \frac{7}{3}$

$Q = 3[(2-2)^2 + (\frac{5}{3} - 2)^2 + (\frac{7}{3} - 2)^2] = \frac{2}{3}$

(b) If there was a significant difference in the players' abilities, the resulting Q-value would have been large. However, as seen in the simulations, $\frac{2}{3}$ is not large at all. Thus, there is not sufficient evidence of a significant difference in the players' abilities.

Theme Two: Collecting and Producing Data

Planning and Conducting Surveys

5

→ **BIAS**
→ **SIMPLE RANDOM SAMPLING**
→ **STRATIFIED SAMPLING**
→ **CLUSTER SAMPLING**
→ **SYSTEMATIC SAMPLING**
→ **SAMPLING FOR HOMOGENEITY**
→ **SAMPLING ERROR**

In the real world, time and cost considerations usually make it impossible to analyze an entire population. Do companies like Apple question every potential consumer before designing a new smart phone? Does a television producer check every household's viewing preferences before deciding whether a pilot program will be continued? By studying statistics, we learn how to estimate a population characteristic (called a population *parameter*) by considering a sample measurement (called a sample *statistic*). For example, later in this review book we will see how to estimate population proportions, means, and slopes by looking at sample proportions, means, and slopes.

To derive conclusions about the larger population, we need to be confident that the sample we have chosen represents the population fairly. Analyzing the data with computers is often easier than gathering the data, but the frequently quoted "garbage in, garbage out" applies here. Nothing can help if the data are collected poorly. Unfortunately, many of the statistics that we are bombarded with in newspapers, radio, and television are based on poorly designed data collection procedures.

There are two important principles in picking a representative sample in order to be able to generalize our findings to the whole population:

1. The use of *randomization* in sample selection increases the chance that the resulting sample exhibits all the known and unknown features of the population.
2. The *sample size*, not the fraction of the population surveyed, is key.

BIAS

The one thing that most quickly invalidates a sample and makes obtaining useful information and drawing meaningful conclusions impossible is *bias*. **A sampling method is biased if in some critical way it consistently results in samples that do not represent the population.**

Voluntary response surveys are based on individuals who choose to participate, typically give too much emphasis to people with strong opinions, and undersample people who

> **TIP**
> You should always be aware to what population you can reasonably generalize the results of your study.

> **NOTE**
> **MYTH:** A random sample will *always* be representative of the population.
>
> **FACT:** A random sample, *simply by chance*, might turn out not to be very representative.

> **NOTE**
> The way data are obtained is crucial. A large sample size cannot make up for a poor survey design or for faulty data collection techniques.

don't care much about a topic. For example, radio call-in programs about controversial topics such as gun control, abortion, and school segregation do not produce meaningful data on what proportion of the population favor or oppose related issues. Online surveys posted to websites are a modern source of voluntary response bias. Although a voluntary response survey may be easy and inexpensive to conduct, the sample is likely to be composed of strongly opinionated people, especially those with negative opinions on a subject.

Convenience surveys, like interviews at shopping malls, are based on choosing individuals who are easy to reach. These surveys tend to produce data highly unrepresentative of the entire population. For example, door-to-door household surveys miss people who do not happen to be at home and typically miss groups such as college students, prison inmates, and the homeless. Although convenience samples may allow a researcher to conduct a survey quickly and inexpensively, generalizing from the sample to a population is almost impossible.

There are many other sources of sampling bias. Although it is **not** necessary to know them by name, it is important to recognize that the sample designs are flawed, to be able to explain why they are flawed, and to understand the *direction* of the bias, that is, whether it will likely cause an under- or an overestimation of a proportion or a mean.

Undercoverage bias happens when there is inadequate representation, and thus some groups in the population are left out of the process of choosing the sample. For example, telephone surveys ignore those who don't have telephones. *Response bias* occurs when the question itself can lead to misleading results because people don't want to be perceived as having unpopular or unsavory views or don't want to admit to having committed crimes. *Nonresponse bias*, where there are low response rates, occurs when individuals chosen for the sample can't be contacted or refuse to participate, and it is often unclear which part of the population is responding. *Quota sampling bias*, where interviewers are given free choice in picking people in the (problematic if not impossible) attempt to pick representatively with no randomization, is a recipe for disaster. *Wording bias* can occur when non-neutral or poorly worded questions lead to very unrepresentative responses or even when the order in which questions are asked makes a difference.

SIMPLE RANDOM SAMPLING

How can we increase our chance of choosing a representative sample? One technique is to write the name of each member of the population on a card, mix the cards thoroughly in a large box, and pull out a specified number of cards. This method gives everyone in the population an equal chance of being selected as part of the sample. Unfortunately, this method is usually too time-consuming, and bias might still creep in if the mixing is not thorough. A *simple random sample* (SRS), **one in which every possible sample of the desired size has an equal chance of being selected**, can be more easily obtained by assigning a number to everyone in the population and using a random digit table or having a computer generate random numbers to indicate choices.

→ EXAMPLE 5.1

Suppose 80 students are taking an AP Statistics course and the teacher wants to pick a sample of 10 students randomly to try out a practice exam. She first assigns the students numbers 01, 02, 03, . . ., 80. While reading off two digits at a time from a random digit table, she ignores any numbers over 80, ignores 00, and ignores repeats, stopping when she has a set of 10. If the table began 75425 56573 90420 48642 27537 61036 15074 84675, she would choose the students numbered 75, 42, 55, 65, 73, 04, 27, 53, 76, and 10. Note that 90 and 86 are ignored because they are over 80. Note that the second and third occurrences of 42 are ignored because they are repeats.

TIP

When using a random digit table, it is important that each label has the same number of digits. It would be wrong to assign the digits 1, 2, 3, . . ., 80 rather than 01, 02, 03, . . ., 80. (This is not necessary when using a random number generator.)

→ EXAMPLE 5.2

An alternative solution to Example 5.1 is to make use of a random number generator on a computer. Assign the students numbers 1, 2, 3, . . ., 80. Use a computer to generate 10 random integers between 1 and 80 *without replacement*, that is, throw out repeats. The sample consists of the students with assigned numbers corresponding to the 10 unique computer-generated numbers.

Advantages of simple random sampling include the following:

- The simplicity of simple random sampling makes it relatively easy to interpret data collected.
- This method requires minimal advance knowledge of the population other than knowing the complete sampling frame.
- Simple random sampling allows us to make generalizations (i.e., statistical inferences) from the sample to the population.

IMPORTANT

To receive full credit, there are 3 steps:

1. Assigning numbers,
2. Using a computer random number generator to generate distinct numbers in a given range, and
3. Linking selected numbers with corresponding individuals selected for the sample.

Disadvantages of simple random sampling include the following:

- Although simple random sampling is a straightforward procedure to understand, it can be difficult to execute, especially if the population is large. For example, how would you obtain a simple random sample of students from the population of all high school students in the United States?
- This method may require you to contact nonrespondents repeatedly, which can be very time-consuming.
- This method may leave out groups you want to be sure are represented.

NOTE

The list of all individuals from which the sample is drawn is called the *sampling frame*.

There are other sampling methods to choose. Each has its own set of advantages and disadvantages.

STRATIFIED SAMPLING

Stratified sampling involves dividing the population into homogeneous groups called *strata*, then picking random samples from each of the strata, and finally combining these individual samples into what is called a *stratified random sample*. For example, we can stratify by age, gender, income level, or race; pick a sample of people from each stratum; and combine to form the final sample. For example, a political pollster might want to be sure to include respondents from various minority groups. Stratifying by religion or race and including people from each group would insure this diversity in the final sample.

Advantages of stratified sampling include the following:

- Samples taken within a stratum have little variability. So the resulting estimates are more precise than when using other sampling methods.
- Important differences among groups can become more apparent.

Disadvantages of stratified sampling include the following:

- Like an SRS, this method might be difficult to implement with large populations.
- Forcing subdivisions when none really exist is meaningless.

We could further do *proportional sampling*, where the sizes of the random samples from each stratum depend on the proportion of the total population represented by the stratum.

CLUSTER SAMPLING

Cluster sampling involves dividing the population into heterogeneous groups called *clusters* and then picking everyone in a random selection of one or more of the clusters. For example, to survey high school seniors, we could randomly pick several senior class homerooms in which to conduct our study and sample all students in those selected homerooms.

Advantages of cluster sampling include the following:

- Clusters are often chosen for ease, convenience, and quickness.
- With limited fixed funds, cluster sampling usually allows for a larger sample size than do other sampling methods.

Disadvantages of cluster sampling include the following:

- With a given sample size, cluster sampling usually provides less precision than will either an SRS or a stratified sample.
- If the population doesn't have natural clusters and the designated clusters are not representative of the population, selection could easily result in a biased sample.

The differences between strata and clusters include the following:

- While all strata are represented in the sample, only a subset of clusters are in the sample.
- Stratified sampling is accurate when each strata consists of *homogeneous* elements, while cluster sampling is accurate when each cluster consists of *heterogeneous* elements.

SYSTEMATIC SAMPLING

Systematic sampling is a relatively simple and quick method. It involves listing the population in some order (for example, alphabetically), choosing a random point to start, and then picking every tenth (or hundredth, or thousandth, or kth) person from the list. This gives a reasonable sample as long as the original order of the list is not in any way related to the variables under consideration.

An advantage of systematic sampling is the following:

- If similar members in the list are grouped together, then we can end up with a kind of stratified sample, only more easily implemented.

A disadvantage of systematic sampling is the following:

- If the list happens to have a periodic structure similar to the period of the sampling, then a very biased sample could result.

NOTE

In picking every kth person, if the list order has no structure, this is in effect cluster sampling from among k clusters!

➡ **EXAMPLE 5.3**

Suppose a sample of 100 high school students from a Chicago school of size 5000 is to be chosen to determine their views on whether they think the Cubs will win another World Series this century. One method would be to have each student write his or her name on a card, put the cards into a box, and have the principal reach in and pull out 100 cards. However, questions could arise regarding how well the cards are mixed. For example, how might the outcome be affected if all students in one PE class toss their names in at the same time so that their cards are clumped together? Another method would be to assign each student a number from 0001 to 5000 and then use a random digit table, picking out four digits at a time, ignoring repeats, 0000, and numbers over 5000, until a unique set of 100 numbers are picked. Then choose the students corresponding to the selected 100 numbers (simple random sampling). What are alternative procedures?

Answer: From a list of the students, the surveyor could simply note every 50th name (systematic sampling). Since students in each class have certain characteristics in common, the surveyor could use a random selection method to pick 25 students from each of the separate lists of freshmen, sophomores, juniors, and seniors (stratified sampling). If each homeroom has a random mix of 20 students from across all grade levels, the surveyor could randomly pick five homerooms with the sample consisting of all the students in these five rooms (cluster sampling).

NOTE

None of these alternative procedures result in a *simple random sample* because every possible sample of size 100 does not have an equal chance of being selected.

TIP

Understand that these other procedures are still proper random samples even if they are not simple random samples.

SAMPLING FOR HOMOGENEITY

Sometimes we randomly pick a single sample, cross-categorize on two variables, and then test to see if there is an association between the variables. However, sometimes we instead randomly pick a sample from each of two or more populations and then compare the distribution of some variable among the different populations. This second method is called sampling for *homogeneity*.

NOTE

We'll revisit this sampling method later when developing chi-square inference, but it's really a design issue and well worth recognizing at this time.

Sampling error is
to be expected and
can be quantified,
while bias is to be
avoided.

SAMPLING ERROR: THE VARIATION INHERENT IN A SURVEY

No matter how well-designed and well-conducted a survey is, it still gives a sample *statistic* as an estimate for a population *parameter*. Different samples give different sample statistics, all of which are estimates for the same population parameter. So error, called *sampling error*, is naturally present. This error can be described using probability; that is, we can say how likely we are to have a certain size error. Generally, the associated probabilities are smaller when the sample size is larger. However, **whenever a sample is taken, sampling error will be present**.

SUMMARY

- Bias is the tendency to favor the selection of certain members of a population.
- Voluntary response bias occurs when individuals choose whether to respond (for example, Internet surveys), and typically give too much emphasis to people with strong opinions (particularly strong negative opinions).
- Convenience surveys, like interviews at shopping malls, are based on choosing individuals who are easy to reach. These surveys tend to produce data highly unrepresentative of the entire population.
- Use of randomization is absolutely critical in selecting an unbiased sample.
- Use of randomization in selecting a sample is crucial in being able generalize from our sample to the population.
- A simple random sample (SRS) is one in which every possible sample of the desired size has an equal chance of being selected.
- Stratified sampling involves dividing the population into homogeneous groups called strata and then picking random samples from each of the strata.
- Cluster sampling involves dividing the population into heterogeneous groups called clusters, randomly selecting one or more of the clusters, and then using everyone in the selected clusters.
- Sampling error is not a negative but, rather, refers to the natural variability among samples.

QUIZ 1

Multiple-Choice Questions

Directions: The questions that follow are each followed by five suggested answers. Choose the response that best answers the question.

1. A school psychologist wants to investigate student depression. She selects, at random, samples of 30 freshmen, 30 sophomores, 30 juniors, and 30 seniors to interview. Which of the following best describes the principal's sampling plan?

   (A) A convenience sample
   (B) A simple random sample
   (C) A stratified random sample
   (D) A cluster sample
   (E) A systematic sample

2. Which of the following is a true statement?

   (A) If bias is present in a sampling procedure, it can be overcome by dramatically increasing the sample size.
   (B) There is no such thing as a "bad sample."
   (C) Sampling techniques that use probability techniques effectively eliminate bias.
   (D) Convenience samples often lead to undercoverage bias.
   (E) Voluntary response samples often underrepresent people with strong opinions.

3. Which of the following is a true statement about sampling error?

   (A) Sampling error can be eliminated only if a survey is both extremely well designed and extremely well conducted.
   (B) Sampling error concerns natural variation between samples, is always present, and can be described using probability.
   (C) Sampling error is generally larger when the sample size is larger.
   (D) Sampling error implies an error, possibly very small, but still an error, on the part of the surveyor.
   (E) Sampling error is higher when bias is present.

4. Many colleges are moving toward coed dormitories. A Residential Life director plans to sample student attitudes toward this arrangement. He randomly selects three of the 12 dorms and sends a questionnaire to all residents living in them. Which of the following best describes the director's sampling plan?

   (A) A convenience sample
   (B) A simple random sample
   (C) A stratified random sample
   (D) A cluster sample
   (E) A systematic sample

5. Ann Landers, who wrote a daily advice column appearing in newspapers across the country, once asked her readers, "If you had it to do over again, would you have children?" Of the more than 10,000 readers who responded, 70% said no. What does this show?

   (A) The survey is meaningless because of voluntary response bias.
   (B) No meaningful conclusion is possible without knowing something more about the characteristics of her readers.
   (C) The survey would have been more meaningful if she had picked a random sample of the 10,000 readers who responded.
   (D) The survey would have been more meaningful if she had used a control group.
   (E) This was a legitimate sample, randomly drawn from her readers and of sufficient size to allow the conclusion that most of her readers who are parents would have second thoughts about having children.

6. Each of the 30 MLB teams has 25 active roster players. A sample of 60 players is to be chosen as follows. Each team will be asked to place 25 cards with its players' names into a hat and randomly draw out two names. The two names from each team will be combined to make up the sample. Will this method result in a simple random sample of the 750 baseball players?

   (A) Yes, because each player has the same chance of being selected.
   (B) Yes, because each team is equally represented.
   (C) Yes, because this is an example of stratified sampling, which is a special case of simple random sampling.
   (D) No, because the teams are not chosen randomly.
   (E) No, because not each group of 60 players has the same chance of being selected.

Free-Response Questions

1. A student is interested in estimating the average length of words in a 600-page textbook, and plans the following three-stage sampling procedure:

   (1) Noting that each of the 12 chapters has a different author, the student decides to obtain a sample of words from each chapter.

   (2) Each chapter is approximately 50 pages long. The student uses a random number generator to pick three pages from each chapter.

   (3) On each chosen page the student notes the length of every tenth word.

      (a) The first stage above represents what kind of sampling procedure? Give an advantage in using it in this context.

      (b) The second stage above represents what kind of sampling procedure? Give an advantage in using it in this context.

      (c) The third stage above represents what kind of sampling procedure? Give a *disadvantage* to using it dependent upon an author's writing style.

2. Two versions of a Pew Research poll were conducted:

   Version I: "Is it more important for the U.S. President to focus on domestic policy or foreign policy?"

   Version II: "Is it more important for the U.S. President to focus on domestic policy or the war on terrorism?"

   The results of one wording were 33% choosing domestic policy, while with the other wording 52% chose domestic policy.

   Which version probably resulted in which result? Explain.

3.

You are supposed to interview the residents of two of the above five houses.

(a) How would you choose which houses to interview in order to obtain an SRS?

(b) You plan to visit the homes at 9 A.M. If someone isn't home, explain the reasons for and against substituting another house.

(c) Are there any differences you might expect to find among the residents based on the above sketch, and noting this, what other sampling procedure might you choose to use?

4. An emotionally charged topic in government these days is immigration reform. Suppose a congresswoman wishes to survey her constituents concerning their opinions on whether the federal government should expand the Dream Act (the process for undocumented immigrants to obtain permanent residency). Discuss possible sources of bias with regard to the following four options: (1) conducting a survey via random telephone dialing into her district, (2) sending out a mailing using a registered voter list, (3) having a pollster interview everyone who walks past her downtown office, and (4) broadcasting a radio appeal urging interested citizens in her district to call in their opinions to her office.

QUIZ 2

Multiple-Choice Questions

Directions: The questions that follow are each followed by five suggested answers. Choose the response that best answers the question.

1. To evaluate their catering service with regard to culinary excellence, airline executives plan to pick a random passenger to start with and then survey every tenth passenger departing from a Beijing to San Francisco flight. Which of the following best describes the executives' sampling plan?

   (A) A convenience sample
   (B) A simple random sample
   (C) A stratified random sample
   (D) A cluster sample
   (E) A systematic sample

2. School uniforms are being adopted by U.S. public schools in increasing numbers. Two possible wordings for a question on whether or not students should have to wear school uniforms are as follows:

   I. Many educators believe in creating a level playing field to reduce socioeconomic disparities. Do you believe that students should have to wear school uniforms?
   II. Many sociologists believe that students have a right to express their individuality. Do you believe that students should have to wear school uniforms?

   One of these questions showed that 18% of the population favors school uniforms, while the other question showed that 23% of the population favors school uniforms. Which question probably produced which result and why?

   (A) The first question probably showed 23% of the population favors school uniforms, and the second question probably showed 18% because of the lack of randomization in the choice of pro-uniform and anti-uniform arguments as evidenced by the wording of the questions.
   (B) The first question probably showed 18% and the second question probably showed 23% because of stratification in the wording of the questions.
   (C) The first question probably showed 23% and the second question probably showed 18% because of the lack of a neutral cluster in the sample.
   (D) The first question probably showed 18% and the second question probably showed 23% because of response bias due to the wording of the questions.
   (E) The first question probably showed 23% and the second question probably showed 18% because of response bias due to the wording of the questions.

3. In a study of successes and failures in adopting Common Core standards, a random sample of high school principals will be selected from each of the 50 states. Selected individuals will be asked a series of evaluative questions. Why is stratification used here?

   (A) To minimize response bias
   (B) To minimize nonresponse bias
   (C) To minimize voluntary response bias
   (D) Because each state is roughly representative of the U.S. population as a whole
   (E) To obtain higher statistical precision because variability of responses within a state is likely less than variability of responses found in the overall population

4. To find out the average occupancy size of student-rented apartments, a researcher picks a simple random sample of 100 such apartments. Even after one follow-up visit, the interviewer is unable to make contact with anyone in 27 of these apartments. Concerned about nonresponse bias, the researcher chooses another simple random sample and instructs the interviewer to continue this procedure until contact is made with someone in a total of 100 apartments. The average occupancy size in the final 100-apartment sample is 2.78. Is this estimate probably too low or too high?

   (A) Too low, because of undercoverage bias.
   (B) Too low, because convenience samples overestimate average results.
   (C) Too high, because of undercoverage bias.
   (D) Too high, because convenience samples overestimate average results.
   (E) Too high, because voluntary response samples overestimate average results.

5. To survey the opinions of bleacher fans at Wrigley Field, a surveyor plans to select every one-hundredth fan entering the bleachers one afternoon. Will this result in a simple random sample of Cub fans who sit in the bleachers?

   (A) Yes, because each bleacher fan has the same chance of being selected.
   (B) Yes, but only if there is a single entrance to the bleachers.
   (C) Yes, because the 99 out of 100 bleacher fans who are not selected will form a control group.
   (D) Yes, because this is an example of systematic sampling, which is a special case of simple random sampling.
   (E) No, because not every sample of the intended size has an equal chance of being selected.

6. A researcher plans a study to examine the depth of belief in God among the adult population. He obtains a simple random sample of 100 adults as they leave church one Sunday morning. All but one of them agree to participate in the survey. Which of the following is a true statement?

(A) Proper use of chance as evidenced by the simple random sample makes this a well-designed survey.
(B) The high response rate makes this a well-designed survey.
(C) Selection bias makes this a poorly designed survey.
(D) The validity of this survey depends on whether or not the adults attending this church are representative of all churches.
(E) The validity of this survey depends upon whether or not similar numbers of those surveyed are male and female.

Free-Response Questions

Directions: You must show all work and indicate the methods you use. You will be graded on the correctness of your methods and on the accuracy of your final answers.

1. A cable company plans to survey potential customers in a small city currently served by satellite dishes. Two sampling methods are being considered. Method A is to randomly select a sample of 25 city blocks and survey every family living on those blocks. Method B is to randomly select a sample of families from each of the five natural neighborhoods making up the city.

(a) What is the statistical name for the sampling technique used in Method A, and what is a possible reason for using it rather than an SRS?

(b) What is the statistical name for the sampling technique used in Method B, and what is a possible reason for using it rather than an SRS?

2. A questionnaire is being designed to determine whether most people are or are not in favor of sending U.S. troops to participate in U.N. peacekeeping operations. Give two examples of poorly worded questions, one biased toward each response.

3. Pelicans are considered an endangered species, and there is a need to protect them from becoming extinct. Unfortunately, the U.S. Fish & Wildlife Service is unable to protect all pelican colonies; however, it plans to protect those colonies identified as most important. One criterion in determining the importance of a geographical region is the density of nesting sites within it.

Given an aerial view of the nests in a geographical region used for pelican nesting, the service plans to divide the rectangular picture into 150 equal size plots as below:

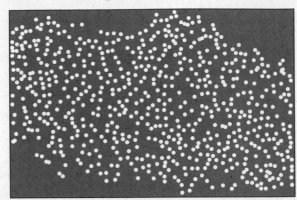

The service will randomly pick 15 plots (10% of the total), count the nests in each of those plots, sum and multiply by 10 to estimate the total number of nests in the region, and then will calculate density. Two methods for counting the number of nests are under consideration.

Method I: Randomly pick one of the above rows and use all the 15 plots in that row.

Method II: Randomly pick one of the plots in each of the 15 columns to obtain the 15 plots.

(a) What sampling procedure is each method striving for, and how can each method be implemented?

(b) Which of these two methods is preferable here? Explain.

4. For each of the following survey designs, identify the problem and the effect it would have on the estimate of interest.

(a) Should spending be increased for the athletic program? A school official decides to sample spectators randomly at the next home basketball game.

(b) Do people believe that eating organic will lower the risk of cancer? A health magazine sends a questionnaire to all their subscribers.

(c) A supermarket quality control inspector opens boxes of strawberries. If no mold is seen among the visible strawberries, he accepts the box.

(d) Are her patients following her diet instructions? A family physician asks each of her patients during their office visits.

(e) Parents are asked to write to a newspaper with a "yes" or "no" response to whether or not they are happy with the education their children are receiving in the local public schools.

ANSWERS EXPLAINED

Quiz 1

MULTIPLE-CHOICE QUESTIONS

1. **(C)** The four classes (freshmen, sophomores, juniors, and seniors) are strata.

2. **(D)** If there is bias, taking a larger sample just magnifies the bias on a larger scale. If there is enough bias, the sample can be worthless. Even when the subjects are chosen randomly, there can be bias due, for example, to non-response or to the wording of the questions. Convenience samples, like shopping mall surveys, are based on choosing individuals who are easy to reach, and they typically miss a large segment of the population. Voluntary response samples, like radio call-in surveys, are based on individuals who offer to participate, and they typically overrepresent persons with strong opinions.

3. **(B)** Different samples give different sample statistics, all of which are estimates of a population parameter. Sampling error relates to natural variation between samples, can never be eliminated, can be described using probability, and is generally smaller if the sample size is larger.

4. **(D)** The dorms are clusters.

5. **(A)** This survey provides a good example of voluntary response bias, which often overrepresents negative opinions. The people who chose to respond were most likely parents who were very unhappy, and so there is very little chance that the 10,000 respondents were representative of the population. Knowing more about her readers, or taking a sample of the sample, would not have helped.

6. **(E)** In a simple random sample, every possible group of the given size has to be equally likely to be selected, and this is not true here. For example, with this procedure it would be impossible for all the Cubs to be together in the final sample. This procedure is an example of stratified sampling, but stratified sampling does not result in simple random samples.

FREE-RESPONSE QUESTIONS

1. (a) This is an example of *stratified sampling*, where the chapters are strata. The advantage is that the student is ensuring that the final sample will represent the 12 different authors, who may well use different average word lengths.

   (b) This is an example of *cluster sampling*, where for each chapter the three chosen pages are clusters. It is reasonable to assume that each page (cluster) resembles the author's overall pattern. The advantage is that using these clusters is much more practical than trying to sample from among all an author's words.

   (c) This is an example of *systematic sampling*, which is quicker and easier than many other procedures. A possible disadvantage is that if ordering is related to the variable under consideration, this procedure will likely result in an unrepresentative sample. For example, in this study if an author's word length is related to word order in sentences, the student could end up with words of particular lengths.

2. Version I probability resulted in 52% choosing domestic policy over foreign policy. The words used to define a category can make a significant difference. In this example, using the words "war on terrorism" instead of the more neutral "foreign policy," probably resonated with people and led to the dramatic difference in results.

3. (a) To obtain an SRS, you might use a random number table and note the first two different numbers between 1 and 5 that appear. Or you could use a calculator to generate numbers between 1 and 5, again noting the first two different numbers that result.

   (b) Time and cost considerations would be the benefit of substitution. However, substitution rather than returning to the same home later could lead to selection bias because certain types of people are not and will not be home at 9 A.M. With substitution the sample would no longer be a simple random sample.

   (c) Corner lot homes like homes 1 and 5 might have different residents (perhaps with higher income levels) than other homes. With this in mind, a stratified sample might be more appropriate, randomly choosing one of the corner lot homes and one of the interior lot homes.

4. The direct telephone and mailing options will both suffer from undercoverage bias. For example, undocumented immigrants themselves and those housing them may be reluctant to put themselves on record by answering either a phone or mail survey. The pollster interviews will result in a convenience sample, which can be highly unrepresentative of the population. In this case, there might be a real question concerning which members of her constituency spend any time in the downtown area where her office is located. The radio appeal will lead to a voluntary response sample, which typically gives too much emphasis to persons with strong opinions.

Quiz 2

MULTIPLE-CHOICE QUESTIONS

1. **(E)** With a random starting point and by picking every nth person for the sample, the method is called systematic sampling.

2. **(E)** The wording "creating a level playing field " and "a right to express their individuality" are nonneutral and clearly leading phrasings.

3. **(E)** Each of the 50 states, with its own long-time past state standards and different regional culture, is considered a homogeneous stratum.

4. **(C)** It is most likely that the apartments at which the interviewer had difficulty finding someone home were apartments with fewer students living in them. Replacing these with other randomly picked apartments most likely replaces smaller-occupancy apartments with larger-occupancy ones.

5. **(E)** In a simple random sample, every possible group of the given size has to be equally likely to be selected, and this is not true here. For example, with this procedure it will be impossible for all the early arrivals to be together in the final sample. This procedure is an example of systematic sampling, but systematic sampling does not result in simple random samples.

6. **(C)** Surveying people coming out of any church results in a very unrepresentative sample of the adult population, especially given the question under consideration. Using chance and obtaining a high response rate will not change the selection bias and make this into a well-designed survey.

FREE-RESPONSE QUESTIONS

1. (a) Method A is an example of *cluster sampling*, where the population is divided into heterogeneous groups called *clusters* and individuals from a random sample of the clusters are surveyed. It is often more practical to simply survey individuals from a random sample of clusters (in this case, a random sample of city blocks) than to try to randomly sample a whole population (in this case the entire city population).

   (b) Method B is an example of *stratified sampling*, where the population is divided into homogeneous groups called *strata* and random individuals from each stratum are chosen. Stratified samples can often give useful information about each stratum (in this case, about each of the five neighborhoods) in addition to information about the whole population (the city population).

2. There are many possible examples, such as "Are you in favor of sending U.S. troops to participate in U.N. peacekeeping operations even though these usually have nothing to do with our national interests?" and "Many people in the world live in conflict zones and experience great difficulty finding adequate housing, food, and medical care. Are you in favor of sending U.S. troops to participate in U.N. peacekeeping operations?" Because of wording bias, the first sentence will result in a lower proportion of people in favor of sending U.S. troops than will the second sentence.

3. (a) Method I results in cluster sampling (the rows are clusters) and can be implemented by numbering the rows 1 through 10, using a random number generator to randomly pick a number between 1 and 10, and using all 15 plots in that row. Method II results in stratified sampling (the columns are strata) and can be implemented by numbering the rows 1 through 10, using a random number generator to pick a number between 1 and 10 to pick a plot in the first column, then repeat this process to pick a random plot in each of the 15 columns, and to use the 15 plots resulting from picking one plot from each column.

   (b) Method II, a stratified sample (with the columns as strata) would be preferable here as each column is somewhat different, so picking a plot from each is meaningful. Method I, an attempt at cluster sampling with the rows as clusters, would work only if each of the rows was heterogeneous (representative of the whole area); this is clearly not true, as for example, the top and bottom rows have far fewer nests than the middle rows.

4. (a) People who attend the school's basketball game are probably more interested in supporting the athletic program than those who do not attend. This study will probably show that a higher proportion of people favor the increased spending than what is the true proportion.

   (b) Subscribers to a health magazine are more likely to believe in the benefits of eating organic than the general population, so the resulting proportion of believers in this study will be greater than it is in the general population.

   (c) The distributer could put any moldy strawberries on the bottom, out of sight! The proportion of accepted boxes will be higher than should be.

   (d) Many patients will be reluctant to admit to their doctor that they are not following her diet instructions. So the proportion who say they are following the instructions will be higher than the true proportion following the instructions.

   (e) Parents who are satisfied with the school system are probably less likely to take the time to respond to this voluntary response survey, while those who are unhappy with the education their kids are receiving are more likely to respond. So the proportion of "unhappy" responses will probably be higher than the true proportion of parents who are unhappy.

Planning and Conducting Experiments

6

→ **EXPERIMENTS VERSUS OBSERVATIONAL STUDIES**
  → **TREATMENTS, CONFOUNDING**
→ **THE LANGUAGE OF EXPERIMENTS**
  → **EXPERIMENTAL UNITS, SUBJECTS, FACTORS, LEVELS**
  → **CONTROL GROUPS, RANDOM ASSIGNMENT**
  → **PLACEBO EFFECTS, BLINDING, DOUBLE BLINDING**
  → **MATCHED PAIR DESIGN, BLOCKING**
→ **REPLICATION AND GENERALIZABILITY OF RESULTS**

Several primary principles deal with the proper planning and conducting of experiments. First, possible confounding variables are controlled by making conditions (other than the treatments) similar for all treatment groups. Second, chance should be used in random assignment of subjects to treatment groups. Third, natural variation in outcomes can be lessened by using more subjects.

NOTE

Recognizing when a study is an experiment versus when it is an observational study is a critical life skill!

EXPERIMENTS VERSUS OBSERVATIONAL STUDIES

In an experiment, we impose some change or treatment and measure the response. In an observational study, we simply observe and measure something that has taken place or is taking place while trying not to cause any changes by our presence. Experiments often suggest causal relationships, while observational studies can show only the existence of associations.

NOTE

Observational studies are generally less expensive and less time-consuming than experiments. Observational studies are often conducted first and suggest experiments to be performed next.

➡ **EXAMPLE 6.1**

A study is to be designed to determine whether a particular commercial review course actually helps raise SAT scores among students attending a particular high school. How could an observational study be performed? An experiment? Which is more appropriate here?

Answer: An observational study would interview students at the school who have taken the SAT Exam, asking whether or not they took the review course, and then the SAT results of those who have and have not taken the review course would be compared.

An experiment performed on students at the school who are planning to take the SAT Exam would use chance to randomly assign some students to take the study course while others to not take the course, and then comparing the SAT results of those who have and have not taken the study course.

The experimental approach is more appropriate here. With the observational study, there could be many explanations for any SAT score difference noted between those who took the course and those who didn't. For example, students who took the course

TIP

It is important to identify the *response variable* (SAT scores in this example) right away. This is critical in answering any experimental design question.

NOTE

Anytime we observe a relationship among two variables, there's always the possibility that some third variable also relates to the two variables and is responsible for "confounding" the apparent relationship between the two original variables.

might be the more serious high school students. Higher score results might be due to their more serious outlook and have nothing to do with taking the course. The experiment tries to control for *confounding* variables by using random assignment to make the groups taking and not taking the course as similar as possible except for whether or not the individuals take the course.

➡ EXAMPLE 6.2

A study is to be designed to examine the GPAs of students who study at least 3 hours every evening and those who don't. Which is more appropriate, an observational study or an experiment?

Answer: Requiring some students to cut their studying to under 3 hours every evening would be unethical. The proper procedure here is an observational study, having students anonymously fill out questionnaires asking about their study habits and GPA.

TIP

You should always be asking whether or not you can reasonably draw cause-and-effect connections between explanatory and response variables.

THE LANGUAGE OF EXPERIMENTS

An experiment is performed on objects called *experimental units*. If the units are people, they are called *subjects*. Experiments involve *explanatory variables*, called *factors*, that are believed to have an effect on *response variables*. A group is treated with some *level* of the explanatory variable, and the outcome of the response variable is measured.

➡ EXAMPLE 6.3

In an experiment to test exercise and blood pressure reduction, volunteers are randomly assigned to participate in either 0, 1, or 2 hours a day of exercise, 5 days a week, for 6 months. What are the explanatory and response variables, and what are the levels?

Answer: The explanatory variable, hours of exercise, is being implemented at three levels: 0, 1, and 2 hours a day. The response variable is not specified but could be the measured change in either systolic or diastolic blood pressure readings after 6 months.

Suppose the volunteers were further randomly assigned to follow either the DASH (Dietary Approaches to Stop Hypertension) or the TLC (Therapeutic Lifestyle Changes) diet for the 6 months. There would then be two factors, one with three levels and one with two levels, and a total of six treatments (DASH diet with 0 hours daily exercise, DASH with 1 hour exercise, DASH with 2 hours exercise, TLC diet with 0 hours daily exercise, TLC with 1 hour exercise, and TLC with 2 hours exercise).

TIP

If asked to list all the treatments (explanatory variables), don't create ones that aren't there!

NOTE

When there is more than one factor, the number of treatments is the product of the number of levels for each factor.

An explanatory variable and another variable are *confounded* when their effects on a response variable cannot be distinguished. For example, in an experiment to determine if different fertilizers cause differences in plant growth, a horticulturist might decide to have many test plots using one or the other of the fertilizers, with equal numbers of sunny and shady plots for each fertilizer, so that fertilizer and sun are not confounded. However, if two fertilizers require different amounts of watering, it might be difficult to determine if the difference in fertilizers or the difference in watering is the real cause of observed differences in plant growth. In this case, fertilizer and water would be confounded.

In an experiment, there is often a *control group* to provide a baseline for comparing the effects of the other treatments. *Random assignment* to the various treatment groups, including to the control group, can help reduce the problem posed by confounding variables. Random assignment to treatments, called *completely randomized design*, is critical in good experimental design. This is especially true if the subjects are not randomly selected, as is the case in most medical/drug experiments.

NOTE

The goal of random assignment is to create treatment groups that are similar in all respects except for the treatment imposed.

NOTE

Random assignment *minimizes*, but can't *eliminate*, the effect of possible confounding variables.

NOTE

Random assignment allows us to make a valid comparison of treatments.

TIP

It is not enough simply to say use "random assignment." You must be able to carefully explain how the random assignment of treatments to subjects is to be accomplished. Don't forget to name which treatment is given to which randomly chosen group!

➡ EXAMPLE 6.4

Sixty patients, ages 5 to 12, all with common warts are enrolled in a study to determine if application of duct tape is as effective as cryotherapy in the treatment of warts. Subjects will receive either cryotherapy (liquid nitrogen applied to each wart for 10 seconds every 2 weeks) for 6 treatments or duct tape occlusion (applied directly to the wart) for 2 months. Describe a completely randomized design.

Answer: Assign each patient a number from 01 to 60. Use a random number table, reading off two digits at a time, ignoring repeats, 00, and numbers over 60, until 30 unique numbers have been selected. (Instead, you can use a random integer generator on a calculator to pick integers between 1 and 60, throwing away repeats, until 30 unique integers have been selected.) The 30 patients corresponding to the 30 selected integers will be given the cryotherapy treatment. The remaining 30 patients will receive the duct tape treatment. At the end of the treatment schedules, compare the proportion of each group that had complete resolution of the warts being studied.

NOTE

When blinding is used, subjects in the treatment and control groups experience the placebo effect equally.

NOTE

Blinding and placebos in experiments are important but are not always feasible. You can still have "experiments" without using these.

NOTE

Of course, with double blinding, someone (the person directing the study) must know who is receiving which treatment!

It is a fact that many people respond to any kind of perceived treatment. This is called the *placebo effect*. For example, when given a sugar pill after surgery but told that it is a strong pain reliever, many people feel immediate relief from their pain. *Blinding* occurs when the subjects don't know which of the different treatments (such as placebos) they are receiving. *Double blinding* is when neither the subjects nor the *response evaluators* know who is receiving which treatment.

➡ EXAMPLE 6.5

There is a pressure point on the wrist that some doctors believe can be used to help control the nausea experienced following certain medical procedures. The idea is to place a band containing a small marble firmly on a patient's wrist so that the marble is located directly over the pressure point. Describe how an experiment might be run on 50 postoperative patients.

Answer: Assign each patient a number from 01 to 50. From a random number table read off two digits at a time, throwing away repeats, 00, and numbers over 50, until 25 unique numbers have been selected. (You can, instead, use a random integer generator on a calculator to pick integers between 1 and 50, ignoring repeats, until 25 unique integers have been selected.) Put wristbands with marbles over the pressure point on the patients with these assigned numbers. Put wristbands with marbles on the remaining patients also, but *not* over the pressure point. Have a researcher check by telephone with all 50 patients at designated time intervals to determine the degree of nausea being experienced. Neither the patients nor the researcher on the telephone should know which patients have the marbles over the correct pressure point.

A *matched pairs design* (also called a *paired comparison design*) is when two treatments are compared based on the responses of paired subjects, one of whom receives one treatment while the other receives the second treatment. Often the paired subjects are really single subjects who are given both treatments, one at a time in random order.

➡ EXAMPLE 6.6

Does seeing pictures of accidents caused by drunk drivers influence one's opinion on penalties for drunk drivers? How could a comparison test be designed for 100 subjects?

Answer: The subjects could be asked questions about drunk driving penalties before and then again after seeing the pictures, and any change in answers noted. This would be a poor design because there is no control group, there is no use of randomization, and subjects might well change their answers because they realize that that is what is expected of them after seeing the pictures.

A better design is to use randomization to split the subjects into two groups, half of whom simply answer the questions while the other half first see the pictures and then answer the questions. For example, number the subjects 00 through 99, and use a random number table, reading off two digits at a time, ignoring repeats, until 50 unique numbers have been selected. The subjects corresponding to these 50 numbers will be the group who sees the pictures.

Another possibility is to use a group of 50 twins as subjects. One of each set of twins is randomly picked (e.g., based on choosing an odd or even digit from a random number table) to answer the questions without seeing the pictures, while the other first sees the pictures and then answers the questions. The answers could be compared from each set of twins. This is a paired comparison test that might help minimize confounding variables due to family environment, heredity, and so on.

Just as stratification is sampling that first divides the population into representative groups called strata, *blocking* in experimental design first divides the subjects into representative groups called *blocks*. Subjects within *each* block are randomly assigned treatments. One can think of blocking as running parallel experiments before combining the results. This technique helps control certain possible confounding variables by bringing them directly into the picture and helps make conclusions more specific. The paired comparison (matched pairs) design is a special case of blocking in which each pair can be considered a block of size 2.

➡ **EXAMPLE 6.7**

There is a rising trend for star college athletes to turn professional without finishing their degrees. A study is performed to assess whether reading an article about professional salaries has an impact on such decisions. Randomization can be used to split the subjects into two groups, and those in one group given the article before answering questions. How can a block design be incorporated into the design of this experiment?

Answer: The subjects can be split into two blocks, underclass and upperclass, before using randomization to assign some to read the article before questioning. With this design, the impact of the salary article on freshmen and sophomores can be distinguished from the impact on juniors and seniors.

Similarly, blocking can be used to separately analyze men and women, those with high GPAs and those with low GPAs, those in different sports, those with different majors, and so on.

TIP

Use proper terminology! The language of experiments is different from the language of observational studies. Don't mix up *blocking* and *stratification*.

TIP

Treatments are not randomly assigned to blocks, and blocks are not formed at random. This experimental design uses randomization *within* blocks.

Completely Randomized Design

Block Design

NOTE

When the results are combined, this estimates the effect, separate from blocks, of the independent variable of interest.

TIP

Replication refers to having multiple subjects, **not** multiple trials of the same experiment.

TIP

Most experiments are not performed on random samples from a population. So don't use *random sample* terminology when you mean *random assignment.*

REPLICATION AND GENERALIZABILITY OF RESULTS

When differences are observed in a comparison test, the researcher must decide whether these differences are *statistically significant* or whether they can be explained by natural variation. One important consideration is the size of the sample: the larger the sample, the more significant the observation. This is the principle of *replication.* In other words, the treatment should be repeated on a sufficient number of subjects so that real response differences are more apparent.

Randomization, in the form of random assignment, is critical to minimize the effect of confounding variables. However, in order to generalize experimental results to a larger population (as we try to do in sample surveys), it would also be necessary that the group of subjects used in the experiment be randomly selected from the population. For example, it is hard to generalize from the effect a television commercial has on students at a private midwestern high school to the effect the same commercial has on retired senior citizens in Florida. If the experiment was performed on those high school students, it can be generalized only to the population of all students at private midwestern high schools.

SUMMARY

- While observational studies involve observing without causing changes, experiments involve actively imposing treatments and observing responses.
- Differences in observed responses are statistically significant when it is unlikely that they can be explained by chance variation.
- Random assignment of subjects to treatment groups is extremely important in handling unknown and uncontrollable differences.
- Random assignment refers to what is done with subjects after they've been picked for a study, whereas random sampling refers to how subjects are selected for a study.
- Variables are said to be confounded when there is uncertainty as to which variable is causing an effect.
- The placebo effect refers to the fact that many people respond to any kind of perceived treatment.
- Blinding refers to subjects not knowing which treatment they are receiving.
- Double-blinding refers to subjects and those evaluating their responses not knowing who received which treatments.
- Completely randomized designs refer to experiments in which everyone has an equal chance of receiving any treatment.
- Blocking is the process of dividing the subjects into representative groups to bring certain differences into the picture (for example, blocking by gender, age, or race).
- Randomized block designs refer to experiments in which the randomization occurs only within blocks.
- Randomized paired comparison designs refer to experiments in which subjects are paired and randomization is used to decide who in each pair receives what treatment.

QUIZ 1

Multiple-Choice Questions

Directions: The questions that follow are each followed by five suggested answers. Choose the response that best answers the question.

1. Suppose you wish to compare the average class size of mathematics classes to the average class size of English classes in your high school. Which is the most appropriate technique for gathering the needed data?

   (A) Census
   (B) Sample survey
   (C) Experiment
   (D) Observational study
   (E) None of these methods is appropriate.

2. Two studies are run to compare the experiences of families living in high-rise public housing to those of families living in townhouse subsidized rentals. The first study interviews 25 families who have been in each government program for at least 1 year, while the second randomly assigns 25 families to each program and interviews them after 1 year. Which of the following is a true statement?

   (A) Both studies are observational studies because of the time period involved.
   (B) Both studies are observational studies because there are no control groups.
   (C) The first study is an observational study, while the second is an experiment.
   (D) The first study is an experiment, while the second is an observational study.
   (E) Both studies are experiments.

3. A study is made to determine whether taking AP Statistics in high school helps students achieve higher GPAs when they go to college. In comparing records of 200 college students, half of whom took AP Statistics in high school, it is noted that the average college GPA is higher for those 100 students who took AP Statistics than for those who did not. Based on this study, guidance counselors begin recommending AP Statistics for college bound students. Which of the following is *incorrect*?

   (A) While this study indicates a relation, it does not prove causation.
   (B) There could well be a confounding variable responsible for the seeming relationship.
   (C) Self-selection here makes drawing the counselors' conclusion difficult.
   (D) A more meaningful study would be to compare an SRS from each of the two groups of 100 students.
   (E) This is an observational study, not an experiment.

4. When the estrogen-blocking drug tamoxifen was first introduced to treat breast cancer, there was concern that it would cause osteoporosis as a side effect. To test this concern, cancer subjects were randomly selected and given tamoxifen, and their bone density was measured before and after treatment. Which of the following is a true statement?

   (A) This study was an observational study.
   (B) This study was a sample survey of randomly selected cancer patients.
   (C) This study was an experiment in which the subjects were used as their own controls.
   (D) With the given procedure, there cannot be a placebo effect.
   (E) Causation cannot be concluded without knowing the survival rates.

5. In designing an experiment, blocking is used

   (A) to reduce bias.
   (B) to reduce variation.
   (C) as a substitute for a control group.
   (D) as a first step in randomization.
   (E) to control the level of the experiment.

6. Consider the following studies being run by three different nursing home establishments.

   I. One nursing home has pets brought in for an hour every day to see if patient morale is improved.
   II. One nursing home allows hourly visits every day by kindergarten children to see if patient morale is improved.
   III. One nursing home administers antidepressants to all patients to see if patient morale is improved.

   Which of the following is true?

   (A) None of these studies uses randomization.
   (B) None of these studies uses control groups.
   (C) None of these studies uses blinding.
   (D) Important information can be obtained from all these studies, but none will be able to establish causal relationships.
   (E) All of the above

7. Is hot oatmeal with fruit or a Western omelet with home fries a more satisfying breakfast? Fifty volunteers are randomly split into two groups. One group is fed oatmeal with fruit, while the other is fed Western omelets with home fries. Each volunteer then rates his/her breakfast on a one to ten scale for satisfaction. If the Western omelet with home fries receives a substantially higher average score, what is a reasonable conclusion?

    (A) In general, people find Western omelets with home fries more satisfying for breakfast than hot oatmeal with fruit.
    (B) There is no reasonable conclusion because the subjects were volunteering rather than being randomly selected from the general population.
    (C) There is no reasonable conclusion because of the small size of the sample.
    (D) There is no reasonable conclusion because blinding was not used.
    (E) There is no reasonable conclusion because there are too many possible confounding variables such as age, race, and ethnic background of the individual volunteers and season when the study was performed.

8. Which of the following is *not* important in the design of experiments?

    (A) Control of confounding variables
    (B) Randomization in assigning subjects to different treatments
    (C) Replication of the experiment using sufficient numbers of subjects
    (D) Care in observing without imposing change
    (E) Isolating variability due to differences between blocks

Free-Response Questions

> **Directions:** You must show all work and indicate the methods you use. You will be graded on the correctness of your methods and on the accuracy of your final answers.

1. Suppose a new drug is developed that appears in laboratory settings to completely prevent people who test positive for human immunodeficiency virus (HIV) from ever developing full-blown acquired immunodeficiency syndrome (AIDS). Putting all ethical considerations aside, design an experiment to test the drug. What ethical considerations might arise during the testing that would force an early end to the experiment?

2. In a study of the effectiveness of arthroscopic knee surgery, researchers randomly assigned patients with knee pain into three groups: one group to receive arthroscopic debridement, one to receive arthroscopic lavage, and the third to receive simulated arthroscopic surgery in which the surgeon made small incisions without inserting instruments. All patients were then followed up by their surgeons to measure improvements in pain and functional ability.

   (a) Was this an observational study or an experiment? Explain.

   (b) Why did the researchers include a group to receive sham surgery?

   (c) Was double blinding used and explain the possible consequences of this decision?

   (d) Explain how which surgeons performed which surgeries could be a possible confounding variable.

3. Explain how you would design an experiment to evaluate whether subliminal advertising (flashing "BUY POPCORN" on the screen for a fraction of a second) results in more popcorn being sold in a movie theater. Show how you will incorporate comparison, randomization, and blinding.

4. A new vegetable fertilizer is to be tested at two different levels (regular concentration and double concentration). Design an experiment, including a control, for 30 test plots, half of which are in shade. Explain carefully how you will use randomization.

5. Two studies are run to measure the extent to which taking zinc lozenges helps to shorten the duration of the common cold. Researchers in the first study send questionnaires to all 5000 employees of a major teaching hospital asking whether they have taken zinc lozenges to fight the common cold and what they perceive to be the benefits, if any. The response rate is 90%. The 500 people who did not respond to the first mailing receive follow-up telephone calls, and eventually responses are obtained from over 99% of the hospital employees. Researchers in the second study take a group of 100 volunteers and randomly select 50 to receive zinc lozenges while the others receive a similar-looking, similar-tasting placebo. The volunteers are not told whether they are taking the zinc lozenges, but their doctors know and are asked to accurately measure the duration of common cold symptoms experienced by the volunteers. Comment on the designs of the two studies, remarking on their good points and on possible sources of error.

QUIZ 2

Multiple-Choice Questions

Directions: The questions that follow are each followed by five suggested answers. Choose the response that best answers the question.

1. Two studies are run to determine the effect of low levels of wine consumption on cholesterol level. The first study measures the cholesterol levels of 100 volunteers who have not consumed alcohol in the past year and compares these values with their cholesterol levels after 1 year, during which time each volunteer drinks one glass of wine daily. The second study measures the cholesterol levels of 100 volunteers who have not consumed alcohol in the past year, randomly picks half the group to drink one glass of wine daily for a year while the others drink no alcohol for the year, and finally measures their levels again. Which of the following is a true statement?

   (A) The first study is an observational study, while the second is an experiment.
   (B) The first study is an experiment, while the second is an observational study.
   (C) Both studies are observational studies, but only one uses both randomization and a control group.
   (D) The first study is a census of 100 volunteers, while the second study is an experiment.
   (E) Both studies are experiments.

2. In a 1927–1932 Western Electric Company study on the effect of lighting on worker productivity, productivity increased with each increase in lighting but then also increased with every decrease in lighting. If it is assumed that the workers knew a study was in progress, this is an example of

   (A) the effect of a treatment unit.
   (B) the placebo effect.
   (C) the control group effect.
   (D) sampling error.
   (E) voluntary response bias.

3. Which of the following is *incorrect*?

   (A) Blocking is to experiment design as stratification is to sampling design.
   (B) By controlling certain variables, blocking can make conclusions more specific.
   (C) The paired comparison design is a special case of blocking.
   (D) Blocking results in increased accuracy because the blocks have smaller size than the original group.
   (E) In a randomized block design, the randomization occurs within the blocks.

4. A consumer product agency tests miles per gallon for a sample of automobiles using each of four different octanes of gasoline. Which of the following is true?

(A) There are four explanatory variables and one response variable.
(B) There is one explanatory variable with four levels of response.
(C) Miles per gallon is the only explanatory variable, but there are four response variables corresponding to the different octanes.
(D) There are four levels of a single explanatory variable.
(E) Each explanatory level has an associated level of response.

5. Which of the following is not true?

(A) In an experiment some treatment is intentionally forced on one group to note the response.
(B) In an observational study information is gathered on an already existing situation.
(C) Sample surveys are observational studies, not experiments.
(D) While observational studies may suggest relationships, it is usually not possible to conclude cause and effect because of the lack of control over possible confounding variables.
(E) A complete census is the only way to establish a cause-and-effect relationship absolutely.

6. Which of the following is a true statement about the design of matched-pair experiments?

(A) Each subject might receive both treatments.
(B) Each pair of subjects receives the identical treatment, and differences in their responses are noted.
(C) Blocking is one form of matched-pair design.
(D) Stratification into two equal sized strata is an example of matched pairs.
(E) Randomization is unnecessary in true matched pair designs.

7. Do teenagers prefer sports drinks colored blue or green? Two different colorings, which have no effect on taste, are used on the identical drink to result in a blue and a green beverage; volunteer teenagers are randomly assigned to drink one or the other colored beverage; and the volunteers then rate the beverage on a one to ten scale. Because of concern that sports interest may affect the outcome, the volunteers are first blocked by whether or not they play on a high school team. Is blinding possible in this experiment?

    (A) No, because the volunteers know whether they are drinking a blue or green drink.
    (B) No, because the volunteers know whether or not they play on a high school team.
    (C) Yes, by having the experimenter in a separate room randomly pick one of two containers and remotely have a drink poured from that container.
    (D) Yes, by having the statistician analyzing the results not knowing which volunteer sampled which drink.
    (E) Yes, by having the volunteers drink out of solid colored thermoses, so that they don't know the color of the drink they are tasting.

8. Some researchers believe that too much iron in the blood can raise the level of cholesterol. The iron level in the blood can be lowered by making periodic blood donations. A study is performed by randomly selecting half of a group of volunteers to give periodic blood donations while the rest do not. Is this an experiment or an observational study?

    (A) An experiment with a single factor
    (B) An experiment with control group and blinding
    (C) An experiment with blocking
    (D) An observational study with comparison and randomization
    (E) An observational study with little, if any, bias

Free-Response Questions

> **Directions:** You must show all work and indicate the methods you use. You will be graded on the correctness of your methods and on the accuracy of your final answers.

1. A new weight-loss supplement is to be tested at three different levels (once, twice, and three times a day). Design an experiment, including a control group and including blocking for gender, for 80 overweight volunteers, half of whom are men. Explain carefully how you will use randomization.

2. Can taking a power nap during the day help adults remember things they just learned? A random sample of college students taking afternoon mathematics classes were asked if they take a short nap back at their dormitory after their last class, and their final grades in their math classes are noted.

   (a) Was this an observational study or an experiment? Explain.

   (b) Analysis of the data showed that among students who take at least a 10 to 20 minute nap after classes, their average math class final grade was higher than that of students who did not take such naps. Based on this result, is it reasonable for college counselors to recommend afternoon naps for students who are struggling in their math classes? Explain.

   (c) Explain a possible confounding variable in this context.

3. A new pain control procedure has been developed in which the patient uses a small battery pack to vary the intensity and duration of electric signals to electrodes surgically embedded in the afflicted area. Putting all ethical considerations aside, design an experiment to test the procedure. What ethical considerations might arise during the testing that would force an early end to the experiment?

4. Explain how you would design an experiment to evaluate whether praying for a hospitalized heart attack patient leads to a speedier recovery. Show how you would incorporate comparison, randomization, and blinding.

5. The computer science department plans to offer three introductory-level CS courses: one using Python, one using C++, and one using Java.

   (a) The department chairperson plans to give all students the same general programming exam at the end of the year and to compare the relative effectiveness of using each of the programming languages by comparing the mean grades of the students from each course. What is wrong, if anything, with the chairperson's plan?

   (b) The chairperson also wishes to determine whether math majors or science majors do better in the courses. Suppose he calculates that the average grade of science majors was higher than the average grade of math majors in each of the courses. Does it follow that the average grade of all the science majors taking the three courses must be higher than the average grade of all the math majors? Explain.

   (c) Suppose 300 students wish to take introductory programming. How would you randomly assign 100 students to each of the three courses?

   (d) How would you randomly assign students to the three courses if you wanted the assignment to be independent from student to student with each student in turn having a one-third probability of taking each of the three classes.

   (e) Name a confounding variable that all the above methods miss.

ANSWERS EXPLAINED

Quiz 1

MULTIPLE-CHOICE QUESTIONS

1. **(A)** The main office at your school should be able to give you the class sizes of every math and English class. If need be, you can check with every math and English teacher.

2. **(C)** In the first study the families were already in the housing units, while in the second study one of two treatments was applied to each family.

3. **(D)** It may well be that very bright students are the same ones who both take AP Statistics and have high college GPAs. If students could be randomly assigned to take or not take AP Statistics, the results would be more meaningful. Of course, ethical considerations might make it impossible to isolate the confounding variable in this way. Only using a sample from the observations gives less information.

4. **(C)** In experiments on people, the subjects can be used as their own controls, with responses noted before and after the treatment. However, with such designs there is always the danger of a placebo effect. Thus the design of choice would involve a separate control group to be used for comparison.

5. **(B)** Blocking divides the subjects into groups, such as men and women, or political affiliations, and thus reduces variation.

6. **(E)** None of the studies has any controls, such as randomization, control groups, or blinding, and so while they may give valuable information, they cannot establish cause and effect.

7. **(A)** There is nothing wrong with using volunteers—what is important is to randomly assign the volunteers into the two treatment groups. There is no way to use blinding in this study—the subjects will clearly know which breakfast they are eating. The main idea behind randomly assigning subjects to the different treatments is to control for various possible confounding variables—it is reasonable to assume that people of various ages, races, ethnic backgrounds, etc., are assigned to receive each of the treatments.

8. **(D)** Control, randomization, and replication are all important aspects of well-designed experiments. Care in observing without imposing change refers to observational studies, not experiments.

FREE-RESPONSE QUESTIONS

1. Ask doctors, hospitals, or blood testing laboratories to make known that you are looking for HIV-positive volunteers. As the volunteers arrive, use a random number table to give each one the drug or a placebo (e.g., if the next digit in the table is odd, the volunteer gets the drug, while if the next digit is even, the volunteer gets a placebo). Use double-blinding; that is, both the volunteers and their doctors should not know if they are receiving the drug or the placebo. Ethical considerations will arise, for example, if the drug is very successful. If volunteers on the placebo are steadily developing full-blown AIDS while no one on the drug is, then ethically the test should be stopped and everyone put on the drug. Or if most of the volunteers on the drug are dying from an unexpected fatal side effect, the test should be stopped and everyone taken off the drug.

2. (a) This was an experiment because the patients were randomly assigned to three different groups and each group received a treatment: arthroscopic debridement, arthroscopic lavage, or simulated arthroscopic surgery.

   (b) They included a group to receive a sham surgery so that in addition to comparing the two actual procedures to each other, they could also see if either of the two actual procedures were any better than no real procedure at all (to combat placebo effect).

   (c) Although the patients may not have been told which surgery they received, there was no double blinding because the patients' surgeons themselves measured improvements in pain and functional ability. This could influence the results as the surgeons could inadvertently treat the patients differently depending upon surgery received or could themselves be subconsciously influenced in their judgments of patients' pain based on knowing which patients received which treatment.

   (d) If a very good surgeon performed one of the real procedures, the results could show a much more significant difference with the other surgeries than if an inferior surgeon performed the real procedure. The procedure itself and the surgeon's competence are confounded.

3. Every day for some specified period of time, look at the next digit on a random number table. If it is odd, flash the subliminal message all day on the screen, while if it is even, don't flash the message that day (randomization). Don't let the customers know what is happening (blinding) and don't let the clerks selling the popcorn know what is happening (double-blinding). Compare the quantity of popcorn bought by the treatment group, that is, by the people who receive the subliminal message, to the quantity bought by the control group, the people who don't receive the message (comparison).

4. To achieve blocking by sunlight, first separate the sunlit and shaded plots. Label the 15 sunlit plots 1 through 15. Use a random integer generator on a calculator to pick integers between 1 and 15, throwing away repeats, until 5 unique integers have been selected. The sunlit plots corresponding to these 5 integers will receive the fertilizer at regular concentration. Continue using the calculator, throwing away repeats, until 5 more unique integers between 1 and 15 have been selected. The sunlit plots corresponding to these 5 new integers will receive the fertilizer at double concentration, while the 5 remaining sunlit plots will be a control group receiving no fertilizer. Now repeat the procedure, this time labeling the shaded plots 1 through 15. Assuming size is the pertinent outcome, weigh all vegetables at the end of the season, compare the average weights among the three sunlit groups, and compare the average weights among the three shaded groups to determine the effect of the fertilizer, if any, at different levels on sunlit plots and separately on shaded plots.

5. The first study, an observational study, does not suffer from nonresponse bias, as do most studies involving mailed questionnaires, because the researchers made follow-up telephone calls and achieved a very high response rate. However, the first study suffers terribly from selection bias. People who work at a teaching hospital are not representative of the general population. One would expect many of them to have heard about how zinc coats the throat to hinder the propagation of viruses. The second study, a controlled experiment, used comparison between a treatment group and a control

group, used randomization in selecting who went into each group, and used blinding to control for a placebo effect on the part of the volunteers. However, they did not use double-blinding; that is, the doctors knew whether their patients were receiving the zinc lozenges, and this could introduce subconscious influence as the doctors make judgments about their patients' health.

Quiz 2

MULTIPLE-CHOICE QUESTIONS

1. **(E)** Both studies apply treatments and measure responses, and so both are experiments.

2. **(B)** The desire of the workers for the study to be successful led to a placebo effect.

3. **(D)** Blocking in experiment design first divides the subjects into representative groups called blocks, just as stratification in sampling design first divides the population into representative groups called strata. This procedure can control certain variables by bringing them directly into the picture, and thus conclusions are more specific. The paired comparison design is a special case of blocking in which each pair can be considered a block. Unnecessary blocking detracts from accuracy because of smaller sample sizes.

4. **(D)** Octane is the only explanatory variable, and it is being tested at four levels. Miles per gallon is the single response variable.

5. **(E)** The first two sentences can be considered part of the definitions of *experiment* and *observational study*. A sample survey does not impose any treatment; it simply counts a certain outcome, and so it is an observational study, not an experiment. A complete census can provide much information about a population, but it doesn't necessarily establish a cause-and-effect relationship among seemingly related population parameters.

6. **(A)** Each subject might receive both treatments, as, for example, in the Pepsi-Coke taste comparison study. The point is to give each subject in a matched pair a different treatment and note any difference in responses. Matched-pair experiments are a particular example of blocking, not vice versa. Stratification refers to a sampling method, not to experimental design. Randomization is used to decide which of a pair gets which treatment or which treatment is given first if one subject is to receive both.

7. **(A)** Blinding does have to do with whether or not the subjects know which treatment (color in this experiment) they are receiving. However, drinking out of solid colored thermoses makes no sense since the beverages are identical except for color and the point of the experiment is the teenager's reaction to color. Blinding has nothing to do with blocking (team participation in this experiment).

8. **(A)** This study is an experiment because a treatment (periodic removal of a pint of blood) is imposed. There is no blinding because the subjects clearly know whether or not they are giving blood. There is no blocking because the subjects are not divided into blocks before random assignment to treatments. For example, blocking would have been used if the subjects had been separated by gender or age before random assignment to give or not give blood donations. There is a single factor—giving or not giving blood.

FREE-RESPONSE QUESTIONS

1. To achieve blocking by gender, first separate the men and women. Label the 40 men 1 through 40. Use a random integer generator on a calculator to pick integers between 1 and 40, ignoring repeats, until 10 unique integers have been selected. The men corresponding to these integers will receive the supplements at the once-a-day level. Continue using the calculator, ignoring repeats, until 10 more unique integers between 1 and 40 have been selected, and the men corresponding to these new integers will receive the supplement at the twice-a-day level. Again ignore repeats and select 10 more unique integers between 1 and 40 with the corresponding men to receive the supplements at the three-a-day level, while the remaining 10 men will be a control group and not receive the supplement. Now repeat the entire procedure, starting by labeling the women 1 through 40. A decision should be made whether or not to use a placebo and have all participants take "something" three times a day. Weigh all 80 overweight volunteers before and after a predetermined length of time. Calculate the change in weight for each individual. Calculate the average change in weight among the ten people in each of the eight groups. Compare the four averages from each block (men and women) to determine the effect, if any, of different levels of the supplement for men and for women.

> **TIP**
>
> Simply saying to "randomly assign" subjects to treatment groups is usually an incomplete response. You need to explain *how* to make the assignments— for example, by using a random number table, or through generating random numbers on a calculator, or picking from a hat.

2. (a) This was an observational study because the researchers applied no treatments. They simply analyzed student responses about taking afternoon naps versus students' final grade in math classes.

    (b) No, it would not be reasonable for college counselors to recommend afternoon naps for students who are struggling in their math classes. An experiment in which students were randomly assigned to take or not take afternoon power naps could establish causation, but this was not done for this study.

    (c) There are many possible answers. One possible confounding variable is that students who are more serious in general take afternoon power naps for health reasons and study harder for academic reasons. If that were true, the higher average final grade in math classes would not be the result of taking naps but rather the result of the overall seriousness of some students, which results in both higher math grades and power naps.

> **NOTE**
>
> The only way to show causation is with a well-designed, well-controlled *experiment*!

3. Ask doctors and hospitals to make known that you are looking for volunteers from among intractable pain sufferers. As the volunteers arrive, use a random integer generator on a calculator to decide which will have the electrodes properly embedded in their pain centers and which will have the electrodes harmlessly embedded in wrong positions. For example, randomly pick integers between 1 and 2. If the next digit chosen is a 1, the volunteer receives the proper embedding, while if the next digit is a 2, the volunteer does not. Use double-blinding; that is, both the volunteers and their doctors should not know if the volunteers are receiving the proper embedding. Ethical considerations will arise, for example, if the procedure is very successful. If volunteers with the wrong embeddings are in constant pain, while everyone with proper embedding is pain-free, then ethically the test should be stopped and everyone given the proper embedding. Or if most of the volunteers with proper embedding develop an unexpected side effect of the pain spreading to several nearby sites, then the test should be stopped and the procedure discontinued for everyone.

4. For each new heart attack patient entering the hospital, look at the next digit from a random number table. If it is odd, give the name to a group of people who will pray for the patient throughout his or her hospitalization, while if it is even, don't ask the group to pray (randomization). Don't let the patients know what is happening (blinding) and don't let the doctors know what is happening (double-blinding). Compare the lengths of hospitalization of patients who receive prayers with those of control group patients who don't receive prayers (comparison).

5. (a) Allowing the students to self-select which class to take leads to confounding that could be significant. For example, perhaps the brighter students all want to learn a certain one of the three languages.

   (b) It is possible for the average score of all science majors to be lower than the average for all math majors even though the science majors averaged higher in each class. For example, suppose that the students taking Java scored much higher than the students in the other two classes. Furthermore, only one science major took Java, and she scored tops in the class. Then the overall average of the math majors could well be higher. This is an example of Simpson's paradox, in which a comparison can be reversed when more than one group is combined to form a single group.

   (c) Number the students 001 through 300. Read off three digits at a time from a random number table, noting all triplets between 001 and 300 and ignoring repeats, until 100 unique such numbers have been selected. Keep reading off three digits, ignoring repeats, until 100 unique new numbers between 001 and 300 are selected. These get C++, while the remaining 100 get Java. Even quicker would be to use a calculator to generate unique random digits between 001 and 300.

   (d) Go through the list of students, flipping a die for each. If a 1 or a 2 shows, the student takes Python, if a 3 or a 4 shows, C++, and if a 5 or a 6 shows, Java.

   (e) Another possible variable are the teachers. For example, perhaps the better teachers teach Java.

Theme Three: Probability

Probability as Relative Frequency

7

→ **LAW OF LARGE NUMBERS**
→ **BASIC PROBABILITY RULES**
→ **MULTISTAGE PROBABILITY CALCULATIONS**
→ **BINOMIAL DISTRIBUTION**
→ **GEOMETRIC PROBABILITIES**
→ **SIMULATION**
→ **DISCRETE RANDOM VARIABLES, MEANS (EXPECTED VALUES), AND STANDARD DEVIATIONS**

In the world around us, unlikely events sometimes take place. At other times, events that seem inevitable do not occur. Chance is everywhere! The cards you are dealt in a poker game, the particular genes you inherit from your parents, and the coin toss at the beginning of a tennis match to determine who serves first are examples of chance behavior that mathematics can help us understand. Even though we may not be able to foretell a specific result, we can sometimes assign what is called a *probability* to indicate the likelihood that a particular event will occur.

Probabilities are always between 0 and 1, with a probability close to 0 meaning that an event is unlikely to occur, and a probability close to 1 meaning that the event is likely to occur. The sum of the probabilities of all the separate outcomes of an experiment is always 1.

THE LAW OF LARGE NUMBERS

The *relative frequency* of an event is the proportion of times the event happened, that is, the number of times the event happened divided by the total number of trials. Relative frequencies may change every time an experiment is performed. The *law of large numbers* states that when an experiment is performed a large number of times, the relative frequency of an event tends to become closer to the probability of the event, that is, *probability is long-term relative frequency.*

The law of large numbers says nothing at all about short-run behavior. There is no such thing as a law of small numbers or a law of averages. Gamblers might say that "red" is due on a roulette table, a basketball player is due to make a shot, a player has a hot hand at the craps table, or a certain number is due to come up in a lottery, but if events are independent, the probability of the outcome of the next trial has nothing to do with what happened in previous trials. Even though casinos and life insurance companies might lose money in the short run, they make long-term profits because of their understanding of the law of large numbers.

TIP
Calculators will express very small probabilities in scientific notation such as 3.4211073E–6. Know what this means, and remember that probabilities are never greater than 1.

EXAMPLE 7.1

There are two games involving flipping a fair coin. In the first game, you win a prize if you can throw between 45% and 55% heads. In the second game, you win if you can throw more than 60% heads. For each game would you rather flip 20 times or 200 times?

Answer: The probability of throwing heads is 0.5. By the law of large numbers, the more times you throw the coin, the more the relative frequency tends to become closer to this probability. With fewer tosses, there is greater chance of wide swings in the relative frequency. Thus, in the first game you would rather have 200 flips, whereas in the second game you would rather have only 20 flips.

BASIC PROBABILITY RULES

 EXAMPLE 7.2

A standard literacy test consists of 100 multiple-choice questions, each with five possible answers. There is no penalty for guessing. A score of 60 is considered passing and a score of 80 is considered superior. When an answer is completely unknown, test takers employ one of three strategies: guess, choose answer (c), or choose the longest answer. The table below summarizes the results of 1000 test takers.

Strategy	Score 0–59	Score 60–79	Score 80–100	
Guess	40	160	100	300
Choose answer (c)	35	170	115	320
Choose longest answer	45	200	135	380
	120	530	350	1000

Note that in analyzing tables such as the one above, it is usually helpful to sum rows and columns.

(a) What is the probability that someone in this group uses the "guess" strategy?

Answer: $P(guess) = \dfrac{300}{1000} = 0.3$

(b) What is the probability that someone in this group scores 60–79?

Answer: $P(score\ 60\text{–}79) = \dfrac{530}{1000} = 0.53$

NOTE

Notation:
$P(A^C) = 1 - P(A)$

(c) What is the probability that someone in this group does not score 60–79?

Answer: $P(does\ not\ score\ 60\text{–}79) = 1 - P(score\ 60\text{–}79) = 1 - 0.53 = 0.47$

[The probability that an event will not occur, that is, the probability of its complement, is equal to 1 minus the probability that the event will occur.]

TIP

The word "and" here means *both* events.

(d) What is the probability that someone in this group chooses strategy "answer (c)" *and* scores 80–100?

Answer: $P(answer\ (c) \cap score\ 80\text{–}100) = \dfrac{115}{1000} = 0.115$

(e) What is the probability that someone in this group chooses strategy "longest answer" *or* scores 0–59?

Answer: $P(\text{longest answer} \cup 0\text{-}59) = \dfrac{40+35+45+200+135}{1000} = \dfrac{455}{1000} = 0.455$

Note that

$P(\text{longest answer} \cup \text{score } 0\text{-}59) =$

$P(\text{longest answer}) + P(\text{score } 0\text{-}59) - P(\text{longest answer} \cap \text{score } 0\text{-}59) =$

$0.380 + 0.120 - 0.045 = 0.455$

[For any pair of events A and B, P(A ∪ B) = P(A) + P(B) − P(A ∩ B).]

(f) What is the probability that someone in this group chooses strategy "guess" *given that* his/her score was 0–59?

Answer: $P(\text{guess}|\text{score } 0\text{-}59) = \dfrac{40}{120} \approx 0.333$

Note that we narrowed our attention to the 120 test takers who scored 0–59 to calculate this *conditional probability*.

(g) What is the probability that someone in this group scored 80–100 *given that* he/she chose strategy "longest answer"?

Answer: $P(\text{score } 80\text{-}100 | \text{longest answer}) = \dfrac{135}{380} \approx 0.355$

Note that we narrowed our attention to the 380 test takers who chose "longest answer" to calculate this *conditional probability*.

[A formula for conditional probability is given by $P(A|B) = \dfrac{P(A \cap B)}{P(B)}$.]

(h) Are the strategy "guess" and scoring 0–59 *independent events*? That is, is whether a test taker used the strategy "guess" unaffected by whether he/she scored 0–59?

Answer: We must check if $P(\text{guess}|\text{score } 0\text{-}59) = P(\text{guess})$. From (f) and (a), we see that these probabilities are not equal (0.333 ≠ 0.3), so the strategy "guess" and scoring 0–59 are *not* independent events.

[Events A and B are *independent* if P(A|B) = P(A), or equivalently, If P(B|A) = P(B). It is also true that A and B are independent if and only if P(A ∩ B) = P(A) P(B), that is, if and only if the probability of both events happening is the product of their probabilities. Yet another insight into independence is that A and B are independent if and only if $P(A|B) = P(A|B^C)$.]

(i) Are the strategy "longest answer" and scoring 80–100 *mutually exclusive events*? That is, are these two events disjoint and cannot simultaneously occur?

Answer: longest answer ∩ score 80–100 ≠ Ø and $P(\text{longest answer} \cap \text{score } 80\text{-}100) \neq 0$, so the strategy "longest answer" and scoring 80–100 are *not* mutually exclusive events.

Note that if two events are mutually exclusive, then the probability that at least one event will occur is equal to the sum of the respective probabilities of the two events:

If P(A ∩ B) = 0, then P(A ∪ B) = P(A) + P(B)

[Whether events are mutually exclusive or are independent are two very different properties! One refers to events being disjoint; the other refers to an event having no effect on whether or not the other occurs. Note that mutually exclusive (disjoint) events are *not* independent (except in the special case that one of the events has probability 0). That is, mutually exclusive gives that P(A ∩ B) = 0, whereas independence gives that P(A ∩ B) = P(A) P(B) (and the only way these are ever simultaneously true is in the very special case when P(A) = 0 or P(B) = 0).]

TIP

The word "or" here means one event or the other event *or both events.*

TIP

Conditional probability shrinks the population of interest.

TIP

If A and B are independent we also have **P(A|not B) = P(A).**

TIP

Don't multiply probabilities unless the events are independent.

TIP

Don't confuse independence with mutually exclusive.

TIP

Don't add probabilities unless the events are mutually exclusive.

MULTISTAGE PROBABILITY CALCULATIONS

➡ EXAMPLE 7.3

On a university campus, 60%, 30%, and 10% of the computers use Windows, Apple, and Linux operating systems, respectively. A new virus affects 3% of the Windows, 2% of the Apple, and 1% of the Linux operating systems. What is the probability a computer on this campus has the virus?

Answer: In such problems it is helpful to start with a tree diagram.

TIP

Tree diagrams can be very useful in working with conditional probabilities.

We then have

$$P(Windows \cap virus) = (0.6)(0.03) = 0.018$$

$$P(Apple \cap virus) = (0.3)(0.02) = 0.006$$

$$P(Linux \cap virus) = (0.1)(0.01) = 0.001$$

At this stage a Venn diagram is helpful in finishing the problem:

IMPORTANT

Naked or *bald* answers will receive little or *no* credit. You must show where answers come from.

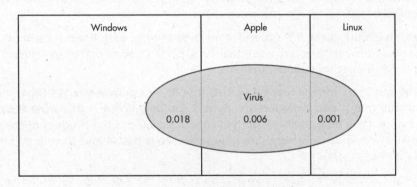

$$P(virus) = P(Windows \cap virus) + P(Apple \cap virus) + P(Linux \cap virus)$$
$$= 0.018 + 0.006 + 0.001 = 0.025$$

We can take the above analysis one stage further and answer such questions as: If a randomly chosen computer on this campus has the virus, what is the probability it is a Windows machine? An Apple machine? A Linux machine?

$$P(Windows \mid virus) = \frac{P(Windows \cap virus)}{P(virus)} = \frac{0.018}{0.025} = 0.72$$

$$P(Apple \mid virus) = \frac{P(Apple \cap virus)}{P(virus)} = \frac{0.006}{0.025} = 0.24$$

$$P(Linux \mid virus) = \frac{P(Linux \cap virus)}{P(virus)} = \frac{0.001}{0.025} = 0.04$$

BINOMIAL DISTRIBUTION

A probability distribution is a listing or formula giving the probability of each outcome.

In many applications, such as coin tossing, there are only *two* possible outcomes. For applications in which a two-outcome situation is repeated a certain number of times, and the probability of each of the two outcomes remains the same for each repetition, the resulting calculations involve what are known as *binomial probabilities*.

➡ EXAMPLE 7.4

Suppose the probability that a lightbulb is defective is 0.1 (so probability of being good is 0.9).

(a) What is the probability that four lightbulbs are all defective?

Answer: Because of independence (i.e., whether one lightbulb is defective is not influenced by whether any other lightbulb is defective), we can multiply individual probabilities of being defective to find the probability that all the bulbs are defective:

$$(0.1)(0.1)(0.1)(0.1) = (0.1)^4 = 0.0001$$

(b) What is the probability that exactly two out of three lightbulbs are defective?

Answer: The probability that the first two bulbs are defective and the third is good is $(0.1)(0.1)(0.9) = 0.009$. The probability the first bulb is good and the other two are defective is $(0.9)(0.1)(0.1) = 0.009$. Finally, the probability that the second bulb is good and the other two are defective is $(0.1)(0.9)(0.1) = 0.009$. Summing, we find that the probability that exactly two out of three bulbs are defective is $0.009 + 0.009 + 0.009 = 0.027$.

(c) What is the probability that exactly three out of eight lightbulbs are defective?

Answer: The probability of any particular arrangement of three defective and five good bulbs is $(0.1)^3(0.9)^5 = 0.00059049$. How many such arrangements are there? The answer is given by combinations: $\binom{8}{3} = \frac{8!}{3!5!} = 56$. Thus, the probability that exactly three out of eight lightbulbs are defective is $56 \times 0.00059049 = 0.03306744$. [On the TI-84, `binompdf(8,.1,3)= 0.03306744`.]

More generally, if an experiment has two possible outcomes, called *success* and *failure*, with the probability of success equal to p, and the probability of failure equal to q (of course, $p + q = 1$), and if the outcome at any particular time has no influence over the outcome at any other time, then if the experiment is repeated n times, the probability of exactly k successes (and thus $n - k$ failures) is

$$\binom{n}{k}p^k q^{n-k} = \frac{n!}{k!(n-k)!}p^k q^{n-k}$$

> **NOTE**
>
> *pdf* stands for "probability density function," that is, it's a function that tells you the probability of certain events occurring.

> **NOTE**
>
> Or use Bpd on the Casio Prizm or BINOMIAL on the HP Prime.

> **NOTE**
>
> On the exam, you can write `binompdf(n=8,p=0.1,x=3) = 0.033` and you will receive full credit.

> That is, "calc-speak" can be used if the parameters are defined.

➡ EXAMPLE 7.5

Super Mario cards were in one-third of cereal boxes advertising enclosed Magic Motion cards. If six boxes of cereal are purchased, what is the probability of exactly two Super Mario cards?

TIP

On the exam it is sufficient to write: Binomial, $n = 6$, $p = 1/3$, $P(X = 2) = 0.329$.

Answer: If the probability of a Mario card is 1/3, then the probability of no Mario card is 1 – 1/3 = 2/3. If two out six boxes have a Mario card, then 6 – 2 = 4 do not.

Thus, the desired probability is $\binom{6}{2}\left(\frac{1}{3}\right)^2\left(\frac{2}{3}\right)^4 = 15\left(\frac{1}{3}\right)^2\left(\frac{2}{3}\right)^4 = \frac{80}{243} \approx 0.329$.

[Or `binompdf(6,1/3,2)` $\approx$ 0.329.]

In some situations it is easier to calculate the probability of the complementary event and subtract this value from 1.

➡ EXAMPLE 7.6

Joe DiMaggio had a career batting average of 0.325. What was the probability that he would get at least one hit in five official times at bat?

NOTE

On the exam it is sufficient to write: Binomial, $n = 5$, $p = 0.325$, $P(X \geq 1) = 0.860$.

Answer: We could sum the probabilities of exactly one hit, two hits, three hits, four hits, and five hits. However, the complement of "at least one hit" is "zero hits." The probability of no hit is

$$\binom{5}{0}(0.325)^0(0.675)^5 = (0.675)^5 \approx 0.140$$

and thus the probability of at least one hit in five times at bat is 1 – 0.140 = 0.860.

[Or binomcdf(5, .675, 4) ≈ 0.860.]

NOTE

Or use Bcd on the Casio Prizm or BINOMIAL_CDF on the HP Prime.

Many, perhaps most, applications of probability involve such phrases as *at least, at most, less than*, and *more than*. In these cases, solutions involve summing two or more cases. For such calculations, the TI-84 `binomcdf` is very useful. `binomcdf(n,p,x)` gives the probability of *x* or fewer successes in a binomial distribution with number of trials *n* and probability of success *p*.

NOTE

cdf stands for "cumulative distribution function," that is, the probability that *X* is less than or equal to a particular value.

➡ EXAMPLE 7.7

A manufacturer has the following quality control check at the end of a production line: If at least eight of ten randomly picked articles meet all specifications, the whole shipment is approved. If, in reality, 85% of a particular shipment meet all specifications, what is the probability that the shipment will make it through the control check?

Answer: The probability of an item meeting specifications is 0.85, and so the probability of it not meeting specifications must be 0.15. We want to determine the probability that at least eight out of ten articles will meet specifications, that is, the probability that exactly eight or exactly nine or exactly ten articles will meet specifications. We sum the three binomial probabilities:

| Exactly 8 of 10 meet specifications | Exactly 9 of 10 meet specifications | Exactly 10 of 10 meet specifications |

$$\binom{10}{8}(0.85)^8(0.15)^2 \quad + \binom{10}{9}(0.85)^9(0.15)^1 \quad + \binom{10}{10}(0.85)^{10}0(.15)^0$$

$$= \frac{10!}{8!2!}(0.85)^8(0.15)^2 + 10(0.85)^9(0.15) + (0.85)^{10} \approx 0.820$$

[One can calculate 1 − binomcdf(10, .85, 7) ≈ 0.820 or binomcdf(10, .15, 2) ≈ 0.820.]

TIP

On the exam it is sufficient to write: Binomial, $n = 10$, $p = 0.85$, $P(X \geq 8) = 0.820$.

➡️ **EXAMPLE 7.8**

For the problem in Example 7.7, what is the probability that a shipment in which only 70% of the articles meet specifications will make it through the control check?

Answer:

$$\binom{10}{8}(0.7)^8(0.3)^2 + \binom{10}{9}(0.7)^9(0.3)^1 + \binom{10}{10}(0.7)^{10}(0.3)^0$$

$$= 45(0.7)^8(0.3)^2 + 10(0.7)^9(0.3) + (0.7)^{10}$$

$$\approx 0.383$$

[Or binomcdf(10, .3, 2) ≈ 0.383.]

TIP

On the exam, it is also sufficient to write: binomcdf ($n = 10$, $p = 0.3$, $x = 2$) = 0.383 .

➡️ **EXAMPLE 7.9**

A grocery store manager notes that 35% of customers who buy a particular product make use of a store coupon to receive a discount. If seven people purchase the product, what is the probability that fewer than four will use a coupon?

Answer: In this situation, "fewer than four" means zero, one, two, or three.

$$\binom{7}{0}(0.35)^0(0.65)^7 + \binom{7}{1}(0.35)^1(0.65)^6 + \binom{7}{2}(0.35)^2(0.65)^5 + \binom{7}{3}(0.35)^3(0.65)^4$$

$$= (0.65)^7 + 7(0.35)(0.65)^6 + 21(0.35)^2(0.65)^5 + 35(0.35)^3(0.65)^4$$

$$\approx 0.800$$

[Or binomcdf(7, .35, 3) ≈ 0.800.]

Sometimes we are asked to calculate the probability of each of the possible outcomes (the results should sum to 1).

➥ EXAMPLE 7.10 _____

If the probability that a male birth will occur is 0.51, what is the probability that a five-child family will have all boys? Exactly four boys? Exactly three boys? Exactly two boys? Exactly one boy? All girls?

Answer:

$$P(5\text{ boys}) = \binom{5}{5}(0.51)^5(0.49)^0 = (0.51)^5 = 0.0345$$

$$P(4\text{ boys}) = \binom{5}{4}(0.51)^4(0.49)^1 = 5(0.51)^4(0.49) = 0.1657$$

$$P(3\text{ boys}) = \binom{5}{3}(0.51)^3(0.49)^2 = 10(0.51)^3(0.49)^2 = 0.3185$$

$$P(2\text{ boys}) = \binom{5}{2}(0.51)^2(0.49)^3 = 10(0.51)^2(0.49)^3 = 0.3060$$

$$P(1\text{ boys}) = \binom{5}{1}(0.51)^1(0.49)^4 = 5(0.51)(0.49)^4 = 0.1470$$

$$P(0\text{ boys}) = \binom{5}{0}(0.51)^0(0.49)^5 = (0.49)^5 = \frac{0.0283}{1.0000}$$

[Or binompdf(5, .49) = {0.0345 0.1657 0.3185 0.3060 0.1470 0.0283}.]

A list such as the one in Example 7.10 shows the entire _probability distribution_, which in this case refers to a listing of all outcomes and their probabilities.

GEOMETRIC PROBABILITIES

Suppose an experiment has two possible outcomes, called _success_ and _failure_, with the probability of success equal to p and the probability of failure equal to $q = 1 - p$, and the trials are independent. Then the probability that the first success is on trial number $X = k$ is

$$q^{k-1}p$$

➥ EXAMPLE 7.11 _____

Suppose only 12% of men in ancient Greece were honest. What is the probability that the first honest man Diogenes encounters will be the third man he meets?

Answer: $(0.88)^2(0.12) = 0.092928$ [or geometpdf (.12, 3) = 0.092928]

What is the probability that the first honest man he encounters will be no later than the fourth man he meets?

Answer: $(0.12) + (0.88)(0.12) + (0.88)^2(0.12) + (0.88)^3(0.12) = 0.40030464$
[or geometcdf (.12, 4) = 0.40030464]

SIMULATION

Instead of algebraic calculations, sometimes we can use simulation to answer probability questions.

➡ EXAMPLE 7.12

If left alone, 70% of birthmarks gradually fade away. If ten children, 5 of each gender, are born with birthmarks, what is the probability that the same number of boys and girls will lose their birthmarks? Answer the question using simulation.

Answer: Let the digits 1–7 represent having a birthmark that fades away, and 8, 9, and 0 represent having a birthmark that doesn't fade away. To simulate the 10 children, select 10 digits from the random number table, with the first 5 representing boys and the next 5 representing girls. Note the number of digits 1–7 in each group and see if there is a match. Underlining the digits 1–7 gives

86961 94141	46633 32552	64452 34882	46266 23371	69214 70203
84873 03236	65618 34476	66335 55856	80396 76708	28303 11388
36392 37904	85601 08943	38324 38221	65492 52566	52057 25970
57161 68039	06748 19239	65683 21773	38557 66359	22908 88647
27941 43337	90190 59437	75107 40518	72389 78958	64819 40361
27492 40058	07181 31446	69345 47300	71402 60588	85664 55760
27827 61501	95435 91684	53217 66012	33702 26371	35333 32652
46511 30857	39655 92730	33859 00386	86309 36997	36997 91524
21567 49374	29800 76389	51752 72124	40673 49251	02103 80901
42505 78167	90045 72843	82718 74608	02434 32440	66180 82562

TIP

Probabilities calculated through simulations should always be referred to as *estimates* or *approximations*.

TIP

Be able to describe simulations so that others can repeat your procedure.

For example, in the first set of 10 digits, there are 3 boys and 4 girls whose birthmarks fade, so no match. In the second set of 10 digits, there are 5 boys and 5 girls whose birthmarks fade, so a match. We have 3-4, 5-5, 5-3, 5-5, 4-3; 3-4, 4-5, 5-4, 2-3, 3-3; 4-3, 3-2, 4-4, 4-5, 4-3; 5-2, 3-3, 4-5, 4-4, 2-3; 4-5, 1-4. 4-3, 3-2, 3-4; 4-2, 3-5, 4-3, 4-2, 4-4; 4-4, 4-3, 5-4, 4-5, 5-5; 5-3, 4-3, 3-2, 2-3, 3-4; 5-4, 1-3, 5-5, 4-4, 3-2; 4-3, 2-4, 3-3, 4-4, 3-4

We count 13 matches out of a possible 50, and so estimate the probability that the same number of boys and girls will lose their birthmarks to be 13/50 = 0.26.

➡ EXAMPLE 7.13

Babe Ruth had a career batting average of 0.342, quite impressive for a home run hitter! Use simulation to estimate the probability that his first hit in a game is on the first at-bat? On the second at-bat? Not until the third at-bat? Fourth at-bat?

Answer: Using the random number table from the previous example, read off 3 digits at a time, with 001–342 representing a hit and anything else, not a hit, and starting all over again every time there is a hit. So, for example, reading off the first line gives

869-619-414-146 is a first hit on the fourth at-bat,

633-325 is a first hit on the second at-bat,

526-445-234 is a first hit on the third at-bat,

882-462-662-337 is a first hit on the fourth at-bat,

169 is a first hit on the first at-bat, and

214 is also a first hit on the first at-bat.

Continuing in this fashion and tabulating the results, we have

Number of the at-bat at which the first hit occurred	Frequency	Estimated probability
1	22	22/64 = 0.344
2	14	14/64 = 0.219
3	13	13/64 = 0.203
4	6	6/64 = 0.094
Over 4	9	9/64 = 0.141
Total	64	

The actual probabilities are 0.342, (0.658)(0.342) = 0.225, $(0.658)^2(0.342)$ = 0.148, $(0.658)^3(0.342)$ = 0.097, and 1 − (0.342 + 0.225 + 0.148 + 0.094) = 0.188. [On the TI-84, one can calculate these *geometric probabilities*: `geometpdf(0.342,1)`, ... , `geometpdf(0.342,4)`, and `1-geometcdf(0.342,4)`.]

In performing a simulation, you must:

1. Set up a correspondence between outcomes and random numbers.
2. Give a procedure for choosing the random numbers (for example, pick three digits at a time from a designated row in a random number table).
3. Give a stopping rule.
4. Note what is to be counted (what is the purpose of the simulation), and give the count if requested.

DISCRETE RANDOM VARIABLES, MEANS (EXPECTED VALUES), AND STANDARD DEVIATIONS

Often each outcome of an experiment has not only an associated probability but also an associated *real number*. For example, the probability may be 0.5 that a student is taking 0 AP classes, 0.3 that he/she is taking 1 AP class, and 0.2 that he/she is taking 2 AP classes. If X represents the different numbers associated with the potential outcomes of some chance situation, we call X a random variable.

While the mean of the set {2, 7, 12, 15} is (2 + 7 + 12 + 15)/4 = 9 or 2(¼) + 7(¼) + 12(¼) + 15(¼) = 9, the *expected value* or mean of a random variable takes into account that the various outcomes may not be equally likely. The expected value or mean of a random variable X is given by $\mu_X = E(X) = \sum xP(x)$ where $P(x)$ is the probability of outcome x. [We also write $\mu_X = \sum x_i p_i$.]

NOTE

The expected value is sometimes referred to as a *weighted average*.

→ EXAMPLE 7.14

In a lottery, 10,000 tickets are sold at $1 each with a prize of $7500 for one winner. What is the average result for each bettor?

Answer: The actual winning payoff is $7499 because the winner paid $1 for a ticket, so we have

Outcome:	Win	Lose
Probability:	$\dfrac{1}{10,000}$	$\dfrac{9,999}{10,000}$
Random variable:	7499	–1

$$\text{Expected Value} = 7499\left(\frac{1}{10,000}\right) + (-1)\left(\frac{9,999}{10,000}\right) = -0.25$$

Thus the *average* result for each person betting on the lottery is a $0.25 loss. Alternatively, we can say that the expected payoff to the lottery system is $0.25 for each ticket sold.

Suppose we have a *binomial random variable,* that is, a random variable whose values are the numbers of "successes" in some binomial probability distribution.

→ EXAMPLE 7.15

Of the automobiles produced at a particular plant, 40% had a certain defect. Suppose a company purchases five of these cars. What is the expected value for the number of cars with defects?

Answer: We might guess that the average or mean or expected value is 40% of 5 = 0.4 × 5 = 2, but let's calculate from the definition. Letting X represent the number of cars with the defect, we have

$$P(0) = \binom{5}{0}(0.4)^0(0.6)^5 = (0.6)^5 = 0.07776$$

$$P(1) = \binom{5}{1}(0.4)^1(0.6)^4 = 5(0.4)(0.6)^4 = 0.25920$$

$$P(2) = \binom{5}{2}(0.4)^2(0.6)^3 = 10(0.4)^2(0.6)^3 = 0.34560$$

$$P(3) = \binom{5}{3}(0.4)^3(0.6)^2 = 10(0.4)^3(0.6)^2 = 0.23040$$

$$P(4) = \binom{5}{4}(0.4)^4(0.6)^1 = 5(0.4)^4(0.6) = 0.07680$$

$$P(5) = \binom{5}{5}(0.4)^5(0.6)^0 = (0.4)^5 = 0.01024$$

Outcome (no. of cars):	0	1	2	3	4	5
Probability:	0.07776	0.25920	0.34560	0.23040	0.07680	0.01024
Random variable:	0	1	2	3	4	5

$$\begin{aligned} E(X) &= 0(0.07776) + 1(0.25920) + 2(0.34560) + 3(0.23040) \\ &\quad + 4(0.07680) + 5(0.01024) \\ &= 2 \end{aligned}$$

Thus the answer turns out to be the same as would be obtained by simply multiplying the probability of a "success" by the number of cases.

The following is true: If we have a binomial probability situation with the probability of a success equal to p and the number of trials equal to n, the expected value or mean number of successes for the n trials is np.

➡ **EXAMPLE 7.16** _____

An insurance salesperson is able to sell policies to 15% of the people she contacts. Suppose she contacts 120 people during a 2-week period. What is the expected value for the number of policies she sells?

Answer: We have a binomial probability with the probability of a success 0.15 and the number of trials 120, and so the mean or expected value for the number of successes is 120 × 0.15 = 18.

_____

We have seen that the mean of a random variable is $\Sigma x_i p_i$. However, not only is the mean important, but also we would like to measure the variability for the values taken on by a random variable. Since we are dealing with chance events, the proper tool is variance (along with standard deviation). *Variance* was defined in Topic Two to be the mean average of the squared deviations $(x - \mu)^2$. If we regard the $(x - \mu)^2$ terms as the values of some random variable (whose probability is the same as the probability of x), the mean of this new random variable is $\Sigma(x_i - \mu_x)^2 p_i$, which is precisely how we define the variance σ^2 of a random variable:

NOTE

It is also true that var(X) = $[\Sigma x_i^2 p_i] - \mu_x^2$

$$\text{var}(X) = \sigma_x^2 = \sum (x_i - \mu_x)^2 p_i$$

As before, the standard deviation σ is the square root of the variance.

➡ **EXAMPLE 7.17** _____

A highway engineer knows that his crew can lay 5 miles of highway on a clear day, 2 miles on a rainy day, and only 1 mile on a snowy day. Suppose the probabilities are as follows:

Outcome:	Clear	Rain	Snow
Probability:	0.6	0.3	0.1
Random variable (miles of highway):	5	2	1

What are the mean (expected value) and the variance?

TIP

In a problem like this, there is no shortcut. You must show your work.

Answer:

$$\mu_x = \sum x_i p_i = 5(0.6) + 2(0.3) + 1(0.1) = 3.7$$

$$\sigma_x^2 = \sum (x_i - \mu_x)^2 p_i = (5 - 3.7)^2 (0.6) + (2 - 3.7)^2 (0.3) + (1 - 3.7)^2 (0.1) = 2.61$$

NOTE

Alternatively, you can calculate $\sigma_x^2 = 5^2(0.6) + 2^2(0.3) + 1^2(0.1) - 3.7^2 = 2.61$

_____

➡ EXAMPLE 7.18 _____

Look again at Example 7.15. We calculated the mean to be 2. What is the variance?

Answer: $\sigma^2 = (0 - 2)^2(0.07776) + (1 - 2)^2(0.2592) + (2 - 2)^2(0.3456) + (3 - 2)^2(0.2304)$
$+ (4 - 2)^2(0.0768) + (5 - 2)^2(0.01024) = 1.2$

Could we have calculated the above result more easily? In this case, we have a binomial random variable and the variance can be simply calculated as $npq = np(1 - p)$.

How can we use this method to calculate the variance in Example 7.15 more simply?

Answer: $np(1 - p) = 5(0.4)(0.6) = 1.2$

_____

Thus, for a random variable X,

> *Mean or expected value:* $\mu_x = \Sigma x_i p_i$
>
> *Variance:* $\sigma^2_x = \Sigma(x_i - \mu_x)^2 p_i$
>
> *Standard deviation:* $\sigma_x = \sqrt{\Sigma(x_i - \mu_x)^2 p_i}$

In the case of a binomial probability distribution with probability of a success equal to p and number of trials equal to n, if we let X be the number of successes in the n trials, the above equations become

> *Mean or expected value:* $\mu_x = np$
>
> *Variance:* $\sigma^2_x = npq = np(1 - p)$
>
> *Standard deviation:* $\sigma_x = \sqrt{npq} = \sqrt{np(1 - p)}$

NOTE

For a geometric probability distribution with probability of success equal to p, if we let X be the number of trials needed to get one success, the mean or expected value is $1/p$ and the variance is $(1 - p)/p^2$.

➡ EXAMPLE 7.19 _____

Sixty percent of all buyers of new cars choose automatic transmissions. For a group of five buyers of new cars, calculate the mean and standard deviation for the number of buyers choosing automatic transmissions.

Answer:

$$\mu_x = np = 5(0.6) = 3.0$$

$$\sigma_x = \sqrt{np(1 - p)} = \sqrt{5(0.6)(0.4)} \approx 1.1$$

_____

- The law of large numbers states that in the long run, a cumulative relative frequency tends closer and closer to what is called the probability of an event.
- The probability of the complement of an event is equal to 1 minus the probability of the event.
- Two events are mutually exclusive if they cannot occur simultaneously.
- If two events are mutually exclusive, the probability that at least one will occur is the sum of their respective probabilities.
- More generally, $P(A \cup B) = P(A) + P(B) - P(A \cap B)$.
- The conditional probability of A given B is given by: $P(A|B) = \dfrac{P(A \cap B)}{P(B)}$; that is, the population of interest is shrunk.
- Two events A and B are independent if $P(A|B) = P(A)$; that is, the probability of one event occurring does not influence whether or not the other occurs.
- If two events A and B are independent, then $P(A \cap B) = P(A) \times P(B)$.
- A random variable takes various numeric values, each with a given probability.
- The mean or expected value of a random variable is calculated by $\mu = \Sigma \, xP(x)$, while the variance is $\sigma^2 = \Sigma(x - \mu)^2 \, P(x)$.
- The binomial distribution arises when there are two possible outcomes, the probability of success is constant, and the trials are independent.
- In a binomial distribution, the mean number of successes is np and the standard deviation is $\sqrt{np(1-p)}$.

QUIZ 1

Multiple-Choice Questions

> **Directions:** The questions that follow are each followed by five suggested answers. Choose the response that best answers the question.

Questions 1–5 refer to the following study: One thousand students at a city high school were classified both according to GPA and whether or not they consistently skipped classes.

	<2.0	GPA 2.0–3.0	>3.0	
Many skipped classes	80	25	5	110
Few skipped classes	175	450	265	890
	255	475	270	

1. What is the probability that a student has a GPA between 2.0 and 3.0?

   (A) 0.025
   (B) 0.227
   (C) 0.450
   (D) 0.475
   (E) 0.506

2. What is the probability that a student has a GPA under 2.0 and has skipped many classes?

   (A) 0.080
   (B) 0.281
   (C) 0.285
   (D) 0.314
   (E) 0.727

3. What is the probability that a student has a GPA under 2.0 or has skipped many classes?

   (A) 0.080
   (B) 0.281
   (C) 0.285
   (D) 0.314
   (E) 0.727

4. What is the probability that a student has a GPA under 2.0 given that he has skipped many classes?

(A) 0.080
(B) 0.281
(C) 0.285
(D) 0.314
(E) 0.727

5. Are "GPA between 2.0 and 3.0" and "skipped few classes" independent?

(A) No, because $0.475 \neq 0.506$.
(B) No, because $0.475 \neq 0.890$.
(C) No, because $0.450 \neq 0.475$.
(D) Yes, because of conditional probabilities.
(E) Yes, because of the product rule.

---

6. The following data are from *The Commissioner's Standard Ordinary Table of Mortality:*

Age	Number Surviving
0	10,000,000
20	9,664,994
40	9,241,359
70	5,592,012

What is the probability that a 20-year-old will survive to be 70?

(A) 0.407
(B) 0.421
(C) 0.559
(D) 0.579
(E) 0.966

7. In a 1974 "Dear Abby" letter, a woman lamented that she had just given birth to her eighth child, and all were girls! Her doctor had assured her that the chance of the eighth child being a girl was only 1 in 100. What was the real probability that the eighth child would be a girl?

(A) 0.01
(B) 0.5
(C) $(0.5)^7$
(D) $(0.5)^8$
(E) $\dfrac{(0.5)^7 + (0.5)^8}{2}$

8. There are two games involving flipping a fair coin. In the first game you win a prize if you can throw between 40% and 60% heads. In the second game you win if you can throw more than 75% heads. For each game would you rather flip the coin 50 times or 500 times?

(A) 50 times for each game
(B) 500 times for each game
(C) 50 times for the first game, and 500 for the second
(D) 500 times for the first game, and 50 for the second
(E) The outcomes of the games do not depend on the number of flips.

9. Suppose that, for any given year, the probabilities that the stock market declines, that women's hemlines are lower, and that both events occur are, respectively, 0.4, 0.35, and 0.3. Are the two events independent?

(A) Yes, because $(0.4)(0.35) \neq 0.3$.
(B) No, because $(0.4)(0.35) \neq 0.3$.
(C) Yes, because $0.4 > 0.35 > 0.3$.
(D) No, because $0.5(0.3 + 0.4) = 0.35$.
(E) There is insufficient information to answer this question.

10. According to a CBS/*New York Times* poll taken in 1992, 15% of the public have responded to a telephone call-in poll. In a random group of five people, what is the probability that exactly two have responded to a call-in poll?

(A) $10(0.15)^2(0.85)^3$
(B) $5(0.15)^2(0.85)^3$
(C) $(0.15)^2(0.85)^3$
(D) $(0.15)^2$
(E) $5(0.15)^2$

11. Suppose that, in a certain part of the world, in any 50-year period the probability of a major plague is 0.39, the probability of a major famine is 0.52, and the probability of both a plague and a famine is 0.15. What is the probability of a famine given that there is a plague?

(A) $0.39 - 0.15$
(B) $\dfrac{0.15}{0.52}$
(C) $0.52 - 0.15$
(D) $\dfrac{0.15}{0.39}$
(E) $0.39 + 0.52 - 0.15$

12. At a warehouse sale 100 customers are invited to choose one of 100 identical boxes. Five boxes contain $700 color television sets, 25 boxes contain $540 camcorders, and the remaining boxes contain $260 cameras. What should a customer be willing to pay to participate in the sale?

(A) $260
(B) $352
(C) $500
(D) $540
(E) $699

13. Suppose you are one of 7.5 million people who send in their name for a drawing with 1 top prize of $1 million, 5 second-place prizes of $10,000, and 20 third-place prizes of $100. Is it worth the $0.49 postage it cost you to send in your name?

(A) Yes, because $\dfrac{1,000,000}{0.49} = 2,040,816$, which is less than 7,500,000.

(B) No, because your expected winnings are only $0.14.

(C) Yes, because $\dfrac{7,500,000}{1+5+20} = 288,642$.

(D) No, because $1,052,000 < 7,500,000$.

(E) Yes, because $\dfrac{1,052,000}{26} = 40,462$.

14. Suppose we have a random variable X where for the values $k = 0, ..., 10$, the associated probabilities are $\binom{10}{k}(0.37)^k(0.63)^{10-k}$. What is the mean of X?

(A) 0.37
(B) 0.63
(C) 3.7
(D) 6.3
(E) None of the above

15. Suppose that 2% of a clinic's patients are known to have cancer. A blood test is developed that is positive in 98% of patients with cancer but is also positive in 3% of patients who do not have cancer. If a person who is chosen at random from the clinic's patients is given the test and it comes out positive, what is the probability that the person actually has cancer?

(A) 0.02
(B) 0.02 + 0.03
(C) (0.02)(0.98)
(D) (0.02)(0.98) + (0.98)(0.03)
(E) $\dfrac{(0.02)(0.98)}{(0.02)(0.98)+(0.98)(0.03)}$

Free-Response Questions

1. A sample of applicants for a management position yields the following numbers with regard to age and experience:

	Years of experience		
	0–5	6–10	>10
Less than 50 years old	80	125	20
More than 50 years old	10	75	50

   (a) What is the probability that a randomly picked applicant from this sample is less than 50 years old? Has more than 10 years' experience? Is more than 50 years old and has five or fewer years' experience?

   (b) What is the probability that a randomly picked applicant from this sample is less than 50 years old given that she has between 6 and 10 years' experience?

   (c) Are the two events "less than 50 years old" and "more than 10 years' experience" independent events? How about the two events "more than 50 years old" and "between 6 and 10 years' experience"? Explain.

2. Explain what is wrong with each of the following statements:

   (a) The probability that a student will score high on the AP Statistics exam is 0.43, while the probability that she will not score high is 0.47.

   (b) The probability that a student plays tennis is 0.18, while the probability that he plays basketball is six times as great.

   (c) The probability that a student enjoys her English class is 0.64, while the probability that she enjoys both her English and her social studies classes is 0.71.

   (d) The probability that a student will be accepted by his first choice for college is 0.38, while the probability that he will be accepted by his first or second choice is 0.32.

   (e) The probability that a student fails AP Statistics and will still be accepted by an Ivy League school is –0.17.

3. Could hands-free, automatic faucets actually be housing more bacteria than the old-fashioned, manual kind? The concern is that decreased water flow may increase the chance that bacteria grows, because the automatic faucets are not being thoroughly flushed through. It is known that 15% of water cultures from older faucets in hospital patient care areas test positive for Legionella bacteria. A recent study at Johns Hopkins Hospital found Legionella bacteria growing in 10 of cultured water samples from 20 electronic faucets.

(a) If the probability of Legionella bacteria growing in a faucet is 0.15, what is the probability that in a sample of 20 faucets, 10 or more have the bacteria growing?

(b) Does the Johns Hopkins study provide sufficient evidence that the probability of Legionella bacteria growing in electronic faucets is greater than 15%? Explain.

4. Consider the following probability distribution for the number of tattoos college students have:

Tattoos	0	1	2	3	4
Probability	0.85	0.11	0.02	0.014	0.006

(a) What is the probability a student has two tattoos given that he has at least one tattoo?

(b) In a random sample of three students, what is the probability that one of the students has at least two tattoos while the other two students have no tattoos?

(c) What is the mean number of tattoos per student?

QUIZ 2

Multiple-Choice Questions

Directions: The questions that follow are each followed by five suggested answers. Choose the response that best answers the question.

Questions 1–4 refer to the following study: Five hundred people used a home test for HIV, and then all underwent more conclusive hospital testing. The accuracy of the home test was evidenced in the following table.

	HIV	Healthy	
Positive test	35	25	60
Negative test	5	435	440
	40	460	

1. What is the *predictive value* of the test? That is, what is the probability that a person has HIV and tests positive?

    (A) $\dfrac{35}{500}$

    (B) $\dfrac{35}{60}$

    (C) $\dfrac{35}{40}$

    (D) $\dfrac{35}{35+25+5}$

    (E) $\dfrac{35+25+5}{500}$

2. What is the *false-positive* rate? That is, what is the probability of testing positive given that the person does not have HIV?

    (A) $\dfrac{25}{460}$

    (B) $\dfrac{25}{60}$

    (C) $\dfrac{35}{40}$

    (D) $\dfrac{25}{500}$

    (E) $\dfrac{35+25+5}{500}$

3. What is the *sensitivity* of the test? That is, what is the probability of testing positive given that the person has HIV?

(A) $\dfrac{35+25+5}{500}$

(B) $\dfrac{35}{35+25+5}$

(C) $\dfrac{35}{500}$

(D) $\dfrac{35}{60}$

(E) $\dfrac{35}{40}$

4. What is the *specificity* of the test? That is, what is the probability of testing negative given that the person does not have HIV?

(A) $\dfrac{5}{40}$

(B) $\dfrac{35}{60}$

(C) $\dfrac{35+5+435}{500}$

(D) $\dfrac{435}{500}$

(E) $\dfrac{435}{460}$

---

5. Suppose you toss a fair coin ten times and it comes up heads every time. Which of the following is a true statement?

(A) By the law of large numbers, the next toss is more likely to be tails than another heads.

(B) By the properties of conditional probability, the next toss is more likely to be heads given that ten tosses in a row have been heads.

(C) Coins actually do have memories, and thus what comes up on the next toss is influenced by the past tosses.

(D) The law of large numbers tells how many tosses will be necessary before the percentages of heads and tails are again in balance.

(E) The probability that the next toss will again be heads is 0.5.

6. Mathematically speaking, casinos and life insurance companies make a profit because of

   (A) their understanding of sampling error and sources of bias.
   (B) their use of well-designed, well-conducted surveys and experiments.
   (C) their use of simulation of probability distributions.
   (D) the central limit theorem.
   (E) the law of large numbers.

7. Given that 52% of the U.S. population are female and 15% are older than age 65, can we conclude that $(0.52)(0.15) = 7.8\%$ are women older than age 65?

   (A) Yes, by the multiplication rule.
   (B) Yes, by conditional probabilities.
   (C) Yes, by the law of large numbers.
   (D) No, because the events are not independent.
   (E) No, because the events are not mutually exclusive.

8. An inspection procedure at a manufacturing plant involves picking three items at random and then accepting the whole lot if at least two of the three items are in perfect condition. If in reality 90% of the whole lot are perfect, what is the probability that the lot will be accepted?

   (A) $3(0.1)^2(0.9)$
   (B) $3(0.9^2)(0.1)$
   (C) $(0.1)^3 + (0.1)^2(0.9)$
   (D) $(0.1)^3 + 3(0.1)^2(0.9)$
   (E) $(0.9)^3 + 3(0.9)^2(0.1)$

9. Sixty-five percent of all divorce cases cite incompatibility as the underlying reason. If four couples file for a divorce, what is the probability that exactly two will state incompatibility as the reason?

   (A) $2(0.65)(0.35)$
   (B) $2(0.65)^2(0.35)^2$
   (C) $4(0.65)^2(0.35)^2$
   (D) $6(0.65)^2(0.35)^2$
   (E) $(0.65)^2$

10. For an advertising promotion, an auto dealer hands out 1000 lottery tickets with a prize of a new car worth $25,000. For someone with a single ticket, what is the standard deviation for the amount won?

    (A) $7.07
    (B) $25.00
    (C) $49.95
    (D) $790.17
    (E) $624,375

11. A city water supply system involves three pumps, the failure of any one of which crashes the system. The probabilities of failure for each pump in a given year are 0.025, 0.034, and 0.02, respectively. Assuming the pumps operate independently of each other, what is the probability that the system does crash during the year?

(A) $0.025 + 0.034 + 0.02$
(B) $1 - (0.025 + 0.034 + 0.02)$
(C) $1 - (0.025)(0.034)(0.02)$
(D) $(1 - 0.025)(1 - 0.034)(1 - 0.02)$
(E) $1 - (1 - 0.025)(1 - 0.034)(1 - 0.02)$

12. An insurance company charges $800 annually for car insurance. The policy specifies that the company will pay $1000 for a minor accident and $5000 for a major accident. If the probability of a motorist having a minor accident during the year is 0.2, and of having a major accident, 0.05, how much can the insurance company expect to make on a policy?

(A) $200
(B) $250
(C) $300
(D) $350
(E) $450

13. Can the function $f(x) = \frac{x+6}{24}$, for $x = 1, 2,$ and 3, be the probability distribution for some random variable?

(A) Yes.
(B) No, because probabilities cannot be negative.
(C) No, because probabilities cannot be greater than 1.
(D) No, because the probabilities do not sum to 1.
(E) Not enough information is given to answer the question.

14. Consider the following table of ages of U.S. senators:

Age (yr):	<40	40–49	50–59	60–69	70–79	>79
Number of senators:	5	30	36	22	5	2

What is the probability that a senator is under 70 years old given that he or she is at least 50 years old?

(A) 0.580
(B) 0.624
(C) 0.643
(D) 0.892
(E) 0.969

15. Suppose that 60% of students who take the AP Statistics exam score 4 or 5, 25% score 3, and the rest score 1 or 2. Suppose further that 95% of those scoring 4 or 5 receive college credit, 50% of those scoring 3 receive such credit, and 4% of those scoring 1 or 2 receive credit. If a student who is chosen at random from among those taking the exam receives college credit, what is the probability that she received a 3 on the exam?

    (A) 0.125
    (B) 0.178
    (C) 0.701
    (D) 0.813
    (E) 0.822

Free-Response Questions

> **Directions:** You must show all work and indicate the methods you use. You will be graded on the correctness of your methods and on the accuracy of your final answers.

1. A manufacturer is considering two options for a quality control check at the end of a production line:

   Option A: If at least five of six randomly picked articles meet all specifications, the day's production is approved.

   Option B: If at least ten of twelve randomly picked articles meet all specifications, the day's production is approved.

   If you are a buyer wanting the most assurance that a day's production will not be accepted if there were only 75% of the articles meeting all specifications, which option would you request the manufacturer to use? Give statistical justification.

2. Assume there is no overlap between the 56% of the population who wear glasses and the 4% who wear contacts. If 55% of those who wear glasses are women and 63% of those who wear contacts are women, what is the probability that the next person you encounter on the street will be a woman with glasses? A woman with contacts? A man with glasses? A man with contacts? A person not wearing glasses or contacts? Explain your reasoning.

3. Suppose USAir accounts for 20% of all U.S. domestic flights. As of mid-1994, USAir was involved in four of the previous seven major disasters. "That's enough to begin getting suspicious but not enough to hang them," said Dr. Brad Efton in *The New York Times* (September 11, 1994, Sec. 4, p. 4). Using a binomial distribution, comment on Dr. Efton's remark.

4. If a car is stopped for speeding, the probability that the driver has illegal drugs hidden in the car is 0.14. A police officer manning a speed trap plans to thoroughly search cars he stops for speeding until he finds one with illegal drugs.

(a) What is the probability that illegal drugs will be found before he stops the tenth car? Show your work.

(b) The results of a 100-trial simulation are shown below. Each trial ends when a car with illegal drugs is found.

Use the above 100-trial results to estimate the mean number of cars searched before finding one with illegal drugs.

ANSWERS EXPLAINED

Quiz 1

MULTIPLE-CHOICE QUESTIONS

1. **(D)** $\frac{475}{1000} = 0.475$

2. **(A)** $\frac{80}{1000} = 0.080$ (probability of an intersection)

3. **(C)** $\frac{80+25+5+175}{1000} = 0.285$ (probability of a union)

4. **(E)** $\frac{80}{110} = 0.727$ (conditional probability)

5. **(A)** $P(2.0 - 3.0 \text{ GPA}) = \frac{475}{1000} = 0.475$; however, $P(2.0 - 3.0 \text{ GPA} \mid \text{few skips}) = \frac{450}{890} = 0.506$. If independent, these would have been equal.

6. **(D)** $\frac{5{,}592{,}012}{9{,}664{,}994} = 0.579$

7. **(B)** The probability of the next child being a girl is independent of the gender of the previous children.

8. **(D)** The probability of throwing heads is 0.5. By the law of large numbers, the more times you flip the coin, the more the relative frequency tends to become closer to this probability. With fewer tosses there is a greater chance of wide swings in the relative frequency.

9. **(B)** If E and F are independent, then $P(E \cap F) = P(E)P(F)$; however, in this problem, $(0.4)(0.35) \neq 0.3$.

10. **(A)** This is a binomial with $n = 5$ and $p = 0.15$. Then, $P(X = 2) = \binom{5}{2}(0.15)^2(0.85)^3$.

11. **(D)** $P(\text{famine}|\text{plague}) = \dfrac{P(\text{famine} \cap \text{plague})}{P(\text{plague})} = \dfrac{0.15}{0.39}$

12. **(B)** $E(X) = \mu_x = \Sigma x_i p_i = 700(0.05) + 540(0.25) + 260(0.7) = 352$

13. **(B)** Your expected winnings are only

$$\frac{1}{7{,}500{,}000}(1{,}000{,}000) + \frac{5}{7{,}500{,}000}(10{,}000) + \frac{20}{7{,}500{,}000}(100) = 0.14$$

14. **(C)** This is a binomial with $n = 10$ and $p = 0.37$, and so the mean is $np = 10(0.37) = 3.7$.

15. **(E)**

$P(\text{positive test}) = P(\text{cancer} \cap \text{positive test})$
$+ P(\text{no cancer} \cap \text{positive test})$
$= (0.02)(0.98) + (0.98)(0.03)$

$$P(\text{cancer}\,|\,\text{positive test}) = \frac{P(\text{cancer} \cap \text{positive test})}{P(\text{positive test})}$$

$$= \frac{(0.02)(0.98)}{(0.02)(0.98)+(0.98)(0.03)}$$

FREE-RESPONSE QUESTIONS

1. It's easiest to first sum the rows and columns:

	Years of experience			
	0–5	6–10	>10	
Less than 50 years old	80	125	20	225
More than 50 years old	10	75	50	135
	90	200	70	360

(a) $P(\text{age} < 50) = \frac{225}{360} = 0.625$

$P(\text{experience} > 10) = \frac{70}{360} = 0.194$

$P(\text{age} > 50 \cap \text{experience } 0\text{--}5) = \frac{10}{360} = 0.028$

(b) $P(\text{age} < 50\,|\,\text{experience } 6\text{--}10) = \frac{125}{200} = 0.625$

(c) $P(\text{age} < 50) = \frac{225}{360} = 0.625$; however, $P(\text{age} < 50\,|\,\text{experience} > 10) = \frac{20}{70}$

$= 0.286$ and so they are not independent.

$P(\text{age} > 50) = \frac{135}{360} = 0.375$ and also $P(\text{age} > 50\,|\,\text{experience } 6\text{--}10) = \frac{75}{200}$

$= 0.375$ and so these two are independent.

2. (a) The probability of the complement is 1 minus the probability of the event, but $1 - 0.43 \neq 0.47$.

   (b) Probabilities are never greater than 1, but $6(0.18) = 1.08$.

   (c) The probability of an intersection cannot be greater than the probability of one of the separate events.

   (d) The probability of a union cannot be less than the probability of one of the separate events.

   (e) Probabilities are never negative.

3. (a) We have a binomial with $n = 20$ and $p = 0.15$, so

   $$P(X \geq 10) = \binom{20}{10}(0.15)^{10}(0.85)^{10} + \binom{20}{11}(0.15)^{11}(0.85)^{9} + \cdots + (0.15) = 0.00025$$

   [The answer comes quickly from a calculator calculation such as 1-binomcdf(20, .15, 9).]

   (b) If the probability of Legionella bacteria growing in an electric faucet is 0.15, then the probability of a result as extreme or more extreme than what was obtained in the Johns Hopkins study is only 0.00025. With such a low probability, there is strong evidence to conclude that the probability of Legionella bacteria growing in electronic faucets is greater than 0.15. That is, there is strong evidence that automatic faucets actually house more bacteria than the old-fashioned, manual kind!

4. (a) $\dfrac{0.02}{0.11 + 0.02 + 0.014 + 0.006} = \dfrac{0.02}{0.15} = \dfrac{2}{15} = 0.133$

   (b) P(at least 2 tattoos) = $0.02 + 0.014 + 0.006 = 0.04$. The student with tattoos can be the first, second, or third student, so we have $3(0.04)(0.85)^2 = 0.0867$.

   (c) $E(X) = 0(0.85) + 1(0.11) + 2(0.02) + 3(0.0014) + 4(0.006) = 0.216$

Quiz 2

1. **(A)** 35 out of the entire population of 500 both have HIV and tested positive.

2. **(A)** Of the 460 healthy people, 25 tested positive.

3. **(E)** Of the 40 people with HIV, 35 tested positive.

4. **(E)** Of the 460 healthy people, 435 tested negative.

5. **(E)** Coins have no memory, and so the probability that the next toss will be heads is .5 and the probability that it will be tails is 0.5. The law of large numbers says that as the number of tosses becomes larger, the percentage of heads tends to become closer to 0.5.

6. **(E)** While the outcome of any single play on a roulette wheel or the age at death of any particular person is uncertain, the law of large numbers gives that the relative frequencies of specific outcomes in the long run tend to become closer to numbers called probabilities.

7. **(D)** $P(E \cap F) = P(E)P(F)$ only if the events are independent. In this case, women live longer than men and so the events are not independent.

8. **(E)** This is a binomial with $n = 3$ and $p = 0.90$. Then $P(X \geq 2) = (0.9)^3 + 3(0.9)^2(0.1)$.

9. **(D)** This is a binomial with $n = 4$ and $p = 0.65$.

$$P(X = 2) = \binom{4}{2}(0.65)^2(0.35)^2 = 6(0.65)^2(0.35)^2$$

10. **(D)** $\bar{x} = E(X) = \sum xP(x) = 0(0.999) + 25{,}000(0.001) = 25$

$$\sigma = \sqrt{\sum (x - \bar{x})^2 P(x)} = \sqrt{(0 - 25)^2(0.999) + (25{,}000 - 25)^2(0.001)} = 790.17$$

[Or on the TI-84, put returns and probabilities into two lists and run `1-Var Stats L1,L2`]

11. **(E)** The probabilities of each pump not failing are $1 - 0.025$, $1 - 0.034$, and $1 - 0.02$, respectively. The probability of none failing is the product of $(1 - 0.025)(1 - 0.034)(1 - 0.02)$. So the probability of at least one failing is $1 - (1 - 0.025)(1 - 0.034)(1 - 0.02)$.

12. **(D)** $1000(0.2) + 5000(0.05) = 450$, and $800 - 450 = 350$.

13. **(A)** The probabilities $\frac{7}{24}$, $\frac{8}{24}$, and $\frac{9}{24}$ are all nonnegative, and they sum to 1.

14. **(D)** $\dfrac{36 + 22}{36 + 22 + 5 + 2} = 0.892$

15. **(B)**

$$P \text{ (credit)} = 0.57 + 0.125 + 0.006 = 0.701$$

$$P\left(3 \mid \text{credit}\right) = \frac{P(3 \cap \text{credit})}{P(\text{credit})} = \frac{0.125}{0.701} = 0.178$$

FREE-RESPONSE QUESTIONS

1. With Option A, if in reality only 75% of the articles meet all specifications, the probability of rejecting the day's production is:

$$P(\text{rejection}) = 1 - P(\text{acceptance}) = 1 - \left[\binom{6}{5}(0.75)^5(0.25) + (0.75)^6 \right] = 0.466$$

[binomcdf(6,.75,4) = .466]

With Option B, if in reality only 75% of the articles meet all specifications, the probability of rejecting the day's production is:

$$P(\text{rejection}) = 1 - P(\text{acceptance}) =$$

$$1 - \left[\binom{12}{10}(0.75)^{10}(0.25)^2 + \binom{12}{11}(0.75)^{11}(0.25) + (0.75)^{12} \right] = 0.609$$

[binomcdf(12,.75,9) = .609]

For the greatest probability of rejecting the day's production if only 75% of the articles meet all specifications, the buyer should request the manufacturer to use Option B with a probability of rejection of 0.609 as opposed to Option A with a probability of rejection of only 0.466.

2.

$$P\left(E \mid F\right) = \frac{P(E \cap F)}{P(F)} \quad \text{or} \quad P\left(E \cap F\right) = P\left(E \mid F\right)P(F),$$

and so

$$P(\text{woman} \cap \text{glasses}) = (0.55)(0.56)$$
$$= 0.308$$

$$P(\text{woman} \cap \text{contacts}) = (0.63)(0.04)$$
$$= 0.0252$$

If 55% of those who wear glasses are women, 45% of those who wear glasses must be men, and if 63% of those who wear contacts are women, 37% of those who wear contacts must be men. Thus we have

$$P(\text{man} \cap \text{glasses}) = (0.45)(0.56)$$
$$= 0.252$$

$$P(\text{man} \cap \text{contacts}) = (0.37)(0.04)$$
$$= 0.0148$$

The probability that you will encounter a person not wearing glasses or contacts is $1 - (0.56 + 0.04) = 0.4$.

3. If USAir accounted for 20% of the major disasters, the chance that it would be involved in at least four of seven such disasters is

$$\binom{7}{4}(0.2)^4(0.8)^3 + \binom{7}{5}(0.2)^5(0.8)^2 + \binom{7}{6}(0.2)^6(0.8)^1 + (0.2)^7$$
$$= 0.029 + 0.004 + 0.000 + 000$$
$$= 0.033$$

[Or binomcdf(7, .8, 3) = 0.033.]

Mathematically, if USAir accounted for only 20% of the major disasters, there is only a 0.033 chance of it being involved in four of seven such disasters. This seems more than enough evidence to be suspicious! There is sufficient evidence that USAir accounts for more than 20% of major disasters in the years leading up to 1994.

4. (a) The probability that the first stopped car will have illegal drugs is 0.14. The probability that the first will be clean but that the second will have drugs is $(0.86)(0.14)$. The probability that two clean cars will be followed by one with drugs is $(0.86)^2(0.14)$. The probability that three clean ones will be followed by one with drugs is $(0.86)^3(0.14)$, and so on. The probability of drugs being found before the tenth car is

$$0.14 + (0.86)(0.14) + (0.86)^2(0.14) + \cdots + (0.86)^8(0.14) = 0.7427$$

Or more simply we could solve by subtracting the complementary probability from 1, that is, $1 - (0.86)^9 = 0.7427$.

(b) $\Sigma x P(x) = 1(0.15) + 2(0.12) + \cdots + 25(0.01) = 6.81$

Combining and Transforming Random Variables 8

→ **MEANS AND VARIANCES FOR SUMS AND DIFFERENCES OF SETS**

→ **MEANS AND VARIANCES FOR SUMS AND DIFFERENCES OF RANDOM VARIABLES**

→ **TRANSFORMING RANDOM VARIABLES**

In many experiments, a pair of numbers is associated with each outcome. This situation leads to a study of pairs of random variables. When the random variables are independent, there is an easy calculation for finding both the mean and standard deviation of the sum (and difference) of the two random variables.

MEANS AND VARIANCES FOR SUMS AND DIFFERENCES OF SETS

➡ **EXAMPLE 8.1**

A casino offers couples at the hotel free chips—each person reaches into a bag and pulls out a card stating the number of free chips—one person's bag has a set of four cards with $X = \{1, 9, 20, 74\}$; the other person's bag has a set of three cards with $Y = \{5, 15, 55\}$. Note that the mean of set X is $\mu_X = \frac{1+9+20+74}{4} = 26$ and that of set Y is $\mu_Y = \frac{5+15+55}{3} = 25$. What is the average amount they should be able to pool together?

Answer: Form the set W of sums: $W = \{1 + 5, 1 + 15, 1 + 55, 9 + 5, 9 + 15, 9 + 55, 20 + 5, 20 + 15, 20 + 55, 74 + 5, 74 + 15, 74 + 55\} = \{6, 16, 56, 14, 24, 64, 25, 35, 75, 79, 89, 129\}$ and

$$\mu_W = \mu_{X+Y} = \frac{6+16+56+14+24+64+25+35+75+79+89+129}{12} = 51$$

We note that $\mu_W = \mu_{X+Y} = \mu_X + \mu_Y = 26 + 25 = 51$.

In general, the mean of a set of sums is equal to the sum of the means of the two original sets. Even more generally, if a sum is formed by picking one element from each of several sets, the mean of such sums is simply the sum of the means of the various sets.

How is the variance of W related to the variances of the original sets?

$$\sigma^2_X = \frac{(1-26)^2 + (9-26)^2 + (20-26)^2 + (74-26)^2}{4} = 813.50$$

$$\sigma^2_Y = \frac{(5-25)^2 + (15-25)^2 + (55-25)^2}{3} = 466.67$$

$$\sigma^2_W = \sigma^2_{X+Y} = \frac{(6-51)^2 + (16-51)^2 + \cdots + (129-51)^2}{12} = 1280.17$$

Note that $\sigma^2_W = \sigma^2_{X+Y} = \sigma^2_X + \sigma^2_Y$.

This is true for the variance of any set of sums. More generally, if a sum is formed by picking one element from each of several sets, the variance of such sums is simply the sum of the variances of the various sets.

We can also calculate the standard deviation $\sigma_W = \sqrt{1280.17} = 35.78$ and conclude the average pooled number of chips received by a couple is 51 with a standard deviation of 35.78.

➡ EXAMPLE 8.2

In Example 8.1, what is the average difference in number of chips received from bag X and bag Y?

Answer: Form the set of differences: Z = {1 – 5, 1 – 15, 1 – 55, 9 – 5, 9 – 15, 9 – 55, 20 – 5, 20 – 15, 20 – 55, 74 – 5, 74 – 15, 74 – 55} = {–4, –14, –54, 4, –6, –46, 15, 5, –35, 69, 59, 19}

A quick calculation gives $\mu_Z = 1$. Note that $\mu_Z = \mu_{X-Y} = \mu_X - \mu_Y = 26 - 25 = 1$. In general, the mean of a set of differences is equal to the difference of the means of the two original sets.

How is the variance of Z related to the variance of the original sets?

A calculation (for example, using `1–Var Stats` on the TI-84) yields $\sigma^2_Z = 1280.17$. Wow, the set of differences has the same variance as the set of sums! We have, $\sigma^2_Z = \sigma^2_{X-Y} = \sigma^2_X + \sigma^2_Y$. More generally, if a total is formed by a procedure that adds or subtracts one element from each of several independent sets, the variance of the resulting totals is simply the sum of the variances of the several sets.

Not only can we sum variances as shown above to calculate total variance, but we can also reverse the process and determine how the total variance is split up among its various sources. For example, we can find what portion of the variance in numbers of sales made by a company's sales representatives is due to the individual salesperson, what portion is due to the territory, what portion is due to the particular products sold by each salesperson, and so on.

➡ EXAMPLE 8.3

A company's sports shoes sell for an average of $100 with a standard deviation of $9. The costs going into their sports shoes are as follows:

Retail store costs and profits: average of $45 with a standard deviation of $5.

Brand company costs and profits: average of $24 with a standard deviation of $4.

Advertising and publicity costs: average of $11 with a standard deviation of $3.

Assuming these various costs are independent of one another, what is the mean and standard deviation for the remaining costs R together (which includes materials, transportation, factory worker wages, etc.)?

Answer: $45 + 24 + 11 + \mu_R = 100$ gives $\mu_R = \$20$; and $5^2 + 4^2 + 3^2 + \sigma^2_R = 9^2$ gives $\sigma_R = 31$ and $\sigma_R = \sqrt{31} = \$5.57$.

MEANS AND VARIANCES FOR SUMS AND DIFFERENCES OF RANDOM VARIABLES

The mean of the sum (or difference) of two random variables is equal to the sum (or difference) of the individual means. *If two random variables are independent*, the variance of the sum (or difference) of the two random variables is equal to the *sum* of the two individual variances.

➡ EXAMPLE 8.4

An insurance salesperson estimates the numbers of new auto and home insurance policies she sells per day as follows:

# of auto policies	0	1	2	3
Probability	0.2	0.4	0.3	0.1

# of home policies	0	1	2
Probability	0.5	0.3	0.2

What is the expected value or mean for the overall number of policies sold per day?

Answer: μ_{auto} = (0)(0.2) + (1)(0.4) + (2)(0.3) + (3)(0.1) = 1.3,
μ_{home} = (0)(0.5) + (1)(0.3) + (2)(0.2) = 0.7, and so
$\mu_{total} = \mu_{auto} + \mu_{home}$ = 1.3 + 0.7 = 2.0.

Assuming the selling of new auto policies is independent of the selling of new home policies (which may not be true if some new customers buy both), what would be the standard deviation in the number of new policies sold per day?

Answer: σ^2_{auto} = (0 − 1.3)^2(0.2) + (1 − 1.3)^2(0.4) + (2 − 1.3)^2(0.3) + (3 − 1.3)^2(0.1) = 0.81,
σ^2_{home} = (0 − 0.7)^2(0.5) + (1 − 0.7)^2(0.3) + (2 − 0.7)^2(0.2) = 0.61, and so,
assuming independence, $\sigma^2_{total} = \sigma^2_{auto} + \sigma^2_{home}$ = 0.81 + 0.61 = 1.42 and
$\sigma_{total} = \sqrt{1.42}$ = 1.192.

➡ EXAMPLE 8.5

Vitamin C megadoses are claimed by alternative medicine advocates to have effects on diseases such as cancer and AIDS, as well as on the common cold. In one long-term study, participants received either a placebo or a 200-mg Vitamin C tablet each day. Suppose that the mean compliance rate was 73% with a standard deviation of 24%. Furthermore, the daily intake of Vitamin C through foods among the adult population is 94 g with a standard deviation of 39 g. Among those receiving the supplement, what is the mean and standard deviation for their daily Vitamin C intake?

Answer: With a compliance of 73%, the mean intake of Vitamin C from the supplement is (0.73)(200) = 146 with a standard deviation of (0.24)(200) = 48. Thus, the total mean intake of Vitamin C is 146 + 94 = 240 g with a standard deviation of $\sqrt{48^2 + 39^2}$ = 61.85 g (with independence, variances add!).

➡ EXAMPLE 8.6

Suppose the mean SAT math, critical reading, and writing scores for students at a particular high school are 515, 501, and 493, respectively, with standard deviations of 65, 40, and 38. What can be said about the mean and standard deviation of the three combined scores for these students?

Answer: The mean combined score is 515 + 501 + 493 = 1509. Can we add the three variances, $65^2 + 40^2 + 38^2$ and then take the square root to obtain the desired standard deviation? No, because we don't have independence—students with high scores on one of the three sections will tend to have high scores on the other two sections. The standard deviation of the combined scores cannot be calculated from the given information.

TRANSFORMING RANDOM VARIABLES

NOTE

This is the same as what we observed in Topic 2 about sets of data.

Adding a constant to every value of a random variable will increase the mean by that constant. However, differences remain the same, so measures of variability like the standard deviation will remain unchanged. Multiplying every value of a random variable by a constant will increase the mean by the same multiple. In this case differences are also increased and the standard deviation will increase by the same multiple.

➡ EXAMPLE 8.7

A carnival game of chance has payoffs of $2 with probability 0.5, of $5 with probability 0.4, and of $10 with probability 0.1. What should you be willing to pay? Let's calculate the mean and standard deviation for the winnings if you play this game.

$$\mu_X = E(X) = \sum xP(x) = 2(0.5) + 5(0.4) + 10(0.1) = 4$$

$$\sigma_X = \sqrt{\sum (x-m)^2 P(x)} = \sqrt{(2-4)^2(0.5) + (5-4)^2(0.4) + (10-4)^2(0.1)} = \sqrt{6}$$

What if $4 is added to each payoff?

$$\mu_{X+4} = 6(0.5) + 9(0.4) + 14(0.1) = 8$$

$$\sigma_{X+4} = \sqrt{(6-8)^2(0.5) + (9-8)^2(0.4) + (14-8)^2(0.1)} = \sqrt{6}$$

Note that $\mu_{X+4} = \mu_X + 4$ and $\sigma_{X+4} = \sigma_X$.

What if instead each payoff is tripled?

$$\mu_{3X} = 6(0.5) + 15(0.4) + 30(0.1) = 12$$

$$\sigma_{3X} = \sqrt{(6-12)^2(0.5) + (15-12)^2(0.4) + (30-12)^2(0.1)} = \sqrt{54} = 3\sqrt{6}$$

Note that $\mu_{3X} = 3\mu_X$ and $\sigma_{3X} = 3\sigma_X$.

For random variables X and Y the generalizations for transforming X and for combining X and Y are sometimes written as

Transforming X: $E(X \pm a) = E(X) \pm a$ $\text{var}(X \pm a) = \text{var}(X)$ $\text{SD}(X \pm a) = \text{SD}(X)$

                        $E(bX) = bE(X)$ $\text{var}(bX) = b^2 \text{var}(X)$ $\text{SD}(bX) = b\text{SD}(X)$

Combining X and Y: $E(X \pm Y) = E(X) \pm E(Y)$

$\text{var}(X \pm Y) = \text{var}(X) + \text{var}(Y)$ if X and Y are independent

$SD(X \pm Y) = \sqrt{[SD(X)]^2 + [SD(X)]^2}$ if X and Y are independent

Does it make sense that variances *add* for both sums and differences of independent random variables? *Ranges* may illustrate this point more clearly: suppose prices for primary residences in a community go from \$210,000 to \$315,000, while prices for summer cottages go from \$50,000 to \$120,000. Someone wants to purchase a primary residence and a summer cottage. The minimum *total* spent will be \$260,000, the maximum possible is \$435,000, and thus the range for the total is \$435,000 – \$260,000 = \$175,000. The minimum *difference* spent between the two homes will be \$90,000, the maximum difference will be \$265,000, and thus the range for the difference is \$265,000 – \$90,000 = \$175,000, the same as the range for the total. (Variances and ranges are both measures of spread and behave the same way with regard to the above rule.)

As can be seen above, the rules from algebra do not follow when combining random variables. For example, with random variables, $2X \neq X + X$. From the above formulas, we have $E(2X) = 2E(X)$ and $SD(2X) = 2SD(X)$; however, while $E(X + X) = E(X) + E(X) = 2E(X)$,

$$SD(X + X) = \sqrt{[SD(X)]^2 + [SD(X)]^2} = \sqrt{2[SD(X)]^2} = \sqrt{2}\, SD(X).$$

➡ EXAMPLE 8.8

Let X be what shows when rolling a single die. What are the mean and the standard deviation of X, $2X$, and $X + X$?

X takes the values $\{1, 2, 3, 4, 5, 6\}$ each with probability 1/6, $\mu_x = \Sigma x P(x) = 3.5$ and $\sigma_x = \sqrt{\Sigma (x - \mu)^2 P(x)} \approx 1.7078$.

$2X$ takes the values $\{2, 4, 6, 8, 10, 12\}$ each with probability 1/6, $\mu_{2x} = 7$, and $\sigma_{2x} \approx 3.4156$. Note that $\mu_{2x} = 2\mu_x$ and $\sigma_{2x} = 2\sigma_x$.

But $X + X$ is throwing a die twice and takes the values $\{2, 3, 4, 5, 6, 7, 8, 9, 10, 11, 12\}$ with probabilities $\left\{\frac{1}{36}, \frac{2}{36}, \frac{3}{36}, \frac{4}{36}, \frac{5}{36}, \frac{6}{36}, \frac{5}{36}, \frac{4}{36}, \frac{3}{36}, \frac{2}{36}, \frac{1}{36}\right\}$, respectively.

Then $\mu_{X+X} = 7$ and $\sigma_{X+X} \approx 2.4152$. Note that $\mu_{X+X} = 2\mu_X$ and $\sigma_{X+X} = \sqrt{2}\,\sigma_X$.

More generally, with n independent copies of X, we have

$$E(X + X + \cdots + X) = E(X) + E(X) + \cdots + E(X) = nE(X)$$

and $\text{var}(X + X + \cdots + X) = \text{var}(X) + \text{var}(X) + \text{var}(X) + \cdots + \text{var}(X) = n\,\text{var}(X)$.

So $SD(X + X + \cdots + X) = \sqrt{n}\, SD(X)$, while $SD(nX) = nSD(X)$.

SUMMARY

- When combining two random variables, $E(X + Y) = E(X) + E(Y)$ and $E(X - Y) = E(X) - E(Y)$.
- When combining two independent random variables, the variances always add: $\text{var}(X \pm Y) = \text{var}(X) + \text{var}(Y)$.

QUIZ 1

Multiple-Choice Questions

> **Directions:** The questions that follow are each followed by five suggested answers. Choose the response that best answers the question.

1. An auto dealer offers discounts averaging $2,450 with a standard deviation of $575. If 50 autos are sold during one month, what is the expected value and standard deviation for the total discounts given?

   (A) $E(Total) = \$17,324$ $SD(Total) = \$170$
   (B) $E(Total) = \$17,324$ $SD(Total) = \$4066$
   (C) $E(Total) = \$122,500$ $SD(Total) = \$170$
   (D) $E(Total) = \$122,500$ $SD(Total) = \$4066$
   (E) $E(Total) = \$122,500$ $SD(Total) = \$28,750$

2. Science majors in college pay an average of $650 per year for books with a standard deviation of $130, whereas English majors pay an average of $465 per year for books with a standard deviation of $90. What is the mean difference and standard deviation between the amounts paid for books by science and English majors?

   (A) $E(Diff) = \$92.50$ $SD(Diff) = \$15$
   (B) $E(Diff) = \$185$ $SD(Diff) = \$110$
   (C) $E(Diff) = \$185$ $SD(Diff) = \$158$
   (D) $E(Diff) = \$185$ $SD(Diff) = \$220$
   (E) $E(Diff) = \$557.50$ $SD(Diff) = \$110$

3. The auditor working for a veterinary clinic calculates that the mean annual cost of medical care for dogs is $98 with a standard deviation of $25 and that the mean annual cost of medical care for pets is $212 with an average deviation of $39 for owners who have one dog and one cat. Assuming expenses for dogs and cats are independent for those owning both a dog and a cat, what is the mean annual cost of medical care for cats, and what is the standard deviation?

   (A) $E(Cats) = \$114$ $SD(Cats) = \$14$
   (B) $E(Cats) = \$114$ $SD(Cats) = \$46$
   (C) $E(Cats) = \$114$ $SD(Cats) = \$30$
   (D) $E(Cats) = \$155$ $SD(Cats) = \$32$
   (E) $E(Cats) = \$188$ $SD(Cats) = \$30$

4. Boxes of 50 "donut holes" weigh an average of 16.0 ounces with a standard deviation of 0.245 ounces. If the empty boxes alone weigh an average of 1.0 ounce with a standard deviation of 0.2 ounces, what are the mean and standard deviation of donut hole weights?

(A) $E(Donut\ hole) = 0.3$ oz $SD(Donut\ hole) = 0.0063$ oz
(B) $E(Donut\ hole) = 0.3$ oz $SD(Donut\ hole) = 0.02$ oz
(C) $E(Donut\ hole) = 0.3$ oz $SD(Donut\ hole) = 0.142$ oz
(D) $E(Donut\ hole) = 15$ oz $SD(Donut\ hole) = 0.142$ oz
(E) $E(Donut\ hole) = 15$ oz $SD(Donut\ hole) = 0.445$ oz

Free-Response Questions

> **Directions:** You must show all work and indicate the methods you use. You will be graded on the correctness of your methods and on the accuracy of your final answers.

1. The two most popular women's spring semester sports at a particular school are softball and lacrosse. The softball team plays 0, 1, or 2 days a week with probabilities 0.2, 0.5, and 0.3, respectively.

   (a) What is the mean number of games played by the softball team per week?

   (b) What is the standard deviation for number of games played by the softball team per week?

   The lacrosse team plays a mean of 1.4 days per week with a standard deviation of 0.6.

   (c) What is the mean total number of softball and lacrosse games played per week?

   (d) Assuming independence, what is the standard deviation for the total number of softball and lacrosse games played per week?

2. Weights of coconuts have a mean of 3.2 pounds with a standard deviation of 0.7 pounds.

   (a) If each pound of coconut produces 5 ounces of fruit, calculate the mean and standard deviation for the amount of fruit per coconut.

   (b) If coconuts are packed five per box for shipment, calculate the mean and standard deviation for the total weight of coconuts per box.

QUIZ 2

Multiple-Choice Questions

> **Directions:** The questions that follow are each followed by five suggested answers. Choose the response that best answers the question.

1. Each hospital/physician internet security intrusion results in the theft of an average of 361 patient records with a standard deviation of 74 records. If there are 45 independent intrusions during a one-year period, what is the expected value and standard deviation for the number of patient records stolen?

   (A) $E(thefts) = \sqrt{45} \times 361$ $SD(thefts) = \sqrt{45} \times 74$

   (B) $E(thefts) = \sqrt{45} \times 361$ $SD(thefts) = 45 \times 74$

   (C) $E(thefts) = 45 \times 361$ $SD(thefts) = \sqrt{45} \times 74$

   (D) $E(thefts) = 45 \times 361$ $SD(thefts) = 45 \times 74$

   (E) $E(thefts) = 45 \times 361$ $SD(thefts) = 45 \times 74^2$

2. An auto dealer pays an average of $8,750 with a standard deviation of $1,200 for used car trade-ins, and sells new cars for an average of $28,500 with a standard deviation of $3,100. Assuming independence of trade-in and new car prices for a customer, what is the standard deviation of the revenue the dealer should expect to make if a customer trades in a used car and buys a new one?

   (A) $\sqrt{3100+1200}$ dollars

   (B) $\sqrt{3100-1200}$ dollars

   (C) $\sqrt{3100^2 + 1200^2}$ dollars

   (D) $\sqrt{3100^2 - 1200^2}$ dollars

   (E) $3100 - 1200$ dollars

3. Suppose a retailer knows that the mean number of broken eggs per carton is 0.3 with a standard deviation of 0.18. In a shipment of 100 cartons, what is the expected number of broken eggs and what is the standard deviation? Assume independence between cartons.

   (A) $E(broken) = 3$ $SD(broken) = 1.8$
   (B) $E(broken) = 30$ $SD(broken) = 1.8$
   (C) $E(broken) = 30$ $SD(broken) = 18$
   (D) $E(broken) = 300$ $SD(broken) = 18$
   (E) $E(broken) = 300$ $SD(broken) = 180$

4. A home theater projector system technician knows that based on past experience, for unpacking, assembly, and fine tuning, the mean total setup time is 5.6 hours with a standard deviation of 0.886 hours. The times for the three steps are independent, the mean and standard deviation for the unpacking time are 1.5 hours and 0.2 hours, and for the assembly time are 2.8 hours and 0.85 hours, respectively. What are the mean and standard deviation for the fine tuning time?

(A) $E(Fine\ tuning) = 1.3$ hr $SD(Fine\ tuning) = 0.15$ hr
(B) $E(Fine\ tuning) = 1.3$ hr $SD(Fine\ tuning) = 1.24$ hr
(C) $E(Fine\ tuning) = 2.15$ hr $SD(Fine\ tuning) = 0.164$ hr
(D) $E(Fine\ tuning) = 2.15$ hr $SD(Fine\ tuning) = 1.24$ hr
(E) $E(Fine\ tuning) = 4.61$ hr $SD(Fine\ tuning) = 0.15$ hr

Free-Response Questions

Directions: You must show all work and indicate the methods you use. You will be graded on the correctness of your methods and on the accuracy of your final answers.

1. A packaging company sells "orchard fruit boxes" containing a variety of in-season fruits. These boxes contain 35, 36, or 37 pieces of fruit with probabilities of 0.1, 0.5, and 0.4, respectively.

   (a) What is the average number of pieces of fruit per box?

   (b) What is the standard deviation for the number of pieces of fruit per box?

   Full boxes of 36 pieces of fruit have an average weight of 10 pounds with a standard deviation of 0.5 pounds, while empty boxes have an average weight of 1 pound with a standard deviation of 0.1 pound.

   (c) What is the mean weight of a piece of fruit in these 36-piece boxes?

   (d) What is the standard deviation for the weight of a piece of fruit in these 36-piece boxes?

2. Weights of pineapples have a mean of 900 grams with a standard deviation of 120 grams.

   (a) If each gram of coconut contains 0.01 grams of fiber, calculate the mean and standard deviation for the amount of fiber per pineapple.

   (b) If pineapples are packed four per box for shipment, calculate the mean and standard deviation for the total weight of pineapples per box.

ANSWERS EXPLAINED

Quiz 1

MULTIPLE-CHOICE QUESTIONS

1. **(D)** Expected values and variances add. Thus,

$$E(Total) = E(X_1) + E(X_2) + \cdots + E(X_{50})$$
$$= 2,450 + 2,450 + \cdots + 2,450 = 50(2,450) = 122,500$$

$$SD(Total) = \sqrt{Var(X_1) + Var(X_2) + \cdots + Var(X_{50})}$$
$$= \sqrt{575^2 + 575^2 + \cdots + 575^2} = \sqrt{50(575^2)} = 4065.86$$

2. **(C)** For a set of differences, means subtract, but variances add. Thus,

$$E(Sci - Eng) = E(Sci) - E(Eng) = 650 - 465 = 185$$

$$SD(Sci - Eng) = \sqrt{Var(Sci) + Var(Eng)} = \sqrt{130^2 + 90^2} = 158.11$$

3. **(C)** Means and variances add. Thus,

$$E(D+C) = E(D) + E(C), \quad 212 = 98 + E(C), \quad E(C) = 114$$

$$Var(D+C) = Var(D) + Var(C), \quad 39^2 = 25^2 + Var(C),$$

$$SD(C) = \sqrt{39^2 - 25^2} = 29.93$$

4. **(B)** Means and variances add. Thus,

$$E(Total) = E(Box) + 50E(Hole), \quad 16.0 = 1.0 + 50E(Hole), \quad E(Hole) = 0.3$$

$$Var(Total) = Var(Box) + 50Var(Hole),$$

$$0.245^2 = 0.2^2 + 50(SD(Hole))^2, \quad SD(Hole) = 0.02$$

FREE-RESPONSE QUESTIONS

1. (a) $E(S) = 0(0.2) + 1(0.5) + 2(0.3) = 1.1$

   (b) $var(S) = (0 - 1.1)^2(0.2) + (1 - 1.1)^2(0.5) + (2 - 1.1)^2(0.3) = 0.49$ and $SD(S) = \sqrt{0.49} = 0.7$

   (c) $E(S + L) = E(S) + E(L) = 1.1 + 1.4 = 2.5$

   (d) $var(S + L) = var(S) + var(L) = (0.7)^2 + (0.6)^2 = 0.85$ and $SD(S + L) = \sqrt{0.85} = 0.922$

2. (a) $F = 5C$ (the weight of coconut is being multiplied by a constant)
   $E(F) = E(5C) = 5E(C) = 5(3.2) = 16$ ounces, and
   $SD(F) = SD(5C) = 5SD(C) = 5(0.7) = 3.5$ ounces

   (b) $T = C + C + C + C + C$ (total weight is the sum of the weights of 5 individual
   coconuts) $E(T) = E(C) + E(C) + E(C) + E(C) + E(C) = 5(3.2) = 16$ pounds and
   $var(T) = var(C) + var(C) + var(C) + var(C) + var(C) = 5(0.7)^2 = 2.45$ with
   $SD(T) = \sqrt{2.45} \approx 1.56552$

Quiz 2

MULTIPLE-CHOICE QUESTIONS

1. **(C)** Expected values and variances add. Thus,

$E(thefts) = E(X_1) + E(X_2) + \cdots + E(X_{45}) = 361 + 361 + \cdots + 361 = 45 \times 361$

and $SD(thefts) = \sqrt{74^2 + 74^2 + \cdots + 74^2} = \sqrt{45 \times 74^2} = \sqrt{45} \times 74$

2. **(C)** For a set of differences, means subtract, but variances add. Thus,

$SD(X - Y) = \sqrt{Var(X) + Var(Y)} = \sqrt{3100^2 + 1200^2}$

3. **(B)** Expected values and variances add. Thus,

$E(broken) = E(X_1) + E(X_2) + \cdots + E(X_{100}) = 0.3 + 0.3 + \cdots + 0.3 = 100(0.3) = 30$

$SD(broken) = \sqrt{Var(X_1) + Var(X_2) + \cdots + Var(X_{100})}$
$$= \sqrt{0.18^2 + 0.18^2 + \cdots + 0.18^2} = \sqrt{100(0.18)^2} = 1.8$$

4. **(A)** Means and variances add. Thus,

$E(Un + As + FT) = E(Un) + E(As) + E(FT)$,
    $5.6 = 1.5 + 2.8 + E(FT)$, $E(FT) = 1.3$

$Var(Un + As + FT) = Var(Un) + Var(As) + Var(FT)$,
    $0.886^2 = 0.2^2 + 0.85^2 + [SD(FT)]^2$,
    $SD(FT) = \sqrt{0.886^2 - 0.2^2 - 0.85^2} = 0.15$

FREE-RESPONSE QUESTIONS

1. (a) $E(X) = 35(0.1) + 36(0.5) + 37(0.4) = 36.3$

   (b) $var(X) = (35 - 36.3)^2(0.1) + (36 - 36.3)^2(0.5) + (37 - 36.3)^2(0.4) = 0.41$ and
       $SD(X) = \sqrt{0.41} = 0.640$

   (c) $E(empty) + 36E(fruit) = E(full)$, $1 + 36E(fruit) = 10$, $E(fruit) = \dfrac{10 - 1}{36} = 0.25$ pounds

   (d) $var(empty) + 36\,var(fruit) = var(full)$, $0.1^2 + 36var(fruit) = 0.5^2$,
       $var(fruit) = \dfrac{0.25 - 0.01}{36} = 0.00667$, and $SD(fruit) = \sqrt{0.0067} = 0.082$ pounds

2. (a) $F = 0.01P$ (the weight of pineapple is being multiplied by a constant)
       $E(F) = E(0.01P) = 0.01E(P) = 0.01(900) = 9$ grams, and
       $SD(F) = SD(0.01P) = 0.01SD(P) = 0.01(120) = 1.2$ grams

   (b) $T = P + P + P + P$ (total weight is the sum of the weights of 4 individual pineapples)
       $E(T) = E(P) + E(P) + E(P) + E(P) = 4(900) = 3600$ grams and
       $var(T) = var(P) + var(P) + var(P) + var(P) = 4(120)^2 = 57,600$ with
       $SD(T) = \sqrt{57,600} = 240$ grams

The Normal Distribution

9

- → **PROPERTIES**
- → **USING TABLES**
- → **USING A CALCULATOR**
- → **A MODEL FOR MEASUREMENT**
- → **COMMONLY USED PROBABILITIES AND *Z*-SCORES**
- → **FINDING MEANS AND STANDARD DEVIATIONS**
- → **NORMAL APPROXIMATION TO THE BINOMIAL**
- → **CHECKING NORMALITY**

Some of the most useful probability distributions are symmetric and bell-shaped:

In this topic we study one such distribution called the *normal distribution* (not *all* symmetric, unimodal distributions are normal). The normal distribution is valuable in describing various natural phenomena, especially those involving growth or decay. However, the real importance of the normal distribution in statistics is that it can be used to describe the results of sampling procedures.

> **IMPORTANT**
>
> On the exam, you should rarely call a distribution *normal*. In general, you should say roughly normal or approximately normal. (Real data is never truly normal.)

The normal distribution can be viewed as a limiting case of the binomial. More specifically, we start with any fixed probability of success *p* and consider what happens as *n* becomes arbitrarily large. To obtain a visual representation of the limit, we draw histograms using an increasingly large *n* for each fixed *p*.

Histograms for *p* = 0.5 and for *n* = 4, 10, and 25, respectively

Histograms for $p = 0.1$ and for $n = 4, 10$, and 25, respectively

Histograms for $p = 0.7$ and for $n = 4, 10$, and 25, respectively

For greater clarity, different scales have been used for the histograms associated with different values of n. From the diagrams above it is reasonable to accept that as n becomes larger without bound, the resulting histograms approach a smooth bell-shaped curve. This is the curve associated with the normal distribution.

PROPERTIES OF THE NORMAL DISTRIBUTION

The normal distribution curve is bell-shaped and symmetric and has an infinite base. Long, flat-looking tails cover many values but only a small proportion of the area. The flat appearance of the tails is deceptive. Actually, far out in the tails, the curve drops proportionately at an ever-increasing rate. In other words, when two intervals of equal length are compared, the one closer to the center may experience a greater numerical drop, but the one further out in the tail experiences a greater drop when measured as a proportion of the height at the beginning of the interval.

The mean here is the same as the median and is located at the center. We want a unit of measurement that applies equally well to any normal distribution, and we choose a unit that arises naturally out of the curve's shape. There is a point on each side where the slope is steepest. These two points are called *points of inflection*, and the distance from the mean to either point is precisely equal to one standard deviation. Thus, it is convenient to measure distances under the normal curve in terms of z-scores (recall from Topic 2 that z-scores are fractions or multiples of standard deviations from the mean).

| Normal curves with the same mean but different standard deviations | Normal curves with the same standard deviation but different means |

Mathematically, the formula for the normal curve turns out to be $y = e^{-z^2/2}$, where y is the relative height above z-score. (By *relative height* we mean proportion of the height above the mean.) However, our interest is not so much in relative heights under the normal curve as it is in *proportionate areas*. The probability associated with any interval under the normal curve is equal to the proportionate area found under the curve and above the interval.

A useful property of the normal curve is that approximately 68% of the area (and thus 68% of the observations) falls within one standard deviation of the mean, approximately 95% of the area falls within two standard deviations of the mean, and approximately 99.7% of the area falls within three standard deviations of the mean.

TIP

The notation N(μ,σ) stands for a normal model with mean μ and standard deviation σ.

NOTE

This "68–95–99.7 Rule" is called the Empirical Rule.

USING TABLES OF THE NORMAL DISTRIBUTION

Table A in the Appendix shows areas under the normal curve. Specifically, this table shows the area to the left of a given point. It shows, for example, that to the left of a z-score of 1.2 there is 0.8849 of the area:

Because the total area is 1, the area to the right of a point can be calculated by calculating 1 minus the area to the left of the point. For example, the area to the right of a z-score of -2.13 is $1 - 0.0166 = 0.9834$. (Alternatively, by symmetry, we could have said that the area to the right of -2.13 is the same as that to the left of $+2.13$ and then read the answer directly from Table A.)

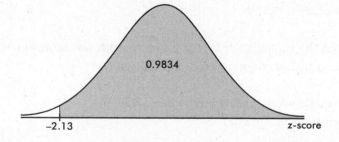

The area between two points can be found by subtraction. For example, the area between the z-scores of 1.23 and 2.71 is equal to $0.9966 - 0.8907 = 0.1059$.

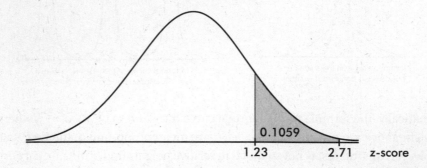

Given a probability, we can use Table A in reverse to find the appropriate z-score. For example, to find the z-score that has an area of 0.8982 to the left of it, we search for the probability closest to 0.8982 in the bulk of the table and read off the corresponding z-score. In this case we find 0.8980 with the z-score 1.27.

To find the z-score with an area of 0.72 to the right of it, we look for $1 - 0.72 = 0.28$ in the bulk of the table. The probability 0.2810 corresponds to the z-score -0.58.

USING A CALCULATOR WITH AREAS UNDER A NORMAL CURVE

On the TI-84, `normalcdf`(lowerbound, upperbound) gives the area (probability) between two z-scores, while `invNorm`(area) gives the z-score with the given area to its left. So we could have calculated above:

> `normalcdf`(–100, 1.2) = 0.8849
> `normalcdf`(–2.13, 100) = 0.9834
> `normalcdf`(1.23, 2.71) = 0.1060
> `invNorm`(.8982) = 1.271
> `invNorm`(.28) = –0.5828

The TI-84 also has the capability of working directly with raw scores instead of z-scores. In this case, the mean and standard deviation must be given:

> `normalcdf`(lowerbound, upperbound, mean, SD)
> `invNorm`(area, mean, SD)

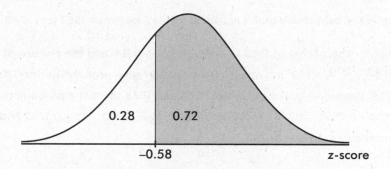

0.28   0.72

−0.58   z-score

THE NORMAL DISTRIBUTION AS A MODEL FOR MEASUREMENT

In solving a problem using the normal distribution as a model, drawing a picture of a normal curve with horizontal lines showing raw scores and z-scores is usually helpful.

➥ **EXAMPLE 9.1**

The life expectancy of a particular brand of lightbulb is roughly normally distributed with a mean of 1500 hours and a standard deviation of 75 hours.

| 1275 | 1350 | 1425 | 1500 | 1575 | 1650 | 1725 | Hours |
| −3 | −2 | −1 | 0 | 1 | 2 | 3 | z-score |

(a) What is the probability that a lightbulb will last less than 1410 hours?

Answer: The z-score of 1410 is $\frac{1410 - 1500}{75}$ = −1.2, and from Table A the probability to the left of −1.2 is 0.1151. [On the TI-84, normalcdf(0, 1410, 1500, 75) = 0.1151 and normalcdf(−10, −1.2) = 0.1151.]

0.1151

1410   1500   Hours

−1.2   0   z-score

(continued)

TIP

Don't use the normal model if the distribution isn't symmetric and unimodal.

TIP

Draw a picture!

TIP

On the exam, the distribution and parameters must be identified. It is sufficient to write: Normal, μ = 1500, σ = 75, $P(X < 1410) = 0.1151$

(b) What is the probability that a lightbulb will last between 1563 and 1648 hours?

Answer: The *z*-score of 1563 is $\frac{1563 - 1500}{75}$ = 0.84. and the *z*-score of 1648 is $\frac{1648 - 1500}{75}$ = 1.97. In Table A, 0.84 and 1.97 give probabilities of 0.7995 and 0.9756, respectively. Thus between 1563 and 1648 there is a probability of 0.9756 − 0.7995 = 0.1761. [normalcdf(1563, 1648, 1500, 75) = 0.1762 and normalcdf(.84, 1.97) = 0.1760.]

TIP

On the exam, it is sufficient to write: Normal, μ = 1500, σ = 75, $P(1563 < X < 1648)$ = 0.1762.

TIP

On the exam, it is also OK to write: normalcdf (lower bd = 1416, upper bd = 1677, μ = 1500, σ = 75) = 0.8595.

(c) What is the probability that a lightbulb will last between 1416 and 1677 hours?

Answer: The *z*-score of 1416 is $\frac{1416 - 1500}{75}$ = −1.2 and the *z*-score of 1677 is $\frac{1677 - 1500}{75}$ = 2.36. In Table A, −1.12 gives a probability of 0.1314, and 2.36 gives a probability of 0.9909. Thus between 1416 and 1677 there is a probability of 0.9909 − 0.1314 = 0.8595. [normalcdf(1416, 1677, 1500, 75) = 0.8595 and normalcdf(−1.12, 2.36) = 0.8595.]

⮕ **EXAMPLE 9.2** _____

A packing machine is set to fill a cardboard box with a mean of 16.1 ounces of cereal. Suppose the amounts per box form an approximately normal distribution with a standard deviation equal to 0.04 ounce.

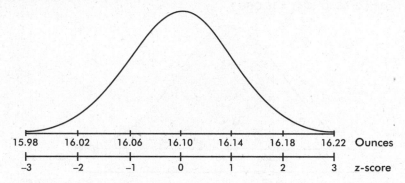

(a) What percentage of the boxes will end up with at least 1 pound of cereal?

Answer: The *z*-score of 16 is $\frac{16 - 16.1}{0.04}$ = –2.5, and –2.5 in Table A gives a probability of 0.0062. Thus the probability of having more than 1 pound in a box is 1 – 0.0062 = 0.9938 or 99.38%. [normalcdf(16, 1000, 16.10, .04) = 0.9938 and normalcdf(–2.5, 10) = 0.9938.]

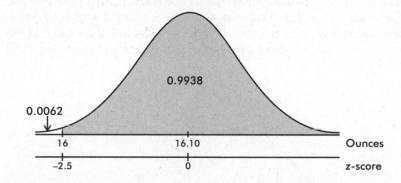

TIP

On the exam, it is sufficient to write:
Normal, μ = 16.1, σ = 0.04, P($X \geq$ 16) = 0.9938 or 99.38%.

(b) Ten percent of the boxes will contain less than what number of ounces?

Answer: In Table A, we note that .1 area (actually 0.1003) is found to the left of a –1.28 *z*-score. [On the TI-84, invNorm(.1) = –1.2816.] Converting the *z*-score of –1.28 into a raw score yields –1.28 = $\frac{x - 16.1}{0.04}$ or 16.1 – 1.28(0.04) = 16.049 ounces. [Or directly on the TI-84, invNorm(.1, 16.10, .04) = 16.049.]

TIP

On the exam, it is also OK to write:
invNorm (area = 0.1, μ = 16.10, σ = 0.04) = 16.049.

(continued)

(c) Eighty percent of the boxes will contain more than what number of ounces?

Answer: In Table A, we note that 1 – 0.8 = 0.2 area (actually 0.2005) is found to the left of a *z*-score of –0.84, and so to the right of a –0.84 *z*-score must be 80% of the area. [Or invNorm(.2) = 0.8416.] Converting the *z*-score of –0.84 into a raw score yields 16.1 – 0.84(0.04) = 16.066 ounces. [Or directly on the TI-84, invNorm(.2, 16.10, .04) = 16.066.]

TIP

On the exam, it is OK to write: invNorm(area = 0.05, μ = 16.10, σ = 0.04) = 16.034 and invNorm(area = 0.95, μ = 16.10, σ = 0.04) = 16.166.

(d) The middle 90% of the boxes will be between what two weights?

Answer: Ninety percent in the middle leaves five percent in each tail. In Table A, we note that .0505 area is found to the left of a *z*-score of –1.64, while 0.495 area is found to the left of a *z*-score of –1.65, and so we use the *z*-score of –1.645. Then 90% of the area is between *z*-scores of –1.645 and 1.645. [Or invNorm(.05) = –1.6449 and invNorm(.95) = 1.6449.] Converting *z*-scores to raw scores yields 16.1 – 1.645(0.04) = 16.034 ounces and 16.1 + 1.645(0.04) = 16.166 ounces, respectively, for the two weights between which we will find the middle 90% of the boxes. [invNorm(.05, 16.10, .04) = 16.034 and invNorm(.95, 16.10, .04) = 16.166.]

COMMONLY USED PROBABILITIES AND *Z*-SCORES

As can be seen from Example 9.2*d*, there is often an interest in the limits enclosing some specified middle percentage of the data. For future reference, the limits most frequently asked for are noted below in terms of *z*-scores.

NOTE

The 2σ corresponding to 95% in the 68–95–99.7 rule is an approximation to the more exact 1.96σ.

Ninety percent of the values are between z-scores of -1.645 and $+1.645$, 95% of the values are between z-scores of -1.96 and $+1.96$, and 99% of the values are between z-scores of -2.576 and $+2.576$.

NOTE

These critical z-scores can be found at the bottom of Table B.

Sometimes the interest is in values with particular percentile rankings. For example,

Ninety percent of the values are below a z-score of 1.282, 95% of the values are below a z-score of 1.645, and 99% of the values are below a z-score of 2.326.

NOTE

These calculations hold true under the assumption that you have a normal distribution.

There are corresponding conclusions for negative z-scores:

Ninety percent of the values are above a z-score of -1.282, 95% of the values are above a z-score of -1.645, and 99% of the values are above a z-score of -2.326.

It is also useful to note the percentages corresponding to values falling between integer z-scores. For example,

Note that 68.26% of the values are between z-scores of -1 and $+1$, 95.44% of the values are between z-scores of -2 and $+2$, and 99.74% of the values are between z-scores of -3 and $+3$.

➡ EXAMPLE 9.3

Suppose that the average height of adult males in a particular locality is 70 inches with a standard deviation of 2.5 inches.

(a) If the distribution is approximately normal, the middle 95% of males are between what two heights?

Answer: As noted above, the critical z-scores in this case are ±1.96, and so the two limiting heights are 1.96 standard deviations from the mean. Therefore, $70 \pm 1.96(2.5) = 70 \pm 4.9$, or from 65.1 to 74.9 inches.

(b) Ninety percent of the heights are below what value?

Answer: The critical z-score is 1.282, and so the value in question is $70 + 1.282(2.5) = 73.205$ inches.

(c) Ninety-nine percent of the heights are above what value?

Answer: The critical z-score is -2.326, and so the value in question is $70 - 2.326(2.5) = 64.185$ inches.

NOTE

The Empirical Rule's 68%, 95%, and 99.7% are approximations to 68.26%, 95.44%, and 99.74%.

(d) What percentage of the heights are between z-scores of ±1? Of ±2? Of ±3?

Answer: 68.26%, 95.44%, and 99.74%, respectively.

FINDING MEANS AND STANDARD DEVIATIONS

If we know that a distribution is normal, we can calculate the mean μ and the standard deviation σ using percentage information from the population.

➡ **EXAMPLE 9.4** _____

Given a normal distribution with a mean of 25, what is the standard deviation if 18% of the values are above 29?

Answer: Looking for a 0.82 probability in Table A (or use invNorm (0.82)), we note that the corresponding z-score is 0.92. Thus 29 – 25 = 4 is equal to 0.92 standard deviations; that is, $0.92\sigma = 4$, and $\sigma = \dfrac{4}{0.92} = 4.35$.

➡ **EXAMPLE 9.5** _____

Given a normal distribution with a standard deviation of 10, what is the mean if 21% of the values are below 50?

Answer: Looking for a 0.21 probability in Table A (or using invNorm) leads us to a z-score of –0.81. Thus 50 is –0.81 standard deviation from the mean, and so $\mu = 50 + 0.81(10) = 58.1$.

EXAMPLE 9.6

Given a normal distribution with 80% of the values above 125 and 90% of the values above 110, what are the mean and standard deviation?

Answer: Table A (or using invNorm) gives critical z-scores of –0.84 and –1.28. Thus we have $\frac{125-\mu}{\sigma}$ = –0.84 and $\frac{110-\mu}{\sigma}$ = –1.28. Solving the system {125 – μ = –0.84σ, 110 – μ = –1.28σ} simultaneously gives μ = 153.64 and σ = 34.09.

NORMAL APPROXIMATION TO THE BINOMIAL

Many practical applications of the binomial involve examples in which n is large. However, for large n, binomial probabilities can be quite tedious to calculate. Since the normal can be viewed as a limiting case of the binomial, it is natural to use the normal to approximate the binomial in appropriate situations.

The binomial takes values only at integers, while the normal is continuous with probabilities corresponding to areas over intervals. Therefore, we establish a technique for converting from one distribution to the other. For approximation purposes we do as follows. Each binomial probability corresponds to the normal probability over a unit interval centered at the desired value. Thus, for example, to approximate the binomial probability of eight successes we determine the normal probability of being between 7.5 and 8.5.

TIP

When n is small, you cannot use the normal model to approximate the binomial model.

EXAMPLE 9.7

Suppose that 15% of the cars coming out of an assembly plant have some defect. In a delivery of 40 cars what is the probability that exactly 5 cars have defects?

Answer: The actual answer is $\left(\frac{40!}{35!5!}\right)(0.15)^5(0.85)^{35}$ [binompdf(40, 0.15, 5) = 0.169].

To approximate the answer using the normal, we first calculate the mean μ and the standard deviation σ as follows:

$$\mu = np = 40(0.15) = 6$$

$$\sigma = \sqrt{np(1-p)} = \sqrt{40(0.15)(0.85)} = 2.258$$

NOTE

To use a normal for a binomial, we would like $\mu - 3\sigma \geq 0$, that is, $np - 3\sqrt{npq} \geq 0$, so $np \geq 3\sqrt{npq}$, $n^2p^2 \geq 9npq$, and $np \geq 9q$. But $9q \leq 9 < 10$, so we ask for $np \geq 10$.

(continued)

$P(4.5 < X < 5.5) = 0.159$

4.5 5.5 6   Cars with defect

−0.66 −0.22 0   z-score

Even more useful are approximations relating to probabilities over intervals.

➥ EXAMPLE 9.8

NOTE

binomcdf (250, .6,155) = 0.760

If 60% of the population support massive federal budget cuts, what is the probability that in a survey of 250 people at most 155 people support such cuts?

Answer: The actual answer is the sum of 156 binomial expressions:

$$(0.4)^{250} + \cdots + \frac{250!}{155!95!}(0.6)^{155}(0.4)^{95}$$

A good approximation can be obtained by using the normal. We calculate μ and σ:

$$\mu = np = 250(0.6) = 150$$

NOTE

normalcdf (−0.5, 155.5, 150, 7.746) = 0.761

$$\sigma = \sqrt{np(1-p)} = \sqrt{250(0.6)(0.4)} = 7.746$$

The binomial of at most 155 successes corresponds to the normal probability of ≤155.5.

$$P\left(z < \frac{155.5 - 150}{7.746}\right) = 0.761$$

Is the normal a good approximation? The answer, of course, depends on the error tolerances in particular situations. A general rule of thumb is that the normal is a good approximation to the binomial whenever both np and $n(1 - p)$ are at least 10. (Some authors use 5 instead of 10 here.)

CHECKING NORMALITY

In the examples in this Topic we have assumed the population has a normal distribution. When you collect your own data, before you can apply the procedures developed above, you must decide whether it is reasonable to assume the data come from a normal population. In later Topics we will have to check for normality before applying certain inference procedures.

The initial check should be to draw a picture. Dotplots, stemplots, and histograms are all useful graphical displays to show that data is unimodal and roughly symmetric.

➡ EXAMPLE 9.9

The ages at inauguration of U.S. presidents from Washington to G. W. Bush were: {57, 61, 57, 57, 58, 57, 61, 54, 68, 51, 49, 64, 50, 48, 65, 52, 56, 46, 54, 49, 51, 47, 55, 55, 54, 42, 51, 56, 55, 51, 54, 51, 60, 61, 43, 55, 56, 61, 52, 69, 64, 46, 54}. Can we conclude that the distribution is roughly normal?

Answer: Entering the 43 data points in a graphing calculator gives the following histogram:

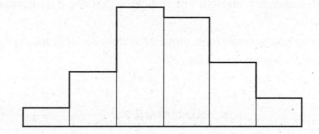

Alternatively, we could have used a dotplot or stemplot.

```
4 | 2 3
4 | 6 6 7 8 9 9
5 | 0 1 1 1 1 1 2 2 4 4 4 4 4
5 | 5 5 5 5 6 6 6 7 7 7 7 8
6 | 0 1 1 1 1 4 4
6 | 5 8 9
```

The histogram, dotplot, and stemplot do indicate a distribution that is *roughly* unimodal and symmetric; note that it is difficult to come to such a conclusion from just a boxplot.

A more specialized graphical display to check normality is the *normal probability plot*. When the data distribution is roughly normal, the plot is roughly a diagonal straight line. While this plot more clearly shows deviations from normality, it is not as easy to understand as a histogram. The normal probability plot is difficult to calculate by hand; however, technology such as the TI-84 readily plots the graph. (On the TI-84, in STATPLOT the sixth choice in TYPE gives the normal probability plot.)

The normal probability plot for the data in Example 9.9 is:

A graph this close to a diagonal straight line indicates that the data have a distribution very close to normal.

Note that alternatively, one could have plotted the data (in this case, inaugural ages) on the *x*-axis and the normal scores on the *y*-axis.

SUMMARY

- The normal distribution is one particular symmetric, unimodal distribution.
- Drawing pictures of normal curves with horizontal lines showing raw scores and *z*-scores is usually helpful.
- Areas (probabilities) under a normal curve can be found using the given table or calculator functions, such as `normalcdf` on TI calculators, Ncd on the Casio Prizm, and NORMAL_CDF on the HP Prime.
- Critical values corresponding to given probabilities can be found using the given table or calculator functions, such as `invNorm` on TI calculators, InvN on the Casio Prizm, and NORMALD_ICDF on the HP Prime.
- The normal probability plot is very useful in gauging whether a given distribution of data is approximately normal: If the plot is nearly straight, the data is nearly normal.

QUIZ 1

Multiple-Choice Questions

> **Directions:** The questions that follow are each followed by five suggested answers. Choose the response that best answers the question.

1. Which of the following is a true statement?

   (A) The area under a normal curve is always equal to 1, no matter what the mean and standard deviation are.
   (B) All bell-shaped curves are normal distributions for some choice of μ and σ.
   (C) The smaller the standard deviation of a normal curve, the lower and more spread out the graph.
   (D) Depending upon the value of the standard deviation, normal curves with different means may be centered around the same number.
   (E) Depending upon the value of the standard deviation, the mean and median of a particular normal distribution may be different.

2. Populations P1 and P2 are normally distributed and have identical means. However, the standard deviation of P1 is twice the standard deviation of P2. What can be said about the percentage of observations falling within two standard deviations of the mean for each population?

   (A) The percentage for P1 is twice the percentage for P2.
   (B) The percentage for P1 is greater, but not twice as great, as the percentage for P2.
   (C) The percentage for P2 is twice the percentage for P1.
   (D) The percentage for P2 is greater, but not twice as great, as the percentage for P1.
   (E) The percentages are identical.

In Questions 3–7, assume the given distributions are roughly normal.

3. A trucking firm determines that its fleet of trucks averages a mean of 12.4 miles per gallon with a standard deviation of 1.2 miles per gallon on cross-country hauls. What is the probability that one of the trucks averages fewer than 10 miles per gallon?

   (A) 0.0082
   (B) 0.0228
   (C) 0.4772
   (D) 0.5228
   (E) 0.9772

4. An electronic product takes an average of 3.4 hours to move through an assembly line. If the standard deviation is 0.5 hour, what is the probability that an item will take between 3 and 4 hours?

(A) 0.2119
(B) 0.2295
(C) 0.3270
(D) 0.3811
(E) 0.6730

5. The average noise level in a restaurant is 30 decibels with a standard deviation of 4 decibels. Ninety-nine percent of the time it is below what value?

(A) 30 + 1.96(4)
(B) 30 − 2.576(4)
(C) 30 + 2.576(4)
(D) 30 − 2.326(4)
(E) 30 + 2.326(4)

6. One company produces movie trailers with mean 150 seconds and standard deviation 40 seconds, while a second company produces trailers with mean 120 seconds and standard deviation 30 seconds. What is the probability that two randomly selected trailers, one produced by each company, will combine to less than three minutes?

(A) 0.000
(B) $P\left(z < \dfrac{180-270}{50}\right)$
(C) $P\left(z < \dfrac{180-270}{70}\right)$
(D) $P\left(z < \dfrac{180-270}{\sqrt{40+30}}\right)$
(E) $P\left(z < \dfrac{180-270}{\sqrt{\dfrac{1}{40}+\dfrac{1}{30}}}\right)$

7. Cucumbers grown on a certain farm have weights with a standard deviation of 2 ounces. What is the mean weight if 85% of the cucumbers weigh less than 16 ounces?

(A) 13.92
(B) 14.30
(C) 14.40
(D) 14.88
(E) 15.70

Free-Response Questions

> **Directions:** You must show all work and indicate the methods you use. You will be graded on the correctness of your methods and on the accuracy of your final answers.

1. Scores on a university entrance exam are approximately normally distributed with a mean of 150 and a standard deviation of 25.

   (a) The university will accept 90% of applicants (based on entrance exam score). What is the cutoff score for acceptance (rounded to nearest integer)?

   (b) Given that a student scores over 100, what is the probability of acceptance?

   (c) Given that three random students all score over 100, what is the probability that at least one is not accepted?

2. The time it takes Steve to walk to school follows an approximately normal distribution with mean 30 minutes and standard deviation 5 minutes, while the time it takes Jan to walk to school follows an approximately normal distribution with mean 25 minutes and standard deviation 4 minutes. Assume their walking times are independent of each other.

   (a) If they leave at the same time, what is the probability that Steve arrives before Jan?

   (b) How much earlier than Jan should Steve leave so that he has a 90% chance of arriving before Jan?

Multiple-Choice Questions

> **Directions:** The questions that follow are each followed by five suggested answers. Choose the response that best answers the question.

1. Which of the following is a true statement?

   (A) The area under the standard normal curve between 0 and 2 is twice the area between 0 and 1.
   (B) The area under the standard normal curve between 0 and 2 is half the area between –2 and 2.
   (C) For the standard normal curve, the interquartile range is approximately 3.
   (D) For the standard normal curve, the range is 6.
   (E) For the standard normal curve, the area to the left of 0.1 is the same as the area to the right of 0.9.

2. Consider the following two normal curves:

   Which has the larger mean and which has the larger standard deviation?

   (A) Larger mean, *a*; larger standard deviation, *a*
   (B) Larger mean, *a*; larger standard deviation, *b*
   (C) Larger mean, *b*; larger standard deviation, *a*
   (D) Larger mean, *b*; larger standard deviation, *b*
   (E) Larger mean, *b*; same standard deviation

In Questions 3–7, assume the given distributions are roughly normal.

3. A factory dumps an average of 2.43 tons of pollutants into a river every week. If the standard deviation is 0.88 tons, what is the probability that in a week more than 3 tons are dumped?

    (A) $P\left(z > \dfrac{3-2.43}{0.88}\right)$

    (B) $P\left(z > \dfrac{2.43-3}{0.88}\right)$

    (C) $P\left(z > \dfrac{\frac{3-2.43}{0.88}}{\sqrt{7}}\right)$

    (D) $P\left(z > \dfrac{\frac{2.43-3}{0.88}}{\sqrt{7}}\right)$

    (E) $P(z > (3 - 2.43))$

4. The mean score on a college placement exam is 500 with a standard deviation of 100. Ninety-five percent of the test takers score above what?

    (A) 260
    (B) 336
    (C) 405
    (D) 414
    (E) 664

5. The mean income per household in a certain state is $9500 with a standard deviation of $1750. The middle 95% of incomes are between what two values?

    (A) $5422 and $13,578
    (B) $6070 and $12,930
    (C) $6621 and $12,379
    (D) $7260 and $11,740
    (E) $8049 and $10,951

6. Jay Olshansky from the University of Chicago was quoted in *Chance News* as arguing that for the average life expectancy to reach 100, 18% of people would have to live to 120. What standard deviation is he assuming for this statement to make sense?

    (A) 21.7
    (B) 24.4
    (C) 25.2
    (D) 35.0
    (E) 111.1

7. A coffee machine can be adjusted to deliver any fixed number of ounces of coffee. If the machine has a standard deviation in delivery equal to 0.4 ounce, what should be the mean setting so that an 8-ounce cup will overflow only 0.5% of the time?

(A) 6.97 ounces
(B) 7.22 ounces
(C) 7.34 ounces
(D) 7.80 ounces
(E) 9.03 ounces

Free-Response Questions

> **Directions:** You must show all work and indicate the methods you use. You will be graded on the correctness of your methods and on the accuracy of your final answers.

1. (a) Two components are in series so the failure of either will cause the system to fail. The time to failure for a new component is approximately normally distributed with a mean of 3000 hours and a standard deviation of 400 hours. One of the components has already run for 2500 hours, while the other has run for 2800 hours. Assuming component failures are independent, what is the probability the system survives for 10 more days (240 hours)?

   (b) If the components are put in parallel, the system will fail only if both components fail. If the two components in (a) are put in parallel, what is the probability that the system survives for 10 more days?

2. Many people have lactose intolerance leading to digestive problems when they eat dairy products. Substantial relief can be obtained by taking dietary supplements of lactase, an enzyme. One product consists of caplets containing a mean of 9000 FCC lactase units with a standard deviation of 590 units and an approximately normal distribution of lactase units. People with severe lactose intolerance may need to take two such tablets whenever they eat dairy products. Tablets with under 8500 FCC lactase units can produce noticeably less relief.

   (a) Determine the probability that a caplet has less than 8500 FCC lactase units. Round off to the nearest tenth.

   (b) Simulations are run to determine the number of tablets a lab technician analyzes before finding two tablets with a specified number of units. The results of two 100-trial simulations, one looking for two tablets each less than 8500 units and one looking for two tablets each greater than 9000 units, are shown below. Which distribution is which? Explain your answer.

ANSWERS EXPLAINED

Quiz 1

MULTIPLE-CHOICE QUESTIONS

1. **(A)** The area under any probability distribution is equal to 1. Many bell-shaped curves are *not* normal curves. The smaller the standard deviation of a normal curve, the higher and narrower the graph. The mean determines the value around which the curve is centered; different means give different centers. Because of symmetry, the mean and median are identical for normal distributions.

2. **(E)** All normal distributions have about 95% of their observations within two standard deviations of the mean.

3. **(B)** [normalcdf(0, 10, 12.4, 1.2) = 0.0228.] (Or: The z-score of 10 is $\frac{10-12.4}{1.2} = -2$. From Table A, to the left of –2 is an area of 0.0228.)

4. **(E)** [normalcdf(3, 4, 3.4, .5) = 0.6731.] (Or: The z-scores of 3 and 4 are $\frac{3-3.4}{0.5} = -0.8$ and $\frac{4-3.4}{0.5} = 1.2$, respectively. From Table A, to the left of –0.8 is an area of 0.2119, and to the left of 1.2 is an area of 0.8849. Thus between 3 and 4 is an area of 0.8849 – 0.2119 = 0.6730.)

5. **(E)** The critical z-score associated with 99% to the left is 2.326.

6. **(B)** $\mu_{X+Y} = 150 + 120 = 270$ and $\sigma_{X+Y} = \sqrt{40^2 + 30^2} = 50$

7. **(A)** The critical z-score associated with 85% to the left is 1.04. Then $\mu + 1.04(2) = 16$ gives $\mu = 13.92$.

FREE-RESPONSE QUESTIONS

1. (a) $P\left(z > \frac{x-150}{25}\right) = 0.90$ so $\frac{x-150}{25} = -1.282$ and $x = 117.95$ for a cutoff score of 118.

   (b) $P(x > 118 \mid x > 100) = \frac{P(x > 118)}{P(x > 100)} = \frac{0.900}{0.9772} = 0.921$

   (c) $1 - (0.921)^3 = 0.219$

2. (a) Let the random variable D be the difference in walking times (Steve–Jan). Then $\mu_D = 30 - 25 = 5$ and $\sigma_D = \sqrt{5^2 + 4^2} = 6.403$. For Steve to arrive before Jan, the difference in walking times must be < 0.

   $P(D < 0) = P\left(z < \frac{0-5}{6.403}\right) = 0.217$

   (b) If m is the number of minutes he should leave early, then the new mean of the differences is $5 - m$, and we want $P\left(z < \frac{0-(5-m)}{6.403}\right) = 0.9$, which gives $\frac{-5+m}{6.403} = 1.282$ and $m = 13.2$ minutes.

Quiz 2

MULTIPLE-CHOICE QUESTIONS

1. **(B)** Statement (B) is true by symmetry of the normal curve, however 0.4772 is not twice 0.3413, 0.67 – (–0.67) is not 3, the range is not finite, and the $P(z < 0.1)$ is more than 0.5 while $P(z > 0.9)$ is less than 0.5.

2. **(C)** Curve a has mean 6 and standard deviation 2, while curve b has mean 18 and standard deviation 1.

3. **(A)** The z-score of 3 is $\frac{3-2.43}{0.88}$.

4. **(B)** [invNorm(.05, 500, 100) = 336.] (Or: If 95% of the area is to the right of a score, 5% is to the left. Looking for 0.05 in the body of Table A, we note the z-score of –1.645. Converting this to a raw score gives 500 – 1.645(100) = 336.)

5. **(B)** The critical z-scores associated with the middle 95% are ±1.96, and 9500 ± 1.96(1750) = 6070 and 12,930. [invNorm(.025, 9500, 1750) = 6070 and invNorm(.975, 9500, 1750) = 12,930.]

6. **(A)** The critical z-score associated with 18% to the right or 82% to the left is 0.92. Then $100 + 0.92\sigma = 120$ gives $\sigma = 21.7$.

7. **(A)** The critical z-score associated with 0.5% to the right (99.5% to the left) is 2.576. Then $c + 2.576(0.4) = 8$ gives $c = 6.97$.

FREE-RESPONSE QUESTIONS

1. (a) $P(2740|2500) = \dfrac{P\left(z > \dfrac{2740-3000}{400}\right)}{P\left(z > \dfrac{2500-3000}{400}\right)} = \dfrac{0.742154}{0.894350} = 0.829825$ and

   $P(3040|2800) = \dfrac{P\left(z > \dfrac{3040-3000}{400}\right)}{P\left(z > \dfrac{2800-3000}{400}\right)} = \dfrac{0.460172}{0.691462} = 0.665506$

   Given independence, the probability that both components last 240 more hours is $(0.829825)(0.665506) = 0.552$.

   (b) The probability that both fail is $(1 - 0.829825)(1 - 0.665506) = 0.056923$.
   The probability that at least one doesn't fail is $1 - 0.056923 \approx 0.943$.

2. (a) $P\left(z < \dfrac{8500-9000}{590}\right) \approx 0.2$

   (b) The probability of a caplet having more than 9000 units is 0.5, and thus one would expect to find two caplets with more than 9000 units much quicker than finding two caplets with less than 8500 units. Thus the first histogram results from looking for two caplets with more than 9000 units and the second histogram results from looking for two caplets with less than 8500 units.

Sampling Distributions

10

→ **SAMPLE PROPORTION**

→ **SAMPLE MEAN**

→ **CENTRAL LIMIT THEOREM**

→ **DIFFERENCE BETWEEN TWO INDEPENDENT SAMPLE PROPORTIONS**

→ **DIFFERENCE BETWEEN TWO INDEPENDENT SAMPLE MEANS**

→ **THE *t*-DISTRIBUTION**

→ **THE CHI-SQUARE DISTRIBUTION**

→ **THE STANDARD ERROR**

→ **BIAS AND ACCURACY**

The *population* is the complete set of items of interest. A *sample* is a part of a population used to represent the population. The population mean μ and population standard deviation σ are examples of *population parameters*. The sample mean $\bar{x}$ and the sample standard deviation s are examples of *statistics*. Statistics are used to make inferences about population parameters. While a population parameter is a fixed quantity, statistics vary depending on the particular sample chosen. The probability distribution showing how a statistic varies is called a *sampling distribution*. The sampling distribution is *unbiased* if its mean is equal to the associated population parameter.

> **IMPORTANT**
>
> A sampling distribution is not the same thing as the distribution of a sample.

We want to know some important truth about a population, but in practical terms this truth is unknowable. What's the average adult human body weight? What proportion of people have high cholesterol? What we can do is carefully collect data from as large and representative a group of individuals as possible and then use this information to estimate the value of the population parameter. How close are we to the truth? We know that different samples would give different estimates, and so sampling error is unavoidable. What is wonderful, and what we will learn in this Topic, is that we can quantify this sampling error! We can make statements like "the average weight must almost surely be within 5 pounds of 176 pounds" or "the proportion of people with high cholesterol must almost surely be within ±3% of 37%."

SAMPLING DISTRIBUTION OF A SAMPLE PROPORTION

Whereas the mean is basically a quantitative measurement, the proportion represents essentially a qualitative approach. The interest is simply in the presence or absence of some attribute. We count the number of yes responses and form a proportion. For example, what proportion of drivers wear seat belts? What proportion of SCUD missiles can be intercepted? What proportion of new smart phones have a certain software problem?

This separation of the population into "haves" and "have-nots" suggests that we can make use of our earlier work on binomial distributions. We also keep in mind that, when n (trials, or in this case sample size) is large enough, the binomial can be approximated by the normal.

In this topic we are interested in estimating a population proportion p by considering a single sample proportion $\hat{p}$. This sample proportion is just one of a whole universe of sample proportions, and to judge its significance we must know how sample proportions vary. Consider the set of proportions from all possible samples of a specified size n. It seems reasonable that these proportions will cluster around the population proportion (the sample proportion is an unbiased statistic of the population proportion) and that the larger the chosen sample size, the tighter the clustering.

How do we calculate the mean and standard deviation of the set of sample proportions? Suppose the sample size is n and the actual population proportion is p. From our work on binomial distributions, we remember that the mean and standard deviation for the number of successes in a given sample are pn and $\sqrt{np(1-p)}$, respectively, and for large n the complete distribution begins to look "normal."

Here, however, we are interested in the proportion rather than in the number of successes. From Topic Two we remember that when we multiply or divide every element by a constant, we multiply or divide both the mean and the standard deviation by the same constant. In this case, to change number of successes to proportion of successes, we divide by n:

$$\mu_{\hat{p}} = \frac{pn}{n} = p \quad \text{and} \quad \sigma_{\hat{p}} = \frac{\sqrt{np(1-p)}}{n} = \sqrt{\frac{p(1-p)}{n}}$$

Furthermore, if each element in an approximately normal distribution is divided by the same constant, it is reasonable that the result will still be an approximately normal distribution.

Thus the principle forming the basis of the following discussion is

---

Start with a population with a given proportion p. Take all samples of size n. Compute the proportion in each of these samples. Then

1. the set of all sample proportions is approximately normally distributed (often stated: the distribution of sample proportions is approximately normal).
2. the mean $\mu_{\hat{p}}$ of the set of sample proportions equals p, the population proportion.
3. the standard deviation $\sigma_{\hat{p}}$ of the set of sample proportions is approximately equal to $\sqrt{\frac{p(1-p)}{n}}$.

Alternatively, we say that the sampling distribution of $\hat{p}$ is approximately normal with mean p and standard deviation $\sqrt{\frac{p(1-p)}{n}}$.

NOTE

That is, the sample size should be large enough so that the expected number of "successes" and the expected number of "failures" in the sample are both at least 10.

Since we are using the normal approximation to the binomial, both np and $n(1-p)$ should be at least 10. Furthermore, in making calculations and drawing conclusions from a specific sample, it is important that the sample be a *random sample*.

Finally, because sampling is usually done without replacement, the sample cannot be too large; the sample size n should be no larger than 10% of the population. (We're actually worried about *independence*, but randomly selecting a relatively small sample allows us to assume independence. Of course, it's always better to have larger samples—it's just that if the sample is large relative to the population, then the proper inference techniques are different from those taught in introductory statistics classes.)

TIP

It is sufficient to write: Normal with $\mu_{\hat{p}}$ = 0.80 and $\sigma_{\hat{p}}$ = 0.0381; $P(\hat{p} < 0.75)$ = 0.0947.

➡ EXAMPLE 10.1

It is estimated that 80% of people with high math anxiety experience brain activity similar to that experienced under physical pain when anticipating doing a math problem. In a simple random sample of 110 people with high math anxiety, what is the probability that less than 75% experience the physical pain brain activity?

Answer: The sample is given to be random, both np = (110)(0.80) = 88 ≥ 10 and $n(1 - p)$ = (110)(0.20) = 22 ≥ 10, and our sample is clearly less than 10% of all people with math anxiety. So the sampling distribution of $\hat{p}$ is approximately normal with mean 0.80 and standard deviation $\sigma_{\hat{p}} = \sqrt{\dfrac{(0.80)(0.20)}{110}} = 0.0381$. With a z-score of $\dfrac{0.75 - 0.80}{0.0381} = -1.312$, the probability that the sample proportion is less than 0.75 is `normalcdf(-1000,-1.312)` = 0.0948.

[Or normalcdf(–1000, .75, .80, .0381) = 0.0947.]

SAMPLING DISTRIBUTION OF A SAMPLE MEAN

Suppose we are interested in estimating the mean μ of a population. For our estimate we could simply randomly pick a single element of the population, but then we would have little confidence in our answer. Suppose instead that we pick 100 elements and calculate their average. It is intuitively clear that the resulting sample mean has a greater chance of being closer to the mean of the whole population than the value for any individual member of the population does.

When we pick a sample and measure its mean $\bar{x}$, we are finding exactly one sample mean out of a whole universe of sample means. To judge the significance of a single sample mean, we must know how sample means vary. Consider the set of means from all possible samples of a specified size. It is both apparent and reasonable that the sample means are clustered around the mean of the whole population; furthermore, these sample means have a tighter clustering than the elements of the original population. In fact, we might guess that the larger the chosen sample size, the tighter the clustering.

How do we calculate the standard deviation $\sigma_{\bar{x}}$ of the set of sample means? Suppose the variance of the population is σ^2 and we are interested in samples of size n. Sample means are obtained by first summing together n elements and then dividing by n. A set of sums has a variance equal to the sum of the variances associated with the original sets. In our case, $\sigma^2_{sums} = \sigma^2 + \cdots + \sigma^2 = n\sigma^2$. When each element of a set is divided by some constant, the new variance is the old one divided by the square of the constant. Since the sample means are obtained by dividing the sums by n, the variance of the sample means is obtained by dividing the variance of the sums by n^2. Thus if $\sigma_{\bar{x}}$ symbolizes the standard deviation of the sample means, we find that

$$\sigma_{\bar{x}}^2 = \frac{\sigma^2_{sums}}{n^2} = \frac{n\sigma^2}{n^2} = \frac{\sigma^2}{n}$$

In terms of standard deviations, we have $\sigma_{\bar{x}} = \frac{\sigma}{\sqrt{n}}$.

We have shown the following:

> Start with a population with a given mean μ and standard deviation σ. Compute the mean of all samples of size n. Then the mean of the set of sample means will equal μ, the mean of the population, and the standard deviation $\sigma_{\bar{x}}$ of the set of sample means will equal $\frac{\sigma}{\sqrt{n}}$, that is, the standard deviation of the whole population divided by the square root of the sample size. Alternatively, we say that the sampling distribution of $\bar{x}$ has mean μ and standard deviation $\frac{\sigma}{\sqrt{n}}$.

Note that the variance of the set of sample means varies directly as the variance of the original population and inversely as the size of the samples, while the standard deviation of the set of sample means varies directly as the standard deviation of the original population and inversely as the square root of the size of the samples.

➥ EXAMPLE 10.2

Emergency room visits after drinking energy drinks is skyrocketing. One particular energy drink has an average of 200 mg of caffeine with a standard deviation of 10 mg. A store sells boxes of six bottles each. What is the mean and standard deviation of the average milligrams of caffeine consumers should expect from the six bottles in each box?

Answer: We have samples of size 6. The mean of these sample means will equal the population mean of 200 mg. The standard deviation of these sample means will equal $\sigma_{\bar{x}} = \frac{10}{\sqrt{6}} = 4.08$ mg.

Note that while giving the mean and standard deviation of the set of sample means, we did not describe the shape of the distribution. If we are also given that the original population is normal, then we can conclude that the set of sample means has a normal distribution.

➡ EXAMPLE 10.3

It is estimated that cats, who live in the wild or as indoor pets allowed to roam outdoors, kill an average of 10 birds a year. Assuming an approximately normal distribution with a standard deviation of 3 birds, what is the probability an SRS of 25 cats will kill an average of between 9 and 12 birds in a year?

Answer: The sampling distribution of the sample means is approximately normal (because the original population is approximately normal) with mean $\mu_{\bar{x}} = 10$ and standard deviation $\sigma_{\bar{x}} = \dfrac{3}{\sqrt{25}} = 0.6$. The z-scores of 9 and 12 are $\dfrac{9-10}{0.6} = -1.667$ and $\dfrac{12-10}{0.6} = 3.333$, and the probability of a sample mean between 9 and 12 is `normalcdf(-1.667,3.333)` $= 0.952$. [Or `normalcdf(9,12,10,0.6)` $= 0.952$.]

NOTE

On the exam, it is sufficient to write: `normalcdf (lower bd = 9, upper bd = 12,` $\mu_{\bar{x}} = 10$, $\sigma_{\bar{x}} = \dfrac{3}{\sqrt{25}}) = 0.952$

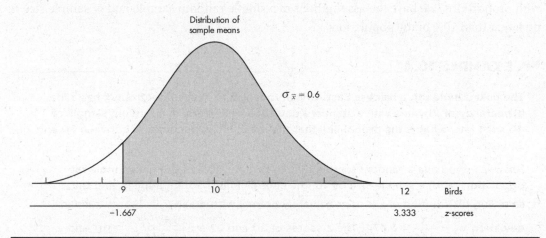

Distribution of sample means

$\sigma_{\bar{x}} = 0.6$

9	10	12 Birds
−1.667		3.333 z-scores

CENTRAL LIMIT THEOREM

We assumed above that the original population had a normal distribution. Unfortunately, few populations are normal, let alone exactly normal. However, it can be shown mathematically that no matter how the original population is distributed, if *n* is large enough, then the set of sample means is approximately normally distributed. For example, there is no reason to suppose that the amounts of money that different people spend in grocery stores are normally distributed. However, if each day we survey 30 people leaving a store and determine the average grocery bill, these daily averages will have a nearly normal distribution.

The following principle forms the basis of much of what we discuss in this topic and in those following. It is a simplified statement of the *central limit theorem* of statistics.

Start with a population with a given mean μ, a standard deviation σ, and any shape distribution whatsoever. Pick n sufficiently large (at least 30) and take all samples of size n. Compute the mean of each of these samples. Then

1. the set of all sample means is approximately normally distributed (often stated: the distribution of sample means is approximately normal).
2. the mean of the set of sample means equals μ, the mean of the population.
3. the standard deviation $\sigma_{\bar{x}}$ of the set of sample means is equal to $\frac{\sigma}{\sqrt{n}}$, that is, equal to the standard deviation of the whole population divided by the square root of the sample size.

Alternatively, we say that for sufficiently large n the sampling distribution of $\bar{x}$ is approximately normal with mean μ and standard deviation $\frac{\sigma}{\sqrt{n}}$.

While we mention $n \geq 30$ as a rough rule of thumb, $n \geq 40$ is often used, and n should be chosen even larger if more accuracy is required or if the original population is far from normal. As with proportions, we have the assumptions of a simple random sample and of sample size n no larger than 10% of the population.

➡ **EXAMPLE 10.4** _____

The naked mole rat, a hairless East African rodent that lives underground, has a life expectancy of 21 years with a standard deviation of 3 years. In a random sample of 40 such rats, what is the probability that the mean life expectancy is between 20 and 22 years?

Answer: We have a random sample that is less than 10% of the naked mole rat population. With a sample size of 40, the central limit theorem applies, and the sampling distribution of $\bar{x}$ is approximately normal with mean $\mu_{\bar{x}} = 21$ and standard deviation $\sigma_{\bar{x}} = \frac{3}{\sqrt{40}} = 0.474$. The z-scores of 20 and 22 are $\frac{20-21}{0.474} = -2.110$ and $\frac{22-21}{0.474} = 2.110$, the probability of sample mean between 20 and 22 is

`normalcdf(-2.110,2.110) = 0.965`. [Or `normalcdf(20,22,21,0.474)= 0.965`.]

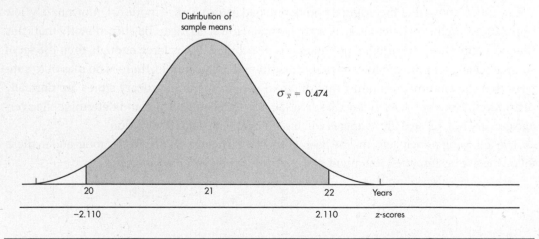

➡ EXAMPLE 10.5

Being bilingual appears to slow the onset of dementia by an average of 4.5 years with a standard deviation of 2 years. In a random sample of 75 bilingual people with dementia, the probability is 0.90 that the average delay in the onset of their dementia was at least how many years?

Answer: We have a random sample that is less than 10% of the total bilingual dementia population. With a sample size of 75, the central limit theorem applies, and the sampling distribution of $\bar{x}$ is approximately normal with mean $\mu_{\bar{x}} = 4.5$ and standard deviation $\sigma_{\bar{x}} = \dfrac{2}{\sqrt{75}} = 0.231$. The critical z-score is `invNorm(0.10)=` -1.282 with a corresponding raw score of $4.5 - 1.282(0.231) = 4.20$ years. [Or `invNorm(.10,4.5,.231)= 4.20`.]

Don't be confused by the several different distributions being discussed! First, there's the distribution of the original population, which may be uniform, bell-shaped, strongly skewed—anything at all. Second, there's the distribution of the data in the sample, and the larger the sample size, the more this will look like the population distribution. Third, there's the distribution of the means of many samples of a given size, and the amazing fact is that this *sampling distribution* can be described by a normal model, regardless of the shape of the original population.

SAMPLING DISTRIBUTION OF A DIFFERENCE BETWEEN TWO INDEPENDENT SAMPLE PROPORTIONS

Numerous important and interesting applications of statistics involve the comparison of two population proportions. For example, is the proportion of satisfied purchasers of American automobiles greater than that of buyers of Japanese cars? How does the percentage of surgeons recommending a new cancer treatment compare with the corresponding percentage of oncologists? What can be said about the difference between the proportion of single parents on welfare and the proportion of two-parent families on welfare?

Our procedure involves comparing two sample proportions. When is a difference between two such sample proportions significant? Note that we are dealing with one difference from the set of all possible differences obtained by subtracting sample proportions of one population from sample proportions of a second population. To judge the significance of one particular difference, we must first determine how the differences vary among themselves. Remember that the variance of a set of differences is equal to the sum of the variances of the individual sets; that is,

$$\sigma_d^2 = \sigma_1^2 + \sigma_2^2$$

Now if

$$\sigma_1 = \sqrt{\frac{p_1(1-p_1)}{n_1}} \quad \text{and} \quad \sigma_2 = \sqrt{\frac{p_2(1-p_2)}{n_2}}$$

then

$$\sigma_d^2 = \frac{p_1(1-p_1)}{n_1} + \frac{p_2(1-p_2)}{n_2} \quad \text{and} \quad \sigma_d = \sqrt{\frac{p_1(1-p_1)}{n_1} + \frac{p_2(1-p_2)}{n_2}}$$

Then we have the following about the sampling distribution of $\hat{p}_1 - \hat{p}_2$:

Start with two populations with given proportions p_1 and p_2. Take all samples of sizes n_1 and n_2, respectively. Compute the difference $\hat{p}_1 - \hat{p}_2$ of the two proportions in each pair of samples. Then

1. the set of all differences of sample proportions is approximately normally distributed (alternatively stated: the distribution of differences of sample proportions is approximately normal).
2. the mean of the set of differences of sample proportions equals $p_1 - p_2$, the difference of population proportions.
3. the standard deviation σ_d of the set of differences of sample proportions is approximately equal to

$$\sqrt{\frac{p_1(1-p_1)}{n_1} + \frac{p_2(1-p_2)}{n_2}}$$

TIP

Whenever doing two-population work, you must check and state that the two populations are independent.

Since we are using the normal approximation to the binomial, $n_1 p_1$, $n_1(1 - p_1)$, $n_2 p_2$, and $n_2(1 - p_2)$ should all be at least 10. Furthermore, in making calculations and drawing conclusions from specific samples, it is important both that the samples be *simple random samples* and that they be taken *independently* of each other. Finally, each sample cannot be too large; the sample sizes should be no larger than 10% of the populations.

➡ EXAMPLE 10.6

In a study of how environment affects our eating habits, scientists revamped one of two close-to-each-other fast-food restaurants with table cloths, candlelight, and soft music. They then noted that at the revamped restaurant customers ate slower and 25% left at least 100 calories of food on their plates. At the unrevamped garish restaurant, customers tended to wolf down their food, and only 19% left at least 100 calories of food on their plates. In a random sample of 110 customers at the revamped restaurant and an independent random sample of 120 customers at the unrevamped restaurant, what is the probability that the difference in the percentages of customers in the revamped setting and unrevamped setting is more than 10% (where the difference is the revamped restaurant percent minus the unrevamped restaurant percent)?

Answer: We have independent random samples, each less than 10% of all fast-food customers, and we note that $n_1 p_1 = 110(0.25) = 27.5$, $n_1(1 - p_1) = 110(0.75) = 82.5$, $n_2 p_2 = 120(0.19) = 22.8$, and $n_2(1 - p_2) = 120(0.81) = 97.2$ are all ≥ 10. Thus, the sampling distribution of $\hat{p}_1 - \hat{p}_2$ is roughly normal with mean $\mu_{\hat{p}_1 - \hat{p}_2} = 0.25 - 0.19 = 0.06$ and

standard deviation $\sigma_{\hat{p}_1 - \hat{p}_2} = \sqrt{\dfrac{(0.25)(0.75)}{110} + \dfrac{(0.19)(0.81)}{120}} = 0.0547$. The *z*-score of

0.10 is $\dfrac{0.10 - 0.06}{0.0547} = 0.731$, and `normalcdf(.731,1000)= 0.232`.

[Or `normalcdf(0.10,1.0,0.06,0.0547)= 0.232`.]

SAMPLING DISTRIBUTION OF A DIFFERENCE BETWEEN TWO INDEPENDENT SAMPLE MEANS

Many real-life applications of statistics involve comparisons of two population means. For example, is the average weight of laboratory rabbits receiving a special diet greater than that of rabbits on a standard diet? Which of two accounting firms pays a higher mean starting salary? Is the life expectancy of a coal miner less than that of a school teacher?

First we consider how to compare the means of samples, one from each population. When is a difference between two such sample means significant? The answer is more apparent when we realize that what we are looking at is one difference from a set of differences. That is, there is the set of all possible differences obtained by subtracting sample means from one set from sample means from a second set. To judge the significance of one particular difference we must first determine how the differences vary among themselves. The necessary key is the fact that the variance of a set of differences is equal to the sum of the variances of the individual sets. Thus,

$$\sigma^2_{\bar{x}_1 - \bar{x}_2} = \sigma^2_{\bar{x}_1} + \sigma^2_{\bar{x}_2}$$

Now if

$$\sigma_{\bar{x}_1} = \frac{\sigma_1}{\sqrt{n_1}} \quad \text{and} \quad \sigma_{\bar{x}_2} = \frac{\sigma_2}{\sqrt{n_2}}$$

then

$$\sigma^2_{\bar{x}_1 - \bar{x}_2} = \frac{\sigma_1^2}{n_1} + \frac{\sigma_2^2}{n_2} \quad \text{and} \quad \sigma_{\bar{x}_1 - \bar{x}_2} = \sqrt{\frac{\sigma_1^2}{n_1} + \frac{\sigma_2^2}{n_2}}$$

TIP

Don't forget to check that the samples are independent!

Then we have the following about the sampling distribution of $\bar{x}_1 - \bar{x}_2$:

> Start with two normal populations with means μ_1 and μ_2 and standard deviations σ_1 and σ_2. Take all samples of sizes n_1 and n_2, respectively. Compute the difference $\bar{x}_1 - \bar{x}_2$ of the two means in each pair of these samples. Then
>
> 1. the set of all differences of sample means is approximately normally distributed (alternatively stated: the distribution of differences of sample means is approximately normal).
> 2. the mean of the set of differences of sample means equals $\mu_1 - \mu_2$, the difference of population means.
> 3. the standard deviation $\sigma_{\bar{x}_1 - \bar{x}_2}$ of the set of differences of sample means is approximately equal to $\sqrt{\dfrac{\sigma_1^2}{n_1} + \dfrac{\sigma_2^2}{n_2}}$.

The more either population varies from normal, the greater should be the corresponding sample size. We also have the assumptions of independent simple random samples, and of sample sizes no larger than 10% of the populations.

➡ EXAMPLE 10.7

It is estimated that 40-year-old men contribute an average of 65 genetic mutations to their new children, whereas 20-year-old men contribute an average of only 25. Assuming standard deviations of 15 and 5 mutations respectively for the 40- and 20-year-olds, what is the probability that the mean number of mutations in a random sample of thirty-five 40-year-old new fathers is between 35 and 45 more than the mean number in a random sample of forty 20-year-old new fathers?

Answer: We have independent random samples, each less than 10% of their age groups, and both sample sizes are over 30, so the sampling distribution of $\bar{x}_1 - \bar{x}_2$ is roughly normal with mean $\mu_{\bar{x}_1 - \bar{x}_2} = 65 - 25 = 40$ and standard deviation

$\sigma_{\bar{x}_1 - \bar{x}_2} = \sqrt{\dfrac{15^2}{35} + \dfrac{5^2}{40}} = 2.656$. The *z*-scores of 35 and 45 are $\dfrac{35 - 40}{2.656} = -1.883$ and $\dfrac{45 - 40}{2.656} = 1.883$, and `normalcdf(-1.883,1.883)= 0.940`.

[Or `normalcdf(35,45,40,2.656)= 0.940`.]

Distribution of differences of sample means

$\sigma_{\bar{x}_1 - \bar{x}_2} = 2.656$

35   40   45

−1.883   1.883   z-scores

➡ EXAMPLE 10.8 _____

The average number of missed school days for students going to public schools is 8.5 with a standard deviation of 4.1, while students going to private schools miss an average of 5.3 with a standard deviation of 2.9. In an SRS of 200 public school students and an SRS of 150 private school students, with a probability of 0.95 the difference in average missed days (public average minus private average) is above what number?

Answer: We have large (n_1 = 200 and n_2 = 150) SRSs that are still smaller than 10% of all public and private school students. The mean of the differences is 8.5 – 5.3 = 3.2, while the standard deviation is $\sqrt{\dfrac{(4.1)^2}{200} + \dfrac{(2.9)^2}{150}} = 0.374$. An area of 0.95 to the right corresponds to a *z*-score of –1.645. The difference in days is 3.2 – 1.645(0.374) = 2.6. Thus there is a 0.95 probability that the difference in average missed days between public and private school samples is over 2.6 days. [invNorm(.05, 3.2, .374) = 2.585.]

THE *t*-DISTRIBUTION

When the population standard deviation σ is unknown, we use the sample standard deviation s as an estimate for σ. But then $\dfrac{\bar{x} - \mu}{s/\sqrt{n}}$ does not follow a normal distribution. If, however, **the original population is normally distributed**, there is a distribution that can be used when working with the $s/\sqrt{n}$ ratios. This *Student t-distribution* was introduced in 1908 by W. S. Gosset, a British mathematician employed by the Guinness Breweries. (When we are working with small samples from a population that is *not* nearly normal, we must use very different "nonparametric" techniques not discussed in this review book.)

Thus, for a sample from a normally distributed population, we work with the variable

$$t = \frac{\bar{x} - \mu}{s/\sqrt{n}}$$

with a resulting *t*-distribution that is bell-shaped and symmetric, but lower at the mean, higher at the tails, and so more spread out than the normal distribution.

Like the binomial distribution, the *t*-distribution is different for different values of *n*. In the tables these distinct *t*-distributions are associated with the values for degrees of freedom (*df*). For this discussion the *df* value is equal to the sample size minus 1. The smaller the *df* value, the larger the dispersion in the distribution. The larger the *df* value, that is, the larger the sample size, the closer the distribution to the normal distribution.

Since there is a separate *t*-distribution for each *df* value, fairly complete tables would involve many pages; therefore, in Table B of the Appendix we list areas and *t*-values for only the more commonly used percentages or probabilities. The last row of Table B is the normal distribution, which is a special case of the *t*-distribution taken when *n* is infinite. For practical purposes, the two distributions are very close for any $n \geq 30$ (some use $n \geq 40$). However, when more accuracy is required, the Student *t*-distribution can be used for much larger values of *n*, and the calculations are easy with technology.

Note that, whereas Table A gives areas under the normal curve to the left of given *z*-values, Table B gives areas to the right of given positive *t*-values. For example, suppose the sample size is 20, and so $df = 20 - 1 = 19$. Then a probability of 0.05 in the tail corresponds with a *t*-value of 1.729, while 0.01 in the tail corresponds to $t = 2.539$.

Thus the *t*-distribution is the proper choice whenever the population standard deviation σ is unknown. In the real world σ is almost always unknown, and so we should almost always use the *t*-distribution. In the past this was difficult because extensive tables were necessary for the various *t*-distributions. However, with calculators today, this problem has diminished. It is no longer necessary to assume that the *t*-distribution is close enough to the *z*-distribution whenever *n* is greater than the arbitrary number 30.

The issue of sample size is refined even further by some statisticians:

Use the
t-distribution
with an SRS
of size *n* and

large *n* ($\geq$40) — Unnecessary to make any assumptions about parent population.

medium *n* (15–40) — Sample should show no extreme values and little, if any, skewness; or assume parent population is normal.

small *n* ($\leq$15) — Sample should show no outliers and no skewness; or assume parent population is normal.

IMPORTANT

For $\dfrac{\bar{x} - \mu}{s/\sqrt{n}}$ to have a *t*-distribution, we actually need that the sampling distribution of $\bar{x}$ is normal. This follows either if the original population has a normal distribution or if the sample size is large enough (from the central limit theorem).

THE CHI-SQUARE DISTRIBUTION

There is a family of sampling distribution models, called the *chi-square models*, which we will use for testing (not confidence intervals). The chi-square distribution has a parameter called degrees of freedom (df). For small df the distribution is skewed to the right; however, for large df it becomes more symmetric and bell-shaped (as does the t-distribution). For one or two degrees of freedom the peak occurs at 0, while for three or more degrees of freedom the peak is at $df-2$.

The chi-square distribution, like the t-distribution, has a separate curve for each df value; therefore, in Table C of the Appendix we list critical chi-square values for only the more commonly used tail probabilities. The chi-square distribution has numerous applications, two of the best known of which are goodness of fit of an observed distribution to a theoretical one, and independence of two criteria of classification of qualitative data. While we will be using chi-square for categorical data, the chi-square distribution as seen above is a *continuous* distribution, and applying it to counting data is just an approximation.

NOTE

Also of interest are that the mean = *df* and the variance = 2*df*.

THE STANDARD ERROR

With proportions and means we typically do not know population parameters. So in calculating standard deviations of the sampling models, we actually estimate using sample statistics. In this case, we use the term *standard error*. That is, for proportions,

$$\sigma_{\hat{p}} = \sqrt{\frac{pq}{n}}, \text{ and we have } SE(\hat{p}) = \sqrt{\frac{\hat{p}\hat{q}}{n}}$$

Similarly, for means,

$$\sigma_{\bar{x}} = \frac{\sigma}{\sqrt{n}}, \text{ and we have } SE(\bar{x}) = \frac{s}{\sqrt{n}}$$

BIAS AND ACCURACY

Understand that a sampling distribution is a theoretical concept that never happens in real data analysis. When all we have is an estimate, know it is just that, only an estimate. Of course it's wrong, but it's the best we've got. This leads to a discussion of *bias* and *accuracy*.

In Topic 5, we discussed *bias* in the context of surveys as a tendency to favor the selection of certain members of a population. In the context of sampling distributions, *bias* means that the sampling distribution is not *centered* on the population parameter. The sampling distributions of proportions and means are *unbiased*, that is, for a given sample size, the set of all sample proportions is centered on the population proportion, and the set of all sample means is centered on the population mean.

The other important, quantifiable judgement of how good an estimate is is *accuracy*, or how narrow the sampling distribution is. For both proportions and means, this is inversely proportional to $\sqrt{n}$, the square root of the sample size.

SUMMARY

- Provided that the sample size n is large enough, the sampling distribution of sample proportions is approximately normal with mean p and standard deviation $\sqrt{\dfrac{p(1-p)}{n}}$ (where p is the population proportion).
- Provided that the sample size n is large enough, the sampling distribution of sample means is approximately normal with mean μ and standard deviation $\dfrac{\sigma}{\sqrt{n}}$ (where μ and σ are the population mean and standard deviation).
- When the population standard deviation σ is unknown (which is almost always the case), and provided that the original population is normally distributed or the sample size n is large enough, the statistic $\dfrac{\bar{x}-\mu}{s/\sqrt{n}}$ follows a *t*-distribution with $n-1$ degrees of freedom.

QUIZ 1

Multiple-Choice Questions

> **Directions:** The questions that follow are each followed by five suggested answers. Choose the response that best answers the question.

1. When selecting a random sample from a population, which of the following combinations of n and p satisfies the conditions for the sampling distribution of $\hat{p}$ to be approximately normal?

   (A) $n = 10$ and $p = 0.67$
   (B) $n = 25$ and $p = 0.32$
   (C) $n = 50$ and $p = 0.18$
   (D) $n = 75$ and $p = 0.15$
   (E) $n = 100$ and $p = 0.08$

2. Which of the following is a true statement?

   (A) The larger the sample, the larger the spread in the sampling distribution.
   (B) Bias has to do with the spread of a sampling distribution.
   (C) Provided that the population size is significantly greater than the sample size, the spread of the sampling distribution does not depend on the population size.
   (D) Sample parameters are used to make inferences about population statistics.
   (E) Statistics from smaller samples have less variability.

3. Which of the following is an *incorrect* statement?

   (A) The sampling distribution of $\bar{x}$ has mean equal to the population mean μ even if the population is not normally distributed.
   (B) The sampling distribution of $\bar{x}$ has standard deviation $\frac{\sigma}{\sqrt{n}}$ even if the population is not normally distributed.
   (C) The sampling distribution of $\bar{x}$ is normal if the population has a normal distribution.
   (D) When n is large, the sampling distribution of $\bar{x}$ is approximately normal even if the population is not normally distributed.
   (E) The larger the value of the sample size n, the closer the standard deviation of the sampling distribution of $\bar{x}$ is to the standard deviation of the population.

4. Which of the following is the best reason that the sample maximum is not used as an estimator for the population maximum?

(A) The sample maximum is biased.
(B) The sampling distribution of the sample maximum is not binomial.
(C) The sampling distribution of the sample maximum is not normal.
(D) The sampling distribution of the sample maximum has too large a standard deviation.
(E) The sample mean plus three sample standard deviations gives a much superior estimate for the population maximum.

5. Thirty-four percent of Americans say that math is the most important subject in school. In a random sample of 400 Americans, what is the probability that between 30% and 35% will say that math is the most important subject?

(A) 0.291
(B) 0.337
(C) 0.382
(D) 0.618
(E) 0.709

6. It is estimated that school CO_2 levels average 1200 ppm with a standard deviation of 300 ppm. In a random sample of 30 schools, what is the probability that the mean CO_2 level is more than 1000, a level at which some researchers feel will cause a drop in academic performance?

(A) $P\left(z > \dfrac{1000-1200}{\frac{300}{\sqrt{30}}}\right)$

(B) $P\left(z > \dfrac{1000-1200}{300}\right)$

(C) $P\left(z > \dfrac{1200-1000}{\frac{300}{\sqrt{30}}}\right)$

(D) $P\left(z > \dfrac{1200-1000}{300}\right)$

(E) $P\left(t > \dfrac{1200-1000}{\frac{300}{\sqrt{30}}}\right)$ with $df = 30 - 1$

7. Binge drinking is a serious problem, killing more than 1700 college students per year. Suppose the incidence of binge drinking is 43% and 37%, respectively, at two large universities. In a random sample of 75 students at the first school and 80 students at the second, what is the probability that the difference (first minus second) between the percentages of binge drinkers is between 5% and 10%?

   (A) 0.176
   (B) 0.245
   (C) 0.473
   (D) 0.527
   (E) 0.755

8. Which of the following statements are true?

   I. Like the normal, t-distributions are always symmetric.
   II. Like the normal, t-distributions are always mound-shaped.
   III. The t-distributions have less spread than the normal, that is, they have less probability in the tails and more in the center than the normal.

   (A) II only
   (B) I and II only
   (C) I and III only
   (D) II and III only
   (E) I, II, and III

9. A study is to be performed to estimate the proportion of voters who believe the economy is "heading in the right direction." Which of the following pairs of sample size and population proportion p will result in the smallest variance for the sampling distribution of $\hat{p}$?

   (A) $n = 100$ and $p = 0.1$
   (B) $n = 100$ and $p = 0.5$
   (C) $n = 100$ and $p = 0.99$
   (D) $n = 1000$ and $p = 0.1$
   (E) $n = 1000$ and $p = 0.5$

10. In a recent study of fatal automobile crashes, for drivers ages 21–24, 32% were legally drunk, while for drivers ages 25–34, 29% were legally drunk. A simulation is conducted in which a random sample of 400 drivers ages 21–24 and a random sample of 400 drivers 25–34, all involved in fatal automobile crashes, are selected. The difference in proportions of those who were legally drunk is calculated, and this simulation is repeated 1000 times. Which of the following is most likely to occur in the simulated sampling distribution of the difference between the two sample proportions?

(A)

(B)

(C)

(D)

(E)

Free-Response Questions

1. Suppose that 16% of teenagers who play video games would enjoy playing a new Super Mario game. The manufacturer doesn't know this actual proportion, and plans on sampling a random sample of 80 teenage game players.

   (a) Calculate $\sigma_{\hat{p}}$, the standard deviation of the sampling distribution of $\hat{p}$.

   (b) Are the conditions for inference met? Explain.

   (c) If a different sample size would result in $\sigma_{\hat{p}} = 0.037$, is the sample size smaller or larger than 80? Explain.

2. It is estimated that 58% of all Americans sleep on their sides.

   (a) What is the probability that a randomly chosen American sleeps on his/her side?

   (b) In a random sample of five Americans, what is the probability that at least three sleep on their sides?

   (c) In a random sample of 350 Americans, what is the probability that at least 50% sleep on their sides?

3. The amount of fuel used by jumbo jets to take off is approximately normally distributed with a mean of 4000 gallons and a standard deviation of 125 gallons.

   (a) What is the probability that a randomly selected jumbo jet will need more than 4250 gallons of fuel to take off?

   (b) What are the shape, mean, and standard deviation of the sampling distribution of the mean gallons of fuel to take off of 40 randomly selected jumbo jets?

   (c) In a random sample of 40 jumbo jets, what is the probability that the mean number of gallons of fuel needed to take off is less than 3950?

   (d) Would the answer to (a), (b), or (c) be affected if the original population of gallons of fuel used by jumbo jets to take off were skewed instead of normal? Explain.

4. Five AP Statistics students want to extend their knowledge beyond the standard course content and read about the *finite population correction factor*, $\sqrt{\dfrac{N-n}{N-1}}$, where N is the population size, and n is the sample size. To study this concept, they each make picks for the 32 games in the first round of the March Madness Men's Division 1 basketball tournament. They note that the number of winners each picked gives a set of size $N = 5$ consisting of the elements 6, 9, 11, 13, and 21.

(a) Determine the population mean μ and the population standard deviation σ for this set.

(b) List all possible samples of size $n = 2$ [there are $C(5, 2) = 10$ of them] and determine the mean $\bar{x}$ of each.

(c) Calculate the mean $\mu_{\bar{x}}$ and standard deviation $\sigma_{\bar{x}}$ of the distribution in (b).

(d) Compare $\mu_{\bar{x}}$ to μ and $\sigma_{\bar{x}}$ to $\dfrac{\sigma}{\sqrt{n}}\sqrt{\dfrac{N-n}{N-1}}$.

(e) More generally, what can be said about $\dfrac{\sigma}{\sqrt{n}}\sqrt{\dfrac{N-n}{N-1}}$ if the population size N is very large compared to the sample size n?

QUIZ 2

Multiple-Choice Questions

Directions: The questions that follow are each followed by five suggested answers. Choose the response that best answers the question.

1. Suppose a population is skewed right. For which of the following sample sizes would the sampling distribution of $\bar{x}$ be closest to normal?

   (A) 10
   (B) 30
   (C) 50
   (D) 100
   (E) Because of the CLT, all give normal sampling distributions.

2. Which of the following is a true statement?

   (A) The sampling distribution of $\hat{p}$ has a mean equal to the population proportion p.

   (B) The sampling distribution of $\hat{p}$ has a standard deviation equal to $\sqrt{np(1-p)}$.

   (C) The sampling distribution of $\hat{p}$ has a standard deviation that becomes larger as the sample size becomes larger.

   (D) The sampling distribution of $\hat{p}$ is considered close to normal provided that $n \geq 30$.

   (E) The sampling distribution of $\hat{p}$ is always close to normal.

3. In a school of 2500 students, the students in an AP Statistics class are planning a random survey of 100 students to estimate the proportion who would rather drop lacrosse rather than band during this time of severe budget cuts. Their teacher suggests instead to survey 200 students in order to

   (A) reduce bias.
   (B) reduce variability.
   (C) increase bias.
   (D) increase variability.
   (E) make possible stratification between lacrosse and band.

4. The distribution of ages of people who died last year in the United States is skewed left. What happens to the sampling distribution of sample means as the sample size goes from $n = 50$ to $n = 200$?

(A) The mean gets closer to the population mean, the standard deviation stays the same, and the shape becomes more skewed left.

(B) The mean gets closer to the population mean, the standard deviation becomes smaller, and the shape becomes more skewed left.

(C) The mean gets closer to the population mean, the standard deviation stays the same, and the shape becomes closer to normal.

(D) The mean gets closer to the population mean, the standard deviation becomes smaller, and the shape becomes closer to normal.

(E) The mean stays the same, the standard deviation becomes smaller, and the shape becomes closer to normal.

5. Which of the following are unbiased estimators for the corresponding population parameters?

I. Sample means
II. Sample proportions
III. Difference of sample means
IV. Difference of sample proportions

(A) None are unbiased.
(B) I and II only
(C) I and III only
(D) III and IV only
(E) All are unbiased.

6. Researchers found that in the aftermath of the 2011 Fukushima nuclear disaster, 12% of the pale grass blue butterfly larvae developed mutations as adults. What is the probability that in a random sample of 300 of these butterfly larvae, more than 15% developed mutations as adults?

(A) $P\left(z > \dfrac{0.12 - 0.15}{\sqrt{(0.12)(0.88)/300}}\right)$

(B) $P\left(z > \dfrac{0.15 - 0.12}{\sqrt{(0.12)(0.88)/300}}\right)$

(C) $P\left(z > \dfrac{0.12 - 0.15}{\sqrt{(0.15)(0.85)/300}}\right)$

(D) $P\left(z > \dfrac{0.15 - 0.12}{(0.12)(0.88)/\sqrt{300}}\right)$

(E) $P\left(z > \dfrac{0.12 - 0.15}{(0.15)(0.85)/\sqrt{300}}\right)$

7. Suppose using accelerometers in helmets, researchers determine that boys playing high school football absorb an average of 355 hits to the head with a standard deviation of 80 hits during a season (including both practices and games). What is the probability on a randomly selected team of 48 players that the average number of head hits per player is between 340 and 360?

(A) 0.236
(B) 0.429
(C) 0.571
(D) 0.614
(E) 0.764

8. Suppose "sleep-trained" babies (allowed to cry themselves to sleep) wake up an average of 1.2 times a night with a standard deviation of 0.3 times, while nontrained babies wake up an average of 1.8 times a night with a standard deviation of 0.5 times. In a random sample of 80 babies, half of which are sleep-trained, what is the probability that the nontrained babies in the sample wake up an average number of times greater than 0.75 more than the average of the sleep-trained babies?

(A) 0.0519
(B) 0.1179
(C) 0.4481
(D) 0.3821
(E) 0.9481

9. Which of the following statements about t-distributions are true?

  I. The greater the number of degrees of freedom, the narrower the tails.
 II. The smaller the number of degrees of freedom, the closer the curve is to the normal curve.
III. Thirty degrees of freedom gives the normal curve.

(A) I only
(B) I and II only
(C) I and III only
(D) II and III only
(E) I, II, and III

10. A random sample of 50 high school girls has a mean of 287 Facebook friends with a standard deviation of 68 friends. An independent random sample of 50 high school boys has a mean of 165 Facebook friends with a standard deviation of 45 friends. What is the standard error of the difference (girls – boys) between the sample means?

(A) $\sqrt{\dfrac{68-45}{100}}$

(B) $\sqrt{\dfrac{68^2-45^2}{50}}$

(C) $\sqrt{\dfrac{68^2-45^2}{100}}$

(D) $\sqrt{\dfrac{68^2+45^2}{50}}$

(E) $\sqrt{\dfrac{68^2+45^2}{100}}$

Free-Response Questions

Directions: You must show all work and indicate the methods you use. You will be graded on the correctness of your methods and on the accuracy of your final answers.

1. In a random sample of men working on Wall Street, the following summary statistics are from a question on the number of minutes spent grooming before leaving for the office in the morning:

$$n = 90,\ \bar{x} = 6.5,\ \text{median} = 3.8$$

(a) What is probably true about the distribution given that the mean is so much greater than the median?

(b) Are conditions for inference met? Explain.

(c) If a different sample size would result in a larger $\sigma_{\bar{x}}$, is the sample size smaller or larger than 90? Explain.

2. Studies have shown that teenage drivers are three times more likely to be involved in a fatal crash than drivers age 20 and older, and so insurance rates for teenagers are higher. The average yearly cost of teenage auto insurance is $2275 with a standard deviation of $650. An SRS is taken of 90 teenage drivers.

(a) Explain why there is insufficient information to determine the probability a randomly chosen teenage driver pays over $2400.

(b) What are the mean and standard deviation for the sampling distribution for $\bar{x}$, the mean amount paid for insurance?

(c) What is the probability that the average amount paid in the sample is more than $2400?

3. It is estimated that a new baby deprives each of its parents of an average of 374 hours of sleep in the first year with a standard deviation of 38.55 hours. Assume an approximately normal distribution of deprived hours.

    (a) What is the interquartile range (IQR) of the distribution of hours deprived?

    (b) In a random sample of three parents of new babies, what is the probability that a majority are deprived of more than 400 hours?

    (c) In a random sample of three parents of new babies, what is the probability that the mean number of deprived hours is more than 400?

4. Administrators of a national writing exam are concerned about any loss of consistency with regard to the graders they use. A random sample of five of their graders is given the same student's paper to grade. Past such evaluations have shown a variance in scores to be around 3 pts^2. The scores for these graders are 70, 71, 71, 73, and 75.

    (a) Calculate the mean $\bar{x}$ and the variance s^2 for this set of $n = 5$ scores.

    It can be shown that the sampling distribution of $\dfrac{(n-1)s^2}{\sigma^2}$ follows a chi-square distribution with $df = n - 1$.

    (b) With $n = 5$ and $\sigma^2 = 3$, calculate $\dfrac{(n-1)s^2}{\sigma^2}$.

    Using simulation, 200 samples, each of size 5, were randomly generated from a population with mean 72 and variance 3. The dotplot below shows the simulated sampling distribution of $Q = \dfrac{(n-1)s^2}{3}$.

Simulated values of Q

    (c) Given your calculation in (b) and given the above dotplot, does it appear that there is a significant loss of consistency with regard to the graders being used? Justify your answer.

ANSWERS EXPLAINED

Quiz 1

MULTIPLE-CHOICE QUESTIONS

1. **(D)** The relevant condition is that both np and $nq \geq 10$, and only Choice (D) with $np = (75)(0.15) = 11.25$ and $nq = (75)(0.85) = 63.75$ satisfies this.

2. **(C)** The larger the sample, the smaller the spread in the sampling distribution. Bias has to do with the center, not the spread, of a sampling distribution. Sample statistics are used to make inferences about population proportions. Statistics from smaller samples have more variability.

3. **(E)** It is always true that the sampling distribution of $\bar{x}$ has mean μ and standard deviation $\frac{\sigma}{\sqrt{n}}$. In addition, the sampling distribution will be normal if the population is normal and will be approximately normal if n is large even if the population is not normal. $\frac{\sigma}{\sqrt{n}}$ and σ are not equal unless $n = 1$.

4. **(A)** The maximum of a sample is never larger than the maximum of the population, so the mean of the sample maximums will not be equal to the population maximum. (A sampling distribution is unbiased if its mean is equal to the population parameter.)

5. **(D)** The sample is given to be random, both $np = (400)(0.34) = 136 \geq 10$ and $n(1 - p) = (400)(0.66) = 264 \geq 10$, and our sample is clearly less than 10% of all people. So the sampling distribution of $\hat{p}$ is approximately normal with mean 0.34 and standard deviation $\sigma_{\hat{p}} = \sqrt{\frac{(0.34)(0.66)}{400}} = 0.0237$. With z-scores of $\frac{0.30 - 0.34}{0.0237} = -1.688$ and $\frac{0.35 - 0.34}{0.0237} = 0.422$, the probability that the sample proportion is between 0.30 and 0.35 is `normalcdf(-1.688,.422)= 0.618`. [Or `normalcdf(.3,.35,.34,.0237)=` 0.618.]

6. **(A)** We have a random sample that is less than 10% of all schools. With a sample size of 30, the central limit theorem applies, and the sampling distribution of $\bar{x}$ is approximately normal with mean $\mu_{\bar{x}} = 1200$ and standard deviation $\sigma_{\bar{x}} = \dfrac{300}{\sqrt{30}}$. The z-score of 1000 is $\dfrac{1000-1200}{\frac{300}{\sqrt{30}}}$, and the probability of a sample mean over 1000 is $P\left(z > \dfrac{1000-1200}{\frac{300}{\sqrt{30}}}\right)$.

Distribution of sample means

$\sigma_{\bar{x}} = \dfrac{300}{\sqrt{30}}$

1000   1200 CO_2 levels

$\dfrac{1000-1200}{300/\sqrt{30}}$ z-scores

7. **(B)** We have two independent random samples, each less than 10% of their respective populations, and we note that $n_1 p_1 = 75(0.43) = 32.25$, $n_1(1-p_1) = 75(0.57) = 42.75$, $n_2 p_2 = 80(0.37) = 29.6$, and $n_2(1-p_2) = 80(0.63) = 50.4$ are all ≥ 10. Thus the sampling distribution of $\hat{p}_1 - \hat{p}_2$ is roughly normal with mean $\mu_{\hat{p}_1 - \hat{p}_2} = 0.43 - 0.37 = 0.06$ and standard deviation $\sigma_{\hat{p}_1 - \hat{p}_2} = \sqrt{\dfrac{(0.43)(0.57)}{75} + \dfrac{(0.37)(0.63)}{80}} = 0.0786$. The z-score of 0.05 is $\dfrac{0.05-0.06}{0.0786} = -0.127$, the z-score of 0.10 is $\dfrac{0.10-0.06}{0.0786} = 0.509$, and

`normalcdf(-.127,.509)= 0.245`. [Or `normalcdf(.05,.10,.06,.0786)=0.245`.]

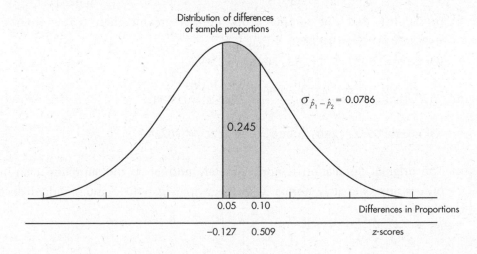

Distribution of differences of sample proportions

$\sigma_{\hat{p}_1 - \hat{p}_2} = 0.0786$

0.245

0.05   0.10   Differences in Proportions

−0.127   0.509   z-scores

8. **(B)** The t-distributions are symmetric and mound-shaped, and they have more, not less, spread than the normal distribution.

9. **(D)** The variance for the sampling distribution of $\hat{p}$ equals $\frac{p(1-p)}{n}$. A larger n in the denominator results in a smaller quotient, and $(0.1)(0.9) < (0.5)(0.5)$.

10. **(C)** The sampling distribution is roughly normal with a mean of $\mu = 0.32 - 0.29 = 0.03$ and a standard deviation of $\sigma = \sqrt{\frac{(0.32)(0.68)}{400} + \frac{(0.29)(0.71)}{400}} = 0.0325$.

FREE-RESPONSE QUESTIONS

1. (a) $\sigma_{\hat{p}} = \sqrt{\frac{p(1-p)}{n}} = \sqrt{\frac{(0.16)(0.84)}{80}} = 0.041$

   (b) Yes, the conditions for inference are met. First, it is given that a random sample is planned. Second, the sample size, $n = 80$, is clearly less than 10% of all game players. Third, $np = (80)(0.16) = 12.8 \geq 10$ and $nq = (80)(0.84) = 67.2 \geq 10$.

   (c) The sample size would be larger than 80, because $0.037 < 0.041$, and in order for $\sigma_{\hat{p}}$ to be smaller, the denominator in $\sigma_{\hat{p}} = \sqrt{\frac{p(1-p)}{n}}$ must be larger.

TIP

On the exam, you can use expressions like binomcdf **only if you note the parameters:** binomcdf $(n = 5, p = .58, 2)$

2. (a) 0.58

   (b) This is a binomial with $n = 3$ and $p = 0.58$.

   $$P(X \geq 3) = \binom{5}{3}(0.58)^3(0.42)^2 + \binom{5}{4}(0.58)^4(0.42) + (0.58)^5 = 0.647$$

   or $1 - \text{binomcdf}(5, .58, 2) = 0.647$.

NOTE

You can also do a direct binomial calculation:
1 − binomcdf
$(350, .58, 174)$ =
0.9989.

   (c) The sample is given to be random, both $np = (350)(0.58) = 203 \geq 10$ and $n(1 - p) = (350)(0.42) = 147 \geq 10$, and our sample is clearly less than 10% of all Americans. So the sampling distribution of $\hat{p}$ is approximately normal with mean 0.58 and standard deviation $\sigma_{\hat{p}} = \sqrt{\frac{(0.58)(0.42)}{350}} = 0.0264$. With a z-score of $\frac{0.5 - 0.58}{0.0264} = -3.030$, the probability that the sample proportion is greater than 0.5 is $\text{normalcdf}(-3.030, 1000) = 0.9988$.
   [Or $\text{normalcdf}(.5, 1, .58, .0264) = 0.9988$.]

3. (a) $P(X > 4250) = P\left(z > \frac{4250 - 4000}{125}\right) = P(z > 2) = 0.0228$.

   [Or $\text{normalcdf}(4250, 10000, 4000, 125) = 0.0228$.]

   (b) The original population is approximately normal so the sampling distribution of $\bar{x}$ is approximately normal with mean $\mu_{\bar{x}} = \mu = 4000$ and standard deviation

   $$\sigma_{\bar{x}} = \frac{\sigma}{\sqrt{n}} = \frac{125}{\sqrt{40}} = 19.764.$$

(c) $P(X < 3950) = P\left(z < \dfrac{3950 - 4000}{19.764}\right) = P(z < -2.530) = 0.0057.$

[Or `normalcdf(0,3950,4000,19.764) = 0.0057`.]

(d) The answer to part (a) would be affected because it assumes an approximately normal population. The other answers would not be affected because for large enough n, the central limit theorem gives that the sampling distribution of $\bar{x}$ is roughly normal regardless of the distribution of the original population.

4. (a) Using a calculator, find $\mu = 12$ and $\sigma = 5.0596$.

   (b)

Sample	{6, 9}	{6, 11}	{6, 13}	{6, 21}	{9, 11}	{9, 13}	{9, 21}	{11, 13}	{11, 21}	{13, 21}
Mean	7.5	8.5	9.5	13.5	10	11	15	12	16	17

   (c) $\mu_{\bar{x}} = 12$ and $\sigma_{\bar{x}} = 3.0984.$

   (d) $\mu_{\bar{x}} = \mu = 12$ and $\dfrac{\sigma}{\sqrt{n}}\sqrt{\dfrac{N-n}{N-1}} = \dfrac{5.0596}{\sqrt{2}}\sqrt{\dfrac{5-2}{5-1}} = 3.0984 = \sigma_{\bar{x}}.$

   (e) If N is very large, $\dfrac{N-n}{N-1}$ is approximately equal to 1, and so the expression is approximately equal to $\dfrac{\sigma}{\sqrt{n}}$.

Quiz 2

MULTIPLE-CHOICE QUESTIONS

1. **(D)** The Central Limit Theorem gives that the larger the sample size, the closer the sampling distribution will be to normal.

2. **(A)** The sampling distribution of $\hat{p}$ has a standard deviation $\sqrt{\dfrac{p(1-p)}{n}}$ which is smaller with larger n. While the sampling distribution of $\hat{p}$ is never exactly normal, it is considered close to normal *provided that* both np and $n(1-p)$ are large enough (greater than 10 is a standard guide).

3. **(B)** Sample proportions are an unbiased estimator for the population proportion, and larger sample sizes lead to reduced variability.

4. **(E)** The sampling distribution of $\bar{x}$ has mean μ, standard deviation $\dfrac{\sigma}{\sqrt{n}}$, and shape, which becomes closer to normal with larger n.

5. **(E)** All are unbiased estimators for the corresponding population parameters; that is, the means of their sampling distributions are equal to the population parameters.

6. **(B)** The sample is given to be random, both $np = (300)(0.12) = 36 \geq 10$ and $n(1 - p) = (300)(0.88) = 264 \geq 10$, and our sample is clearly less than 10% of all butterfly larvae. So the sampling distribution of $\hat{p}$ is approximately normal with mean $\mu_{\hat{p}} = 0.12$ and standard deviation $\sigma_{\hat{p}} = \sqrt{\dfrac{(0.12)(0.88)}{300}}$. With a z-score of $\dfrac{0.15 - 0.12}{\sqrt{\dfrac{(0.12)(0.88)}{300}}}$, the probability that the sample proportion is more than 15% is $P\left(z > \dfrac{0.15 - 0.12}{\sqrt{\dfrac{(0.12)(0.88)}{300}}}\right)$.

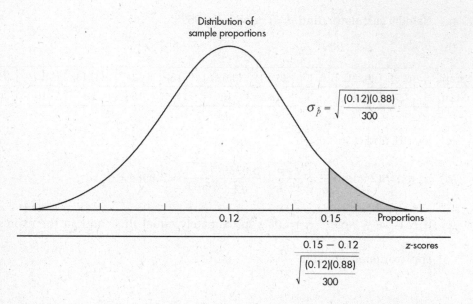

7. **(C)** We have a random sample that is less than 10% of the high school football population. With a sample size of 48, the central limit theorem applies, and the sampling distribution of $\bar{x}$ is approximately normal with mean $\mu_{\bar{x}} = 355$ and standard deviation $\sigma_{\bar{x}} = \dfrac{80}{\sqrt{48}} = 11.547$. The z-scores of 340 and 360 are $\dfrac{340 - 355}{11.547} = -1.299$ and $\dfrac{360 - 355}{11.547} = 0.433$, the probability of a sample mean between 340 and 360 is `normalcdf(-1.299,.433)= 0.571`. [Or `normalcdf(340,360,355,11.547)= 0.571`.]

8. **(A)** We have independent random samples, each less than 10% of babies, and both sample sizes are $40 \geq 30$, so the sampling distribution of $\bar{x}_1 - \bar{x}_2$ is roughly normal with mean $\mu_{\bar{x}_1 - \bar{x}_2} = 1.8 - 1.2 = 0.6$ and standard deviation $\sigma_{\bar{x}_1 - \bar{x}_2} = \sqrt{\dfrac{(0.5)^2}{40} + \dfrac{(0.3)^2}{40}} = 0.0922$. The z-score of 0.75 is $\dfrac{0.75 - 0.6}{0.0922} = 1.627$, and `normalcdf(1.627,1000)=0.0519`.

[Or `normalcdf(.75,1000,.6,.0922)=0.0519`.]

9. **(A)** The larger the number of degrees of freedom, the closer the curve to the normal curve. While around the 30 level is often considered a reasonable approximation to the normal curve, it is not the normal curve.

10. **(D)** Variances add, and so $\sqrt{\left(\dfrac{68}{\sqrt{50}}\right)^2 + \left(\dfrac{45}{\sqrt{50}}\right)^2} = \sqrt{\dfrac{68^2}{50} + \dfrac{45^2}{50}} = \sqrt{\dfrac{68^2 + 45^2}{50}}$.

FREE-RESPONSE QUESTIONS

1. (a) When the mean is much greater than the median, the distribution is probably skewed right (toward the greater values).

   (b) Yes, the conditions for inference are met. First, it is given that a random sample was planned. Second, the sample size, $n = 90$, is clearly less than 10% of all men working on Wall Street. Third, while the population distribution is probably not normal (it's probably skewed right), the sample size is larger enough, $n = 90 \geq 30$, so that the CLT gives that the sampling distribution of $\bar{x}$ is approximately normal.

   (c) The sample size would be smaller than 90, because in order for $\sigma_{\bar{x}}$ to be larger, the denominator in $\sigma_{\bar{x}} = \dfrac{\sigma}{\sqrt{n}}$ must be smaller.

2. (a) We do not know the shape of the distribution of the amount individual teenage drivers pay for insurance, so there is no way of calculating the probability a randomly chosen teenage driver pays over $2400 a year for auto insurance.

(b) $\mu_{\bar{x}} = \mu = 2275$ and $\sigma_{\bar{x}} = \dfrac{\sigma}{\sqrt{n}} = \dfrac{650}{\sqrt{90}} = 68.52$

(c) With this large a sample size ($90 \geq 30$), the central limit theorem tells us that the sampling distribution of $\bar{x}$ is approximately normal. We calculate

$$P(\bar{x} > 2400) = P\left(z > \dfrac{2400 - 2275}{68.52}\right) = P(z > 1.824) = 0.0341$$

3. (a) The z-scores corresponding to cumulative probabilities of 0.25 and 0.75 are ± 0.6745. Thus $Q_1 = 374 - (0.6745)(38.55) = 348.00$, $Q_3 = 374 + (0.6745)(38.55) = 400.00$, and IQR $= Q_3 - Q_1 = 400 - 348 = 52$ hours.

(b) From part (a), we have that the probability of less than 400 deprived hours is 0.75, and thus the probability of being more than 400 deprived hours is 0.25. Now we have a binomial with $n = 3$ and $p = 0.25$. $P(\text{majority} > 400) = P(2 \text{ or } 3 \text{ are} > 400) = 3(0.25)^2(0.75) + (0.25)^3 = 0.15625$.

(c) The sampling distribution of the sample means is approximately normal (because the original population is normal) with mean $\mu_{\bar{x}} = \mu = 374$ and standard deviation $\sigma_{\bar{x}} = \dfrac{\sigma}{\sqrt{n}} = \dfrac{38.55}{\sqrt{3}} = 22.257$. The z-score of 400 is $\dfrac{400 - 374}{22.257} = 1.168$, and the probability of a sample mean over 400 is `normalcdf(1.168,1000)= 0.121`. [Or `normalcdf(400,1000,374,22.257)=0.121`.]

4. (a) A calculator quickly gives $\bar{x} = 72$ and $s^2 = 4$.

(b) $\dfrac{(n-1)s^2}{\sigma^2} = \dfrac{(5-1)(4)}{3} = 5.33$

(c) Even though the sample variance of 4 is greater than the old variance of 3, looking at where 5.33 is found in the dotplot (the dotplot generated from a population with variance 3), it is very plausible that the given sample came from a population with variance 3. It does not appear that there is evidence of a significant loss of consistency with regard to the graders being used.

Theme Four: Statistical Inference

Confidence Intervals

11

→ **DEFINITION**

→ **CONDITIONS FOR INFERENCE**

→ **FOR A PROPORTION**

→ **FOR A DIFFERENCE OF TWO PROPORTIONS**

→ **FOR A MEAN**

→ **FOR A DIFFERENCE BETWEEN TWO MEANS**

→ **FOR THE SLOPE OF A LEAST SQUARES REGRESSION LINE**

U sing a measurement from a sample, we are never able to say *exactly* what a population proportion or mean is; rather we always say we have a certain *confidence* that the population proportion or mean lies in a particular *interval*. The particular interval is centered around a sample proportion or mean (or other statistic) and can be expressed as the sample estimate plus or minus an associated *margin of error*.

THE MEANING OF A CONFIDENCE INTERVAL

Using what we know about sampling distributions, we are able to establish a certain confidence that a sample proportion or mean lies within a specified interval around the population proportion or mean. However, we then have the same confidence that the population proportion or mean lies within a specified interval around the sample proportion or mean (e.g., the distance from Missoula to Whitefish is the same as the distance from Whitefish to Missoula).

Typically we consider 90%, 95%, and 99% confidence interval estimates, but any percentage is possible. The percentage is the percentage of samples that would pinpoint the unknown p or μ within plus or minus a certain margin or error. We do *not* say there is a 0.90, 0.95, or 0.99 probability that p or μ is within a certain margin of error of a given sample proportion or mean. For a given sample proportion or mean, p or μ either is or isn't within the specified interval, and so the probability is either 1 or 0.

As will be seen, there are two aspects to this concept. First, there is the *confidence interval*, usually expressed in the form:

$$\text{estimate} \pm \text{margin of error}$$

Second, there is the *success rate for the method*, that is, the proportion of times repeated applications of this method would capture the true population parameter.

TIP

Of course, to obtain better estimates, it's always better to have larger samples—it's just that if the sample size is large relative to the population, then the proper inference techniques are different from those taught in the AP Statistics curriculum.

CONDITIONS FOR INFERENCE

Following are the three standard assumptions/conditions for our inference procedures and the *ideal* way they are met. Of course, we seldom, if ever, meet ideals!

1. **RANDOMIZATION ASSUMPTION:** For samples, the sample should be a simple random sample. For experiments, the subjects should be randomly assigned to treatments.

2. **INDEPENDENCE ASSUMPTION:** Individuals in a sample must be independent of each other, and this is obtained through random sampling/selection. Independence across samples is obtained by selecting two (or more) separate random samples. Always examine how the data were collected to check if the assumption of independence is reasonable. Sample size can also affect independence. Because sampling is usually done without replacement, if the sample is too large, lack of independence becomes a concern. So we typically require that the sample size n be no larger than 10% of the population (the 10% Rule).

3. **NORMALITY ASSUMPTION:** Inference for proportions is based on a normal model for the sampling distribution of $\hat{p}$, but actually we have a binomial distribution. Fortunately, the binomial is approximately normal if both np and $nq \geq 10$. Inference for means is based on a normal model for the sampling distribution of $\bar{x}$; this is true if the population is normal and is approximately true (thanks to the CLT) if the sample size is large enough (typically we accept $n \geq 30$). With regard to means, this is referred to as the *Normal/Large Sample* condition.

➡ EXAMPLE 11.1

If we pick an SRS of size 80 from a large population, which of the following values of the population proportion p would allow use of the normal model for the sampling distribution of $\hat{p}$?

(A) 0.10
(B) 0.15
(C) 0.90
(D) 0.95
(E) 0.99

Answer: (B) The relevant condition is that both np and $nq \geq 10$. In (A), $np = (80)(0.10) = 8$; in (C), $nq = (80)(0.10) = 8$; in (D), $nq = (80)(0.05) = 4$; and in (E), $nq = (80)(0.01) = 0.8$. However, in (B), $np = (80)(0.15) = 12$ and $nq = (0.85)(80) = 68$ are both ≥ 10.

➡ EXAMPLE 11.2

It is hypothesized that the ratio $\dfrac{\text{outer edges of lips}}{\text{upper ridges of lips}}$ is most pleasing when equal to the golden ratio, $\varphi = 1.618$. A student would like to calculate the mean ratio for students in her high school and plans to randomly sample 50 out of the 780 students at the school. Does this satisfy the 10% Rule, and what is the purpose of the rule?

Answer: 10% of 780 = 78, and 50 < 78, so, yes, the 10% Rule is satisfied. In picking the sample of 50 students, the sampling will obviously be done without replacement, because the same student will not be used twice. If the sample was too large compared to the size of the population, lack of independence would have become a concern.

➡ EXAMPLE 11.3

A human resources director is interested in the proportion of employees in an 85-member department who believe they are saving enough for retirement. She randomly selects 12 employees to interview. Explain whether conditions for inference are satisfied.

Answer: Even though it may or may not have been the ideal method of sampling, it is stated that a random sample was taken, satisfying the randomization condition. The 10% condition (related to independence) is violated because 10% of 85 is 8.5, and 12 is greater than 8.5. The normality assumption is violated because np and nq total 12, so both cannot be ≥ 10.

➡ EXAMPLE 11.4

An English teacher is interested in the mean length of words in a Hemingway novel. She picks a random sample of 25 words. Explain whether conditions for inference are satisfied.

Answer: Even though it may or may not have been the ideal method of sampling, it is stated that a *random* sample was taken, satisfying the randomization condition. The 10% condition (related to independence) is also satisfied because it seems clear that 25 is less than 10% of the number of words in an entire novel. However, the normal/large sample condition is not satisfied because it is not given that the distribution of all words in the novel is approximately normal, and the sample size is not large, that is, $n = 25 < 30$.

CONFIDENCE INTERVAL FOR A PROPORTION

We are interested in estimating a population proportion p by considering a single sample proportion $\hat{p}$. This sample proportion is just one of a whole universe of sample proportions, and from Topic 10 we remember the following:

1. The set of all sample proportions is approximately normally distributed.
2. The mean $\mu_{\hat{p}}$ of the set of sample proportions equals p, the population proportion.
3. The standard deviation $\sigma_{\hat{p}}$ of the set of sample proportions is approximately equal to $\sqrt{\dfrac{p(1-p)}{n}}$.

Remember that we are really using a normal approximation to the binomial, so $n\hat{p}$ and $n(1 - \hat{p})$ should both be at least 10. Furthermore, in making calculations and drawing conclusions from a specific sample, it is important that the sample be a *simple random sample*. Finally, the population should be large, typically checked by the assumption that the sample is less than 10% of the population. (If the population is small and the sample exceeds 10% of the population, then models other than the normal are more appropriate.)

In finding confidence interval estimates of the population proportion p, how do we find $\sqrt{\dfrac{p(1-p)}{n}}$ since p is unknown? The reasonable procedure is to use the sample proportion $\hat{p}$:

$$\sigma_{\hat{p}} \approx \sqrt{\frac{\hat{p}(1-\hat{p})}{n}}$$

When the standard deviation is estimated in this way (using the sample), we use the term *standard error*. That is,

$$SE(\hat{p}) = \sqrt{\frac{\hat{p}(1-\hat{p})}{n}}$$

➡ EXAMPLE 11.5

If 64% of an SRS of 550 people leaving a shopping mall claim to have spent over $25, determine a 99% confidence interval estimate for the proportion of shopping mall customers who spend over $25.

Answer: We check that $n\hat{p}$ = 550(0.64) = 352 ≥ 10 and $n(1 - \hat{p})$ = 550(0.36) = 198 ≥ 10, we are given that the sample is an SRS, and it is reasonable to assume that 550 is less than 10% of all mall shoppers. Since $\hat{p}$ = 0.64, the standard deviation of the set of sample proportions is

$$\sigma_{\hat{p}} \approx \sqrt{\frac{(0.64)(0.36)}{550}} = 0.0205$$

(continued)

From Topic 9 we know that 99% of the sample proportions should be within 2.576 standard deviations of the population proportion. Equivalently, we are 99% certain that the population proportion is within 2.576 standard deviations of any sample proportion.[1] Thus the 99% confidence interval estimate for the population proportion is 0.64 ± 2.576(0.0205) = 0.64 ± 0.053. We say that the *margin of error* is ±0.053. We are 99% certain that the proportion of shoppers spending over $25 is between 0.587 and 0.693.

We can also say, using the definition of confidence *level*, that if the interviewing procedure were repeated many times, about 99% of the resulting confidence intervals would contain the true proportion (thus we're 99% confident that the method worked for the interval we got).[1]

TIP

The margin of error accounts for variation inherent in sampling, but it does not account for and is not a redress for bias in sampling.

➡ EXAMPLE 11.6

In a simple random sample of machine parts, 18 out of 225 were found to have been damaged in shipment. Establish a 95% confidence interval estimate for the proportion of machine parts that are damaged in shipment.

IMPORTANT

After finding a confidence interval, you must always interpret the interval in context.

Answer: We check that $n\hat{p} = 18 \geq 10$ and $n(1 - \hat{p}) = 225 - 18 = 207 \geq 10$, we are given that the sample is an SRS, and it is reasonable to assume that 225 is less than 10% of all shipped machine parts.

Calculator software (such as `1-PropZInt` on the TI-84 or 1-Prop ZInterval on the Casio Prizm) gives (0.04455,0.11545).

[For instructional purposes in this review book, we note that $\hat{p} = \frac{18}{225} = 0.08$,

$SE(\hat{p}) = \sqrt{\frac{(0.08)(0.92)}{225}} = 0.0181$, and 0.08 ± 1.96(0.0181) = 0.08 ± 0.035.]

TIP

On the TI-84, for a confidence interval of a proportion, use 1-PropZInt, *not* ZInterval; and *x* must be an integer.

Thus we are 95% certain that the proportion of all machine parts damaged in shipment is between 0.045 and 0.115.

On the TI-Nspire the result shows as:

zInterval_1Prop 18,225,0.95: *stat.results*	"Title"	"1-Prop z Interval"
	"CLower"	0.044552
	"CUpper"	0.115448
	"p̂"	0.08
	"ME"	0.035448
	"n"	225.

Suppose there are 50,000 parts in the entire shipment. We can translate from proportions to actual numbers:

$$0.045(50,000) = 2250 \quad \text{and} \quad 0.115(50,000) = 5750$$

So we can be 95% confident that there are between 2250 and 5750 defective parts in the whole shipment.

TIP

Finding confidence intervals *using formulas* is not necessary on free-response questions, but answer choices using formulas may well appear on multiple-choice questions.

---

[1] Note that we cannot say there is a 0.99 probability that the population proportion is within 2.576 standard deviations of a given sample proportion. For a given sample proportion, the population proportion either is or isn't within the specified interval, and so the probability is either 1 or 0.

➥ EXAMPLE 11.7 _____

A telephone survey of 1000 adults was taken shortly after the United States began bombing Iraq.

(a) If 832 voiced their support for this action, with what confidence can it be asserted that 83.2% ± 3% of the adult U.S. population supported the decision to go to war?

Answer: We check that $n\hat{p} = 832 \geq 10$ and $n(1 - \hat{p}) = 1000 - 832 = 168 \geq 10$, we assume an SRS, and clearly 1000 is <10% of the adult U.S. population.

$$\hat{p} = \frac{832}{1000} = 0.832 \quad \text{and so} \quad SE(\hat{p}) = \sqrt{\frac{(0.832)(0.168)}{1000}} = 0.0118$$

The relevant *z*-scores are $\pm\frac{0.03}{0.0118} = \pm 2.54$. Normalcdf(–2.54, 2.54) = 0.9889.

In other words, 83.2% ± 3% is a 98.89% confidence interval estimate for U.S. adult support of the war decision.

(b) If the adult U.S. population is 191 million, estimate the actual numerical support.

Answer: Since

$$0.802(191,000,000) \approx 153,000,000$$

while

$$0.862(191,000,000) \approx 165,000,000$$

we can be 98.90% confident that between 153 and 165 million adults supported the initial bombing decision.

NOTE

On the exam, you can write:

normalcdf$\Big($lower bd = .802,

upper bd = .862, mean = .832,

SD = $\sqrt{\frac{(.832)(.168)}{1000}}\Big)$ = .98898

or 98.9%

➥ EXAMPLE 11.8 _____

A U.S. Department of Labor survey of 6230 unemployed adults classified people by marital status, gender, and race. The raw numbers are as follows:

	White, 16 yr and older		
	Married	Widow/Div.	Single
Men	1090	337	1168
Women	952	423	632

	Nonwhite, 16 yr and older		
	Married	Widow/Div.	Single
Men	266	135	503
Women	189	186	349

(a) Find a 90% confidence interval estimate for the proportion of unemployed men who are married.

Answer: Totaling the first row across both tables, we find that there are 3499 men in the survey; 1090 + 266 = 1356 of them are married. We check that $n\hat{p} = 1356 \geq 10$ and $n(1 - \hat{p}) = 3499 - 1356 = 2143 \geq 10$, we assume an SRS, and clearly the sample is <10% of the total unemployed adult population.

A calculator gives (0.37399, 0.40109). [For instructional purposes we note that $\hat{p} = \dfrac{1356}{3499} = 0.3875$ and $SE(\hat{p}) = \sqrt{\dfrac{(0.3875)(0.6125)}{3499}} = 0.008236.$

Thus the 90% confidence interval estimate is $0.3875 \pm 1.645(0.008236) =$ $0.3875 \pm 0.0135.$] We are 90% confident that the true proportion of unemployed men who are married is between 0.374 and 0.401.

(b) Find a 98% confidence interval estimate of the proportion of unemployed single persons who are women.

Answer: There are $1168 + 632 + 503 + 349 = 2652$ singles in the survey. Of these, $632 + 349 = 981$ are women [$n\hat{p} = 981 \geq 10$ and $n(1 - \hat{p}) = 2652 - 981 = 1671 \geq 10$]. A calculator gives (0.3481, 0.39172).

[For instructional purposes we note that $\hat{p} = \dfrac{981}{2652} = 0.3699$ and

$$SE(\hat{p}) = \sqrt{\dfrac{(0.3699)(0.6301)}{2652}} = 0.009375$$

The 98% confidence interval estimate is $0.3699 \pm 2.326(0.009375) =$ $0.3699 \pm 0.0218.$] We are 98% confident that the true proportion of unemployed single persons who are women is between 0.348 and 0.392.

As we have seen, there are two types of statements that come out of confidence intervals. First, we can interpret the confidence interval and say we are 90% confident that between 60% and 66% of all voters favor a bond issue. Second, we can interpret the confidence level and say that if this survey were conducted many times, then about 90% of the resulting confidence intervals would contain the true proportion of voters who favor the bond issue.

Incorrect statements include "The percentage of all voters who support the bond issue is between 60% and 66%," "There is a 0.90 probability that the true percentage of all voters who favor the bond issue is between 60% and 66%," and "If this survey were conducted many times, then about 90% of the sample proportions would be in the interval (0.60, 0.66)."

One important consideration in setting up a survey is the choice of sample size. To obtain a smaller, more precise interval estimate of the population proportion, we must either decrease the degree of confidence or increase the sample size. Similarly, if we want to increase the degree of confidence, we can either accept a wider interval estimate or increase the sample size. Again, while choosing a larger sample size may seem desirable, in the real world this decision involves time and cost considerations.

In setting up a survey to obtain a confidence interval estimate of the population proportion, what should we use for $\sigma_{\hat{p}}$? To answer this question, we first must consider how large $\sqrt{p(1-p)}$ can be. We plot various values of p:

p:	0.1	0.2	0.3	0.4	0.5	0.6	0.7	0.8	0.9
$\sqrt{p(1-p)}$:	0.3	0.4	0.458	0.490	0.5	0.490	0.458	0.4	0.3

This results in 0.5 as the intuitive answer. Thus $\sqrt{\dfrac{p(1-p)}{n}}$ is at most $\dfrac{0.5}{\sqrt{n}}$. We make use of this fact to determine sample sizes in problems such as Examples 11.9 and 11.0.

TIP

Noting the
inequality, the
result must be
rounded *up* if it
is a decimal.

EXAMPLE 11.9

An Environmental Protection Agency (EPA) investigator wants to know the proportion of fish that are inedible because of chemical pollution downstream of an offending factory. If the answer must be within ±0.03 at the 96% confidence level, how many fish should be in the sample tested?

Answer: We want $2.05\sigma_{\hat{p}} \le 0.03$. From the above remark, $\sigma_{\hat{p}}$ is at most $\dfrac{0.5}{\sqrt{n}}$ and so it is

sufficient to consider $2.05\left(\dfrac{0.5}{\sqrt{n}}\right) \le 0.03$. Algebraically, we have $\sqrt{n} \ge \dfrac{2.05(0.5)}{0.03} = 34.17$

and $n \ge 1167.4$. Therefore, choosing a sample of 1168 fish gives the inedible proportion to within ±0.03 at the 96% level.

NOTE

For 96%
confidence,
there is 0.02 in
each tail and
invNorm(.02)
= –2.05 or
invNorm(.98) =
2.05

Note that the accuracy of the estimate does *not* depend on what fraction of the whole population we have sampled. What is critical is the *absolute size* of the sample. Is some minimal value of n necessary for the procedures we are using to be meaningful? Since we are using the normal approximation to the binomial, both np and $n(1-p)$ should be at least 10 (see Topic 9).

EXAMPLE 11.10

A study is undertaken to determine the proportion of industry executives who believe that workers' pay should be based on individual performance. How many executives should be interviewed if an estimate is desired at the 99% confidence level to within ±0.06? To within ±0.03? To within ±0.02?

NOTE

For 99%
confidence,
there is 0.005
in each tail and
invNorm(.005)
= –2.576 or
invNorm(.995)
= 2.576

Answer: Algebraically, $2.576\left(\dfrac{0.5}{\sqrt{n}}\right) \le 0.06$ gives $\sqrt{n} \ge \dfrac{2.576(0.5)}{0.06} = 21.5$, so

$n \ge 462.25$. Similarly, $2.576\left(\dfrac{0.5}{\sqrt{n}}\right) \le 0.03$ gives $\sqrt{n} \ge \dfrac{2.576(0.5)}{0.03} = 42.9$, so $n \ge 1840.4$.

Finally, $2.576\left(\dfrac{0.5}{\sqrt{n}}\right) \le 0.02$ gives $\sqrt{n} \ge \dfrac{2.576(0.5)}{0.02} = 64.4$, so $n \ge 4147.4$. Thus 463, 1841,

or 4148 executives should be interviewed, depending on the accuracy desired.

Note that to cut the interval estimate in half (from ±0.06 to ±0.03), we would have to increase the sample size fourfold, and to cut the interval estimate to a third (from ±0.06 to ±0.02), a ninefold increase in the sample size would be required (answers are not exact because of round-off error.)

More generally, to divide the interval estimate by d without affecting the confidence level, we must increase the sample size by a multiple of d^2.

A formula for the calculations in Examples 11.5 and 11.6 is

$$n = \left[\frac{z}{2(\text{error})}\right]^2$$

(Note: this formula is not given on the AP exam formula page.)

CONFIDENCE INTERVAL FOR A DIFFERENCE OF TWO PROPORTIONS

From Topic 10, we have the following information about the sampling distribution of $\hat{p}_1 - \hat{p}_2$:

1. The set of all differences of sample proportions is approximately normally distributed.
2. The mean of the set of differences of sample proportions equals $p_1 - p_2$, the difference of population proportions.
3. The standard deviation σ_d of the set of differences of sample proportions is approximately equal to

$$\sqrt{\frac{p_1(1-p_1)}{n_1} + \frac{p_2(1-p_2)}{n_2}}.$$

Remember that we are using the normal approximation to the binomial, so $n_1\hat{p}_1$, $n_1(1 - \hat{p}_1)$, $n_2\hat{p}_2$, and $n_2(1 - \hat{p}_2)$ should all be at least 10. In making calculations and drawing conclusions from specific samples, it is important both that the samples be *simple random samples* and that they be taken *independently* of each other. Finally, the original populations should be large compared to the sample sizes.

➡ EXAMPLE 11.11

Suppose that 84% of an SRS of 125 nurses working 7 a.m. to 3 p.m. shifts in city hospitals express positive job satisfaction, while only 72% of an SRS of 150 nurses on 11 p.m. to 7 a.m. shifts express similar fulfillment. Establish a 90% confidence interval estimate for the difference.

Answer: Note that $n_1\hat{p}_1 = (125)(0.84) = 105$, $n_1(1 - \hat{p}_1) = (125)(0.16) = 20$, $n_2\hat{p}_2 = (150)(0.72) = 108$, and $n_2(1 - \hat{p}_2) = (150)(0.28) = 42$ are all ≥ 10, we are given independent SRSs, and the population of city hospital nurses is assumed to be large. `2-PropZInt` on the TI-84 or 2-Prop ZInterval on the Casio Prizm give (0.0391, 0.2009). [For instructional purposes, we note that

$$n_1 = 125 \quad n_2 = 150$$

$$\hat{p}_1 = 0.84 \quad \hat{p}_2 = 0.72$$

$$\sigma_d \approx \sqrt{\frac{(0.84)(0.16)}{125} + \frac{(0.72)(0.28)}{150}} = 0.0492$$

The observed difference is 0.84 – 0.72 = 0.12, and the critical z-scores are ±1.645. The confidence interval estimate is 0.12 ± 1.645(0.0492) = 0.12 ± 0.081.]

We can be 90% certain that the proportion of satisfied nurses on 7 a.m. to 3 p.m. shifts is between 0.039 and 0.201 higher than the proportion for nurses on 11 p.m. to 7 a.m. shifts.

TIP

When interpreting the confidence interval for a difference, you must indicate which direction the difference is taken.

➡ EXAMPLE 11.12

A grocery store manager notes that in an SRS of 85 people going through the express checkout line, only 10 paid with checks, whereas, in an SRS of 92 customers passing through the regular line, 37 paid with checks. Find a 95% confidence interval estimate for the difference between the proportion of customers going through the two different lines who use checks.

Answer: Note that $n_1 \hat{p}_1 = 10$, $n_1(1 - \hat{p}_1) = 75$, $n_2 \hat{p}_2 = 37$, and $n_2(1 - \hat{p}_2) = 55$ are all at least 10, we are given independent SRSs, and the total number of customers is large. `2-PropZInt` gives (–0.4059, –0.1632). [For instructional purposes, we note that

$$n_1 = 85 \qquad n_2 = 92$$

$$\hat{p}_1 = \frac{10}{85} = 0.118 \quad \hat{p}_2 = \frac{37}{92} = 0.402$$

$$\sigma_d \approx \sqrt{\frac{(0.118)(0.882)}{85} + \frac{(0.402)(0.598)}{92}} = 0.0619$$

The observed difference is 0.118 – 0.402 = – 0.284, and the critical z-scores are ±1.96. Thus, the confidence interval estimate is – 0.284 ± 1.96(0.0619) = –0.284 ± 0.121.] The manager can be 95% sure that the proportion of all customers passing through the express line who use checks is between 0.163 and 0.405 lower than the proportion going through the regular line who use checks.

With regard to choosing a sample size, $\sqrt{p(1-p)}$ is at most 0.5. Thus

$$\sqrt{p(1-p)\left(\frac{1}{n_1} + \frac{1}{n_2}\right)} \le (0.5)\sqrt{\frac{1}{n_1} + \frac{1}{n_2}}$$

Now, if we simplify by insisting that $n_1 = n_2 = n$, the above statement can be reduced as follows:

$$(0.5)\sqrt{\frac{1}{n} + \frac{1}{n}} = (0.5)\sqrt{\frac{2}{n}} = \frac{0.5\sqrt{2}}{\sqrt{n}}$$

(Note: this formula is not given on the AP exam formula page.)

➡ EXAMPLE 11.13

A pollster wants to determine the difference between the proportions of high-income voters and low-income voters who support a decrease in the capital gains tax. If the answer must be known to within ±0.02 at the 95% confidence level, what size samples should be taken?

Answer: Assuming we will pick the same size samples for the two sample proportions, we have $\sigma_d \le \frac{0.5\sqrt{2}}{\sqrt{n}}$ and $1.96\sigma_d \ge 0.02$. Thus $\frac{1.96(0.5)\sqrt{2}}{\sqrt{n}} \ge 0.02$.

Algebraically we find that $\sqrt{n} \ge \frac{1.96(0.5)\sqrt{2}}{0.02} = 69.3$. Therefore, $n \ge 69.3^2 = 4802.5$, and the pollster should use 4803 people for each sample.

CONFIDENCE INTERVAL FOR A MEAN

We are interested in estimating a population mean μ by considering a single sample mean $\bar{x}$. This sample mean is just one of a whole universe of sample means, and from Topic 10 we remember that if n is sufficiently large,

1. the set of all sample means is approximately normally distributed.
2. the mean of the set of sample means equals μ, the mean of the population.
3. the standard deviation $\sigma_{\bar{x}}$ of the set of sample means is approximately equal to $\frac{\sigma}{\sqrt{n}}$, that is, equal to the standard deviation of the whole population divided by the square root of the sample size.

Frequently we do not know σ, the population standard deviation. In such cases, we must use s, the *standard deviation of the sample*, as an estimate of σ. In this case $\frac{s}{\sqrt{n}}$ is called the *standard error*, $SE(\bar{x})$, and is used as an estimate for $\sigma_{\bar{x}} = \frac{\sigma}{\sqrt{n}}$. (Note that we use t-distributions instead of the standard normal curve whenever σ is unknown, no matter what the sample size.)

Remember that in making calculations and drawing conclusions from a specific sample, it is important that the sample be a *simple random sample* and be no more than 10% of the population.

➡ EXAMPLE 11.14 _____

A bottling machine is operating with a standard deviation of 0.12 ounce. Suppose that in an SRS of 36 bottles the machine inserted an average of 16.1 ounces into each bottle.

(a) Estimate the mean number of ounces in all the bottles this machine fills. More specifically, give an interval within which we are 95% certain that the mean lies.

Answer: For samples of size 36, the sample means are approximately normally distributed with a standard deviation of $\sigma_{\bar{x}} = \frac{\sigma}{\sqrt{n}} = \frac{0.12}{\sqrt{36}} = 0.02$. From Topic 9 we know that 95% of the sample means should be within 1.96 standard deviations of the population mean. Equivalently, we are 95% certain that the population mean is within 1.96 standard deviations of any sample mean. In our case, $16.1 \pm 1.96(0.02) = 16.1 \pm 0.0392$, and we are 95% sure that the mean number of ounces in all bottles is between 16.0608 and 16.1392. This is called a *95% confidence interval estimate*. [On the TI-84, under STAT and then TESTS, go to ZInterval. With Inpt:Stats, σ:.12, $\bar{x}$:16.1, n:36, and C-Level:.95, Calculate gives (16.061, 16.139).]

NOTE

On the Casio Prizm, under STATISTICS and then INTR, go to *t* and 1-Sample.

(b) How about a 99% confidence interval estimate?

Answer: Here, $16.1 \pm 2.576(0.02) = 16.1 \pm 0.0515$, and we are 99% sure that the mean number of ounces in all bottles is between 16.0485 and 16.1515. [ZInterval gives (16.048, 16.152).] We can also say, using the definition of confidence *level*, that if the sampling procedure were repeated many times, about 99% of the resulting confidence intervals would contain the true population mean (thus we're 99% confident that the method worked for the interval we got).

Note that when we wanted a higher certainty (99% instead of 95%), we had to settle for a larger, less specific interval (± 0.0515 instead of ± 0.0392).

➡ EXAMPLE 11.15

At a certain plant, batteries are being produced with a life expectancy that has a variance of 5.76 months squared. Suppose the mean life expectancy in an SRS of 64 batteries is 12.35 months.

(a) Find a 90% confidence interval estimate of life expectancy for all the batteries produced at this plant.

Answer: We are given an SRS, and *n* = 64 is less than 10% of all batteries. The standard deviation of the population is $\sigma = \sqrt{5.76} = 2.4$, and the standard deviation of the sample means is $\sigma_{\bar{x}} = \frac{\sigma}{\sqrt{n}} = \frac{2.4}{\sqrt{64}} = 0.3$. The 90% confidence interval estimate for the population mean is 12.35 ± 1.645(0.3) = 12.35 ± 0.4935. Thus we are 90% certain that the mean life expectancy of the batteries is between 11.8565 and 12.8435 months. [A calculator gives (11.857, 12.843).]

(b) What would the 90% confidence interval estimate be if the sample mean of 12.35 had come from a sample of 100 batteries?

Answer: The standard deviation of the sample means would then have been $\sigma_{\bar{x}} = \frac{\sigma}{\sqrt{n}} = \frac{2.4}{\sqrt{100}} = 0.24$, and the 90% confidence interval estimate would be 12.35 ± 1.645(0.24) = 12.35 ± 0.3948. [A calcualtor gives (11.955, 12.745).]

Note that when the sample size increased (from 64 to 100), the same sample mean resulted in a narrower, more specific interval (±0.3948 versus ±0.4935).

➡ EXAMPLE 11.16

A new drug results in lowering the heart rate by varying amounts with a standard deviation of 2.49 beats per minute.

(a) Find a 95% confidence interval estimate for the mean lowering of the heart rate in all patients if a 50-person SRS averages a drop of 5.32 beats per minute.

Answer: We are given an SRS, and *n* = 50 is less than 10% of all people. The standard deviation of sample means is $\sigma_{\bar{x}} = \frac{\sigma}{\sqrt{n}} = \frac{2.49}{\sqrt{50}} = 0.352$.

We are 95% certain that the mean lowering of the heart rate is 5.32 ± 1.96(0.352) = 5.32 ± 0.69 or between 4.63 and 6.01 heartbeats per minute. [A calculator gives (4.6298, 6.0102).]

NOTE

normalcdf(-2.13,2.13) = 0.9668

(b) With what certainty can we assert that the new drug lowers the heart rate by a mean of 5.32 ± 0.75 beats per minute?

Answer: Converting ±0.75 to *z*-scores yields $\frac{\pm 0.75}{0.352} = \pm 2.13$. From Table A, these *z*-scores give probabilities of 0.0166 and 0.9834, respectively, so our answer is 0.9834 – 0.0166 = 0.9668. In other words, 5.32 ± 0.75 beats per minute is a 96.68% confidence interval estimate of the mean lowering of the heart rate effected by this drug.

NOTE

You cannot conclude "nearly normal" or even "roughly symmetric" from a boxplot.

Remember, when σ is unknown (which is almost always the case), we use the *t*-distribution instead of the *z*-distribution. This follows if the parent population is normal (or at least nearly normal, which we can roughly check using a dotplot, stemplot, boxplot, histogram, or normal probability plot of the sample data), or if the sample size is large enough for the central limit theorem to apply. If we are not given that the parent population is normal, then for small samples ($n \leq 15$) the sample data should be unimodal and symmetric with no outliers and no skewness; for medium samples ($15 < n < 40$) the sample data should be unimodal and reasonably symmetric with no extreme values and little, if any, skewness; while for large samples ($n \geq 40$) the t-methods can be used no matter what the sample data show.

TIP

Understand that you are not trying to prove that the sample data are normal, but rather trying to infer something about the parent population.

➡ **EXAMPLE 11.17**

When a random sample of 10 cars of a new model were tested for gas mileage, the results showed a mean of 27.2 miles per gallon with a standard deviation of 1.8 miles per gallon. What is a 95% confidence interval estimate for the gas mileage achieved by this model? (Assume that the population of mpg results for all the new model cars is approximately normally distributed.)

Answer: The sample is given to be random, 10 cars should be less than 10% of all cars of the new model, the population is approximately normal so the distribution of sample means is approximately normal, and a t-interval may be found. Calculator software (such as `TInterval` on the TI-84 or 1-Sample tInterval on the Casio Prizm) gives (25.912, 28.488). [For instructional purposes we note that the standard error of the sample means is $SE(\bar{x}) = \dfrac{1.8}{\sqrt{10}} = 0.569$. With $10 - 1 = 9$ degrees of freedom and 2.5% in each tail, the appropriate t-scores are ± 2.262 (from Table B or InvT on the TI-84 or Inverse Student-t on the Casio Prizm). Then $27 \pm 2.262(0.569) = 27.2 \pm 1.3$.] We are 95% confident that the gas mileage of the new model is between 25.9 and 28.5 miles per gallon.

NOTE

df here is the sample size minus the number of parameters being estimated (just the mean), so *df = n − 1*

➡ **EXAMPLE 11.18**

A new process for producing synthetic gems yielded six randomly selected stones weighing 0.43, 0.52, 0.46, 0.49, 0.60, and 0.56 carats, respectively, in its first run. Find a 90% confidence interval estimate for the mean carat weight from this process. (Assume that the population of carats of all gems produced by this new process is approximately normally distributed.)

Answer: The sample is given to be random, six stones is assumed to be less than 10% of all stones produced by this process, the population is approximately normal so the distribution of sample means is approximately normal, and a t-interval may be found. Calculator software (such as `TInterval` on the TI-84 with the data put in a List) gives (0.45797, 0.56203). [For instructional purposes, we note that

$$\bar{x} = \frac{\Sigma x}{n} = \frac{0.43 + 0.52 + 0.46 + 0.49 + 0.60 + 0.56}{6} = \frac{3.06}{6} = 0.51$$

$$s = \sqrt{\frac{\Sigma (x - \bar{x})^2}{n - 1}}$$
$$= \sqrt{\frac{(0.08)^2 + (0.01)^2 + (0.05)^2 + (0.02)^2 + (0.09)^2 + (0.05)^2}{5}}$$
$$= 0.0632$$

and

$$SE(\bar{x}) = \frac{s}{\sqrt{n}} = \frac{0.0632}{\sqrt{6}} = 0.0258$$

NOTE

Or, Inverse
Student-t on the
Casio Prizm, or
STUDENT_ICDF
on the HP Prime.

With $df = 6 - 1 = 5$ and 5% in each tail, the t-scores are ±2.015 (from Table B or InvT on the TI-84). Then $0.51 ± 2.015(0.0258) = 0.51 ± 0.052.$] We are 90% confident that the mean carat weight for all stones produced by this procedure is between 0.458 and 0.562 carats.

On the TI-Nspire the result shows as:

A carats		tInterval *carats*,1,0.9: *stat.results*	
		"Title"	"t Interval"
1	0.43	"CLower"	0.457972
2	0.52	"CUpper"	0.562028
3	0.46	"$\bar{x}$"	0.51
4	0.49	"ME"	0.052028
5	0.6	"df"	5.
6	0.56	"sx := s$_{n-1}$x"	0.063246
		"n"	6.

TIP

Read the
question
carefully!
Be sure you
understand
exactly what
you are being
asked to do or
find or explain.

➡ **EXAMPLE 11.19** _____

A survey was conducted involving 250 out of 125,000 families living in a city. The average amount of income tax paid per family in the sample was $3540 with a standard deviation of $1150. Establish a 99% confidence interval estimate for the total taxes paid by all the families in the city. (Assume that the population of all income taxes paid is approximately normally distributed.)

Answer: $n = 250$ is less than 10% of 125,000, but we must assume that the 250 families was an SRS of the 125,000 families. We are given $\bar{x} = 3540$ and $s = 1150$. We use s as an estimate for σ and calculate the standard error of the sample means to be $SE(\bar{x}) = \frac{s}{\sqrt{n}} = \frac{1150}{\sqrt{250}} = 72.73$.

Since σ is unknown, we use a t-distribution. In this case we obtain $3540 ± 2.596(72.73) = 3540 ± 188.8$ and then $125,000(3540 ± 188.8) = 442,500,000 ± 23,600,000$. That is, we are 99% confident that the total tax paid by all families in the city is between $418,900,000 and $466,100,000.

_____

Statistical principles are useful not only in analyzing data but also in setting up experiments, and one important consideration is the choice of a sample size. In making interval estimates of population means, we have seen that each inference must go hand in hand with an associated confidence level statement. Generally, if we want a smaller, more precise interval estimate, we either decrease the degree of confidence or increase the sample size. Similarly, if we want to increase the degree of confidence, we either accept a wider interval estimate or increase the sample size. Thus, choosing a larger sample size always seems desirable; in the real world, however, time and cost considerations are involved.

➡ **EXAMPLE 11.20** _____

Ball bearings are manufactured by a process that results in a standard deviation in diameter of 0.025 inch. What sample size should be chosen if we wish to be 99% sure of knowing the diameter to within ±0.01 inch?

Answer: We have

$$\sigma_{\bar{x}} = \frac{\sigma}{\sqrt{n}} = \frac{0.025}{\sqrt{n}} \quad \text{and} \quad 2.576\sigma_{\bar{x}} \le 0.01$$

Thus

$$2.576\left(\frac{0.025}{\sqrt{n}}\right) \leq 0.01$$

Algebraically, we find that

$$\sqrt{n} \geq \frac{2.576(0.025)}{0.01} = 6.44$$

so $n \geq 41.5$. We choose a sample size of 42.

A formula for the above calculation is

$$n = \left(\frac{z\sigma}{error}\right)^2$$

(Note: This formula is not given on the AP exam formula page.)

Note that we can obtain a better estimate by recalculating using a t-score obtained from $df = n - 1$ where n is the value obtained from using the z-score. For example, in the above problem, a 99% confidence interval with $df = 42 - 1 = 41$ gives $t = \pm 2.701$. Then $2.701\left(\frac{0.025}{\sqrt{n}}\right) \leq 0.01$ results in $n = 46$.

TIP

Understand that you use a critical *z*-value here because to use a *t*-value you would need to know *df*, which is unknown without starting with a sample size.

CONFIDENCE INTERVAL FOR A DIFFERENCE BETWEEN TWO MEANS

We have the following information about the sampling distribution of $\bar{x}_1 - \bar{x}_2$:

1. The set of all differences of sample means is approximately normally distributed.
2. The mean of the set of differences of sample means equals $\mu_1 - \mu_2$, the difference of population means.
3. The standard deviation $\sigma_{\bar{x}_1-\bar{x}_2}$ of the set of differences of sample means is approximately equal to $\sqrt{\dfrac{\sigma_1^2}{n_1} + \dfrac{\sigma_2^2}{n_2}}$.

In making calculations and drawing conclusions from specific samples, it is important both that the samples be *simple random samples* and that they be taken *independently* of each other.

➡ **EXAMPLE 11.21** _____

A 30-month study is conducted to determine the difference in the numbers of accidents per month occurring in two departments in an assembly plant. Suppose the first department averages 12.3 accidents per month with a standard deviation of 3.5, while the second averages 7.6 accidents with a standard deviation of 3.4. Determine a 95% confidence interval estimate for the difference in the numbers of accidents per month. (Assume that the two populations are independent and approximately normally distributed.)

Answer: We are not given that the sample is random so must assume that the 30 months are representative. We are given that the two populations are independent and approximately normal, so the distributions of sample means are approximately normal, and a *t*-interval may be found. Calculator software (such as `2-SampTInt` on the TI-84 or 2-Sample tInterval on the Casio Prizm) gives (2.9167, 6.4833). [For instructional purposes we note

$$n_1 = 30 \qquad n_2 = 30$$
$$\bar{x}_1 = 12.3 \qquad \bar{x}_2 = 7.6$$
$$s_1 = 3.5 \qquad s_2 = 3.4$$

$$\sigma_{\bar{x}_1 - \bar{x}_2} = \sqrt{\frac{\sigma_1^2}{n_1} + \frac{\sigma_2^2}{n_2}} \approx \sqrt{\frac{s_1^2}{n_1} + \frac{s_2^2}{n_2}} = \sqrt{\frac{(3.5)^2}{30} + \frac{(3.4)^2}{30}} = 0.89$$

The observed difference is 12.3 – 7.6 = 4.7, and with $df = (n_1 - 1) + (n_2 - 1) = 58$, the critical *t*-scores are ±2.00. Thus the confidence interval estimate is 4.7 ± 2.00(0.89) = 4.7 ± 1.78.]

We are 95% confident that the first department has between 2.92 and 6.48 more accidents per month than the second department.

➥ EXAMPLE 11.22

A hardware store owner wishes to determine the difference between the drying times of two brands of paint. Suppose the standard deviation between cans in each population is 2.5 minutes. How large a sample (same number) of each must the store owner use if he wishes to be 98% sure of knowing the difference to within 1 minute?

Answer:

$$\sigma_{\bar{x}_1 - \bar{x}_2} = \sqrt{\frac{(2.5)^2}{n} + \frac{(2.5)^2}{n}} = \frac{3.536}{\sqrt{n}}$$

With a critical *z*-score of 2.326, we have $2.326 \left(\frac{3.536}{\sqrt{n}} \right) \leq 1$, so $\sqrt{n} \geq 8.22$ and $n \geq 67.6$.

Thus the owner should test samples of 68 paint patches from each brand.

In the case of independent samples, there is also a procedure based on the assumption that both original populations not only are normally distributed but also have equal variances. Then we can get a better estimate of the common population variance by *pooling* the two sample variances. But it's never wrong not to pool.

The analysis and procedure discussed in this section require that the two samples being compared be independent of each other. However, many experiments and tests involve comparing two populations for which the data naturally occur in pairs. In this case of *paired differences*, the proper procedure is to run a one-sample analysis on the single variable consisting of the differences from the paired data.

➡️ EXAMPLE 11.23

An SAT preparation class of 30 randomly selected students produces the following improvement in scores:

	First Score	Second Score	Improvement Second Score–First Score
Mean	1640.17	1703.37	63.2
Standard Deviation	131.64	128.59	41.887

Find a 90% confidence interval estimate of the mean improvement in test scores.

Answer: It would be wrong to calculate a confidence interval for a difference between two means using the means and standard deviations of the first and second scores. The condition/assumption of independence between the two samples is violated! The proper procedure is a one-sample t-interval on the set of differences (improvement) between the scores for each of the 30 students.

We are given a random sample, and $n = 30$ is large enough so that by the CLT, the distribution of sample means is approximately normal. So a *t*-interval may be found. Using $\bar{x} = 63.2$ and $s = 41.887$, calculator software (such as T-Interval on the TI-84 or 1-Sample tInterval on the Casio Prizm) gives (50.206, 76.194). We are 90% confident that the mean improvement in test scores is between 50.2 and 76.2.

When to do a two-sample analysis versus a matched pair analysis can be confusing. The key idea is whether or not the two groups are independent. The difference has to do with design. What is the average reaction time of people who are sober compared to that of people who have had two beers? If our experimental design calls for randomly assigning half of a group of volunteers to drink two beers and then comparing reaction times with the remaining sober volunteers, then a two-sample analysis is called for. However, if our experimental design calls for testing reaction times of all the volunteers and then giving all the volunteers two beers and retesting, then a matched pair analysis is called for. Matching may come about as above because you have made two measurements on the same person, or measurements might be made on sets of twins, or between salaries of president and provost at a number of universities, etc. The key is whether the two sets of measurements are independent, or related in some way relevant to the question under consideration.

Note the
similarities
between
$b_1 \pm t^*SE(b_1)$,

$\hat{p} \pm z^*SE(\hat{p})$,

and

$\bar{x} \pm t^*SE(\bar{x})$.

CONFIDENCE INTERVAL FOR THE SLOPE OF A LEAST SQUARES REGRESSION LINE

In Topic 3, we discussed the least squares line

$$\hat{y} = \bar{y} + b_1 (x - \bar{x})$$

where b_1 is the slope of the line. This slope is readily found using the statistical software on a calculator. It is an estimate of the slope β of the true regression line. A confidence interval estimate for β can be found using t-scores, where $df = n - 2$, and the appropriate standard deviation, the standard error of the slope, written s_{b_1} or $SE(b_1)$, is

$$s_{b_1} = \frac{\sqrt{\dfrac{\Sigma(y_i - \hat{y}_i)^2}{n-2}}}{\sqrt{\Sigma(x_i - \bar{x})^2}}$$

where the sum in the numerator is the sum of the squared residuals (see Topic 3) and the sum in the denominator is the sum of the squared deviations from the mean. The value of s_{b_1} is found at the same time that the statistical software on your calculator finds the value

of b_1. Some calculators calculate the numerator $s = \sqrt{\dfrac{\Sigma(y_i - \hat{y}_i)^2}{n-2}}$, and then the denominator

can be calculated by $\sqrt{\Sigma(x_i - \bar{x})^2} = s_x \sqrt{n-1}$.

   Assumptions here include: (1) the sample must be randomly selected; (2) the scatterplot should be approximately linear; (3) there should be no apparent pattern in the residuals plot (On the TI-84, after Stat → Calc → LinReg, the list of residuals is stored in RESID under LIST NAMES); and (4) the distribution of the residuals should be approximately normal. (The fourth point can be checked by a histogram, dotplot, stemplot, or a normal probability plot of the residuals.)

⮕ **EXAMPLE 11.24** _____

A random sample of ten high school students produced the following results for number of hours of television watched per week and GPA.

TV hours	12	21	8	20	16	16	24	0	11	18
GPA	3.1	2.3	3.5	2.5	3.0	2.6	2.1	3.8	2.9	2.6

Determine the least squares line and give a 95% confidence interval estimate for the true slope.

Answer: Checking the assumptions:

We are told that the data come from a *random* sample of students.

Scatterplot is approximately linear.

No apparent pattern is evident in residuals plot.

Distribution of residuals is approximately normal.

Checked by histogram.

Checked by normal probability plot.

Now using the statistics software on a calculator gives

$$\hat{y} = 3.892 - 0.07202x$$

Putting the data into two Lists and using calculator software (such as `LinRegTInt` on the TI-84) gives (−0.0887, −0.0554).

(continued)

TIP

If you refer to a graph, whether it is a histogram, boxplot, stemplot, scatterplot, residuals plot, normal probability plot, or some other kind of graph, you should *roughly draw it*. It is not enough to simply say, "I did a normal probability plot of the residuals on my calculator and it looked linear."

NOTE

Remember from the Checking Normality section in Topic Nine that if the normal probability plot is nearly straight, the data is nearly normal.

We are thus 95% confident that each additional hour before the television each week is associated with between a 0.055 and a 0.089 drop in GPA.

On the TI-Nspire one can show the following:

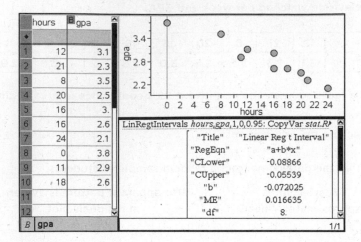

Remember we must be careful about drawing conclusions concerning cause and effect; it is possible that students who have lower GPAs would have these averages no matter how much television they watched per week.

Note that $s = \sqrt{\dfrac{\sum(y-\hat{y})^2}{n-2}}$, called the *standard deviation of the residuals*, is sometimes written s_e. It is a measure of the spread about the regression line, and is almost always in the regression output (usually simply labeled as s or S). Note that s_{b_1} is sometimes written $SE(b_1)$ and is also referred to as the *standard error of the slope*. So we can also write $SE(b_1) = \dfrac{s_e}{s_x\sqrt{n-1}}$. We see that increasing s_e (that is, increasing spread about the line) *increases* the slope's standard error, while increasing s_x (that is, increasing the range of x-values) or increasing n (the sample size) *decreases* the slope's standard error.

TIP

Practice how to read *generic computer output*, and note that you usually will *not* need to use all the information given.

➡ **EXAMPLE 11.25**

Information concerning SAT verbal scores and SAT math scores was collected from 15 randomly selected subjects. A linear regression performed on the data using a statistical software package produced the following printout:

```
Dependent variable: Math
Variable        Coef          SE Coef          T           Prob
Constant        92.5724       31.75            2.92        0.012
Verbal          0.763604      0.05597          13.6        0.000
S = 16.69    R-Sq = 93.5%    R-Sq(adj) = 93.0%
```

(continued)

What is the regression equation?

Answer: Assuming that all assumptions for regression are met (we are given that the sample is random, and the scatterplot is approximately linear), the *y*-intercept and slope of the equation are found in the `Coef` column of the above printout.

$$\widehat{Math} = 92.57 + 0.764 \text{ Verbal}$$

What is a 95% confidence interval estimate for the slope of the regression line?

Answer: The standard deviation of the residuals is S = 16.69 and the standard error of the slope is S_{b_1} = 0.05597. With 15 data points, *df* = 15 − 2 = 13, and the critical *t*-values are ±2.160. The 95% confidence interval of the true slope is:

$$b_1 \pm t\, s_{b_1} = 0.764 \pm 2.160(0.05597) = 0.764 \pm 0.121$$

We are 95% confident that for every 1-point increase in verbal SAT score, the average increase in math SAT score is between 0.64 and 0.89.

➡ EXAMPLE 11.26

Consider the following scatterplot and regression analysis of the 15 data points:

Scatterplot of y versus x

95 percent confidence interval for the slope β: (0.325,0.430)

Suppose the data point (35,14) is added. Would the confidence interval change, and, if so, would the margin of error increase or decrease?

Answer: (35,14) appears to follow the linear trend of the rest of the data. The slope of the regression line will change little, if at all. However, with more points following the same trend, the standard error of the slope will be smaller, and thus the margin of error will decrease. In this example, the 95% confidence interval for the slope with the additional data point is actually (0.339,0.415).

What would happen if instead the data point (12,1) is added?

Answer: With the *x*-coordinate, 12, appearing close to the mean of all the *x*-coordinates, the slope of the regression line will change little, if at all; however, the *y*-intercept will change, and both the standard error of the slope and the margin of error of the confidence interval will increase. In this example, the 95% confidence interval for the slope with the additional data point is actually (0.272,0.479).

SUMMARY

- Important assumptions/conditions always must be checked before calculating confidence intervals.
- We are never able to say exactly what a population parameter is; rather, we say that we have a certain confidence that it lies in a certain interval.
- If we want a narrower interval, we must either decrease the confidence level or increase the sample size.
- If we want a higher level of confidence, we must either accept a wider interval or increase the sample size.
- The level of confidence refers to the percentage of samples that produce intervals that capture the true population parameter.
- When scoring a confidence interval free-response question on the exam, readers will look for whether you:
  1. Identify the procedure by name (such as one-sample z-interval for a population proportion) or formula (such as $\hat{p} \pm z^* \sqrt{\dfrac{\hat{p}(1-\hat{p})}{n}}$) and check the conditions.
  2. Compute a proper interval (it is sufficient to simply copy the interval from your calculator calculation).
  3. Interpret the interval in context (the conclusion must mention the confidence level and refer to the population, not to the sample).

QUIZ 1

Multiple-Choice Questions

Directions: The questions that follow are each followed by five suggested answers. Choose the response that best answers the question.

1. For a class project, an AP Statistics student wishes to determine a 95% confidence interval for the proportion of students in the school building who have Facebook pages. The total number of students in the building is 3520. Which of the following combinations of n and $\hat{p}$ will satisfy the conditions for inference?

   (A) $n = 15$ and $\hat{p} = 0.67$
   (B) $n = 25$ and $\hat{p} = 0.35$
   (C) $n = 50$ and $\hat{p} = 0.17$
   (D) $n = 75$ and $\hat{p} = 0.15$
   (E) $n = 100$ and $\hat{p} = 0.09$

2. Changing from a 95% confidence interval estimate for a population proportion to a 99% confidence interval estimate, with all other things being equal,

   (A) increases the interval size by 4%.
   (B) decreases the interval size by 4%.
   (C) increases the interval size by 31%.
   (D) decreases the interval size by 31%.
   (E) This question cannot be answered without knowing the sample size.

3. Most recent tests and calculations estimate at the 95% confidence level that the maternal ancestor to all living humans called mitochondrial Eve lived 273,000 ±177,000 years ago. What is meant by "95% confidence" in this context?

   (A) A confidence interval of the true age of mitochondrial Eve has been calculated using z-scores of ±1.96.
   (B) A confidence interval of the true age of mitochondrial Eve has been calculated using t-scores consistent with $df = n - 1$ and tail probabilities of ±.025.
   (C) There is a 0.95 probability that mitochondrial Eve lived between 96,000 and 450,000 years ago.
   (D) If 20 random samples of data are obtained by this method, and a 95% confidence interval is calculated from each, then the true age of mitochondrial Eve will be in 19 of these intervals.
   (E) 95% of all random samples of data obtained by this method will yield intervals that capture the true age of mitochondrial Eve.

4. A confidence interval estimate is determined from the GPAs of a simple random sample of n students. All other things being equal, which of the following will result in a smaller margin of error?

(A) A smaller confidence level
(B) A larger sample standard deviation
(C) A smaller sample size
(D) A larger population size
(E) A smaller sample mean

5. In an SRS of 50 teenagers, two-thirds said they would rather text a friend than call. What is the 98% confidence interval for the proportion of all teens who would rather text than call a friend?

(A) $\dfrac{2}{3} \pm 1.96 \sqrt{\dfrac{(2/3)(1/3)}{50}}$

(B) $\dfrac{2}{3} \pm 2.326 \sqrt{\dfrac{(2/3)(1/3)}{50}}$

(C) $\dfrac{2}{3} \pm 2.405 \sqrt{\dfrac{(2/3)(1/3)}{50}}$

(D) $\dfrac{2}{3} \pm 1.96 \dfrac{\sqrt{(2/3)(1/3)}}{50}$

(E) $\dfrac{2}{3} \pm 2.326 \dfrac{\sqrt{(2/3)(1/3)}}{50}$

6. A 1993 *Los Angeles Times* poll of 1703 adults revealed that only 17% thought the media was doing a "very good" job. With what degree of confidence can the newspaper say that 17% ± 2% of all adults believe the media is doing a "very good" job?

(A) 72.9%
(B) 90.0%
(C) 95.0%
(D) 97.2%
(E) 98.6%

7. Donald Trump makes heavy use of Twitter to broadcast his agenda. A market analyst wants to know what percentage of corporate CEOs make similar use of Twitter. What size sample should be obtained to determine with 98% confidence the use level to within 2.5%?

   (A) $n \le \sqrt{\dfrac{0.025}{2.326 \times 0.5}}$

   (B) $n \le \sqrt{\dfrac{2.326 \times 0.5}{0.025}}$

   (C) $n \le \left(\dfrac{2.326 \times 0.5}{0.025}\right)^2$

   (D) $n \ge \sqrt{\dfrac{2.326 \times 0.5}{0.025}}$

   (E) $n \ge \left(\dfrac{2.326 \times 0.5}{0.025}\right)^2$

8. One gallon of gasoline is put in each of 30 test autos, and the resulting mileage figures are tabulated with $\bar{x} = 28.5$ and $s = 1.2$. Determine a 95% confidence interval estimate of the mean mileage of all comparable autos.

   (A) (28.46, 28.54)
   (B) (28.42, 28.58)
   (C) (28.1, 28.9)
   (D) (27.36, 29.64)
   (E) (27.3, 29.7)

9. A company owns 335 trucks. For an SRS of 30 of these trucks, the average yearly road tax paid is $9540 with a standard deviation of $1205. What is a 99% confidence interval estimate for the total yearly road taxes paid for the 335 trucks?

   (A) $9540 ± $103
   (B) $9540 ± $567
   (C) $3,196,000 ± $606
   (D) $3,196,000 ± $35,000
   (E) $3,196,000 ± $203,000

10. What sample size should be chosen to find the mean number of absences per month for school children to within ±0.2 at a 95% confidence level if it is known that the standard deviation is 1.1?

(A) $n \le \sqrt{\dfrac{0.2}{1.96 \times 1.1}}$

(B) $n \le \sqrt{\dfrac{1.96 \times 1.1}{0.2}}$

(C) $n \le \left(\dfrac{1.96 \times 1.1}{0.2}\right)^2$

(D) $n \ge \sqrt{\dfrac{1.96 \times 1.1}{0.2}}$

(E) $n \ge \left(\dfrac{1.96 \times 1.1}{0.2}\right)^2$

11. The National Research Council of the Philippines reported that 210 of 361 members in biology are women, but only 34 of 86 members in mathematics are women. Establish a 95% confidence interval estimate of the difference in proportions of all women in biology and all women in mathematics in the Philippines.

(A) 0.186 ± 0.115
(B) 0.186 ± 0.154
(C) 0.395 ± 0.103
(D) 0.543 ± 0.154
(E) 0.582 ± 0.051

12. A researcher plans to investigate the difference between the proportion of psychiatrists and the proportion of psychologists who believe that most emotional problems have their root causes in childhood. How large a sample should be taken (same number for each group) to be 90% certain of knowing the difference to within ±0.03?

(A) 39
(B) 376
(C) 752
(D) 1504
(E) 3007

13. In a study aimed at reducing developmental problems in low-birth-weight (under 2500 grams) babies (*Journal of the American Medical Association*, June 13, 1990, page 3040), 347 infants were exposed to a special educational curriculum while 561 did not receive any special help. After 3 years the children exposed to the special curriculum showed a mean IQ of 93.5 with a standard deviation of 19.1; the other children had a mean IQ of 84.5 with a standard deviation of 19.9. Find a 95% confidence interval estimate for the difference in mean IQs of all low-birth-weight babies who receive special intervention and those who do not.

(A) $(93.5 - 84.5) \pm 1.97 \sqrt{\dfrac{(19.1)^2}{347} + \dfrac{(19.9)^2}{561}}$

(B) $(93.5 - 84.5) \pm 1.97 \left(\dfrac{19.1}{\sqrt{347}} + \dfrac{19.9}{\sqrt{561}} \right)$

(C) $(93.5 - 84.5) \pm 1.65 \sqrt{\dfrac{(19.1)^2}{347} + \dfrac{(19.9)^2}{561}}$

(D) $(93.5 - 84.5) \pm 1.65 \left(\dfrac{19.1}{\sqrt{347}} + \dfrac{19.9}{\sqrt{561}} \right)$

(E) $(93.5 - 84.5) \pm 1.65 \sqrt{\dfrac{(19.1)^2 + (19.9)^2}{347 + 561}}$

14. Nine subjects, 87 to 96 years old, were given 8 weeks of progressive resistance weight training (*Journal of the American Medical Association*, June 13, 1990, page 3032). Strength before and after training for each individual was measured as maximum weight (in kilograms) lifted by left knee extension:

Before:	3	3.5	4	6	7	8	8.5	12.5	15
After:	7	17	19	12	19	22	28	20	28

Find a 95% confidence interval estimate for the strength gain of all 87- to 96-year olds who are given the training.

(A) 11.61 ± 3.03
(B) 11.61 ± 3.69
(C) 11.61 ± 3.76
(D) 19.11 ± 1.25
(E) 19.11 ± 3.69

15. In a random sample of 25 professional baseball players, their salaries (in millions of dollars) and batting averages result in the following regression analysis:

```
The regression equation is    Batting = 0.234 + 0.00805 Salary

Predictor         Coef    SE Coef      T      P
Constant      0.233577   0.005883   39.70  0.000
Salary        0.008051   0.001058    7.61  0.000

S = 0.0169461    R-Sq = 71.6%    R-Sq(adj) = 70.3%
```

Which of the following gives a 98% confidence interval for the slope of the regression line?

(A) $0.008051 \pm 2.326(0.001058)$

(B) $0.008051 \pm 2.326(0.0169461)$

(C) $0.008051 \pm 2.485\left(\dfrac{0.001058}{\sqrt{25}}\right)$

(D) $0.008051 \pm 2.492\left(\dfrac{0.0169461}{\sqrt{25}}\right)$

(E) $0.008051 \pm 2.500(0.001058)$

Free-Response Questions

> **Directions:** You must show all work and indicate the methods you use. You will be graded on the correctness of your methods and on the accuracy of your final answers.

1. An SRS of 1000 voters finds that 57% believe that competence is more important than character in voting for president of the United States.

   (a) Determine a 95% confidence interval estimate for the percentage of all voters who believe competence is more important than character.

   (b) If your parents know nothing about statistics, how would you explain to them why you can't simply say that 57% of voters believe that competence is more important.

   (c) Also explain to your parents what is meant by the 95% confidence level.

2. In a random sample of 500 new births in the United States, 41.2% were to unmarried women, while in a random sample of 400 new births in the United Kingdom, 46.5% were to unmarried women.

    (a) Calculate a 95% confidence interval for the difference in the proportions of new births to unmarried women in the United States and United Kingdom.

    (b) Does the confidence interval support the belief by a UN health care statistician that the proportions of new births to unmarried women is different in the United States and United Kingdom? Explain.

3. In a July 2008 study of 1050 randomly selected smokers, 74% said they would like to give up smoking.

    (a) At the 95% confidence level, what is the margin of error?

    (b) Explain the meaning of "95% confidence interval" in this example.

    (c) Explain the meaning of "95% confidence level" in this example.

    (d) Give an example of possible response bias in this example.

    (e) If we want instead to be 99% confident, would our confidence interval need to be wider or narrower?

    (f) If the sample size were greater, would the margin of error be smaller or greater?

4. An SRS of 40 inner city gas stations shows a mean price for regular unleaded gasoline to be $3.45 with a standard deviation of $0.05, while an SRS of 120 suburban stations shows a mean of $3.38 with a standard deviation of $0.08.

    (a) Construct 95% confidence interval estimates for the mean price of regular gas in inner city and in suburban stations.

    (b) The confidence interval for the inner city stations is wider than the interval for the suburban stations even though the standard deviation for inner city stations is less than that for suburban stations. Explain why this happened.

    (c) Based on your answer in part (a), are you confident that the mean price of inner city gasoline is less than $3.50? Explain.

5. An SRS of ten brands of breakfast cereals is tested for the number of calories per serving. The following data result: 185, 190, 195, 200, 205, 205, 210, 210, 225, 230.

    Establish a 95% confidence interval estimate for the mean number of calories for servings of all breakfast cereals. Be sure to check assumptions.

6. Information with regard to the assessed values (in $1000) and the selling prices (in $1000) of a random sample of homes sold in an NE market yields the following computer output:

Dependent variable is   **Price**

R squared = 89.8%   R squared (adjusted) = 89.2%
s = 9.508   with   20 – 2 = 18   degrees of freedom

Source	SS	df	MS	F-ratio
Regression	14271.2	1	14271.2	158
Residual	1627.31	18	90.4063	

Variable	Coeff	s.e. of coeff	t-ratio	prob
Constant	0.890087	16.16	0.055	0.956
Assessed	1.0292	0.08192	12.6	0.000

(a) Determine the equation of the least squares regression line.

(b) Construct a 99% confidence interval estimate for the slope of the regression line, and interpret this in context.

QUIZ 2

Multiple-Choice Questions

Directions: The questions that follow are each followed by five suggested answers. Choose the response that best answers the question.

1. A consumer testing agency plans to calculate a 99% confidence interval for the mean mpg for all cars on the road in 2019. Suppose the mpg measurements for the population of interest is actually sharply skewed right. For which of the sample sizes, $n = 30$, 50, or 70, would the sampling distribution of $\bar{x}$ be closest to normal?

   (A) 30
   (B) 50
   (C) 70
   (D) Because of skewness of the population, none of sampling distributions can be approximately normal.
   (E) Because of the CLT, all sampling distributions with $n \geq 30$ are equally approximately normal.

2. In general, how does doubling the sample size change the confidence interval size?

   (A) Doubles the interval size
   (B) Halves the interval size
   (C) Multiplies the interval size by 1.414
   (D) Divides the interval size by 1.414
   (E) This question cannot be answered without knowing the sample size.

3. Suppose (25, 30) is a 90% confidence interval estimate for a population mean μ. Which of the following are true statements?

   (A) There is a 0.90 probability that $\bar{x}$ is between 25 and 30.
   (B) 90% of the sample values are between 25 and 30.
   (C) There is a 0.90 probability that μ is between 25 and 30.
   (D) If 100 random samples of the given size are picked and a 90% confidence interval estimate is calculated from each, then μ will be in 90 of the resulting intervals.
   (E) If 90% confidence intervals are calculated from all possible samples of the given size, μ will be in 90% of these intervals.

4. One month the actual unemployment rate in France was 13.4%. If during that month you took an SRS of 100 Frenchmen and constructed a confidence interval estimate of the unemployment rate, which of the following would have been true?

   (A) The center of the interval was 13.4.
   (B) The interval contained 13.4.
   (C) A 99% confidence interval estimate contained 13.4.
   (D) The z-score of 13.4 was between ±2.576.
   (E) None of the above are true statements.

5. The margin of error in a confidence interval estimate using z-scores covers which of the following?

   (A) Sampling variability
   (B) Errors due to undercoverage and nonresponse in obtaining sample surveys
   (C) Errors due to using sample standard deviations as estimates for population standard deviations
   (D) Type I errors
   (E) Type II errors

6. In a survey funded by Burroughs-Welcome, 750 of 1000 adult Americans said they didn't believe they could come down with a sexually transmitted disease (STD). Construct a 95% confidence interval estimate of the proportion of all adult Americans who don't believe they can contract an STD.

   (A) $0.750 \pm 1.645 \sqrt{1000(0.750)(0.250)}$

   (B) $0.750 \pm 1.645 \sqrt{\dfrac{(0.750)(0.250)}{1000}}$

   (C) $0.750 \pm 1.96 \sqrt{1000(0.750)(0.250)}$

   (D) $0.750 \pm 1.96 \sqrt{\dfrac{(0.750)(0.250)}{1000}}$

   (E) $0.750 \pm 2.576 \sqrt{1000(0.750)(0.250)}$

7. A politician wants to know what percentage of all voters support her position on the issue of forced busing for integration. What size voter sample should be obtained to determine with 90% confidence the support level to within 4%?

   (A) 21
   (B) 25
   (C) 423
   (D) 600
   (E) 1691

8. In an SRS of 80 teenagers, the average number of texts handled in a day was 50 with a standard deviation of 15. What is the 96% confidence interval for the average number of texts handled by all teens daily?

   (A) $50 \pm 2.054(15)$

   (B) $50 \pm 2.054 \dfrac{15}{\sqrt{79}}$

   (C) $50 \pm 2.054 \dfrac{15}{\sqrt{80}}$

   (D) $50 \pm 2.088 \dfrac{15}{\sqrt{79}}$

   (E) $50 \pm 2.088 \dfrac{15}{\sqrt{80}}$

9. The number of accidents per day at a large factory is noted for each of 64 days with $\bar{x} = 3.58$ and $s = 1.52$. With what degree of confidence can we assert that the mean number of accidents per day at the factory is between 3.20 and 3.96?

   (A) 48%
   (B) 63%
   (C) 90%
   (D) 95%
   (E) 99%

10. Hospital administrators wish to learn the average length of stay of all surgical patients. A statistician determines that, for a 95% confidence level estimate of the average length of stay to within ±0.5 days, 50 surgical patients' records will have to be examined. How many records should be looked at to obtain a 95% confidence level estimate to within ±0.25 days?

    (A) 25
    (B) 50
    (C) 100
    (D) 150
    (E) 200

11. In a simple random sample of 300 elderly men, 65% were married, while in an independent simple random sample of 400 elderly women, 48% were married. Determine a 99% confidence interval estimate for the difference between the proportions of all elderly men and women who are married.

(A) $(0.65 - 0.48) \pm 2.326 \sqrt{\dfrac{(0.65)(0.35)}{300} + \dfrac{(0.48)(0.52)}{400}}$

(B) $(0.65 - 0.48) \pm 2.576 \sqrt{\dfrac{(0.65)(0.35)}{300} + \dfrac{(0.48)(0.52)}{400}}$

(C) $(0.65 - 0.48) \pm 2.576 \left(\dfrac{(0.65)(0.35)}{\sqrt{300}} + \dfrac{(0.48)(0.52)}{\sqrt{400}} \right)$

(D) $\left(\dfrac{0.65 + 0.48}{2} \right) \pm 2.576 \sqrt{\dfrac{(0.65)(0.35)}{300} + \dfrac{(0.48)(0.52)}{400}}$

(E) $\left(\dfrac{0.65 + 0.48}{2} \right) \pm 2.807 \sqrt{(0.565)(0.435)\left(\dfrac{1}{300} + \dfrac{1}{400} \right)}$

12. Does socioeconomic status relate to age at time of HIV infection? For 274 high-income HIV-positive individuals the average age of infection was 33.0 years with a standard deviation of 6.3, while for 90 low-income individuals the average age was 28.6 years with a standard deviation of 6.3 (*The Lancet*, October 22, 1994, page 1121). Find a 90% confidence interval estimate for the difference in ages of all high- and low-income people at the time of HIV infection.

(A) 4.4 ± 0.963

(B) 4.4 ± 1.27

(C) 4.4 ± 2.51

(D) 30.8 ± 2.51

(E) 30.8 ± 6.3

13. An engineer wishes to determine the difference in life expectancies of two brands of batteries. Suppose the standard deviation of each brand is 4.5 hours. How large a sample (same number) of each type of battery should be taken if the engineer wishes to be 90% certain of knowing the difference in life expectancies to within 1 hour?

(A) 10

(B) 55

(C) 110

(D) 156

(E) 202

14. A catch of five fish of a certain species yielded the following ounces of protein per pound of fish: 3.1, 3.5, 3.2, 2.8, and 3.4. What is a 90% confidence interval estimate for ounces of protein per pound of this species of fish?

(A) 3.2 ± 0.202
(B) 3.2 ± 0.247
(C) 3.2 ± 0.261
(D) 4.0 ± 0.202
(E) 4.0 ± 0.247

15. Smoking is a known risk factor for cardiovascular diseases and cancer. A study analyzing smoking levels (seven levels from 0.5 to 3.0 packs per day) and risk of dementia (as measured by the Cox hazard ratio) resulted in the following regression analysis:

```
The regression equation is
Hazard ratio = 0.774 + 0.434 Packs

Predictor      Coef      SE Coef        T         P
Constant    0.77372    0.07833       9.88     0.000
Packs       0.43413    0.04292      10.11     0.000

S = 0.0981745  R-Sq = 95.3%  R-Sq(adj) = 94.4%
```

Which of the following gives a 99% confidence interval for the slope of the regression line?

(A) $0.43413 \pm 2.576(0.04292)$
(B) $0.43413 \pm 2.576(0.0981745)$
(C) $0.43413 \pm 3.499\left(\dfrac{0.04292}{\sqrt{7}} \right)$
(D) $0.43413 \pm 3.707\left(\dfrac{0.0981745}{\sqrt{7}} \right)$
(E) $0.43413 \pm 4.032(0.04292)$

Free-Response Questions

1. During the H1N1 pandemic, one published study concluded that if someone in your family had H1N1, you had a 1 in 8 chance of also coming down with the disease. A state health officer tracks a random sample of new H1N1 cases in her state and notes that 129 out of a potential 876 family members later come down with the disease.

   (a) Calculate a 90% confidence interval for the proportion of family members who come down with H1N1 after an initial family member does in this state.

   (b) Based on this confidence interval, is there evidence that the proportion of family members who come down with H1N1 after an initial family member does in this state is different from the 1 in 8 chance concluded in the published study? Explain.

   (c) Would the conclusion in (b) be any different with a 99% confidence interval? Explain.

2. In a simple random sample of 30 subway cars during rush hour, the average number of riders per car was 83.5 with a standard deviation of 5.9. Assume the sample data are unimodal and reasonably symmetric with no extreme values and little, if any, skewness.

   (a) Establish a 90% confidence interval estimate for the average number of riders per car during rush hour. Show your work.

   (b) Assuming the same standard deviation of 5.9, how large a sample of cars would be necessary to determine the average number of riders to within ±1 at the 90% confidence level? Show your work.

3. In a sample of ten basketball players the mean income was $196,000 with a standard deviation of $315,000.

   (a) Assuming all necessary assumptions are met, find a 95% confidence interval estimate of the mean salary of basketball players.

   (b) What assumptions are necessary for the above estimate? Do they seem reasonable here?

4. Bisphenol A (BPA), a synthetic estrogen found in packaging materials, has been shown to leach into infant formula and beverages. Most recently it has been detected in high concentrations in cash register receipts. Random sample biomonitoring data to see if retail workers carry higher amounts of BPA in their bodies than non-retail workers is as follows:

		BPA concentration (μg/L)	
	Sample size	Mean	Standard deviation
Non-retail workers	528	2.43	0.45
Retail workers	197	3.28	0.48

(a) Calculate a 99% confidence interval for the difference in mean BPA body concentrations of non-retail and retail workers.

(b) Does the confidence interval support the belief that retail workers carry higher amounts of BPA in their bodies than non-retail workers? Explain.

5. As of January 1, 2014, Serena Williams had career prize money earnings of $54,380,000 during her 18 year professional career. Among the top active women tennis players, is there a linear relationship between career prize money earnings and years as a pro? Computer output of earnings (millions of $) versus professional years for 10 top active women players are shown below:

Regression Analysis: Earnings versus Pro Years

Predictor	Coef	SE Coef	T	P
Constant	2.69	10.78	0.25	0.809
Pro Years	1.6752	0.8264	2.03	0.077

S = 10.5317   R-Sq = 33.9%   R-Sq(adj) = 25.7%

Analysis of Variance

Source	DF	SS	MS	F	P
Regression	1	455.8	455.8	4.11	0.077
Residual Error	8	887.3	110.9		
Total	9	1343.1			

(a) Determine the equation of the least squares regression line.

(b) Comment on the strength of the linear relationship.

(c) Predict the career earnings for Years = 15 and Years = 30 for active women tennis players, and explain your confidence in both of your answers.

(d) Find the 90% confidence interval for the slope of the regression line, and interpret in context.

(e) A second linear model is found: $\hat{y} = 1.8712(Years)$. Explain why you might prefer this model to the one in (a).

6. A new drug is tested for relief of allergy symptoms. In a double-blind experiment, 12 patients are given varying doses of the drug and report back on the number of hours of relief. The following table summarizes the results:

Dosage (mg)	2	3	4	4	5	6	6	7	8	8	9	10
Duration of relief (hrs)	3	7	6	8	10	8	13	16	15	21	23	24

(a) Determine the equation of the least squares regression line.

(b) Construct a 90% confidence interval estimate for the slope of the regression line, and interpret this in context.

ANSWERS EXPLAINED

Quiz 1

MULTIPLE-CHOICE QUESTIONS

1. **(D)** The relevant condition is that both $n\hat{p}$ and $n\hat{q} \geq 10$, and only Choice (D) with $n\hat{p} = (75)(0.15) = 11.25$ and $n\hat{q} = (75)(0.85) = 63.75$ satisfies this.

2. **(C)** The critical z-scores will go from ± 1.96 to ± 2.576, resulting in an increase in the interval size: $\frac{2.576}{1.96} = 1.31$ or an increase of 31%.

3. **(E)** The 95% refers to the method: 95% of all intervals obtained by this method will capture the true population parameter. Nothing is certain about any particular set of 20 intervals. For any particular interval, the probability that it captures the true parameter is 1 or 0 depending upon whether the parameter is or isn't in it.

4. **(A)** The margin of error varies directly with the critical z-value and directly with the standard deviation of the sample, but inversely with the square root of the sample size. The value of the sample mean and the population size do not affect the margin of error.

5. **(B)** The critical z-score with 0.01 in the tails is 2.326 (`invNorm(0.01)` on the TI-84), and $SE(\hat{p}) = \sqrt{\frac{\hat{p}(1-\hat{p})}{n}} = \sqrt{\frac{(2/3)(1/3)}{50}}$.

6. **(D)** $SE(\hat{p}) = \sqrt{\frac{(0.17)(0.83)}{1703}} = 0.0091$

   $z(0.0091) = 0.02$ $z = 2.20$, $0.9861 - 0.0139 = 97.2\%$ (or normalcdf(–2.20,2.20) = 0.972)

7. **(E)** 0.01 is in each tail, `invNorm(0.99) = 2.326`, $\sqrt{p(1-p)} \leq 0.5$ and $\frac{2.326 \times 0.5}{\sqrt{n}} \leq 0.025$.

8. **(C)** Calculator software gives (28.052, 28.948).

9. **(E)** Using t-scores:

$$335\left[9540 \pm 2.756\left(\frac{1205}{\sqrt{30}}\right)\right] = 335(9540 \pm 606.3) = \$3{,}196{,}000 \pm \$203{,}000$$

   Or TInterval gives (8933.6,10146) and multiplying by 335 gives (2992756,3398910) = 3,195,833 ± 203,077

10. **(E)** $1.96\left(\frac{1.1}{\sqrt{n}}\right) \leq 0.2$

11. **(A)** Calculator software gives (0.07119, 0.30155) or 0.18637 ± 0.11518.

12. **(D)** $1.645\left(\frac{0.5\sqrt{2}}{\sqrt{n}}\right) \leq 0.03, \sqrt{n} \geq 38.77$, and $n \geq 1503.3$; the researcher should choose a sample size of at least 1504.

13. **(A)** $\sigma_d = \sqrt{\frac{(19.1)^2}{347} + \frac{(19.9)^2}{561}}$ and with $df = \min(347 - 1, 561 - 1)$, critical t-scores are ±1.97.

14. **(C)** TInterval on the set of differences gives (7.8514, 15.371) or 11.6112 ± 3.7598.

15. **(E)** The critical t-values with $df = 25 - 2 = 23$ are ± 2.500. Thus, we have $b_1 \pm t^* \times SE(b_1)$ $= 0.008051 \pm (2.500)(0.001058)$.

FREE-RESPONSE QUESTIONS

1. (a) First, identify the confidence interval and check the conditions: This is a one-sample z-interval for the proportion of all voters who believe competence is more important than character. It is given that this is a random sample, we calculate $n\hat{p} = 1000(0.57) = 570 \geq 10$ and $n\hat{q} = 1000(0.43) = 430 \geq 10$, and $n = 1000$ is less than 10% of all voters.

Second, calculate the interval:

Calculator software (such as 1-PropZInt on the TI-84) gives (0.53932, 0.60068). [For instructional purposes in this review book, we note that $\hat{p} = 0.57$ and $\sigma_{\hat{p}} = \sqrt{\dfrac{(0.57)(0.43)}{1000}} = 0.0157$. The critical z-scores associated with the 95% level are ± 1.96. Thus the confidence interval is $0.57 \pm 1.96(0.0157) = 0.57 \pm 0.031$, or between 53.9% and 60.1%.]

Third, interpret in context:

We are 95% confident that between 53.9% and 60.1% of all voters believe competence is more important than character in voting for president.

(b) Explain to your parents that by using a measurement from a sample we are never able to say *exactly* what a population proportion is; rather we are only able to say we are confident that it is within some range of values, in this case between 53.9% and 60.1%.

(c) In 95% of all possible samples of 1000 voters, the method used gives an estimate that is within three percentage points of the true answer.

2. (a) First, identify the confidence interval and check the conditions:

This is a two-sample z-interval for $p_{US} - p_{UK}$, the difference in population proportions of new births to unmarried women in the United States and United Kingdom, that is,

$$(\hat{p}_{US} - \hat{p}_{UK}) \pm z^* \sqrt{\dfrac{\hat{p}_{US}(1 - \hat{p}_{US})}{n_{US}} + \dfrac{\hat{p}_{UK}(1 - \hat{p}_{UK})}{n_{UK}}}$$

It is given that these are random samples, they are clearly independent, and we calculate $n_{US}\hat{p}_{US} = (500)(0.412) = 206 \geq 10$,
$n_{US}(1 - \hat{p}_{US}) = (500)(0.588) = 294 \geq 10$,
$n_{UK}\hat{p}_{UK} = (400)(0.465) = 186 \geq 10$, and
$n_{UK}(1 - \hat{p}_{UK}) = (400)(0.535) = 214 \geq 10$.

The sample sizes, $n_{US} = 500$ and $n_{UK} = 400$, are less than 10% of the numbers of new births in the United States and United Kingdom, respectively.

Second, calculate the interval:

Calculator software (such as `2-PropZInt` on the TI-84) gives (–0.1182, 0.01219). [For instructional purposes in this review book, we note that:

$$(0.412 - 0.465) \pm 1.96 \sqrt{\frac{(0.412)(0.588)}{500} + \frac{(0.465)(0.535)}{400}} = -0.053 \pm 0.065 \text{ or } (-0.118, 0.012).]$$

Third, interpret in context:

We are 95% confident that the difference in proportions, $p_{US} - p_{UK}$, of new births to unmarried women in the United States and United Kingdom is between –0.118 and 0.012.

(b) Because 0 is in the interval of plausible values for the difference of population proportions, this confidence interval does not support the belief by the UN health care statistician that the proportions of new births to unmarried women is different in the United States and United Kingdom.

3. (a) $1.96 \sqrt{\frac{(0.74)(0.26)}{1050}} = 0.0265$. So the margin of error is ±2.65%.

(b) We are 95% confident that between 71.35% and 76.65% of all smokers would like to give up smoking.

(c) If this survey were conducted many times, we would expect about 95% of the resulting confidence intervals to contain the true proportion of all smokers who would like to give up smoking.

(d) Smoking is becoming more undesirable in society as a whole, so some smokers may untruthfully say they would like to stop.

(e) To be more confident we must accept a *wider* interval.

(f) All other things being equal, the greater the sample size, the *smaller* the margin of error.

4. (a) We are calculating one-sample *t*-intervals for the mean price of gas in inner city stations and for the mean price of gas in suburban stations. In each case, we are given random samples and the sample sizes of 40 and 120 are large enough so that by the CLT the distributions of sample means are approximately normal and *t*-intervals may be found. We assume that the sample sizes are less than 10% of the populations of all gas stations. Calculator software gives (3.434, 3.466) for inner city stations and (3.3655, 3.3945) for suburban stations. That is, we are 95% confident that the mean price for gas in all inner city stations is between $3.434 and $3.466, and are 95% confident that the mean price for gas in all suburban stations is between $3.3655 and $3.3945.

(b) Because the sample size of inner city stations is smaller and $SE(\bar{x}) = \frac{s}{\sqrt{n}}$.

(c) $3.50 is greater than anything in the interval (3.434, 3.466), so based on this confidence interval, we are confident that the mean price of inner city gasoline is less than $3.50. Or, you could reason as follows:

Yes, because the standard error of the set of sample means is $\frac{0.05}{\sqrt{40}} = 0.0079$, and so $3.50 is more than three standard deviations away from $3.45.

Thus the probability that the true mean is this far from the sample mean is extremely small.

5. This is a one-sample *t*-interval for the mean number of calories of all breakfast cereals. We are given that we have a random sample, and the nearly normal condition seems reasonable from, for example, either a stemplot or a normal probability plot:

TIP

Linearity in a normal probability plot indicates normality in the data.

Under these conditions the mean calories can be modeled by a *t*-distribution with $n - 1 = 10 - 1 = 9$ degrees of freedom.

Calculator software (such as `TInterval` using `Data` on the TI-84) gives (195.32, 215.68). [For instructional purposes in this review book, we note that a one-sample *t*-interval for the mean gives:

$$\bar{x} \pm t \frac{s}{\sqrt{n}} = 205.5 \pm 2.262 \frac{14.23}{\sqrt{10}} = 205.5 \pm 10.2 \,]$$

We are 95% confident that the true mean number of calories of all breakfast cereals is between 195.3 and 215.7.

6. (a) Assuming that all assumptions for regression are met, the *y*-intercept and slope of the equation are found in the `Coeff` column of the computer printout.

$$\widehat{Selling\ price} = 0.890 + 1.029\,(Assessed\ value)$$

Both the selling price and the assessed value are in $1000s.

(b) From the printout, the standard error of the slope is $s_{b_1} = 0.08192$. With $df = 18$ and 0.005 in each tail, the critical *t*-values are ±2.878. The 99% confidence interval of the true slope is:

$$b_1 \pm t\,s_{b_1} = 1.029 \pm 2.878(0.08192) = 1.029 \pm 0.236$$

We are 99% confident that for every $1 increase in assessed value, the average increase in selling price is between $0.79 and $1.27 (or for every $1000s increase in assessed value, the average increase in selling price is between $790 and $1270).

Quiz 2

MULTIPLE-CHOICE QUESTIONS

1. **(C)** The Central Limit Theorem gives that no matter what the population distribution, the larger the sample size, the closer the sampling distribution of $\bar{x}$ will be to normal.

2. **(D)** Increasing the sample size by a multiple of d divides the interval estimate by $\sqrt{d}$.

3. **(E)** The 90% refers to the method; 90% of all intervals obtained by this method will capture μ. Nothing is sure about any particular set of 100 intervals. For any particular interval, the probability that it captures μ is either 1 or 0 depending on whether μ is or isn't in it.

4. **(E)** There is no guarantee that 13.4 is anywhere near the interval, so none of the statements are true.

5. **(A)** The margin of error has to do with measuring chance variation but has nothing to do with faulty survey design. As long as n is large, s is a reasonable estimate of σ; however, again this is not measured by the margin of error. (With t-scores, there is a correction for using s as an estimate of σ.)

6. **(B)** 1-PropZInt on the TI-84 gives (0.72316, 0.77684).

7. **(C)** $1.645\left(\dfrac{0.5}{\sqrt{n}}\right) \le 0.04, \sqrt{n} \ge 20.563, n \ge 422.8$, so choose $n = 423$.

8. **(E)** LOL, OMG, the critical t-score with 0.02 in the tails is 2.088 (Table B or `invt`(0.02,79) on the TI-84), and $SE(\bar{x}) = \dfrac{s}{\sqrt{n}} = \dfrac{15}{\sqrt{80}}$.

9. **(D)** $SE(\bar{x}) = \dfrac{1.52}{\sqrt{64}} = 0.19, \dfrac{0.38}{0.19} = 2$, and $0.4772 + 0.4772 = 0.9544 \approx 95\%$

10. **(E)** To divide the interval estimate by d without affecting the confidence level, multiply the sample size by a multiple of d^2. In this case, 4(50) = 200.

11. **(B)**
$$n_1 = 300 \quad n_2 = 400$$
$$\hat{p}_1 = 0.65 \quad \hat{p}_2 = 0.48$$
$$\sigma_d = \sqrt{\dfrac{(0.65)(0.35)}{300} + \dfrac{(0.48)(0.52)}{400}} = 0.0372$$
$$(0.65 - 0.48) \pm 2.576(0.0372) = 0.17 \pm 0.096$$

12. **(B)** Calculator software gives (3.1333, 5.6667) or 4.4 ± 1.2667.

13. **(C)** $\sigma_{\bar{x}_1 - \bar{x}_2} = \sqrt{\dfrac{(4.5)^2}{n} + \dfrac{(4.5)^2}{n}} = \dfrac{6.364}{\sqrt{n}}$

$1.645\left(\dfrac{6.364}{\sqrt{n}}\right) \le 1, \sqrt{n} \ge 10.47, n \ge 109.6$; choose $n = 110$.

14. **(C)** TInterval on the TI-84 gives (2.9389, 3.4611) or 3.2 ± 0.2611.

15. **(E)** The critical t-values with $df = 7 - 2 = 5$ are ± 4.032. Thus, we have $b \pm t^* SE(b) = 0.43413 \pm 4.032(0.04292)$.

FREE-RESPONSE QUESTIONS

1. (a) First, identify the confidence interval and check the conditions:

   This is a one-sample z-interval for the proportion of family members who come down with H1N1 after an initial family member does in this state, that is, $\hat{p} \pm z^* \sqrt{\dfrac{\hat{p}(1-\hat{p})}{n}}$.

   It is given that this is a random sample, and we calculate $n\hat{p} = 129 \geq 10$ and $n(1-\hat{p}) = 747 \geq 10$. $n = 876$ is assumed to be less than 10% of the number of all family members of H1N1 patients in the state.

   Second, calculate the interval:

   Calculator software gives $(0.12757, 0.16695)$.

   [For instructional purposes in this review book, we note that $\hat{p} = \dfrac{129}{876} = 0.147$

   and $0.147 \pm 1.645 \sqrt{\dfrac{(0.147)(0.853)}{876}} = 0.147 \pm 0.020$ or $(0.127, 0.167)$]

   Third, interpret in context:

   We are 90% confident that the proportion of all family members who come down with H1N1 after an initial family member does in this state is between 0.127 and 0.167.

   (b) Because $1/8 = 0.125$ is not in the interval of plausible values for the population proportion, there *is* evidence that the proportion of all family members who come down with H1N1 after an initial family member does in this state is different from the 1 in 8 chance concluded in the published study.

   (c) $0.147 \pm 2.576 \sqrt{\dfrac{(0.147)(0.853)}{876}} = 0.147 \pm 0.031$ or $(0.116, 0.178)$ which does include 0.125, so using a 99% confidence interval there is *not* evidence that the proportion of all family members who come down with H1N1 after an initial family member does in this state is different from the 1 in 8 chance concluded in the published study.

2. (a) We are calculating a one-sample t-interval for the mean number of riders per car during rush hour. We are given a random sample, the sample size of 30 is assumed to be less than 10% of all subway cars, and assuming that the sample data are unimodal and reasonably symmetric with no extreme values, the sample size of 30 is large enough so that by the CLT the distribution of sample means is approximately normal and a t-interval may be found. Calculator software gives $(81.67, 85.33)$. [For instructional purposes in this review book, we note that: With $df = n - 1 = 29$, the critical t-scores from Table B (or from InvT on the TI-84) are ± 1.699, and $\bar{x} \pm \dfrac{s}{\sqrt{n}} = 83.5 \pm 1.699 \left(\dfrac{5.9}{\sqrt{30}} \right) = 83.5 \pm 1.83$.] We are 90% confident

NOTE

You can refine
this answer by
recalculating
using a *t*-score
obtained from
df = 95 – 1 = 94:

$1.661\left(\dfrac{5.9}{n}\right) \leq 1,$

and *n* ≥ 97

that the mean number of riders per car during rush hour is between 81.67 and 85.33.

(b) $1.645\left(\dfrac{5.9}{\sqrt{n}}\right) \leq 1, \sqrt{n} \geq 9.7055,$ and $n \geq 94.2,$ so you must choose a random sample of at least 95 subway cars.

3. (a) We are calculating a one-sample *t*-interval for the mean income of some population of basketball players. We are told to assume that all necessary assumptions are met. Calculator software gives (–29,337, 421,337), or we are 95% confident that the mean salary of all basketball salaries from the population from which this sample was taken is between –$29,337 and $421,337. [For instructional purposes in this review book, we note that with *df* = 9,

$$\bar{x} \pm t\frac{s}{\sqrt{n}} = 196{,}000 \pm 2.262\frac{315{,}000}{\sqrt{10}} = 196{,}000 \pm 225{,}000.]$$

(b) We must assume the sample is an SRS and that basketball salaries are normally distributed. This does *not* seem reasonable—the salaries are probably strongly skewed to the right by a few high ones. With the given mean and standard deviation, and noting that salaries are not negative, clearly we do not have a normal distribution, and furthermore, the sample size, *n* = 10, is too small to invoke the CLT. The calculated confidence interval is not meaningful.

4. (a) First, identify the confidence interval and check the conditions:

This is a two-sample *t*-interval for $\mu_{NRW} - \mu_{RW}$, the difference in population means of BPA body concentrations in non-retail workers and retail workers, that is,

$$(\bar{x}_{NRW} - \bar{x}_{RW}) \pm t^* \sqrt{\frac{(s_{NRW})^2}{n_{NRW}} + \frac{(s_{NR})^2}{n_{RW}}}$$

It is given that these are random samples, it is reasonable to assume the samples are independent, and both samples sizes (528 and 197) are large enough so that by the CLT, the distributions of sample means are approximately normal and a *t*-interval may be found.

Second, calculate the interval:

Calculator software gives (–0.9521, –0.7479) with *df* = 332.3.

Third, interpret in context:

We are 99% confident that the difference in means of BPA body concentrations in all non-retail and retail workers (non-retail mean minus retail mean) is between –0.75 and –0.95 μg/L.

(b) Because 0 is not in the interval of plausible values for the difference of population means, and the entire interval is negative, the interval does support the belief that retail workers carry higher amounts of BPA in their bodies than non-retail workers.

5. (a) $\widehat{\text{Earnings}} = 2.69 + 1.6752 \,(\text{Years})$ where Earnings is in millions of dollars.

(b) 33.9% of the variability in career earnings is associated with years as a pro. However, looking at the scatterplot, if Serena's data point, (18, 54.38), is removed, there appears to be no linear relationship at all. (Serena's data point is called an *influential point.*)

(c) 2.69 + 1.6752(15) = \$27.82 million, and 2.69 + 1.6752(30) = \$52.95 million. The estimate for 15 years might be reasonable, but the one for 30 years is not; it is an extrapolation far from the given sample data points.

(d) We note that the scatterplot looks very roughly linear, there is no major pattern in the residual plot, and the histogram of residuals is very roughly unimodal and symmetric. With $df = n - 2 = 8$ and 90% confidence, Table B, the t-distribution table, gives critical t-scores of ±1.860. Then 1.6752 ± 1.860(0.8264) = 1.6752 ± 1.537. We are 90% confident that the true slope of the regression line linking career earnings to years as a pro is between \$0.138 million and \$3.212 million per year. The interpretation in context: We are 90% confident that for each additional year as a tennis professional, the career earnings go up by between \$0.138 million and \$3.212 million, on average.

(e) The model with 0 intercept makes more sense in this situation. When there are 0 years as a pro, the career earnings should be 0.

6. We first check conditions: we must assume use of randomization. The scatterplot is roughly linear, and there is no apparent pattern in the residuals plot.

The distribution of the residuals is very roughly normal.

(a) Now using the statistics software on a calculator gives

$$\widehat{Hours} = -3.225 + 2.676 \, Dosage$$

(b) Using calculator software (such as LinRegTInt on the TI-84 with the data in Lists) gives (2.1631, 3.1898). We are 90% confident that each additional gram of the new drug is associated with an average of between 2.16 and 3.19 more hours of allergy relief.

Tests of Significance— Proportions and Means 12

→ **LOGIC OF SIGNIFICANCE TESTING**

→ **NULL AND ALTERNATIVE HYPOTHESES**

→ *P*-**VALUES**

→ **TYPE I AND TYPE II ERRORS**

→ **CONCEPT OF POWER**

→ **HYPOTHESIS TESTING**

→ **SIMULATIONS AND *P*-VALUES**

Closely related to the problem of estimating a population proportion or mean is the problem of testing a hypothesis about a population proportion or mean. For example, a travel agency might determine an interval estimate for the proportion of sunny days in the Virgin Islands or, alternatively, might test a tourist bureau's claim about the proportion of sunny days. A major stockholder in a construction company might ascertain an interval estimate for the proportion of successful contract bids or, alternatively, might test a company spokesperson's claim about the proportion of successful bids. A consumer protection agency might determine an interval estimate for the mean nicotine content of a particular brand of cigarettes or, alternatively, might test a manufacturer's claim about the mean nicotine content of its cigarettes. An agricultural researcher could find an interval estimate for the mean productivity gain caused by a specific fertilizer or, alternatively, might test the developer's claimed mean productivity gain. In each of these cases, the experimenter must decide whether the interest lies in an interval estimate of a population proportion or mean or in a hypothesis test of a claimed proportion or mean.

LOGIC OF SIGNIFICANCE TESTING, NULL AND ALTERNATIVE HYPOTHESES, *P*-VALUES, ONE- AND TWO-SIDED TESTS, TYPE I AND TYPE II ERRORS, AND THE CONCEPT OF POWER

The general testing procedure is to choose a specific hypothesis to be tested, called the *null hypothesis*, pick an appropriate random sample, and then use measurements from the sample to determine the likelihood of the null hypothesis. If the sample statistic is far enough away from the claimed population parameter, we say that there is sufficient evidence to reject the null hypothesis. We attempt to show that the null hypothesis is unacceptable by showing that it is improbable.

Consider the context of the population proportion. The null hypothesis H_0 is stated in the form of an equality statement about the *population* proportion (for example, H_0: $p = 0.37$). There is an *alternative hypothesis*, stated in the form of an inequality (for example, H_a: $p < 0.37$ or H_a: $p > 0.37$ or H_a: $p \neq 0.37$). The strength of the sample statistic $\hat{p}$ can be gauged through its associated *P-value*, which is the probability of obtaining a sample statistic as extreme (or more extreme) as the one obtained if the null hypothesis is assumed to be true. The smaller the *P*-value, the more significant the difference between the null hypothesis and the sample results.

Note that only population parameter symbols appear in H_0 and H_a. Sample statistics like $\bar{x}$ and $\hat{p}$ do NOT appear in H_0 or H_a.

There are two types of possible errors: the error of mistakenly rejecting a true null hypothesis and the error of mistakenly failing to reject a false null hypothesis. The α-risk, also called the *significance level* of the test, is the probability of committing a *Type I error* and mistakenly rejecting a true null hypothesis. A *Type II error*, a mistaken failure to reject a false null hypothesis, has associated probability β. There is a different value of β for each possible correct value for the population parameter *p*. For each β, $1 - \beta$ is called the power of the test against the associated correct value. The *power* of a hypothesis test is the probability that a Type II error is not committed. That is, given a true alternative, the power is the probability of rejecting the false null hypothesis. Increasing the sample size and increasing the significance level are both ways of increasing the power.

		Population truth	
		H_0 true	H_0 false
Decision based on sample	Reject H_0	Type I error	Correct decision
	Fail to reject H_0	Correct decision	Type II error

A simple illustration of the difference between a Type I and a Type II error is as follows: Suppose the null hypothesis is that all systems are operating satisfactorily with regard to a NASA liftoff. A Type I error would be to delay the liftoff mistakenly thinking that something was malfunctioning when everything was actually OK. A Type II error would be to fail to delay the liftoff mistakenly thinking everything was OK when something was actually malfunctioning. The power is the probability of recognizing a particular malfunction. (Note the complimentary aspect of power, a "good" thing, with Type II error, a "bad" thing.)

Our justice system provides another often quoted illustration. If the null hypothesis is that a person is innocent, then a Type I error results when an innocent person is found guilty, while a Type II error results when a guilty person is not convicted. We try to minimize Type I errors in criminal trials by demanding unanimous jury guilty verdicts. In civil suits, however, many states try to minimize Type II errors by accepting simple majority verdicts.

It should be emphasized that with regard to calculations, questions like "What is the *power* of this test?" and "What is the probability of a *Type II error* in this test?" cannot be answered without reference to a specific alternative hypothesis. Furthermore, AP students are not required to know how to calculate these probabilities. However, they are required to understand the concepts and interactions among the concepts.

HYPOTHESIS TEST FOR A PROPORTION

We must check that we have a simple random sample, that both np_0 and $n(1-p_0)$ are at least 10, and that the sample size is less than 10% of the population.

➡ EXAMPLE 12.1 _____

A union spokesperson claims that 75% of union members will support a strike if their basic demands are not met. A company negotiator believes the true percentage is lower and runs a hypothesis test. What is the conclusion if 87 out of an SRS of 125 union members say they will strike?

Answer: We note that $np_0 = (125)(0.75) = 93.75$ and $n(1-p_0) = (125)(0.25) = 31.25$ are both ≥ 10 and that we have an SRS, and we assume that 125 is less than 10% of the total union membership.

$$H_0: \quad p = 0.75$$
$$H_a: \quad p < 0.75$$

Calculator software (such as `1-PropZTest` on the TI-84 or the Casio Prizm) gives $z = -1.394$ and $P = 0.0816$. [For instructional purposes in this review book, we note that the observed sample proportion is $\hat{p} = \frac{87}{125} = 0.696$, and using the claimed proportion, 0.75, we have $\sigma_{\hat{p}} = \sqrt{\frac{(0.75)(0.25)}{125}} = 0.03873$ and $z = \frac{0.696 - 0.75}{0.03873} = -1.394$, with a resulting P-value of 0.0817.]

There are two possible answers: With this large a P-value, $0.0816 > 0.05$, there is not sufficient evidence at the 5% significance level to reject H_0, that is, there is not sufficient evidence that the true percentage of union members who support a strike is less than 75%.

OR   With this small a P-value, $0.0816 < 0.10$, there is sufficient evidence to reject H_0, that is, there is sufficient evidence at the 10% significance level that the true percentage of union members who support a strike is less than 75%.

> **IMPORTANT**
>
> We never "accept" a null hypothesis; we either do or do not have sufficient evidence to reject it.

On the TI-Nspire the result shows as:

zTest_1Prop 0.75,87,125,-1: *stat.results*	"Title"	"1-Prop z Test"
	"Alternate Hyp"	"prop < p0"
	"z"	-1.39427
	"PVal"	0.081617
	"p̂"	0.696
	"n"	125.

TIP

If the question
does not suggest
a direction for the
alternative, then
use a two-sided
test.

➡ EXAMPLE 12.2

A cancer research group surveys a random sample of 500 women more than 40 years old to test the hypothesis that 28% of women in this age group have regularly scheduled mammograms. Should the hypothesis be rejected at the 5% significance level if 151 of the women respond affirmatively?

Answer: Note that $np_0 = (500)(0.28) = 140$ and $n(1 - p_0) = (500)(0.72) = 360$ are both ≥ 10, we are given a random sample, and clearly 500 < 10% of all women over 40 years old. Since no suspicion is voiced that the 28% claim is low or high, we run a two-sided *z*-test for proportions:

$$H_0: \quad p = 0.28$$
$$H_a: \quad p \neq 0.28$$

Calculator software gives $z = 1.0956$ and $P = 0.2732$.

TIP

Remember: when
we fail to reject
H_0, we are *not*
accepting it; it
simply means
that there is
not sufficient
evidence against
H_0, that is, we
cannot rule
out random
chance as the
explanation.

[For instructional purposes, we note that the observed $\hat{p} = \dfrac{151}{500} = 0.302$, $\sigma_{\hat{p}} = \sqrt{\dfrac{(0.28)(0.72)}{500}} = 0.02008$, $z = \dfrac{0.302 - 0.28}{0.02008} = 1.0956$, which corresponds to a tail probability of 0.1366. Doubling this value (because the test is two-sided) gives a *P*-value of 0.2732.] With this large a *P*-value, 0.2732 > 0.05, there is not sufficient evidence to reject H_0, that is, the cancer research group should not dispute the 28% claim.

HYPOTHESIS TEST FOR A DIFFERENCE BETWEEN TWO PROPORTIONS

TIP

The samples must
be independent
to use these two
sample proportion
methods.

As with confidence intervals for the difference of two proportions, $n_1\hat{p}_1$, $n_1(1 - \hat{p}_1)$, $n_2\hat{p}_2$, and $n_2(1 - \hat{p}_2)$ should all be at least 10, it is important both that the samples be *simple random samples* and that they be taken *independently* of each other, and the original populations should be large compared to the sample sizes.

The fact that a sample proportion from one population is greater than a sample proportion from a second population does not automatically justify a similar conclusion about the population proportions themselves. Two points need to be stressed. First, sample proportions from the same population can vary from each other. Second, what we are really comparing are confidence interval estimates, not just single points.

For many problems the null hypothesis states that the population proportions are equal or, equivalently, that their difference is 0:

$$H_0: p_1 - p_2 = 0$$

The alternative hypothesis is then

$$H_a: p_1 - p_2 < 0 \quad H_a: p_1 - p_2 > 0 \quad \text{or} \quad H_a: p_1 - p_2 \neq 0$$

where the first two possibilities lead to one-sided tests and the third possibility leads to two-sided tests.

Since the null hypothesis is that $p_1 = p_2$, we call this common value p and use this pooled value in calculating σ_d:

$$\sigma_d = \sqrt{\frac{p(1-p)}{n_1} + \frac{p(1-p)}{n_2}} = \sqrt{p(1-p)\,\frac{1}{n_1} + \frac{1}{n_2}}$$

In practice, if $\hat{p}_1 = \frac{x_1}{n_1}$ and $\hat{p}_2 = \frac{x_2}{n_2}$, we use $\hat{p} = \frac{x_1 + x_2}{n_1 + n_2}$ as an estimate of p in calculating σ_d.

TIP

➡ **EXAMPLE 12.3**_____

This is the one situation where it is necessary to use a pooled proportion.

Suppose that early in an election campaign a telephone poll of 800 randomly chosen registered voters shows 460 in favor of a particular candidate. Just before election day, a second poll shows only 520 of 1000 randomly chosen registered voters expressing the same preference. Is there sufficient evidence that the candidate's popularity has decreased?

Answer: Note that $n_1\hat{p}_1 = 460$, $n_1(1 - \hat{p}_1) = 340$, $n_2\hat{p}_2 = 520$, and $n_2(1 - \hat{p}_2) = 480$ are all at least 10 (where n_1 and $\hat{p}_1$ refer to the first poll, ,and n_2 and $\hat{p}_2$ refer to the second poll), we are given that the samples are random and independent by design, and clearly the total population of registered voters is large.

$$H_0: p_1 - p_2 = 0 \quad \text{or} \quad H_0: p_1 = p_2$$
$$H_a: p_1 - p_2 > 0 \quad \text{or} \quad H_a: p_1 > p_2$$

Calculator software (such as `2-PropZTest` on the TI-84 or the Casio Prizm) gives $z = 2.328$ and $P = 0.00995$. [For instructional purposes, we note that the

observed difference is $0.575 - 0.520 = 0.055$, $\hat{p} = \frac{460 + 520}{800 + 1000} = 0.544$,

$\sigma_d = \sqrt{(0.544)(0.456)\,\frac{1}{800} + \frac{1}{1000}} = 0.0236$, $z = \frac{0.055 - 0}{0.0236} = 2.33$, and the tail

probability gives $P = 0.0099$.] With this small a P-value, $0.00995 < 0.05$, there is sufficient evidence to reject H_0, that is, there is sufficient evidence that the candidate's popularity has dropped.

TIP

Along with the P-value, you should also give the test statistic, as we have been doing.

➡ EXAMPLE 12.4 _____

An automobile manufacturer tries two distinct assembly procedures. In a random sample of 350 cars coming off the line using the first procedure there are 28 with major defects, while a random sample of 500 autos from the second line shows 32 with defects. Is the difference significant at the 10% significance level?

Answer: Note that $n_1\hat{p}_1 = 28$, $n_1(1 - \hat{p}_1) = 322$, $n_2\hat{p}_2 = 32$, and $n_2(1 - \hat{p}_2) = 468$ are all at least 10 (where n_1 and $\hat{p}_1$ refer to the sample of cars using the first procedure, and n_2 and $\hat{p}_2$ refer to the sample of cars using the second procedure), we are given that the samples are random and independent by design, and the total population of autos manufactured at the plant is large. Since there is no mention that one procedure is believed to be better or worse than the other, this is a two-sided test.

$$H_0: p_1 - p_2 = 0 \quad \text{or} \quad H_0: p_1 = p_2$$

$$H_a: p_1 - p_2 \neq 0 \quad \text{or} \quad H_a: p_1 \neq p_2$$

Calculator software gives $z = 0.8963$ and $P = 0.3701$. [For instructional purposes, we note that the observed difference is $0.080 - 0.064 = 0.016$,

$$\hat{p} = \frac{28 + 32}{350 + 500} = 0.0706 , \quad \sigma_d = \sqrt{(0.0706)(0.9294)\left(\frac{1}{350} + \frac{1}{500}\right)} = 0.01785 ,$$

$$z = \frac{0.016 - 0}{0.01785} = 0.8964 ,$$ which corresponds to a tail probability of 0.1850. Doubling this value (because the test is two-sided) gives a *P*-value of 0.3700.] With this large a *P*-value, $0.3700 > 0.05$, there is not sufficient evidence to reject H_0, that is, there is not sufficient evidence of a significant difference in proportions of all cars with major defects coming from the two assembly procedures.

HYPOTHESIS TEST FOR A MEAN

We must check that we have a simple random sample, that the sample size is less than 10% of the population, and that the distribution of sample means is approximately normal, which follows if either the sample size is large enough for the CLT to apply or the population has an approximately normal distribution (if we are given the sample data, we check the normality assumption with a graph such as a histogram or a normal probability plot: the histogram should be unimodal and reasonably symmetric, although this condition can be relaxed the larger the sample size.)

➡ EXAMPLE 12.5

A manufacturer claims that a new brand of air-conditioning unit uses only 6.5 kilowatts of electricity per day. A consumer agency believes the true figure is higher and runs a test on a random sample of size 50. If the sample mean is 7.0 kilowatts with a standard deviation of 1.4, should the manufacturer's claim be rejected at a significance level of 5%? of 1%?

Answer: We are given a random sample, $n = 50$ is large enough for the CLT to apply so the distribution of sample means is approximately normal, and we can assume that 50 is less than 10% of all of the new AC units. The population SD is unknown, so a *t*-test is called for with $H_0: \mu = 6.5$ and $H_a: \mu > 6.5$. Calculator software (such as T-Test on the TI-84 or 1-Sample zTest on the Casio Prizm) gives $t = 2.525$ and $P = 0.0074$.

[For instructional purposes, we note that $\sigma_{\bar{x}} = \dfrac{1.4}{\sqrt{50}} = 0.198$, and $t = \dfrac{7.0 - 6.5}{0.198} = 2.525$

with a resulting *P*-value of 0.0074.] With this small a *P*-value, $0.0074 < 0.05$, there is sufficient evidence to reject H_0, that is, there is sufficient evidence for the consumer agency to reject the manufacturer's claim that the new unit uses a mean of only 6.5 kilowatts of electricity per day.

```
T-Test
 Inpt:Data Stats
 μ0:6.5
 x̄:7
 Sx:1.4
 n:50
 μ:≠μ0  <μ0  >μ0
 Calculate  Draw
```

➡ EXAMPLE 12.6

A local chamber of commerce claims that the mean sale price for homes in the city is $90,000. A real estate salesperson notes a random sample of eight recent sales of $75,000, $102,000, $82,000, $87,000, $77,000, $93,000, $98,000, and $68,000. How strong is the evidence to reject the chamber of commerce claim?

Answer: We are given a random sample, we can assume $n = 8$ is less than 10% of all home sales, and we check for normality with a histogram, which yields the roughly unimodal, roughly symmetric:

60,000 70,000 80,000 90,000 100,000 110,000

The population SD is unknown, so a *t*-test is called for with $H_0: \mu = 90,000$ and $H_a: \mu \neq 90,000$. Putting the data into a List, calculator software gives $t = -1.131$ and $P = 0.2953$. With such a large *P*-value, $0.2953 > 0.05$, there is not sufficient evidence to reject H_0, that is, there is not sufficient evidence to reject the commerce claim of a mean sale price of $90,000 for homes in the city.

On the TI-Nspire the result shows as:

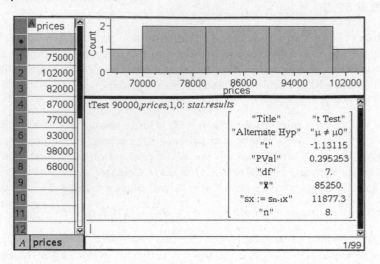

HYPOTHESIS TEST FOR A DIFFERENCE BETWEEN TWO MEANS (UNPAIRED AND PAIRED)

We assume that we have two independent simple random samples, each from an approximately normally distributed population (the normality assumptions can be relaxed if the sample sizes are large).

In this situation the null hypothesis is usually that the means of the populations are the same or, equivalently, that their difference is 0:

$$H_0: \mu_1 - \mu_2 = 0$$

The alternative hypothesis is then

$$H_a: \mu_1 - \mu_2 < 0 \quad H_a: \mu_1 - \mu_2 > 0 \quad \text{or} \quad H_a: \mu_1 - \mu_2 \neq 0$$

The first two possibilities lead to one-sided tests, and the third possibility leads to two-sided tests.

➡ EXAMPLE 12.7

A sales representative believes that his company's computer has more average downtime per week than a similar computer sold by a competitor. Before taking this concern to his director, the sales representative gathers data and runs a hypothesis test. He determines that in a simple random sample of 40 week-long periods at different firms using his company's product, the average downtime was 125 minutes with a standard deviation of 37 minutes. However, 35 week-long periods involving the competitor's computer yield an average downtime of only 115 minutes with a standard deviation of 43 minutes. What conclusion should the sales representative draw?

TIP

Avoid "calculator speak." For example, do not simply write "2-SampTTest..." or "binomcdf..." There are lots of calculators out there, each with its own abbreviations.

Answer: We are given independent SRSs, the sample sizes can be assumed to be less than 10% of all possible weeks and are large enough to relax the normality assumptions. The population SDs are unknown, so a *t*-test is called for with

$H_0: \mu_1 - \mu_2 = 0$ (where μ_1 = mean down time of company's computers
$H_a: \mu_1 - \mu_2 > 0$ and μ_2 = mean down time of competitor's computers)

Calculator software gives $t = 1.0718$ and $P = 0.1438$. With this large a *P*-value, $0.1438 > 0.05$, there is not sufficient evidence to reject H_0, that is, the sales representative does not have sufficient evidence that his company's computers have greater mean downtime than that of the competitor's computers.

TIP

While the normal probability plot is not part of the AP Stat curriculum, it is useful and is accepted on the exam.

➡ EXAMPLE 12.8

A city council member claims that male and female officers wait equal times for promotion in the police department. A women's spokesperson, however, believes women must wait longer than men. If a random sample of five men waited 8, 7, 10, 5, and 7 years, respectively, for promotion while a random sample of four women waited 9, 5, 12, and 8 years, respectively, what conclusion should be drawn?

Answer: We are given random samples and must assume they are independent and less than 10% of total numbers of male and female officers. These samples are too small for histograms to show anything other than no outliers, but the normal probability plots are fairly linear indicating that it is not unreasonable to assume the data come from roughly normal distributions:

Male officers   Female officers

The population SDs are unknown, so a *t*-test is called for with $H_0: \mu_M = \mu_F$ and $H_a: \mu_M < \mu_F$. Putting the data into Lists, calculator software gives $t = -0.6641$ and $P = 0.2685$. With this large a *P*-value, $0.2685 > 0.05$, there is not sufficient evidence to reject H_0, that is, there is not sufficient evidence to dispute the council member's claim that male and female officers wait equal times for promotion. (There is not sufficient evidence of a difference in the mean waiting times for promotion of male and female officers.)

The analysis and procedure described above require that the two samples being compared be independent of each other. However, many experiments and tests involve comparing two populations for which the data naturally occur in pairs. In this case, the proper procedure is to run a one-sample test on a single variable consisting of the differences from the paired data.

→ EXAMPLE 12.9 _____

Does a particular drug slow reaction times? If so, the government might require a warning label concerning driving a car while taking the medication. An SRS of 30 people who might benefit from the drug are tested before and after taking the drug, and their reaction times (in seconds) to a standard testing procedure are noted. The resulting histograms and summary statistics are as follows:

Before   After   Differences
                               (Before–After)

	n	Mean	SD
Before	30	1.46433	0.26267
After	30	1.50200	0.20942
Differences	30	–0.037667	0.11755

TIP

When the data comes in pairs, you cannot use a two sample *t*-test.

If we performed a two-sample test, we would calculate the *P*-value to be 0.2708 and would conclude that with such a large *P*, the observed rise is *not* significant. However, this would not be the proper test or conclusion! The two-sample test works for *independent* sets. However, in this case, there is a clear relationship between the data, in pairs, and this relationship is completely lost in the procedure for the two-sample test. The proper procedure is to form the set of 30 differences, being careful with signs, {–0.06, 0.12, 0.03, –0.24, . . .}, and to perform a single sample hypothesis test as follows:

Name the test:

We are using a *paired t-test*, that is, a single sample hypothesis test on the set of differences.

State the hypothesis:

H_0: The reaction times of individuals to a standard testing procedure are the same before and after they take a particular drug; the mean difference is zero: $\mu_d = 0$.

H_a: The reaction time is greater after they take the drug; the mean difference (Before minus After) is less than zero: $\mu_d < 0$.

Check the conditions:

TIP

Use proper terminology! In a 2-sample *t*-test, the hypotheses are about the difference of means. In a paired *t*-test, the hypotheses are about the mean difference.

1.   The *data are paired* because they are measurements on the same individuals before and after taking the drug.
2.   The reaction times of any individual are independent of the reaction times of the others, so the *differences are independent*.
3.   A *random sample* of people are tested.
4.   The histogram of the differences looks *nearly normal* (roughly unimodal and symmetric).
5.   The sample size, *n* = 30, is less than 10% of the population of all people who may take the drug.

Perform the mechanics:

T-Test immediately gives $t = -1.755$ with $P = 0.0449$. [Or we can calculate

$$SE(\bar{x}) = \frac{s}{\sqrt{n}} = \frac{0.11755}{\sqrt{30}} = 0.021462 \quad \text{and} \quad t = \frac{\bar{x} - 0}{\sigma_{\bar{x}}} = \frac{-0.037667}{0.021462} = -1.755$$

With $df = n - 1 = 29$, the P-value is $P(t < -1.755) = 0.0449$.]

Give a conclusion in context:

With this small a P-value, $0.0449 < 0.05$, there is sufficient evidence to reject H_0, that is, there is sufficient evidence that the mean observed rise in reaction times after taking the drug is significant.

SIMULATIONS AND *P*-VALUES

When our learned test procedures do not apply, either because conditions are not met or because we are working with a parameter for which we have not learned a test procedure, we may be able to proceed with a simulation. We can use a simulation to determine what values of a test statistic are likely to occur by random chance alone, assuming the null hypothesis is true. Then looking at where our test statistic falls, we can estimate a P-value.

➡ EXAMPLE 12.10 _____

One of the most surprising teams in the 2015 MLB season was the Chicago Cubs, who had 97 wins and only 65 losses. Kyle Schwarber made his MLB debut with the Cubs late in the 2015 season, and the Cubs went 22–10 with Schwarber in the lineup. Thus the Cubs had a winning proportion of $\frac{22}{32} = 0.6875$ with Schwarber in the lineup and $\frac{75}{130} \approx 0.5769$ when he was not in the lineup. Is the difference $0.6875 - 0.5769 = 0.1106$ significant?

Answer: We conduct a simulation as follows. The season has 162 games. We let the integers 1 through 97 represent wins and 98 through 162 represent loses. We use a random number generator to pick 32 integers between 1 and 162. These 32 integers represent the simulated results for Schwarber's games, assuming the null hypothesis that the Cubs would do equally well with and without Schwarber in the lineup. We then note the difference between the winning proportion in the simulated 32 game results and the corresponding winning proportion in the remaining 130 game results. Suppose we run this simulation 200 times with the following resulting dotplot:

Simulated Difference in Winning Proportion (With and Without Schwarber)

The dotplot shows differences in winning proportions with and without Schwarber that we would expect to see by random chance if the null hypothesis of no difference with and without Schwarber was true. In 18 + 11 + 6 + 3 + 1 + 1 = 40 out of the 200 simulations, the difference was equal to or greater than 0.1106. The estimated *P*-value is $\frac{40}{200}$ = 0.20. With a *P*-value this large, 0.20 > 0.05, there is not enough evidence to reject the null hypothesis, that is, there is not enough evidence that the Cubs do better with Schwarber than without him.

➡ EXAMPLE 12.11 _____

One measure of variability is the median absolute deviation (MAD). It is defined as the median deviation from the median, that is, as the median of the absolute values of the deviations from the median. A particular industrial product has the following quality control check. Random samples of size 6 are gathered periodically, and measurements are taken of the dimension under observation. If the MAD calculation is significantly greater than what is expected during proper operation of the machinery, a recalibration is necessary. In a simulation of 100 such samples of size 6 from when the machinery is working properly, the resulting MAD calculations are summarized in the following dotplot:

**MAD Calculations from Random Samples of Size 6
When Machinery Is Operating Properly**

Suppose in a random sample of 6 products the measurements are {8.04, 8.06, 8.10, 8.14, 8.18, 8.19}. Is there sufficient evidence to necessitate a recalibration of the machinery?

Answer: The median is $\frac{8.10 + 8.14}{2}$ = 8.12. The absolute deviations from the median are {0.08, 0.06, 0.02, 0.02, 0.06, 0.07} with median $\frac{0.06 + 0.06}{2}$ = 0.06 (the MAD calculation). In the simulation, there were 3 values out of 100 that were 0.06 or greater. This gives an estimated P-value of 0.03. With this small a *P*-value, 0.03 < 0.05, there is sufficient evidence to necessitate a recalibration of the machinery.

MORE ON POWER AND TYPE II ERRORS

Given a specific alternative hypothesis, a Type II error is a mistaken failure to reject the false null hypothesis, while the *power* is the probability of rejecting that false null hypothesis.

➡ **EXAMPLE 12.12** _____

A candidate claims to have the support of 70% of the people, but you believe that the true figure is lower. You plan to gather an SRS and will reject the 70% claim if in your sample shows 65% or less support. What if in reality only 63% of the people support the candidate?

TIP

You will not have to calculate actual probabilities of Type II errors or power.

The upper graph shows the null hypothesis model with the claim that $p_0 = 0.70$, and the plan to reject H_0 if $\hat{p} < 0.65$. The lower graph shows the true model with $p = 0.63$. When will we fail to pick up that the null hypothesis is incorrect? Answer: precisely when the sample proportion is greater than 0.65. This is a Type II error with probability β. When will we rightly conclude that the null hypothesis is incorrect? Answer: when the sample proportion is less than 0.65. This is the power of the test and has probability $1 - \beta$.

TIP

While calculations are not called for, you must understand the interactions among the errors, confidence level, power, *effect*, and sample size.

The following points should be emphasized:

- Power gives the probability of avoiding a Type II error.
- Power has a different value for different possible correct values of the population parameter; thus it is actually a *function* where the independent variable ranges over specific alternative hypotheses.
- Choosing a smaller α (that is, a tougher standard to reject H_0) results in a higher risk of Type II error and a lower power—observe in the above graphs how making α smaller (in this case moving the critical cutoff value to the left) makes the power less and β more!
- The greater the difference between the null hypothesis p_0 and the true value p, the smaller the risk of a Type II error and the greater the power—observe in the above picture how moving the lower graph to the left makes the power greater and β less. (The difference between p_0 and p is sometimes called the *effect*—thus the greater the effect, the greater is the power to pick it up.)
- A larger sample size n will reduce the standard deviations, making both graphs narrower resulting in smaller α, smaller β, and larger power!

TIP

Both blocking in experiments and stratification in sampling reduce variability and so can improve power.

CONFIDENCE INTERVALS VERSUS HYPOTHESIS TESTS

A question sometimes arises as to whether a problem calls for calculating a confidence interval or performing a hypothesis test. Generally, a claim about a population parameter indicates a hypothesis test, while an estimate of a population parameter asks for a confidence interval. However, some confusion may arise because sometimes it is possible to conduct a hypothesis test by constructing a confidence interval estimate of the parameter and then checking whether or not the null parameter value is in the interval. While, when possible, this alternative approach is accepted on the AP exam, it does require very special care in still remembering to state hypotheses, in dealing with one-sided versus two-sided, and in how standard errors are calculated in problems involving proportions. So the recommendation is to conduct a hypothesis test like a hypothesis test. Carefully read the question. If it asks whether or not there is evidence, then this is a hypothesis test; if it involves how much or how effective, then this is a confidence interval calculation. However, it should be noted that there have been AP free-response questions where part (a) calls for a confidence interval calculation and part (b) asks if this calculation provides evidence relating to a hypothesis test.

SUMMARY

- Important assumptions/conditions always must be checked before proceeding with a hypothesis test.
- The null hypothesis is stated in the form of an equality statement about a population parameter, while the alternative hypothesis is in the form of an inequality.
- We attempt to show that a null hypothesis is unacceptable by showing it is improbable.
- The P-value is the probability of obtaining a sample statistic as extreme (or more extreme) as the one obtained if the null hypothesis is assumed to be true.
- When the P-value is small (typically, less than 0.05 or 0.10), we say we have sufficient evidence to reject the null hypothesis.
- A Type I error is the probability of mistakenly rejecting a true null hypothesis.
- A Type II error is the probability of mistakenly failing to reject a false null hypothesis.
- The power of a hypothesis test is the probability that a Type II error is not committed.
- When our learned test procedures do not apply, we may be able to proceed with a simulation.
- When scoring a hypothesis test free-response question on the exam, readers will look for whether you:

   1. State the hypotheses (both H_0 and H_a) and define the parameter of interest. (The hypotheses must refer to the population, not to the sample.)
   2. Identify the procedure by name (such as one-sample z-test for a population proportion) or formula (such as $z = \dfrac{\hat{p} - p_0}{\sqrt{\dfrac{p_0(1-p_0)}{n}}}$) and check the conditions.
   3. Compute the test statistic, P-value, and degrees of freedom if appropriate. (It is sufficient to simply copy all of these from your calculator calculation.)
   4. Give a conclusion in context with linkage to the P-value. (The conclusion must refer to the parameter and to the population, not to the sample.)

QUIZ 1

Multiple-Choice Questions

> **Directions:** The questions that follow are each followed by five suggested answers. Choose the response that best answers the question.

1. Which of the following is a true statement?

   (A) A well-planned hypothesis test should result in a statement either that the null hypothesis is true or that it is false.
   (B) The alternative hypothesis is stated in terms of a sample statistic.
   (C) If a sample is large enough, the necessity for it to be a simple random sample is diminished.
   (D) When the null hypothesis is rejected, it is because it is not true.
   (E) Hypothesis tests are designed to measure the strength of evidence against the null hypothesis.

Questions 2–3 refer to the following:

Video arcades provide a vehicle for competition and status within teenage peer groups and are recognized as places where teens can hang out and meet their friends. The national PTA organization, concerned about the time and money being spent in video arcades by middle school students, commissions a statistical study to investigate whether or not middle school students are spending an average of over two hours per week in video arcades. Twenty communities are randomly chosen, five middle schools are randomly picked in each of the communities, and ten students are randomly interviewed at each school.

2. What is the parameter of interest?

   (A) Video arcades
   (B) A particular concern of the national PTA organization
   (C) 1000 randomly chosen middle school students
   (D) The mean amount of money spent in video arcades by middle school students
   (E) The mean number of hours per week spent in video arcades by middle school students

3. What are the null and alternative hypotheses which the PTA is testing?

   (A) $H_0: \mu = 2$ $\qquad H_a: \mu < 2$
   (B) $H_0: \mu = 2$ $\qquad H_a: \overline{x} < 2$
   (C) $H_0: \mu = 2$ $\qquad H_a: \mu > 2$
   (D) $H_0: \mu = 2$ $\qquad H_a: \overline{x} > 2$
   (E) $H_0: \mu = 2$ $\qquad H_a: \mu \neq 2$

4. A company manufactures a synthetic rubber (jumping) bungee cord with a braided covering of natural rubber and a minimum breaking strength of 450 kg. If the mean breaking strength of a sample drops below a specified level, the production process is halted and the machinery inspected. Which of the following would result from a Type I error?

(A) Halting the production process when too many cords break.
(B) Halting the production process when the breaking strength is below the specified level.
(C) Halting the production process when the breaking strength is within specifications.
(D) Allowing the production process to continue when the breaking strength is below specifications.
(E) Allowing the production process to continue when the breaking strength is within specifications.

5. One ESP test asks the subject to view the backs of cards and identify whether a circle, square, star, or cross is on the front of each card. If p is the proportion of correct answers, this may be viewed as a hypothesis test with H_0: $p = 0.25$ and H_a: $p > 0.25$. The subject is recognized to have ESP when the null hypothesis is rejected. What would a Type II error result in?

(A) Correctly recognizing someone has ESP
(B) Mistakenly thinking someone has ESP
(C) Not recognizing that someone really has ESP
(D) Correctly realizing that someone doesn't have ESP
(E) Failing to understand the nature of ESP

6. A coffee-dispensing machine is supposed to deliver 12 ounces of liquid into a large paper cup, but a consumer believes that the actual amount is less. As a test he plans to obtain a sample of 5 cups of the dispensed liquid and if the mean content is less than 11.5 ounces, to reject the 12-ounce claim. If the machine operates with a known standard deviation of 0.9 ounces, what is the probability that the consumer will mistakenly reject the 12-ounce claim even though the claim is true? (Assume that all conditions for inference are met.)

(A) $P\left(t > \dfrac{11.5 - 12}{0.9/\sqrt{5}}\right)$

(B) $P\left(t < \dfrac{11.5 - 12}{0.9/\sqrt{5}}\right)$

(C) $P\left(z > \dfrac{11.5 - 12}{0.9/\sqrt{5}}\right)$

(D) $P\left(z < \dfrac{11.5 - 12}{0.9/\sqrt{5}}\right)$

(E) $2P\left(t < \dfrac{11.5 - 12}{0.9/\sqrt{5}}\right)$

7. A fast food chain advertises that their large bag of french fries has a weight of 150 grams. Some high school students, who enjoy french fries at every lunch, suspect that they are getting less than the advertised amount. With a scale borrowed from their physics teacher, they weigh a random sample of 15 bags. What is the conclusion if the sample mean is 145.8 g and standard deviation is 12.81 g? (Assume that all conditions for inference are met.)

   (A) There is sufficient evidence to prove the fast food chain advertisement is true.
   (B) There is sufficient evidence to prove the fast food chain advertisement is false.
   (C) The students have sufficient evidence to reject the fast food chain's claim.
   (D) The students do not have sufficient evidence to reject the fast food chain's claim.
   (E) There is not sufficient data to reach any conclusion.

8. Is Internet usage different in the Middle East and Latin America? In a random sample of 500 adults in the Middle East, 151 claimed to be regular Internet users, while in a random sample of 1000 adults in Latin America, 345 claimed to be regular users. What is the P-value for the appropriate hypothesis test?

   (A) $P\left(z < \dfrac{0.302 - 0.345}{\sqrt{\dfrac{(0.302)(0.698)}{500} + \dfrac{(0.345)(0.655)}{1000}}} \right)$

   (B) $2P\left(z < \dfrac{0.302 - 0.345}{\sqrt{\dfrac{(0.302)(0.698)}{500} + \dfrac{(0.345)(0.655)}{1000}}} \right)$

   (C) $P\left(z < \dfrac{0.302 - 0.345}{\sqrt{(0.331)(0.669)\left(\dfrac{1}{500} + \dfrac{1}{1000}\right)}} \right)$

   (D) $2P\left(z < \dfrac{0.302 - 0.345}{\sqrt{(0.331)(0.669)\left(\dfrac{1}{500} + \dfrac{1}{1000}\right)}} \right)$

   (E) $2(0.095167)$

9. A research dermatologist believes that cancers of the head and neck will occur most often of the left side, the side next to a window when a person is driving. In a review of 565 cases of head/neck cancers, 305 occurred on the left side. What is the resulting P-value?

(A) $P\left(z > \dfrac{0.54 - 0.50}{\sqrt{(0.5)(0.5)/565}}\right)$

(B) $2P\left(z > \dfrac{0.54 - 0.50}{\sqrt{(0.5)(0.5)/565}}\right)$

(C) $P\left(z > \dfrac{0.54 - 0.50}{\sqrt{(0.54)(0.46)/565}}\right)$

(D) $P\left(z \geq \dfrac{0.54 - 0.50}{\sqrt{(0.54)(0.46)/565}}\right)$

(E) $2P\left(z > \dfrac{0.54 - 0.50}{\sqrt{(0.54)(0.46)/565}}\right)$

10. Given an experiment with H_0: $\mu = 35$, H_a: $\mu < 35$, and a possible correct value of 32, you obtain a sample statistic of $\bar{x} = 33$. After doing analysis, you realize that the sample size n is actually larger than you first thought. Which of the following results from reworking with the increase in sample size?

(A) Decrease in probability of a Type I error; decrease in probability of a Type II error; decrease in power.
(B) Increase in probability of a Type I error; increase in probability of a Type II error; decrease in power.
(C) Decrease in probability of a Type I error; decrease in probability of a Type II error; increase in power.
(D) Increase in probability of a Type I error; decrease in probability of a Type II error; decrease in power.
(E) Decrease in probability of a Type I error; increase in probability of a Type II error; increase in power.

11. Thirty students volunteer to test which of two strategies for taking multiple-choice exams leads to higher average results. Each student flips a coin, and if heads, uses Strategy A on the first exam and then Strategy B on the second, while if tails, uses Strategy B first and then Strategy A. The average of all 30 Strategy A results is then compared to the average of all 30 Strategy B results. What is the conclusion at the 5% significance level if a two-sample hypothesis test, $H_0: \mu_1 = \mu_2$, $H_a: \mu_1 \neq \mu_2$, results in a P-value of 0.18?

    (A) The observed difference in average scores is significant.
    (B) The observed difference in average scores is not significant.
    (C) A conclusion is not possible without knowing the average scores resulting from using each strategy.
    (D) A conclusion is not possible without knowing the average scores and the standard deviations resulting from using each strategy.
    (E) A two-sample hypothesis test should not be used here.

12. The greater the difference between the null hypothesis claim and the true value of the population parameter,

    (A) the smaller the risk of a Type II error and the smaller the power.
    (B) the smaller the risk of a Type II error and the greater the power.
    (C) the greater the risk of a Type II error and the smaller the power.
    (D) the greater the risk of a Type II error and the greater the power.
    (E) the greater the probability of no change in Type II error or in power.

Free-Reponse Questions

> **Directions:** You must show all work and indicate the methods you use. You will be graded on the correctness of your methods and on the accuracy of your final answers.

1. No vaccinations are 100% risk free, and the theoretical risk of rare complications always have to be balanced against the severity of the disease. Suppose the CDC decides that a risk of one in a million is the maximum acceptable risk of GBS (Guillain-Barre syndrome) complications for a new vaccine for a particularly serious strain of influenza. A large sample study of the new vaccine is conducted with the following hypotheses:

    H_0: The proportion of GBS complications is 0.000001 (one in a million)
    H_a: The proportion of GBS complications is greater than 0.000001 (one in a million)

    The P-value of the test is 0.138.

    (a) Interpret the P-value in context of this study.

    (b) What conclusion should be drawn at the $\alpha = 0.10$ significance level?

    (c) Given this conclusion, what possible error, Type I or Type II, might be committed, and give a possible consequence of committing this error.

2. A particular wastewater treatment system aims at reducing the most probable number per ml (MPN/ml) of *E. coli* to 1000 MPN/100 ml. A random study of 40 of these systems in current use is conducted with the data showing a mean of 1002.4 MPN/100 ml and a standard deviation of 7.12 MPN/100 ml. A test of significance is conducted with:

H_0: The mean concentration of *E. coli* after treatment under this system is 1000 MPN/100 ml.

H_a: The mean concentration of *E. coli* after treatment under this system is greater than 1000 MPN/100 ml.

The resulting *P*-value is 0.0197 with $df = 39$ and $t = 2.132$.

(a) Interpret the *P*-value in context of this study.

(b) What conclusion should be drawn at a 5% significance level?

(c) Given this conclusion, what possible error, Type I or Type II, might be committed, and give a possible consequence of committing this error.

3. A 20-year study of 5000 British adults noted four bad habits: smoking, drinking, inactivity, and poor diet. The study looked to show that there is a higher death rate (proportion who die in a 20-year period) among people with all four bad habits than among people with none of the four bad habits.

(a) Was this an experiment or observational study? Explain.

(b) What are the null and alternative hypotheses?

(c) What would be the result of a Type I error?

(d) What would be the result of a Type II error?

Of the 314 people who had all four bad habits, 91 died during the study, while of the 387 people with none of the four bad habits, 32 died during the study.

(e) Calculate and interpret the *P*-value in context of this study.

4. A behavior study of high school students looked at whether a higher proportion of boys than girls meet a recommended level of physical activity (increased heart rate for 60 minutes/day for at least 5 days during the 7 days before the survey). What is the proper conclusion if 370 out of a random sample of 850 boys and 218 out of an independent random sample of 580 girls met the recommended level of activity?

5. A study is proposed to compare two treatments for patients with significantly narrowed neck arteries. Some patients will be treated with surgery to remove built-up plaque, while others will be treated with stents to improve circulation. The response variable will be the proportion of patients who suffer a major complication such as a stroke or heart attack within one month of the treatment. The researchers decide to block on whether or not a patient has had a mini stroke in the previous year.

   (a) There are 1000 patients available for this study, half of whom have had a mini stroke in the previous year. Explain a block design to assign patients to treatments.

   (b) Give two methods other than blocking to increase the power of detecting a difference between using surgery versus stents for patients with this medical condition. Explain your choice of methods.

6. A car simulator was used to compare effect on reaction time between DWI (driving while intoxicated) and DWT (driving while texting). Ten volunteers were instructed to drive at 50 mph and then hit the brakes in response to the sudden image of a child darting into the road. A baseline stopping distance was established for each driver. Then one day each driver was tested for stopping distance while driving while texting, and another day the driver was tested after consuming a quantity of alcohol. For each driver, which test was done on the first day was decided by coin toss. The following table gives the extra number of feet necessary to stop at 50 mph for each driver for DWI and DWT.

DWI	30	26	28	35	42	33	36	28	27	37
DWT	30	31	25	39	45	32	38	30	28	38

The sample means are $\bar{x}_{DWI} = 32.2$, $\bar{x}_{DWT} = 33.6$, and a two-sample t-test, $H_0: \mu_{DWI} = \mu_{DWT}$, $H_a: \mu_{DWI} \neq \mu_{DWT}$, gives a P-value of 0.590, and a conclusion that there is no evidence of a difference between the effect on reaction time between DWI and DWT. Explain why this is not the proper hypothesis test, and then perform the proper test.

QUIZ 2

Multiple-Choice Questions

Directions: The questions that follow are each followed by five suggested answers. Choose the response that best answers the question.

1. Which of the following is a true statement?

   (A) The *P*-value of a test is the probability of obtaining a result as extreme (or more extreme) as the one obtained assuming the null hypothesis is false.

   (B) If the *P*-value for a test is 0.015, the probability that the null hypothesis is true is 0.015.

   (C) The larger the *P*-value, the more evidence there is against the null hypothesis.

   (D) If possible, always examine your data before deciding whether to use a one-sided or two-sided hypothesis test.

   (E) The alternative hypothesis is one-sided if there is interest in deviations from the null hypothesis in only one direction.

2. A hypothesis test comparing two population proportions results in a *P*-value of 0.032. Which of the following is a proper conclusion?

   (A) The probability that the null hypothesis is true is 0.032.

   (B) The probability that the alternative hypothesis is true is 0.032.

   (C) The difference in sample proportions is 0.032.

   (D) The difference in population proportions is 0.032.

   (E) None of the above are proper conclusions.

3. A pharmaceutical company claims that a medicine will produce a desired effect for a mean time of 58.4 minutes. A government researcher runs a hypothesis test of 40 patients and calculates a mean of $\bar{x} = 59.5$ with a standard deviation of $s = 8.3$. What is the *P*-value?

   (A) $P\left(t > \dfrac{59.5 - 58.4}{8.3/\sqrt{40}}\right)$ with $df = 39$

   (B) $P\left(t > \dfrac{59.5 - 58.4}{8.3/\sqrt{40}}\right)$ with $df = 40$

   (C) $2P\left(t > \dfrac{59.5 - 58.4}{8.3/\sqrt{40}}\right)$ with $df = 39$

   (D) $2P\left(t > \dfrac{59.5 - 58.4}{8.3/\sqrt{40}}\right)$ with $df = 40$

   (E) $2P\left(z > \dfrac{59.5 - 58.4}{8.3/\sqrt{40}}\right)$

4. You plan to perform a hypothesis test with a level of significance of $\alpha = 0.05$. What is the effect on the probability of committing a Type I error if the sample size is increased?

   (A) The probability of committing a Type I error decreases.
   (B) The probability of committing a Type I error is unchanged.
   (C) The probability of committing a Type I error increases.
   (D) The effect cannot be determined without knowing the relevant standard deviation.
   (E) The effect cannot be determined without knowing if a Type II error is committed.

5. A researcher believes a new diet should improve weight gain in laboratory mice. If ten control mice on the old diet gain an average of 4 ounces with a standard deviation of 0.3 ounces, while the average gain for ten mice on the new diet is 4.8 ounces with a standard deviation of 0.2 ounces, what is the P-value? (Assume that all conditions for inference are met.)

   (A) $P\left(t < \dfrac{4 - 4.8}{\sqrt{\dfrac{(0.3)^2}{10} + \dfrac{(0.2)^2}{10}}} \right)$

   (B) $P\left(z < \dfrac{4 - 4.8}{\sqrt{\dfrac{(0.3)^2}{10} + \dfrac{(0.2)^2}{10}}} \right)$

   (C) $P\left(t < \dfrac{4 - 4.8}{\dfrac{0.3}{\sqrt{10}} + \dfrac{0.2}{\sqrt{10}}} \right)$

   (D) $P\left(z < \dfrac{4 - 4.8}{\dfrac{0.3}{\sqrt{10}} + \dfrac{0.2}{\sqrt{10}}} \right)$

   (E) $P\left(t < \dfrac{4 - 4.8}{\dfrac{0.3}{\sqrt{10-1}} + \dfrac{0.2}{\sqrt{10-1}}} \right)$

6. A pharmaceutical company claims that 8% or fewer of the patients taking their new statin drug will have a heart attack in a 5-year period. In a government-sponsored study of 2300 patients taking the new drug, 198 have heart attacks in a 5-year period. Is this strong evidence against the company claim?

   (A) Yes, because the P-value is 0.005657.
   (B) Yes, because the P-value is 0.086087.
   (C) No, because the P-value is only 0.005657.
   (D) No, because the P-value is only 0.086087.
   (E) No, because the P-value is over 0.10.

7. What is the probability of mistakenly failing to reject a false null hypothesis when a hypothesis test is being conducted at the 5% significance level ($\alpha = 0.05$)?

(A) 0.025
(B) 0.05
(C) 0.10
(D) 0.95
(E) There is insufficient information to answer this question.

8. Suppose you do five independent tests of the form $H_0: \mu = 38$ versus $H_a: \mu > 38$, each at the $\alpha = 0.01$ significance level. What is the probability of committing a Type I error and incorrectly rejecting a true null hypothesis with at least one of the five tests?

(A) 0.01
(B) 0.049
(C) 0.05
(D) 0.226
(E) 0.951

9. Choosing a smaller level of significance, that is, a smaller α-risk, results in

(A) a lower risk of Type II error and lower power.
(B) a lower risk of Type II error and higher power.
(C) a higher risk of Type II error and lower power.
(D) a higher risk of Type II error and higher power.
(E) no change in risk of Type II error or in power.

10. Suppose $H_0: p = 0.4$, and the power of the test for the alternative hypothesis $p = 0.35$ is 0.75. Which of the following is a valid conclusion?

(A) The probability of committing a Type I error is 0.05.
(B) The probability of committing a Type II error is 0.65.
(C) If the alternative $p = 0.35$ is true, the probability of failing to reject H_0 is 0.25.
(D) If the null hypothesis is true, the probability of rejecting it is 0.25.
(E) If the null hypothesis is false, the probability of failing to reject it is 0.65.

11. A factory is located close to a city high school. The manager claims that the plant's smokestacks spew forth an average of no more than 350 pounds of pollution per day. As an AP Statistics project, the class plans a one-sided hypothesis test with a critical value of 375 pounds. Suppose the standard deviation in daily pollution poundage is known to be 150 pounds and the true mean is 385 pounds. If the sample size is 100 days, what is the probability that the class will mistakenly fail to reject the factory manager's false claim?

(A) 0.0475
(B) 0.2525
(C) 0.7475
(D) 0.7514
(E) 0.9525

12. Do high school girls apply to more colleges than high school boys? A two-sample t-test of the hypotheses $H_0: \mu_{girls} = \mu_{boys}$ versus $H_a: \mu_{girls} > \mu_{boys}$ results in a P-value of 0.02.

Which of the following statements must be true?

I.   A 90% confidence interval for the difference in means contains 0.
II. A 95% confidence interval for the difference in means contains 0.
III. A 99% confidence interval for the difference in means contains 0.

(A) I only
(B) III only
(C) I and II only
(D) II and III only
(E) I, II, and III

Free-Reponse Questions

Directions: You must show all work and indicate the methods you use. You will be graded on the correctness of your methods and on the accuracy of your final answers.

1. Next to good brakes, proper tire pressure is the most crucial safety issue on your car. Incorrect tire pressure compromises cornering, braking, and stability. Both underinflation and overinflation can lead to problems. The number on the tire is the maximum allowable air pressure—not the recommended pressure for that tire. At a roadside vehicle safety checkpoint, officials randomly select 30 cars for which 35 psi is the recommended tire pressure and calculate the average of the actual tire pressure in the front right tires. What is the parameter of interest, and what are the null and alternative hypotheses that the officials are testing?

2. It is estimated that 17.4% of all U.S. households own a Roth IRA. The American Association of University Professors (AAUP) believes this figure is higher among their members and commissions a study. If 150 out of a random sample of 750 AAUP members own Roth IRAs, is this sufficient evidence to support the AAUP belief?

3. A long accepted measure of the discharge rate (in 1000 ft^3/sec) at the mouth of the Mississippi River is 593. To test if this has changed, ten measurements at random times are taken: 590, 596, 592, 588, 589, 594, 590, 586, 591, 589. Is there statistical evidence of a change?

4. In a random sample of 35 NFL games the average attendance was 68,729 with a standard deviation of 6,110, while in a random sample of 30 Big 10 Conference football games the average attendance was 70,358 with a standard deviation of 9,139. Is there evidence that the average attendance at Big 10 Conference football games is greater than that at NFL games?

5. An engineer wishes to test which of two drills can more quickly bore holes in various materials. He assembles a random sampling of ten materials of various hardnesses and thicknesses.

   (a) Given that a drill's efficiency is influenced by how long and how hard it has been operating, the engineer decided to randomly choose the order in which the materials will be tested. Design and implement a scheme to place the materials in random order using the following random number table:

   51844 73424 84380 82259 28273 58102 18727 69708

   (b) Suppose the drilling times (in seconds) are summarized as follows:

Material	A	B	C	D	E	F	G	H	I	J
Drill 1	4.2	5.1	8.8	1.5	0.8	7.4	3.4	4.7	6.2	4.8
Drill 2	4.2	5.4	8.7	1.7	0.9	7.8	3.4	4.4	6.3	4.9

   What is the mean drilling time for each drill? Is the difference significant? Justify your answer.

6. Paul the octopus, who lives in a tank at Sea Life Centre in Oberhausen, Germany, correctly predicted the winner in 12 out of 14 soccer matches (by choosing to eat mussels from the boxes labeled with national flags of the eventual winning teams). Assume these 14 matches represent a random sample of matches that Paul could predict, and assume that all matches end in a win for one of the teams (using sudden death overtime or penalty kicks to settle tie scores from regulation time).

   (a) Why would it be incorrect to perform a one-sample z-test on whether the above is sufficient evidence that Paul can correctly predict the results of more than 50% of soccer matches?

   (b) Perform a proper hypothesis test for the question in (a).

ANSWERS EXPLAINED

Quiz 1

MULTIPLE-CHOICE QUESTIONS

1. **(E)** We attempt to show that the null hypothesis is unacceptable by showing that it is improbable; however, we cannot show that it is definitely true or false. Both the null and alternative hypotheses are stated in terms of a population parameter, not a sample statistic. These hypotheses tests assume simple random samples.

2. **(E)** The parameter of interest is μ = the mean number of hours per week which middle school students spend in video arcades.

3. **(C)** The alternative hypothesis is always an inequality, either <, or >, or ≠. In this case, the concern is whether middle school students are spending an average of *more than* two hours per week in video arcades. The hypotheses are about parameters, not statistics.

4. **(C)** This is a hypothesis test with H_0: breaking strength is within specifications, and H_a: breaking strength is below specifications. A Type I error is committed when a true null hypothesis is mistakenly rejected.

5. **(C)** A Type II error is a mistaken failure to reject a false null hypothesis or, in this case, a failure to realize that a person really does have ESP.

6. **(D)** This is a one-sided test, and because the population standard deviation is known (and all conditions for inference are met), we use z instead of t.

7. **(D)** $t = \dfrac{145.8 - 150}{12.81/\sqrt{15}} = -1.270$, and with $df = 14$, $P = 0.112$. With this high P-value $0.112 > 0.10$, the students do not have sufficient evidence to reject the fast food chain's claim. (T-Test gives $P = 0.112423$.)

8. **(D)** $\hat{p}_1 = \dfrac{151}{500} = 0.302$, $\hat{p}_2 = \dfrac{345}{1000} = 0.345$, $\hat{p} = \dfrac{151 + 345}{500 + 1000} = 0.331$

   This is a two-sided test (H_0: $p_1 - p_2 = 0$, H_a: $p_1 - p_2 \neq 0$) with

   $$\sigma_d = \sqrt{p(1-p)\left(\frac{1}{n_1} + \frac{1}{n_2}\right)}$$

9. **(A)** This is a one-sided z-test, H_a: $p > 0.5$, with $\sigma_{\hat{p}} = \sqrt{(0.5)(0.5)/565}$.

10. **(C)** A larger sample size n reduces the standard deviation of the sampling distributions resulting in narrower sampling distributions so that for the given sample statistic, the P-value is smaller, and the probabilities of mistakenly rejecting a true null hypothesis or mistakenly failing to reject a false null hypothesis are both decreased. Furthermore, a lower Type II error results in higher power.

11. **(E)** The two-sample hypothesis test is not the proper one and can only be used when the two sets are independent. In this case, there is a clear relationship between the data, in pairs, one pair for each student, and this relationship is completely lost in the proce-

dure for the two-sample test. The proper procedure is to run a one-sample test on the single variable consisting of the differences from the paired data.

12. **(B)** If the null hypothesis is far off from the true parameter value, there is a greater chance of rejecting the false null hypothesis and thus a smaller risk of a Type II error. Power is the probability that a Type II error is not committed, so a lower Type II error results in higher power.

FREE-RESPONSE QUESTIONS

1. (a) The P-value of 0.138 gives the probability of observing a sample proportion of GBS complications as great or greater as the proportion found in the study if in fact the proportion of GBS complications is 0.000001.

   (b) Since $0.138 > 0.10$, there is not sufficient evidence to reject H_0, that is, there is not sufficient evidence that under the new vaccine the proportion of GBS complications is greater than 0.000001 (one in a million).

   (c) The null hypothesis is not rejected, so there is the possibility of a Type II error, that is of mistakenly failing to reject a false null hypothesis. A possible consequence is continued use of vaccine with a higher rate of serious complications than is acceptable.

2. (a) The P-value of 0.0197 gives the probability of observing a sample mean of 1002.4 or greater if in fact this system results in a mean *E. coli* concentration of 1000 MPN/100 ml.

   (b) Since $0.0197 < 0.05$, there is sufficient evidence to reject H_0, that is, there is sufficient evidence that the mean *E. coli* concentration is greater than 1000 MPN/100 ml, and the system is not working properly.

   (c) With rejection of the null hypothesis, there is the possibility of a Type I error, that is of mistakenly rejecting a true null hypothesis. Possible consequences are that sales of the system drop even though the system is doing what it claims, or that the company performs an overhaul to fix the system even though the system is operating properly.

3. (a) This was an observational study as no treatments were imposed. It would have been highly unethical to impose treatments, that is, to instruct randomly chosen volunteers to smoke, drink, skip exercise, and eat poorly.

   (b) $H_0: p_4 = p_0$, $H_a: p_4 > p_0$ where p_4 is the proportion of adults with all four bad habits who die during a 20-year period, and p_0 is the proportion of adults with none of the four bad habits who die during a 20-year period. (Note that the hypotheses are about the population of all adults with and with none of the four bad habits, not about the volunteers who took part in the study.)

   (c) A Type I error, that is, a mistaken rejection of a true null hypothesis, would result in people being encouraged to not smoke, not drink, exercise, and eat well, when these actions actually will not help decrease 20-year death rates.

(d) A Type II error, that is, a mistaken failure to reject a false null hypothesis, would result in people thinking that smoking, drinking, inactivity, and poor diet don't increase 20-year death rates, when actually they do contribute to higher 20-year death rates.

(e) Calculator software (such as `2-PropZTest` on the TI-84 or on the Casio Prizm) gives $P = 0.000$.

[For instructional purposes in this review book, we note that:

$\hat{p}_1 = \dfrac{91}{314} = 0.290$, $\hat{p}_2 = \dfrac{32}{387} = 0.083$, $\hat{p} = \dfrac{91+32}{314+387}$ and $= 0.175$, and

$$P\left(z > \dfrac{0.290-0.083}{\sqrt{(0.175)(0.825)\left(\dfrac{1}{314}+\dfrac{1}{387}\right)}}\right) = 0.000.]$$

If the null hypothesis were true, that is, if there was no difference in the 20-year death rates between people with all four bad habits and people with none of the bad habits, then the probability of sample proportions with a difference as extreme or more extreme than observed is 0.000 (to three decimals).

4. First, state the hypotheses: H_0: $p_B - p_G = 0$ (or $p_B = p_G$) and H_a: $p_B - p_G > 0$ (or $p_B > p_G$). (If asked to state the parameters, then state "where p_B is the proportion of all high school boys who meet the recommended level of physical activity, and p_G is the proportion of all high school girls who meet the recommended level of physical activity.")

Second, identify the test by name or formula and check the assumptions:

This is a two-sample z-test for proportions. Checks: Independent random samples (given), and $n_B \hat{p}_B = 370 \geq 10$, $n_B (1 - \hat{p}_B) = 480 \geq 10$, $n_G \hat{p}_G = 218 \geq 10$, and $n_G (1 - \hat{p}_G) = 362 \geq 10$.

Third, calculate the test statistic z and the P-value. Calculator software (such as `2-PropZTest` on the TI-84) gives $z = 2.2427$ and $P = 0.012$. [For instructional purposes in this review book, we note that

$$\hat{p}_B = \dfrac{370}{850} \approx 0.4353, \quad \hat{p}_G = \dfrac{218}{580} \approx 0.3759, \quad \hat{p} = \dfrac{370+218}{850+580} = \dfrac{588}{1430} \approx 0.4112,$$

$$z = \dfrac{0.4353-0.3759}{\sqrt{(0.4112)(0.5888)\left(\dfrac{1}{850}+\dfrac{1}{580}\right)}} \approx 2.2427, \text{ and } P = P(z > 2.2427) = 0.012.]$$

Fourth, linking to the P-value, give a conclusion in context. With a P-value this small ($0.012 < 0.05$), there is sufficient evidence that the proportion of all high school boys who meet the recommended level of physical activity is greater than the proportion of all high school girls who meet the recommended level of physical activity.

5. (a) One block consists of the 500 patients who have had a mini stroke in the previous year, and a second block consists of the 500 patients who have not. In one block, number the patients 1 through 500, and then using a random number generator, pick numbers between 1 and 500, throwing out repeats, until 250 patients are picked. Assign surgery to these patients and stents to the remaining 250. Repeat for the other block.

   (b) One method is to increase the sample size, resulting in a reduction in the standard error of the sampling distribution, which in turn increases the probability of rejecting the null hypothesis if it is false. A second method is to increase α, the significance level, which in turn also increases the probability of rejecting the null hypothesis if it is false. In either case, there is increased probability of detecting any difference in the proportions of patients suffering major complications between the surgery and stent recipients.

6. The data come in pairs, and the two-sample test does not apply the knowledge of what happened to each individual driver (the condition of independence of the two samples is violated). The appropriate test is a one-population, small-sample hypothesis test on the set of differences: {0, –5, 3, –4, –3, 1, –2, –2, –1, –1}. We proceed as follows:

   First, state the hypotheses: H_0: $\mu_D = 0$, H_a: $\mu_D \neq 0$ (If asked to state the parameter, then state "where μ_D is the mean difference in reaction times between DWI and DWT.")

   Second, identify the test and check the assumptions: This is a paired t-test, that is, a single-sample hypothesis test on the set of differences.

   The *data are paired* because they are measurements on the same individuals under DWI and DWT.

   The reaction times of any individual are assumed independent of the reaction times of the others, so the *differences are independent*.

   We must assume that the volunteers are a *representative sample*.

   Either a dotplot of the sample is roughly unimodal and symmetric or the normal probability plot is roughly linear:

   The sample size, $n = 10$, is less than 10% of all possible drivers.

   Third, calculate the test statistic t and the P-value: Calculator software (such as T-Test and Data on the TI-84) gives $t = -1.871$ and $P = 0.094174$. [For instructional purposes in this review book, we note that $\bar{x} = -1.4$ and $s = 2.366$ giving $t = \dfrac{-1.4 - 4}{2.366/\sqrt{10}} = -1.871$, and with $df = 10 - 1 = 9$, $P(t < -1.871) = 0.0471$. Since this is a two-sided test, we double this value to find the P-value to be $P = 0.0942$.]

Fourth, linking to the P-value, give a conclusion in context: With a P-value this small, $0.0942 < 0.10$, there is sufficient evidence to reject H_0, that is, there is sufficient evidence at the 10% significance level of a difference between the mean effect on reaction time between DWI and DWT. Or, with a P-value this large, $0.0942 > 0.05$, there is not sufficient evidence to reject H_0, that is, there is not sufficient evidence at the 5% significance level of a difference between the mean effect on reaction time between DWI and DWT.

Quiz 2

MULTIPLE-CHOICE QUESTIONS

1. **(E)** The P-value of a test is the probability of obtaining a result as extreme (or more extreme) as the one obtained assuming the null hypothesis is *true*. Small P-values are evidence against the null hypothesis. The null and alternative hypotheses are decided upon *before* the data come in.

2. **(E)** The P-value is a conditional probability; in this case, there is a 0.032 probability of an observed difference in sample proportions as extreme (or more extreme) as the one obtained if the null hypothesis is assumed to be true.

3. **(C)** Medications having an effect shorter or longer than claimed should be of concern, so this is a two-sided test: H_a: $\mu \neq 58.4$, and $df = n - 1 = 40 - 1 = 39$.

4. **(B)** The level of significance is defined to be the probability of committing a Type I error, that is, of mistakenly rejecting a true null hypothesis.

5. **(A)** With unknown population standard deviations, the t-distribution must be used,

   and $\sigma_{\bar{x}_1 - \bar{x}_2} = \sqrt{\dfrac{s_1^2}{n_1} + \dfrac{s_2^2}{n_2}} = \sqrt{\dfrac{(0.3)^2}{10} + \dfrac{(0.2)^2}{10}}$.

6. **(E)** With $\hat{p} = \dfrac{198}{2300} = 0.086087$, get $z = \dfrac{0.086087 - 0.08}{\sqrt{(0.08)(0.92)/2300}} = 1.076$ and $P = 0.141$.

   With a P-value this high $0.141 > 0.10$, the government does not have sufficient evidence to reject the company's claim. (1PropZTest gives $P = 0.140956$.)

7. **(E)** There is a different answer for each possible correct value for the population parameter.

8. **(B)** $P(\text{at least one Type I error}) = 1 - P(\text{no Type I errors}) = 1 - (0.99)^5 = 0.049$

9. **(C)** With a smaller α, that is, with a tougher standard to reject H_0, there is a greater chance of failing to reject a false null hypothesis, that is, there is a greater chance of committing a Type II error. Power is the probability that a Type II error is not committed, so a higher Type II error results in lower power.

10. **(C)** If the alternative is true, the probability of failing to reject H_0 and thus committing a Type II error is 1 minus the power, that is, $1 - 0.75 = 0.25$.

11. **(B)** $P\left(z < \dfrac{375-385}{150/\sqrt{100}}\right) = 0.2525$

[normalcdf(–100, –10/15) = 0.252492.]

12. **(B)**

With 0.02 in the tail, a confidence over $0.48 + 0.48 = 0.96$ will contain 0.

FREE-RESPONSE QUESTIONS

1. The parameter of interest is μ = the mean tire pressure in the front right tires of cars with recommended tire pressure of 35 psi. H_0: $\mu = 35$ and H_a: $\mu \neq 35$.

2. First, state the hypotheses:

H_0: $p = 0.174$ and H_a: $p > 0.174$. (If asked to state the parameter, then state "where p is the proportion of AAUP members who own Roth IRAs.")

Second, identify the test by name or formula and check the assumptions.

This is a one-sample z-test for a population proportion.

Assumptions: Random sample (given), $np = 750(0.174) = 130.5 \geq 10$, $n(1 - p) = 750(0.826) = 619.5 \geq 10$, and the sample size, $n = 750$, is less than 10% of all AAUP members.
Third, calculate the test statistic z and the P-value. Calculator software (such as 1-PropZTest on the TI-84 or on the Casio Prizm) gives $z = 1.878$ and $P = 0.030$.
[For instructional purposes in this review book, we note that

$$\hat{p} = \frac{150}{750} = 0.2, \quad z = \frac{0.2-0.174}{\sqrt{\dfrac{(0.174)(0.826)}{750}}} = 1.878, \text{ and } P = P\,(z > 1.878) = 0.030.]$$

Fourth, linking to the P-value, give a conclusion in context.

With a P-value this small , $0.030 < 0.05$, there is sufficient evidence to reject H_0, that is, there is sufficient evidence that more than 17.4% of all AAUP members own Roth IRAs.

3. First, state the hypotheses: H_0: $\mu = 593$ and H_a: $\mu \neq 593$. (If asked to state the parameter, then state "where μ is the mean discharge rate (in 1000 ft^3/sec) at the mouth of the Mississippi River.")

Second, identify the test by name or formula and check the assumptions:

This is a one-sample t-test for the mean with $t = \dfrac{\bar{x} - x_0}{s/\sqrt{n}}$. Checks: The measurements were taken at random times, the sample size, $n = 10$, is less than 10% of all possible measurements, and either a dotplot of the sample is roughly unimodal and symmetric or the normal probability plot is roughly linear:

Third, calculate the test statistic t and the P-value:

Calculator software (such as T-Test using Data on the TI-84) gives $t = -2.7116$ and $P = 0.0239$. [For instructional purposes in this review book, we note that $\bar{x} = 590.5$ and $s = 2.9155$, which gives

$$t = \frac{590.5 - 593}{2.9155/\sqrt{10}} = -2.7116 \text{ and with } df = 10 - 1 = 9,$$

$$P(t < -2.7116) = 0.0120.$$

Since this is a two-sided test, we double this value to find the P value to be $P = 0.0240$.]

Fourth, linking to the P-value, give a conclusion in context: With a P-value this small, $0.0239 < 0.05$, there is sufficient evidence to reject H_0, that is, there is sufficient evidence that the long accepted measure of the discharge rate at the mouth of the Mississippi River has changed.

4. First, state the hypotheses: H_0: $\mu_{NFL} - \mu_{10} = 0$ (or $\mu_{NFL} = \mu_{10}$) and H_a: $\mu_{NFL} - \mu_{10} < 0$ (or $\mu_{NFL} < \mu_{10}$) (or ">" depending on the order of variables). [If asked to state the parameters, then state "where μ_{NFL} is the mean attendance at NFL games and μ_{10} is the mean attendance at Big 10 football games."]

Second, identify the test by name or formula and check the assumptions: This is a two-sample t-test for means. Checks: Independent random samples (given), both samples sizes, 35 and 30, are large enough so that by the CLT, the distribution of sample means is approximately normal and a t-test may be run, and the sample sizes, $n_{NFL} = 35$ and $n_{10} = 30$, are less than 10% of all NFL and Big 10 football games.

Third, calculate the test statistic t and the P-value. Calculator software (such as 2-SampTTest on the TI-84 or 2-Sample tTest on the Casio Prizm) gives $t = -0.8301$, $df = 49.3$, and $P = 0.2052$. [For instructional purposes in this review book, we note that:

$$t = \frac{68,729 - 70,358}{\sqrt{\dfrac{6110^2}{35} + \dfrac{9139^2}{30}}} = -0.8301, \text{ and with the conservative}$$

$$df = \min(35 - 1, 30 - 1) = 29, \ P = P(t < -0.8301) = 0.2066.]$$

Fourth, linking to the P-value, give a conclusion in context. With a P-value this large, $0.2052 > 0.05$, there is not sufficient evidence to reject H_0, that is, there is not sufficient evidence that the average attendance at Big 10 Conference football games is greater than that at NFL games.

5. (a) Different schemes are possible. For example, assign each material a single-digit number, such as A-0, B-1, C-2, D-3, E-4, F-5, G-6, H-7, I-8, J-9. Then read off the digits from the random number list, one at a time, throwing away any repeats, until each of the materials have been picked (or nine have been picked, as the last one left will go last). The order of picking then gives the order of being tested. Using this scheme we would get the following order:

```
FBIE   HD C       A    J                       G
51844 73424 84380 82259 28273 58102 18727 69708
```

(b) The mean drilling times in the ten materials are 4.69 seconds for Drill 1 and 4.77 seconds for Drill 2.

The proper hypothesis test is a matched pairs t-test on the set of differences, $\{0, -0.3, 0.1, -0.2, -0.1, -0.4, 0, 0.3, -0.1, -0.1\}$

First, state the hypotheses: H_0: $\mu_d = 0$, H_a: $\mu_d \neq 0$. (If asked to state the parameter, then state "where μ_d is the difference in mean drilling times of the two drills through different materials.")

Second, identify the test by name or formula and check the assumptions: This is a one-sample t-test for the mean of paired differences.

Checks: Random samples (given), $n = 10$ is less than 10% of all possible materials, hardnesses, and thicknesses, and either a dotplot of the difference sample is roughly unimodal and symmetric, or the normal probability plot is roughly linear:

Third, calculate the test statistic t and the P-value: Calculator software (such as T-Test with Data on the TI-84) gives $t = -1.272$ and $P = 0.2353$. [For instructional purposes in this review book, we note that:

$\bar{x} = -0.08$ and $s = 0.1989$ which gives $t = \dfrac{-0.08 - 0}{0.1989 / \sqrt{10}} = -1.272$, and with

$df = 10 - 1 = 9$, $P(t < -1.272) = 0.1176$. Since this is a two-sided test, we double this value to find the P-value to be $P = 0.2352$.]

Fourth, linking to the P-value, give a conclusion in context. With a P-value this large, $0.2353 > 0.05$, there is not sufficient evidence to reject H_0, that is, there is not sufficient evidence of a difference in mean drilling times of the two drills through different materials.

6. (a) The conditions for a one-sample z-test on a population proportion include that both $np \geq 10$ and $n(1 - p) \geq 10$. In this case, np and $n(1 - p)$ both $= 14(0.5) = 7 < 10$.

   (b) First, state the hypotheses: H_0: $p = 0.5$ and H_a: $p > 0.5$, [If asked to state the parameter, then state "where p is the proportion of soccer matches for which Paul the octopus can correctly predict the winning team."]

   Second, identify the test and check the assumptions:

   This is a test of a binomial model.

   For each trial there are two outcomes (Paul has a choice of two boxes of mussels), the trials are independent, and the probability of Paul picking a winner can be assumed to be the same for each trial. We are given that the 14 matches are a random sample of all matches.

   Third, calculate the P-value: With $p = 0.5$,

   $$P(x \geq 12) = \binom{14}{12} (0.5)^{12}(0.5)^2 + \binom{14}{13} (0.5)^{13} (0.5) + (0.5)^{14} = 0.00647$$

   Fourth, linking to the P-value, give a conclusion in context: With a P-value this small, $0.00647 < 0.05$, there is sufficient evidence to reject H_0, that is, there is sufficient evidence that Paul can correctly predict the results of more than 50% of soccer matches.

Tests of Significance— Chi-Square and Slope of Least Squares Line

13

→ **CHI-SQUARE TEST FOR GOODNESS OF FIT**
→ **CHI-SQUARE TEST FOR INDEPENDENCE**
→ **CHI-SQUARE TEST FOR HOMOGENEITY OF PROPORTIONS**
→ **HYPOTHESIS TEST FOR SLOPE OF LEAST SQUARES LINE**
→ **INDEPENDENCE**

In this topic we continue our development of tools to analyze data. We learn about inference on distributions of counts using chi-square models. This can be used to solve such problems as "Do test results support Mendel's genetic principles?" (goodness-of-fit test); "Was surviving the Titanic sinking independent of a passenger's status?" (independence test); and "Do students, teachers, and staff show the same distributions in types of cars driven?" (homogeneity test). We then learn about inference with regard to linear association of two variables. This can be used to solve such problems as "Is there a linear relationship between the grade received on a term paper and the number of pages turned in?"

TIP

Unless you have counts, you cannot use χ^2 for these particular problems.

CHI-SQUARE TEST FOR GOODNESS OF FIT

A critical question is often whether or not an observed pattern of data fits some given distribution. A perfect fit cannot be expected, and so we must look at discrepancies and make judgments as to the *goodness of fit.*

One approach is similar to that developed earlier. There is the null hypothesis of a good fit, that is, the hypothesis that a given theoretical distribution correctly describes the situation, problem, or activity under consideration. Our observed data consist of one possible sample from a whole universe of possible samples. We ask about the chance of obtaining a sample with the observed discrepancies if the null hypothesis is really true. Finally, if the chance is too small, we reject the null hypothesis and say that the fit is not a good one.

How do we decide about the significance of observed discrepancies? It should come as no surprise that the best information is obtained from squaring the discrepancy values, as this has been our technique for studying variances from the beginning. Furthermore, since, for example, an observed difference of 23 is more significant if the original values are 105 and 128 than if they are 10,602 and 10,625, we must appropriately *weight* each difference. Such weighting is accomplished by dividing each difference by the expected values. The sum of these weighted differences or discrepancies is called *chi-square* and is denoted as χ^2 (χ is the lowercase Greek letter chi):

$$\chi^2 = \sum \frac{(obs - exp)^2}{exp}$$

The smaller the resulting χ^2-value, the better the fit. The *P*-value is the probability of obtaining a χ^2 value as extreme or more extreme than the one obtained if the null hypothesis is

TIP

In Topic 12, the tests were about population parameters; however, with χ^2, the tests will be about distributions or relationships involving categorical variables.

assumed true. If the χ^2 value is large enough, that is, if the *P*-value is small enough, we say there is sufficient evidence to reject the null hypothesis and to claim that the fit is poor.

To decide how large a calculated χ^2-value must be to be significant, that is, to choose a critical value, we must understand how χ^2-values are distributed. A χ^2-distribution is not symmetric and is always skewed to the right. There are distinct χ^2-distributions, each with an associated number of degrees of freedom (*df*). The larger the *df* value, the closer the χ^2-distribution to a normal distribution. Note, for example, that squaring the often-used *z*-scores 1.645, 1.96, and 2.576 results in 2.71, 3.84, and 6.63, respectively, which are entries found in the first row of the χ^2-distribution table.

A large χ^2-value may or may not be significant—the answer depends on which χ^2-distribution we are using. A table is given of critical χ^2-values for the more commonly used percentages or probabilities. To use the χ^2-distribution for approximations in goodness-of-fit problems, the individual expected values cannot be too small. An often-used rule of thumb is that no expected value should be less than 5. Finally, as in all hypothesis tests we've looked at, the sample should be randomly chosen from the given population.

➡ **EXAMPLE 13.1** _____

In a recent year, at the 6 p.m. time slot, television channels 2, 3, 4, and 5 captured the entire audience with 30%, 25%, 20%, and 25%, respectively. During the first week of the next season, 500 viewers are interviewed.

(a) If viewer preferences have not changed, what number of persons is expected to watch each channel?

Answer: 0.30(500) = 150, 0.25(500) = 125, 0.20(500) = 100, and 0.25(500) = 125, so we have

	Channel			
	2	3	4	5
Expected number	150	125	100	125

(b) Suppose that the actual observed numbers are as follows:

Channel

	2	3	4	5
Observed number	139	138	112	111

Do these numbers indicate a change? Are the differences significant?

Answer: Check the conditions:

1. *Randomization*: We must assume that the 500 viewers are a representative sample.
2. We note that the expected values (150, 125, 100, 125) are all ≥ 5.

H_0: The television audience is distributed over channels 2, 3, 4, and 5 with percentages 30%, 25%, 20%, and 25%, respectively.

H_a: The audience distribution is not 30%, 25%, 20%, and 25%, respectively.

We calculate

$$\chi^2 = \sum \frac{(obs - exp)^2}{exp}$$

$$= \frac{(139 - 150)^2}{150} + \frac{(138 - 125)^2}{125} + \frac{(112 - 100)^2}{100}$$

$$+ \frac{(111 - 125)^2}{125}$$

$$= 5.167$$

Then the P-value is $P = P(\chi^2 > 5.167) = 0.1600$. [If n is the number of classes, $df = n - 1 = 3$, and χ^2cdf(5.167, 1000, 3) = 0.1600.] We also note that putting the observed and expected numbers in Lists, calculator software (such as χ^2GOF-Test on the TI-84 or on the Casio Prizm) gives $\chi^2 = 5.167$ and $P = 0.1600$.

Conclusion:

With this large a P-value (0.1600) there is not sufficient evidence to reject H_0. That is, there is not sufficient evidence that viewer preferences have changed.

NOTE

We check that the *expected* values are all ≥ 5.

NOTE

Sometimes it is useful to note which terms in the summation are larger and which are smaller, that is, which terms contribute the most to the final sum.

TIP

With the *t*-distribution, the *df* depends on the sample size, but here, *df* depends on the number of classes or categories.

Remember: the
hypotheses are
always about
the population,
never about
the sample; it
would be *wrong*
to say H_0: the
sample data
are uniformly
distributed
over the five
locations.

TIP

Even though the
observed values
are counts and
thus integers,
the expected
values might
not be integers
and should not
be rounded to
integers.

➡ EXAMPLE 13.2

A grocery store manager wishes to determine whether a certain product will sell equally well in any of five locations in the store. Five displays are set up, one in each location, and the resulting numbers of the product sold are noted.

	Location				
	1	2	3	4	5
Actual number sold	43	29	52	34	48

Is there enough evidence that location makes a difference? Test at both the 5% and 10% significance levels.

Answer:

H_0: Sales of the product are uniformly distributed over the five locations.
H_a: Sales are not uniformly distributed over the five locations.

A total of 43 + 29 + 52 + 34 + 48 = 206 units were sold. If location doesn't matter, we would expect $\frac{206}{5}$ = 41.2 units sold per location (uniform distribution).

	Location				
	1	2	3	4	5
Expected number sold	41.2	41.2	41.2	41.2	41.2

Check the conditions:

1. *Randomization:* We must assume that the 206 units sold are a representative sample.
2. We note that the expected values (all 41.2) are all ≥ 5.

Putting the observed and expected numbers in Lists, and with *df* = 5 – 1 = 4, calculator software (such as χ^2GOF-Test on the TI-84 or on the Casio Prizm) gives $\chi^2 = 8.903$ and $P = 0.0636$. [For instructional purposes, we note that

$$\chi^2 = \frac{(43-41.2)^2}{41.2} + \frac{(29-41.2)^2}{41.2} + \frac{(52-41.2)^2}{41.2} + \frac{(34-41.2)^2}{41.2} + \frac{(48-41.2)^2}{41.2}$$

$$= 8.903$$

With $P = 0.0636$ there is sufficient evidence to reject H_0 at the 10% level but not at the 5% level. If the grocery store manager is willing to accept a 10% chance of committing a Type I error, there is enough evidence to claim location makes a difference.

CHI-SQUARE TEST FOR INDEPENDENCE

In the goodness-of-fit problems above, a set of expectations was based on an assumption about how the distribution should turn out. We then tested whether an observed sample distribution could reasonably have come from a larger set based on the assumed distribution.

There is a class of real-world problems in which we want to determine if there is a significant association between two categorical variables. The procedure is called a *test of independence*. For example, do nonsmokers, light smokers, and heavy smokers all have the same likelihood of being eventually diagnosed with cancer, heart disease, or emphysema? Is there a relationship (association) between smoking status and being diagnosed with one of these diseases?

We classify our observations in two ways and then ask whether the two ways are independent of each other. For example, we might consider several age groups and within each group ask how many employees show various levels of job satisfaction. The null hypothesis is that age and job satisfaction are independent, that is, that the proportion of employees expressing a given level of job satisfaction is the same no matter which age group is considered.

Analysis involves calculating a table of *expected* values, assuming the null hypothesis about independence is true. We then compare these expected values with the observed values and ask whether the differences are reasonable if H_0 is true. The significance of the differences is gauged by the same χ^2-value of weighted squared differences. The smaller the resulting χ^2-value, the more reasonable the null hypothesis of independence. If the χ^2-value is large enough, that is, if the P-value is small enough, we can say that the evidence is sufficient to reject the null hypothesis and to claim that there *is* some relationship between the two variables or methods of classification.

NOTE

Again, expected values should all be ≥ 5.

In this type of problem,

$$df = (r-1)\,(c-1)$$

where df is the number of degrees of freedom, r is the number of rows, and c is the number of columns.

A point worth noting is that even if there is sufficient evidence to reject the null hypothesis of independence, we cannot necessarily claim any direct *causal* relationship. In other words, although we can make a statement about some link or relationship between two variables, we are *not* justified in claiming that one causes the other. For example, we may demonstrate a relationship between salary level and job satisfaction, but our methods would not show that higher salaries cause higher job satisfaction. Perhaps an employee's higher job satisfaction impresses his superiors and thus leads to larger increases in pay. Or perhaps there is a third variable, such as training, education, or personality, that has a direct causal relationship to both salary level and job satisfaction.

➡ **EXAMPLE 13.3**

In a nationwide telephone poll of 1000 randomly selected adults representing Democrats, Republicans, and Independents, respondents were asked two questions: their party affiliation and if their confidence in the U.S. banking system had been shaken by the savings and loan crisis. The answers, cross-classified by party affiliation, are given in the following *contingency table*.

Confidence Shaken

Observed	Yes	No	No opinion
Democrats	175	220	55
Republicans	150	165	35
Independents	75	105	20

Test the null hypothesis that shaken confidence in the banking system is independent of party affiliation. Use a 10% significance level.

Answer:

H_0: Party affiliation and shaken confidence in the banking system are independent.
H_a: Party affiliation and shaken confidence in the banking system are not independent.

Putting the observed data into a Matrix, calculator software (such as χ^2-Test on the TI-84, Casio Prizm, or HP Prime) gives $\chi^2 = 3.024$ and $P = 0.5538$, and stores the expected values in a second matrix:

180	220.5	49.5
140	171.5	38.5
80	98	22

TIP

It is not enough to just say that all expected cells are ≥ 5; you must list the expected counts.

We are given a random sample and note that all expected cells are ≥ 5.

With this large a *P*-value, 0.5538 > 0.05, there is not sufficient evidence to reject H_0, that is, there is not sufficient evidence of any relationship between party affiliation and shaken confidence in the banking system. [For instructional purposes, we note that the expected values, χ^2, and *P* can be calculated as follows:

Row totals: 175 + 220 + 55 = 450,
150 + 165 + 35 = 350,
75 + 105 + 20 = 200.

Column totals: 175 + 150 + 75 = 400,
220 + 165 + 105 = 490,
55 + 35 + 20 = 110.

	Yes	No	No opinion	
Democrats				450
Republicans				350
Independents				200
	400	490	110	

To calculate, for example, the expected value in the upper left box, we can proceed in any of several equivalent ways. First, we could note that the proportion of Democrats is $\frac{450}{1000}$ = 0.45; and so, if independent, the expected number of Democrat *yes* responses is 0.45(400) = 180. Instead, we could note that the proportion of *yes* responses is $\frac{400}{1000}$ = 0.4; and so, if independent, the expected number of Democrat *yes* responses is 0.4(450) = 180. Finally, we could note that both these calculations simply involve $\frac{(450)(400)}{1000}$ = 180.

In other words, *the expected value of any box can be calculated by multiplying the corresponding row total by the appropriate column total and then dividing by the grand total*. Thus, for example, the expected value for the middle box, which corresponds to Republican *no* responses, is $\frac{(350)(490)}{1000}$ = 171.5.

Continuing in this manner, we fill in the table as follows:

	Yes	No	No opinion	
Democrats	180	220.5	49.5	450
Republicans	140	171.5	38.5	350
Independents	80	98	22	200
	400	490	110	

Expected

[An appropriate check at this point is that each expected cell count is at least 5.]

Next we calculate the value of chi-square:

$$\chi^2 = \frac{(175 - 180)^2}{180} + \frac{(220 - 220.5)^2}{220.5} + \frac{(55 - 49.5)^2}{49.5} + \frac{(150 - 140)^2}{140} + \frac{(165 - 171.5)^2}{171.5}$$

$$+ \frac{(35 - 38.5)^2}{38.5} + \frac{(75 - 80)^2}{80} + \frac{(100 - 98)^2}{98} + \frac{(20 - 22)^2}{22}$$

$$= 3.024$$

$$df = (r - 1)(c - 1) = (3 - 1)(3 - 1) = 4$$

P is then calculated using calculator software such as χ^2cdf(3.024, 1000, 4) of the TI-84 or Ccd on the Casio Prizm or CHISQUARE_CDF on the HP Prime.]

On the TI-Nspire the result shows as:

$\chi^2$2way $\begin{bmatrix} 175 & 220 & 55 \\ 150 & 165 & 35 \\ 75 & 105 & 20 \end{bmatrix}$: *stat.results*	"Title"	"χ^2 2–way Test"
	"χ^2"	3.02428
	"PVal"	0.553771
	"df"	4.
	"ExpMatrix"	"[...]"
	"CompMatrix"	"[...]"
stat.ExpMatrix		$\begin{bmatrix} 180. & 220.5 & 49.5 \\ 140. & 171.5 & 38.5 \\ 80. & 98. & 22. \end{bmatrix}$

As for conditions to check for chi-square tests for independence, we should check that the sample is randomly chosen and that the expected values for all cells are at least 5. If a category has one of its expected cell count less than 5, we can combine categories that are logically similar (for example "disagree" and "strongly disagree"), or combine numerically small categories collectively as "other."

➡ **EXAMPLE 13.4**

To determine whether men with a combination of childhood abuse and a certain abnormal gene are more likely to commit violent crimes, a study is run on a simple random sample of 575 males in the 25 to 35 age group. The data are summarized in the following table:

	Not abused, normal gene	Abused, normal gene	Not abused, abnormal gene	Abused, abnormal gene
Criminal behavior	48	21	32	26
Normal behavior	201	79	118	50

(a) Is there evidence of a relationship between the four categories (based on childhood abuse and abnormal genetics) and behavior (criminal versus normal)? Explain.

(b) Is there evidence that among men with the normal gene, the proportion of abused men who commit violent crimes is greater than the proportion of nonabused men who commit violent crimes? Explain.

(c) Is there evidence that among men who were not abused as children, the proportion of men with the abnormal gene who commit violent crimes is greater than the proportion of men with the normal gene who commit violent crimes? Explain.

(d) Is there a contradiction in the above results? Explain.

Answers:

(a) A chi-square test for independence is indicated. The expected cell counts are as follows:

55.0	22.1	33.1	16.8
194.0	77.9	116.9	59.2

The condition that all cell counts are greater than 5 is met.

H_0: The four categories (based on childhood abuse and abnormal genetics) and behavior (criminal versus normal) are independent.

H_a: The four categories (based on childhood abuse and abnormal genetics) and behavior (criminal versus normal) are not independent (there is a relationship).

Running a chi-square test gives $\chi^2 = 7.752$. With $df = (r - 1)(c - 1) = 3$, we get $P = 0.0514$. Since $0.0514 < 0.10$, the data do provide some evidence (at least at the 10% significance level) to reject H_0 and conclude that there is evidence of a relationship between the four categories (based on childhood abuse and abnormal genetics) and behavior (criminal versus normal).

(b) A two-proportion z-test is indicated. We must check that n is large enough:
$n_1\hat{p}_1 = 21 \geq 10$, $n_1(1 - \hat{p}_1) = 79 \geq 10$, $n_2\hat{p}_2 = 48 \geq 10$, $n_2(1 - \hat{p}_2) = 201 \geq 10$. We are given SRSs that are less than 10% of all males in the 25 to 35 age group.

$H_0: p_1 - p_2 = 0$ (where p_1 is the proportion of abused men with normal genetics who commit violent crimes and p_2 is the proportion of nonabused men with normal genetics who commit violent crimes)

$H_a: p_1 - p_2 > 0$ (the proportion of abused men with normal genetics who commit violent crimes is greater than the proportion of nonabused men with normal genetics who commit violent crimes)

Calculator software (such as 2-PropZTest on the TI-84 or on the Casio Prizm) gives $z = 0.3653$ and $P = 0.3574$. With this large a P-value, $0.3574 > 0.05$, there is not sufficient evidence to reject H_0, that is, there is not sufficient evidence that the proportion of abused men with normal genetics who commit violent crimes is greater than the proportion of nonabused men with normal genetics who commit violent crimes.

(c) A two-proportion z-test is indicated. We must check that n is large enough:
$n_1\hat{p}_1 = 32 \geq 10$, $n_1(1 - \hat{p}_1) = 118 \geq 10$, $n_2\hat{p}_2 = 48 \geq 10$, $n_2(1 - \hat{p}_2) = 201 \geq 10$. We are given SRSs that are less than 10% of all males in the 25 to 35 age group.

$H_0: p_1 - p_2 = 0$ (where p_1 is the proportion of nonabused men with abnormal genetics who commit violent crimes and p_2 is the proportion of nonabused men with normal genetics who commit violent crimes)

$H_a: p_1 - p_2 > 0$ (the proportion of nonabused men with abnormal genetics who commit violent crimes is greater than the proportion of nonabused men with normal genetics who commit violent crimes)

Calculator software (such as 2-PropZTest on the TI-84 or on the Casio Prizm) gives $z = 0.4969$ and $P = 0.3096$. With this large a P-value, $0.3096 > 0.05$, there is not sufficient evidence to reject H_0, that is, there is not sufficient evidence that the proportion of nonabused men with the abnormal gene who commit violent crimes is greater than the proportion of nonabused men with the normal gene who commit violent crimes.

(d) There is no contradiction. It is possible to have evidence of an overall relationship without significant evidence showing in a subset of the categories.

CHI-SQUARE TEST FOR HOMOGENEITY OF PROPORTIONS

In chi-square goodness-of-fit tests we work with a single variable in comparing a single sample to a population model. In chi-square independence tests we work with a single sample classified on two variables. Chi-square procedures can also be used with a single variable to compare samples from two or more populations. It is important that the samples be *simple random samples*, that they be taken *independently* of each other, that the original populations be large compared to the sample sizes, and that the expected values for all cells be at least 5. The contingency table used has a row for each sample.

➡ EXAMPLE 13.5

In a large city, a group of AP Statistics students work together on a project to determine which group of school employees has the greatest proportion who are satisfied with their jobs. In independent simple random samples of 100 teachers, 60 administrators, 45 custodians, and 55 secretaries, the numbers satisfied with their jobs were found to be 82, 38, 34, and 36, respectively. Is there evidence that the proportion of employees satisfied with their jobs is different in different school system job categories?

Answer:

H_0: The proportion of employees satisfied with their jobs is the same across the various school system job categories.

H_a: At least two of the job categories differ in the proportion of employees satisfied with their jobs.

The observed counts are as follows:

	Satisfied	Not satisfied
Teachers	82	18
Administrators	38	22
Custodians	34	11
Secretaries	36	19

Putting the observed data into a matrix, calculator software (such as χ^2-Test on the TI-84 or on the Casio Prizm) gives $\chi^2 = 8.707$ and $P = 0.0335$, and stores the expected values in a second matrix:

73.1	26.9
43.8	16.2
32.9	12.1
40.2	14.8

We are given an SRS and note that all expected cells are ≥ 5. With this small a P-value, $0.0335 < 0.05$, there is sufficient evidence to reject H_0, that is, there is sufficient evidence that the proportion of employees satisfied with their jobs is not the same across all the school system job categories.

The difference between the test for independence and the test for homogeneity can be confusing. When we're simply given a two-way table, it's not obvious which test is being performed. The crucial difference is in the *design* of the study. Did we pick samples from each of two or more populations to compare the distribution of some variable among the different populations? If so, we are doing a test for homogeneity. Did we pick one sample from a single population and cross-categorize on two variables to see if there is an association between the variables? If so, we are doing a test for independence.

For example, if we separately sample Democrats, Republicans, and Independents to determine whether they are for, against, or have no opinion with regard to stem cell research (several samples, one variable), then we do a test for homogeneity with a null hypothesis that the distribution of opinions on stem cell research is the same among Democrats, Republicans, and Independents. However, if we sample the general population, noting the political preference and opinions on stem cell research of the respondents (one sample, two variables), then we do a test for independence with a null hypothesis that political preference is independent of opinion on stem cell research.

Finally, it should be remembered that while we used χ^2 for categorical data, the χ^2 distribution is a *continuous* distribution, and applying it to counting data is just an approximation.

HYPOTHESIS TEST FOR SLOPE OF LEAST SQUARES LINE

In addition to finding a confidence interval for the true slope, one can also perform a hypothesis test for the value of the slope. Often one uses the null hypothesis H_0: $\beta = 0$, that is, that there is no linear relationship between the two variables. Assumptions for inference for the slope of the least squares line include the following: (1) the sample must be randomly selected; (2) the scatterplot should be approximately linear; (3) there should be no apparent pattern in the residuals plot; and (4) the distribution of the residuals should be approximately normal. (This third point can be checked by a histogram, dotplot, stemplot, or normal probability plot of the residuals.)

> **TIP**
>
> *Be careful* about abbreviations. For example, your teacher might use LOBF (line of best fit), but the grader may have no idea what this refers to.

Note that a low *P*-value tells us that if the two variables did not have some linear relationship, it would be highly unlikely to find such a random sample; however, strong evidence that there is some association does not mean the association is strong.

➡ EXAMPLE 13.6

The following table gives serving speeds in mph (using a flat or "cannonball" serve) of ten randomly selected professional tennis players before and after using a newly developed tennis racquet.

| With old racquet | 125 | 133 | 108 | 128 | 115 | 135 | 125 | 117 | 130 | 121 |
| With new racquet | 133 | 134 | 112 | 139 | 123 | 142 | 140 | 129 | 139 | 126 |

> **TIP**
>
> If the data don't look straight, do not fit a straight line.

(a) Is there evidence of a *straight line* relationship with positive slope between serving speeds of professionals using their old and the new racquets?

(continued)

Answer: Checking the assumptions:

We are told that the data come from a *random* sample of professional players.

<u>Scatterplot is approximately linear.</u>   <u>No apparent pattern in residuals plot.</u>

<u>Distribution of residuals is approximately normal.</u>

<u>Checked by histogram.</u>   *or*   <u>Checked by normal probability plot.</u>

We proceed with the hypothesis test for the slope of the regression line.

$$H_0: \beta_1 = 0 \qquad H_a: \beta_1 > 0$$

Using the statistics software on a calculator (for example, `LinRegTTest` on the TI-84 or LinearReg tTest or on the Casio Prizm), gives:

$\overset{\frown}{\text{New speed}}$ = 8.76 + 0.99 (Old speed)   with $t = 5.853$ and $P = 0.00019$

With such a small P value, $0.00019 < 0.001$, there is very strong evidence to reject H_0, that is, there is very strong evidence of a straight line relationship with positive slope between serving speeds of professionals using their old and the new racquets.

(b) Interpret in context the least squares line.

Answer: With a slope of approximately 1 and a *y*-intercept of 8.76, the regression line indicates that use of the new racquet increases serving speed on the average by 8.76 mph regardless of the old racquet speed. That is, players with lower and higher old racquet speeds experience on the average the same numerical (rather than percentage) increase when using the new racquet.

(continued)

On the TI-Nspire the result shows as:

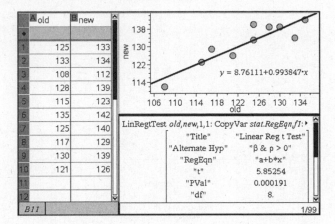

It is important to be able to interpret computer output, and appropriate computer output is helpful in checking the assumptions for inference.

➡ EXAMPLE 13.7

A company offers a 10-lesson program of study to improve students' SAT scores. A survey is made of a random sampling of 25 students. A scatterplot of improvement in total (Math + Evidence-Based Reading and Writing) SAT score versus number of lessons taken is as follows:

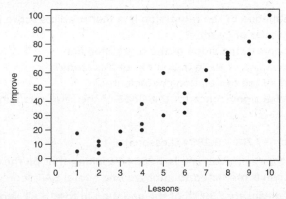

Some computer output of the regression analysis follows, along with a plot and a histogram of the residuals.

Regression Analysis: Improvement Versus Lessons

Dependent variable: Improvement

Predictor	Coef	SE Coef	T	P
Constant	−7.718	4.272	−1.81	0.084
Lessons	9.2854	0.6789	13.68	0.000

S = 9.744   R—Sq = 89.1%

Analysis of Variance

Source	DF	SS	MS	F	P
Regression	1	17761	17761	187.06	0.000
Residual Error	23	2184	95		
Total	24	19945			

(continued)

(a) What is the equation of the regression line that predicts score improvement as a function of number of lessons?

(b) Interpret, in context, the slope of the regression line.

(c) What is the meaning of R–Sq in context of this study?

(d) Give the value of the correlation coefficient.

(e) Perform a test of significance for the slope of the regression line.

Answers:

(a) $\widehat{\text{Improvement}} = -7.718 + 9.2854 \text{ (Lessons)}$

(b) The slope of the regression line is 9.2854, meaning that on the average, each additional lesson is predicted to improve one's total SAT score by 9.2854.

(c) R–Sq = 89.1% means that 89.1% of the variation in total SAT score improvement can be explained by variation in the number of lessons taken.

(d) $r = +\sqrt{.891} = 0.944$, where the sign is positive because the slope is positive.

(e) A test of significance of the slope of the regression line with $H_0: \beta = 0$, $H_a: \beta \neq 0$ (test of significance of correlation) is as follows:

Assumptions: We are told the data came from a random sample, the scatterplot appears to be approximately linear, there is no apparent pattern in the residuals plot, and the histogram of residuals appears to be approximately normal.

$$t = \frac{9.2854 - 0}{0.6789} = 13.68, \text{ and with } df = 25 - 2 = 23, P = 0.000$$

(Or t and P can simply be read off the computer output!)

With this small a P-value, 0.000 < any reasonable significance level, there is very strong evidence to reject H_0 and conclude that there is strong evidence of a linear relationship between improvement in total SAT score and number of lessons taken.

INDEPENDENCE

Throughout this review book the concept of *independence* has arisen many times, with different meanings in different contexts, and a quick review is worthwhile.

1. If the outcome of one event, E, doesn't affect the outcome of another event, F, we say the events are *independent*, and we have $P(E|F) = P(E)$, $P(F|E) = P(F)$, and $P(E \cap F) = P(E)P(F)$.

2. When the distribution of one variable in a contingency table is the same for all categories of another variable, we say the two variables are *independent*.

3. When two random variables are *independent*, we can use the rule for adding variances with regard to the random variables $X + Y$ and $X - Y$.

4. In a binomial distribution, the trials must be *independent;* that is, the probability of success must be the same on every trial, irrespective of what happened on a previous trial. However, when we pick a sample from a population, we do change the size and makeup of the remaining population. We typically allow this as long as the sample is not too large—the usual rule of thumb is that the sample size should be smaller than 10% of the population.

5. In inference we wish to make conclusions about a population parameter by analyzing a sample. A crucial assumption is always that the sampled values are *independent* of each other. This typically involves checking if there was proper randomization in the gathering of data. Also, to minimize the effect on independence from samples being drawn without replacement, we check that less than 10% of the population is sampled.

6. Furthermore, if inference involves the comparison of two groups, the two groups must be *independent* of each other.

7. With paired data, while the observations in each pair are not independent, the differences must be *independent* of each other.

8. With regression, the errors in the regression model must be *independent,* and one may check this by confirming there are no patterns in the residual plot.

SUMMARY

- Chi-square analysis is an important tool for inference on distributions of counts.
- The chi-square statistic is found by summing the weighted squared differences between observed and expected counts.
- A chi-square goodness-of-fit test compares an observed distribution to some expected distribution.
- A chi-square test of independence tests for evidence of an association between two categorical variables from a single sample.
- A chi-square test of homogeneity compares samples from two or more populations with regard to a single categorical variable.
- The sampling distribution for the slope of a regression line can be modeled by a t-distribution with $df = n - 2$.

QUIZ 1

Multiple-Choice Questions

> **Directions:** The questions that follow are each followed by five suggested answers. Choose the response that best answers the question.

1. To test the claim that dogs bite more or less depending upon the phase of the moon, a university hospital counts admissions for dog bites and classifies with moon phase.

	New moon	First quarter	Full moon	Last quarter
Dog bite admissions	32	27	47	38

Which of the following is the proper conclusion?

(A) The data prove that dog bites occur equally during all moon phases.
(B) The data prove that dog bites do not occur equally during all moon phases.
(C) The data give sufficient evidence that dog bites occur equally during all moon phases.
(D) The data give sufficient evidence that dog bites do not occur equally during all moon phases.
(E) The data do not give sufficient evidence to conclude that dog bites are related to moon phases.

2. For a project, a student randomly picks 100 fellow AP Statistics students to survey on whether each has either a PC or Apple at home (all students in the school have a home computer) and what score (1, 2, 3, 4, 5) each expects to receive on the AP exam. A chi-square test of independence results in a test statistic of 8. How many degrees of freedom are there?

(A) 1
(B) 4
(C) 7
(D) 9
(E) 99

3. A random sample of 100 former student-athletes are picked from each of the 16 colleges that are members of the Big East conference. Students are surveyed about whether or not they feel they received a quality education while participating in varsity athletics. A 2×16 table of counts is generated. Which of the following is the most appropriate test to determine whether there is a difference among these schools as to the student-athlete perception of having received a quality education.

(A) A chi-square goodness-of-fit test for a uniform distribution
(B) A chi-square test of independence
(C) A chi-square test of homogeneity of proportions
(D) A multiple-sample z-test of proportions
(E) A multiple-population z-test of proportions

4. A disc jockey wants to determine whether middle school students and high school students have similar music tastes. Independent random samples are taken from each group, and each person is asked whether he/she prefers hip-hop, pop, or alternative. A chi-square test of homogeneity of proportions is performed, and the resulting P-value is below 0.05. Which of the following is a proper conclusion?

(A) There is sufficient evidence that for all three music choices the proportion of middle school students who prefer each choice is equal to the corresponding proportion of high school students.
(B) There is sufficient evidence that the proportion of middle school students who prefer hip-hop is different from the proportion of high school students who prefer hip-hop.
(C) There is sufficient evidence that for all three music choices the proportion of middle school students who prefer each choice is different from the corresponding proportion of high school students.
(D) There is sufficient evidence that for at least one of the three music choices the proportion of middle school students who prefer that choice is equal to the corresponding proportion of high school students.
(E) There is sufficient evidence that for at least one of the three music choices the proportion of middle school students who prefer that choice is different from the corresponding proportion of high school students.

5. A food biologist surveys people at an ice cream parlor, noting their taste preferences and cross-classifying against the presence or absence of a particular marker in a saliva swab test.

	Presence	Absence
Vanilla	32	12
Chocolate	15	7
Strawberry	24	19

Is there sufficient evidence of a relationship between taste preference and the marker presence?

(A) At the 10% significance level, the data prove that there is a relationship between taste preference and the presence of the marker.
(B) At the 10% significance level, the data prove that there is no relationship between taste preference and the presence of the marker.
(C) There is sufficient evidence at the 5% significance level of a relationship between taste preference and the presence of the marker.
(D) There is sufficient evidence at the 10% significance level, but not at the 5% significance level, of a relationship between taste preference and the presence of the marker.
(E) There is not sufficient evidence at the 10% significance level of a relationship between taste preference and the presence of the marker.

6. Can dress size be predicted from a woman's height? In a random sample of 20 female high school students, dress size versus height (cm) gives the following regression results:

```
The regression equation is
Size = -48.8 + 0.374 Height

Predictor     Coef     SE Coef        T         P
Constant     -48.81      30.57     -1.60     0.128
Height        0.3736      0.1898     1.97     0.065

S = 4.46720       R-Sq = 17.7%      R-Sq(adj) = 13.1%
```

Is there statistical evidence of a linear relationship between dress size and height?

(A) No, because r^2, the coefficient of determination, is too small.
(B) No, because 0.128 is above any reasonable significance level.
(C) Yes, because by any reasonable observation, taller women tend to have larger dress sizes.
(D) Yes, because the computer printout does give a regression equation.
(E) There is sufficient evidence at the 10% significance level, but not at the 5% level.

7. A 95% confidence interval for the slope of a regression line is calculated to be (–0.783, 0.457). Which of the following must be true?

(A) The slope of the regression line is 0.
(B) The slope of the regression line is –0.326.
(C) A scatterplot of the data would show a linear pattern.
(D) A residual plot would show no pattern.
(E) The correlation coefficient is negative.

Free-Response Questions

> **Directions:** You must show all work and indicate the methods you use. You will be graded on the correctness of your methods and on the accuracy of your final answers.

1. A candy manufacturer advertises that their fruit-flavored sweets have hard sugar shells in five colors with the following distribution: 35% cherry red, 10% vibrant orange, 10% daffodil yellow, 25% emerald green, and 20% royal purple. A random sample of 300 sweets yielded the counts in the following table:

	Cherry red	Vibrant orange	Daffodil yellow	Emerald green	Royal purple
Observed counts	94	34	22	77	73

Is there sufficient evidence that the distribution is different from what is claimed by the manufacturer?

2. You want to study whether smokers and non-smokers have equal fitness levels (low, medium, high) and are considering two survey designs:

I. Take a random sample of 200 people, asking each whether or not they smoke, and measuring the fitness level of each.

II. Take a random sample of 100 smokers and measure the fitness level of each, and take a random sample of 100 non-smokers and measure the fitness level of each.

(a) Which of the designs results in a test of independence and which results in a test of homogeneity? Explain.

(b) If we are interested in whether the proportion of people who have various fitness levels differs among smokers and non-smokers, which design should be used? Explain.

(c) If we are interested in the conditional distribution of people with given fitness levels who are smokers or are not smokers, which design should be used? Explain.

3. Is there a difference in happiness levels between busy and idle people? In one study, after filling out a survey, 175 randomly chosen high school students were told they could either sit 15 minutes while the survey was being tabulated, or they could walk 15 minutes to another building where the survey was being tabulated. Then they were given a questionnaire asking how good they felt during the past 15 minutes (on a scale of 1, "not good," to 5, "very good"). The results of the questionnaire were as follows:

	Happiness level				
	1	2	3	4	5
Busy (walking)	8	15	24	26	25
Idle (sitting)	18	20	15	10	14

(a) Does the above data give statistical evidence of a relationship between happiness level and busy/idle choice of the students?

(b) If the answer above is positive for a relationship, is it reasonable to conclude that encouraging high school students to keep more busy will lead to higher happiness levels?

4. A study of 100 randomly selected teenagers, ages 13–17, looked at number of texts per waking hour versus age, yielding the computer regression output below:

```
Predictor      Coef     SE Coef          T        P
Constant     -1.055       2.815      -0.37    0.709
Age          0.4577       0.1866       2.45    0.016

S = 2.66501       R-Sq = 5.8%       R-Sq(adj) = 4.8%
```

(a) Interpret the slope in context.

(b) What three graphs should be checked with regard to conditions for a test of significance for the slope of the regression line?

(c) Assuming all conditions for inference are met, perform this test of significance.

(d) Give a conclusion in context, taking into account both your answer to the hypothesis test as well as the value of R-Sq.

QUIZ 2

Multiple-Choice Questions

Directions: The questions that follow are each followed by five suggested answers. Choose the response that best answers the question.

1. A random sample of mice is obtained, and each mouse is timed as it moves through a maze to a reward treat at the end. After several days of training, each mouse is timed again. The data should be analyzed using

   (A) a z-test of proportions.
   (B) a two-sample test of means.
   (C) a paired t-test.
   (D) a chi-square test.
   (E) a regression analysis.

2. According to theory, blood types in the general population occur in the following proportions: 46% O, 40% A, 10% B, and 4% AB. Anthropologists come upon a previously unknown civilization living on a remote island. A random sampling of blood types yields the following counts: 77 O, 85 A, 23 B, and 15 AB. Is there sufficient evidence to conclude that the distribution of blood types found among the island population differs from that which occurs in the general population?

   (A) The data prove that blood type distribution on the island is different from that of the general population.
   (B) The data prove that blood type distribution on the island is not different from that of the general population.
   (C) The data give sufficient evidence at the 1% significance level that blood type distribution on the island is different from that of the general population.
   (D) The data give sufficient evidence at the 5% significance level, but not at the 1% significance level, that blood type distribution on the island is different from that of the general population.
   (E) The data do not give sufficient evidence at the 5% significance level that blood type distribution on the island is different from that of the general population.

3. Is there a relationship between education level and sports interest? A study cross-classified 1500 randomly selected adults in three categories of education level (not a high school graduate, high school graduate, and college graduate) and five categories of major sports interest (baseball, basketball, football, hockey, and tennis). The χ^2-value is 13.95. Is there evidence of a relationship between education level and sports interest?

   (A) The data prove there is a relationship between education level and sports interest.
   (B) The evidence points to a cause-and-effect relationship between education and sports interest.
   (C) There is sufficient evidence at the 5% significance level of a relationship between education level and sports interest.
   (D) There is sufficient evidence at the 10% significance level, but not at the 5% significance level, of a relationship between education level and sports interest.
   (E) The P-value is greater than 0.10, so there is not sufficient evidence of a relationship between education level and sports interest.

4. A geneticist claims that four species of fruit flies should appear in the ratio 1:3:3:9. Suppose that a sample of 2000 flies contained 110, 345, 360, and 1185 flies of each species, respectively. Is there sufficient evidence to reject the geneticist's claim?

   (A) The data prove the geneticist's claim.
   (B) The data prove the geneticist's claim is false.
   (C) The data do not give sufficient evidence to reject the geneticist's claim.
   (D) The data give sufficient evidence to reject the geneticist's claim.
   (E) The evidence from this data is inconclusive.

5. Random samples of 25 students are chosen from each high school class level, students are asked whether or not they are satisfied with the school cafeteria food, and the results are summarized in the following table:

	Freshmen	Sophomores	Juniors	Seniors
Satisfied	15	12	9	7
Dissatisfied	10	13	16	18

   Is there sufficient evidence of a difference in cafeteria food satisfaction among the class levels?

   (A) The data prove that there is a difference in cafeteria food satisfaction among the class levels.
   (B) There is sufficient evidence of a linear relationship between food satisfaction and class level.
   (C) There is sufficient evidence at the 1% significance level of a difference in cafeteria food satisfaction among the class levels.
   (D) There is sufficient evidence at the 5% significance level, but not at the 1% significance level, of a difference in cafeteria food satisfaction among the class levels.
   (E) With $P = 0.1117$ there is not sufficient evidence of a difference in cafeteria food satisfaction among the class levels.

6. To study the relationship between calories (kcal) and fat (g) in pizza, slices of 14 randomly selected major brand pizzas are chemically analyzed. Computer printout for regression:

```
Predictor        Coef    SE Coef       T        P
Constant        1.593      1.422    1.12    0.285
Calories     0.035881   0.004896    7.33    0.000

S = 1.27515      R-Sq = 81.7%     R-Sq(adj) = 80.2%
```

What is measured by $S = 1.27515$?

(A) Variability in calories among slices
(B) Variability in fat among slices
(C) Variability in the slope (g/kcal) of the regression line
(D) Variability in the y-intercept of the regression line
(E) Variability in the residuals

7. Can Points Per Game (PPG) be predicted based on players' salaries? For the Chicago Bulls 2015–2016 season, a linear association study yields the following computer output:

Regression Analysis: PPG versus Salary (in Millions of $)

```
Variable      N     Mean    SE Mean     StDev
Salary       14     6.00       1.71      6.39
PPG          14     8.33       1.63      6.08
```

The regression equation is PPG = 4.39 + 0.656 Salary

```
Predictor      Coef       SE Coef
Constant      4.390         1.709
Salary       0.6564        0.1988

S = 4.58077    R-Sq = 47.6%
```

Which of the following is an appropriate test statistic for testing the null hypothesis that the slope of the regression line is greater than 0? (Assume all conditions for inference are met.)

(A) $\dfrac{8.33}{1.63}$

(B) $\dfrac{8.33}{6.08}$

(C) $\dfrac{4.390}{1.709}$

(D) $\dfrac{0.6564}{0.1988}$

(E) $\dfrac{\frac{0.6564}{0.1988}}{\sqrt{14}}$

Free-Response Questions

> **Directions:** You must show all work and indicate the methods you use. You will be graded on the correctness of your methods and on the accuracy of your final answers.

1. A poll, asking a random sample of adults whether or not they eat breakfast and to rate their morning energy level, results in the table:

	Morning energy level		
	Low	Medium	High
Breakfast	22%	24%	24%
No breakfast	12%	10%	8%

   (a) If the sample size was $n = 500$, is there sufficient evidence of a relationship between eating breakfast and morning energy level?

   (b) Does the answer to part (a) change if $n = 1000$ instead? Explain.

2. A survey on acne treatments randomly selected 100 teenagers using topical treatments, 100 using oral medications, and 50 using laser therapy, and asked each subject about satisfaction level. The resulting counts were:

	Topical	Oral	Laser
Very satisfied	61	54	24
Somewhat satisfied	28	21	10
Unsatisfied	11	25	16

   (a) Do the different treatments lead to different satisfaction levels? Perform an appropriate hypothesis test.

   (b) What is a possible confounding variable that could lead you to jump to a misleading conclusion? Explain.

3. Can the average women's life expectancy in a country be predicted from the average fertility rate (children/woman) in that country? Following are the most recent data from 11 randomly selected countries:

Country	Fertility rate (children/woman)	Life expectancy (women)
Afghanistan	6.25	45.5
Angola	5.33	51.4
Argentina	2.16	80.0
Australia	1.85	84.4
France	1.85	85.1
Liberia	4.69	61.5
Nepal	2.66	69.0
Netherlands	1.77	82.6
Pakistan	3.58	68.3
Poland	1.29	80.4
Singapore	1.29	83.4

Perform a test of significance for the slope of the regression line that relates fertility rate (children/woman) to life expectancy (women).

4. A heavy backpack can cause chronic shoulder, neck, and back pain. Wide, padded shoulder straps, a waist belt, and avoidance of single-strap bags all help, but weight is the main problem. The recommendation for a safe weight for school backpacks is no more than 10% of body weight. In a study of 500 randomly selected high school students, the weights of their backpacks give rise to the following:

Weight (lb)	Below 12.5	12.5–17.5	17.5–22.5	22.5–27.5	Above 27.5
Observed #	28	134	182	112	44

(a) In a normal distribution with $\mu = 20$ and $\sigma = 5$, what are the probabilities of $x < 12.5$, $12.5 < x < 17.5$, $17.5 < x < 22.5$, $22.5 < x < 27.5$, and $x > 27.5$?

(b) Given 500 students, if the data follow a normal distribution with $\mu = 20$ and $\sigma = 5$, what are the expected values for the numbers of backpacks in each of the indicated weight ranges?

(c) Test the null hypothesis that the data follow a normal distribution with $\mu = 20$ and $\sigma = 5$.

ANSWERS EXPLAINED

Quiz 1

MULTIPLE-CHOICE QUESTIONS

1. **(E)** The expected counts if dog bites occur equally during all moon phases are each
$\frac{1}{4}(32 + 27 + 47 + 38) = 36$. A chi-square goodness-of-fit test gives
$$\chi^2 = \sum \frac{(obs-exp)^2}{exp} = \frac{(32-36)^2}{36} + \frac{(27-36)^2}{36} + \frac{(47-36)^2}{36} + \frac{(38-36)^2}{36} = 6.167,$$
and with $df = 4 - 1 = 3$, $P(\chi^2 > 6.167) = 0.1038$. With this large a P-value $(0.1038 > 0.10)$, there is not sufficient evidence to conclude that dog bites are related to moon phases.

2. **(B)** $df = (\text{rows} - 1)(\text{columns} - 1) = (2 - 1)(5 - 1) = 4$

3. **(C)** Picking separate samples from each of 16 populations and classifying according to one variable (perception of quality education) is a survey design that is most appropriately analyzed using a chi-square test of homogeneity of proportions.

4. **(E)** With a P-value this small (less than 0.05), there is sufficient evidence in support of the alternative hypothesis H_a: the distributions of music preferences are different, that is, they differ for at least one of the proportions.

5. **(E)** A chi-square test of independence gives $\chi^2 = 2.852$, and with $df = 2$, we find $P = 0.2403$, and since $0.2403 > 0.10$, there is not sufficient evidence at the 10% significance level of a relationship between taste preference and the presence of the marker.

6. **(E)** The relevant P-value is 0.065, which is less than 0.10 but greater than 0.05.

7. **(E)** The slope of the regression line is -0.163, which is the middle of the confidence interval. The correlation coefficient has the same sign as the slope.

FREE-RESPONSE QUESTIONS

NOTE

χ^2GOF-Test on the TI-84; CHI-GOF on the Casio Prizm, or χ^2Test; Goodness-of-fit on the HP Prime.

1. First, state the hypotheses: H_0: The colors of the sugar shells are distributed according to 35% cherry red, 10% vibrant orange, 10% daffodil yellow, 25% emerald green, and 20% royal purple, and H_a: The colors of the sugar shells are not distributed as claimed by the manufacturer. [Or H_0: $P_{CR} = 0.35$, $P_{VO} = 0.10$, $P_{DY} = 0.10$, $P_{EG} = 0.25$, $P_{RP} = 0.20$, and H_a: at least one proportion is different from this distribution.]

Second, identify the test and check the assumptions: χ^2 goodness-of-fit test. We are given a random sample, and calculate that all expected cells are at least 5: 35% of $300 = 105$, 10% of $300 = 30$, 25% of $300 = 75$, and 20% of $300 = 60$.

Third, calculate the test statistic χ^2 and the P-value: A calculator gives
$$\chi^2 = \sum \frac{(obs-exp)^2}{exp} = 6.689, \text{ and with } df = 5 - 1 = 4, P = 0.153.$$

Fourth, linking to the P-value, give a conclusion in context: With a P-value this large, $0.153 > 0.10$, there is not sufficient evidence to reject H_0, that is, there is not sufficient evidence that the distribution is different from what is claimed by the manufacturer.

2. (a) Design I, with a single sample from one population classified on two variables (smoking and fitness), will result in a test of independence. Design II, with independent samples from two populations each with the single variable (fitness), will result in a test of homogeneity.

   (b) Design II, with its test of homogeneity, and using an equal sample size from each of the two populations (smokers and non-smokers), is best for comparing proportions of smokers who have different fitness levels with proportions of non-smokers who have different fitness levels.

   (c) Design I, which classifies one population on the two variables, smoking and fitness, is the only one of these two designs which will give data on the conditional distribution of people with given fitness levels who are smokers or are not smokers.

3. (a) First, state the hypotheses: H_0: Happiness level is independent of busy/idle choice for high school students and H_a: Happiness level is not independent of busy/idle choice for high school students.

   Second, identify the procedure and check the conditions: This is a chi-square test of independence. It is given that there is a random sample, the data are measured as "counts," and the expected counts are all at least 5 (put the observed counts in a matrix; then χ^2-Test on the TI-84 gives expected counts of

14.6	19.6	21.8	20.2	21.8
11.4	15.4	17.2	15.8	17.2

NOTE

χ^2-Test on the TI-84; CHI-2WAY on the Casio Prizm, or χ^2-Test; 2-way test on the HP Prime.

   Third, calculate the test statistic and the P-value: Calculator software gives $\chi^2 = 14.54$ with $P = 0.0058$ and $df = 4$.

   Fourth, give a conclusion in context with linkage to the P-value: With a P-value this small, $0.0058 < 0.01$, there is strong evidence to reject H_0, that is, there is strong evidence of a relationship between happiness level and busy/idle choice for high school students.

   (b) No, it is not reasonable to conclude that encouraging high school students to keep more busy will lead to higher happiness levels. This was not an experiment with students randomly chosen to sit or walk. The students themselves chose whether or not to sit or walk so no cause-and-effect conclusion is possible. For example, it could well be that the happier students choose to walk, whereas the less happy students choose to sit.

4. (a) Each additional year in age of teenagers is associated with an average of 0.4577 more texts per waking hour.

   (b) The scatterplot of texts per hour versus age should be roughly linear, there should be no apparent pattern in the residual plot, and a histogram of the residuals should be approximately normal.

   (c) First, state the hypotheses: H_0: $\beta = 0$, H_a: $\beta \neq 0$, where β is the slope of the regression line that relates average texts per hour to age.

Second, identify the test and check the assumptions: This is a test of significance for the slope of the regression line, and we are given that all conditions for inference are met.

Third, calculate the test statistic t and the P-value: The computer printout gives that $t = 2.45$ and $P = 0.016$.

Fourth, linking to the P-value, give a conclusion in context: With this small a P-value, $0.016 < 0.05$, there is sufficient evidence to reject H_0, that is, there is sufficient evidence of a linear relationship between average texts per hour and age for teenagers ages 13–17.

(d) R-Sq = 5.8%, so even though there is evidence of a linear relationship between average texts per hour and age for teenagers ages 13–17, only 5.8% of the variability in average texts per hour is explained by this regression model (or "is accounted for by the variation in age.").

Quiz 2

MULTIPLE-CHOICE QUESTIONS

1. **(C)** There are two observations of each mouse, a *before* time and an *after* time. These two times are dependent so a paired t-test is appropriate, *not* a two-sample test.

2. **(D)** With $77 + 85 + 23 + 15 = 200$ samples, the expected counts if the blood type distribution on the island is the same as that of the general population are 46% of $200 = 92$, 40% of $200 = 80$, 10% of $200 = 20$, and 4% of $200 = 8$. A chi-square goodness-of-fit test gives

$$\chi^2 = \sum \frac{(obs - exp)^2}{exp} = \frac{(77 - 92)^2}{92} + \frac{(85 - 80)^2}{80} + \frac{(23 - 20)^2}{20} + \frac{(15 - 8)^2}{8} = 9.333,$$

and with $df = 4 - 1 = 3$, $P(\chi^2 > 9.333) = 0.0252$. With a P-value this small ($0.0252 < 0.05$), there is sufficient evidence at the 5% significance level that blood type distribution on the island is different from that of the general population.

3. **(D)** With $df = (3 - 1)(5 - 1) = 8$, $P(\chi^2 > 13.95) = 0.083$. Since $0.05 < 0.083 < 0.10$, there is sufficient evidence at the 10% significance level, but not at the 5% significance level, of a relationship between education level and sports interest.

4. **(D)** With $1 + 3 + 3 + 9 = 16$, according to the geneticist the expected number of fruit flies of each species is $\frac{1}{16}(2000) = 125$, $\frac{3}{16}(2000) = 375$, $\frac{3}{16}(2000) = 375$, $\frac{9}{16}(2000) = 1125$. A chi-square goodness-of-fit test gives

$$\chi^2 = \sum \frac{(obs - exp)^2}{exp} = \frac{(110 - 125)^2}{125} + \frac{(345 - 375)^2}{375} + \frac{(360 - 375)^2}{375} + \frac{(1185 - 1125)^2}{1125} = 8,$$

and with $df = 4 - 1 = 3$, $P(\chi^2 > 8) = 0.0460$. With a P-value this small ($0.0460 < 0.05$), there is sufficient evidence at the 5% significance level to reject the geneticist's claim.

5. **(E)** A chi-square test of homogeneity gives $\chi^2 = 5.998$, and with $df = 3$, the P-value is 0.1117. With a P-value this large ($0.1117 > 0.10$) there is not sufficient evidence of a difference in cafeteria food satisfaction among the class levels.

6. **(E)** In computer printouts of regression analysis, "S" typically gives the standard deviation of the residuals.

7. **(D)** With $H_0: \beta = 0$ and $H_a: \beta > 0$, $t = \dfrac{0.6564 - 0}{0.1988}$.

FREE-RESPONSE QUESTIONS

1. (a) First, state the hypotheses: H_0: Eating breakfast and morning energy level are independent and H_a: Eating breakfast and morning energy level are not independent.

   Second, identify the test and check the assumptions: This is a χ^2 test of independence on

110	120	120
60	50	40

   where we are given a random sample, and a calculator gives that all expected cells are at least 5:

119	119	112
51	51	48

   Third, calculate the test statistic χ^2 and the P-value: Calculator software gives $\chi^2 = 4.202$ and $P = 0.1224$.

   Fourth, linking to the P-value, give a conclusion in context: With this large a P-value, $0.1224 > 0.05$, there is not sufficient evidence to reject H_0, that is, there is not sufficient evidence of a relationship between eating breakfast and morning energy level.

   (b) Yes, the conclusion changes. With $n = 1000$, the observed numbers are:

220	240	240
120	100	80

   with $\chi^2 = 8.403$ and $P = 0.0150$. With a P-value this small, $0.0150 < 0.05$, now there is sufficient evidence of a relationship between eating breakfast and morning energy level.

2. (a) First, state the hypotheses: H_0: The different treatments lead to the same satisfaction levels and H_a: The different treatments lead to different satisfaction levels.

   Second, identify the test and check the assumptions: χ^2 test of homogeneity. We are given a random sample, and a calculator gives that all expected cells are at least 5:

55.6	55.6	27.8
23.6	23.6	11.8
20.8	20.8	10.4

Third, calculate the test statistic χ^2 and the P-value: A calculator gives $\chi^2 = 10.9521$, and with $df = 4$, $P = 0.0271$.

Fourth, linking to the P-value, give a conclusion in context: With this small a P-value, $0.0271 < 0.05$, there is sufficient evidence to reject H_0, that is, there is sufficient evidence that the different treatments do lead to different satisfaction levels.

(b) An example of a possible confounding variable is severity of the acne outbreak. It could be that those with more severe cases have less satisfaction no matter what the treatment and are also the ones who are encouraged to use oral medications or laser therapy. So it would be wrong to conclude that oral medications or laser therapy are the causes of less satisfaction.

3. First, state the hypotheses: H_0: $\beta = 0$, H_a: $\beta \neq 0$. (If asked to state the parameter, then state "where β is the slope of the regression line that relates average fertility rate to women's life expectancy.")

Second, identify the test and check the assumptions: This is a test of significance for the slope of the regression line. We are given that the data come from a random sample of countries.

Scatterplot is approximately linear

No apparent pattern in residual plot

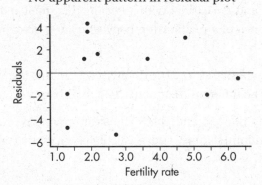

Distribution of residuals is approximately normal
(Checked by normal probability plot is roughly linear)

Residual

Third, calculate the test statistic t and the P-value: Using a calculator gives that $t = -12.53$ and $P = 0.0000$.

Fourth, linking to the P-value, give a conclusion in context: With a P-value this small, $0.0000 < 0.05$, there is sufficient evidence to reject H_0, that is, there is sufficient evidence of a linear relationship between average fertility rate (children/woman) and life expectancy (women).

> **NOTE**
>
> **LinRegTTest on the TI-84; LinearReg tTest on the Casio Prizm; or linear t-test on the HP Prime.**

4. (a) The z-scores for 22.5 and 27.5 are, respectively, $\dfrac{22.5 - 20}{5} = 0.5$ and $\dfrac{27.5 - 20}{5} = 1.5$.

Similarly, 17.5 and 12.5 have z-scores of -0.5 and -1.5, respectively. Using the normal probability table or a calculator gives the probabilities: $P(z < -1.5) = 0.0668$, $P(-1.5 < z < -0.5) = 0.2417$, $P(-0.5 < z < 0.5) = 0.3830$, $P(0.5 < z < 1.5) = 0.2417$, $P(z > 1.5) = 0.0668$.

(b) Multiplying each probability by 500: $0.0668(500) = 33.4$, $0.2417(500) = 120.85$, $0.3830(500) = 191.5$, $0.2417(500) = 120.85$, and $0.0668(500) = 33.4$.

Weight (lb)	Below 12.5	12.5–17.5	17.5–22.5	22.5–27.5	Above 27.5
Expected#	33.4	120.85	191.5	120.85	33.4

(c) First, state the hypotheses: H_0: The weights of student backpacks follow a normal distribution with $\mu = 20$ and $\sigma = 5$ and H_a: The weights of student backpacks do not follow a normal distribution with $\mu = 20$ and $\sigma = 5$.

Second, identify the test and check the assumptions: This is a chi-square test of goodness-of-fit. We are given that the weights are from a random sample of students. We note from above that all expected cells are ≥ 5:

Third, calculate the test statistic χ^2 and the P-value: A calculator gives

$$\chi^2 = \sum \frac{(obs - exp)^2}{exp} = 6.7874, \text{ and with } df = 5 - 1 = 4, P = 0.1476.$$

Fourth, linking to the P-value, give a conclusion in context: With a P-value this large, $0.1476 > 0.05$, there is not sufficient evidence to reject H_0, that is, there is not sufficient evidence that the data do not follow a normal distribution with $\mu = 20$ and $\sigma = 5$.

Theme Five: More Practice and Review

The Investigative Task

14

→ **THREE WORKED-OUT EXAMPLES**
→ **SEVEN PRACTICE QUESTIONS**

The last question on the exam is called the Investigative Task. This free-response question counts for one-eighth of the total grade on the AP Statistics exam. The purpose of this question is to test your ability to take what you've learned throughout the course and apply that knowledge in a novel way. The topic of this question often includes material not only unfamiliar to students but also to their teachers.

This question always has several parts. Very often the first parts are straightforward, coming directly out of standard concepts of the curriculum. These are followed by parts that will challenge you to think creatively and require you to use learned concepts in new ways. This "investigative part" is written in such a way as to be accessible to students who have learned sound statistical thinking, not just memorized formulas and procedures.

There is no reason to panic and absolutely no reason to leave a blank response. All students should be able to receive some credit by answering the more standard beginning to this question. Following this with something "reasonable" will earn you even more credit.

It is not possible to list all topics that you might encounter on the Investigative Task. However, working through the following examples will help you understand how to approach the task on the exam. This should give you the confidence to do well on the Investigative Task you will see on the May exam.

Just as the other free-response questions, the Investigative Task is scored on a 0 to 4 scale with 1 point for a *minimal* response, 2 points for a *developing* response, 3 points for a *substantial* response, and 4 points for a *complete* response. Individual parts of this question are scored as E for *essentially* correct, P for *partially* correct, and I for *incorrect*. Note that *essentially* correct does not mean *perfect*. Work is also graded *holistically*. That is, your complete response is considered as a whole whenever scores do not fall precisely on an integral value on the 0 to 4 scale.

Just as for the other free-response questions, the scoring rubrics here are not consistent as to what the minimal requirement is. These rubrics tend to be tough on the things the Investigative Task is intended to address and lenient on peripheral issues. Choosing appropriate methods and demonstrating good communication are the keys to high scores.

Following are three Investigative Tasks with carefully explained solutions illustrating suggestive thought processes for approaching such problems. Following these are seven more Investigative Tasks for which you're on your own. (Of course, there are answers and grading rubrics at the end of this topic!)

➡ EXAMPLE 14.1 _____

The R&D division of an automobile company has developed a new fuel additive that they hope will increase gas mileage. For a pilot study, they send 30 cars using the new additive on a road trip from Detroit to Los Angeles. From previous studies, they know that without the additive, the cars will average 32.0 mpg with a standard deviation of σ = 2.5 mpg. (Assume σ remains the same with or without the additive.)

(a) What is the parameter of interest, and what are the null and alternative hypotheses?

Think: The interest is in gas mileage, and there is mention of the "average" mpg. So the parameter is the mean mpg using the new fuel additive. The parameter is always something about a whole population, so be sure to use μ, not $\bar{x}$. The null hypothesis is some specific claim about this parameter, and the old mean mpg is given to be a specific 32.0 mpg. The alternative hypothesis will have an inequality. Since the R&D division hopes to increase gas mileage, the alternative must be > 32.0.

> *Answer:* The parameter of interest is μ, the mean gas mileage in mpg for cars using the new fuel additive.
>
> $$H_0: \mu = 32.0 \quad \text{and} \quad H_a: \mu > 32.0$$

(b) What are Type I and Type II errors in this context? Give a possible consequence of each.

Think: A Type I error is when you mistakenly reject a true null hypothesis. In this case, if the null hypothesis was true, that would mean the cars really get a mean of 32.0 mpg. If you reject this, you mistakenly think the mean mpg is > 32.0. A consequence is that you'll waste money on this new additive, thinking it will improve gas mileage when in reality it won't. So *Answer:*

> *Answer:* Type I error: The additive doesn't improve gas mileage, but the company mistakenly thinks it does. Consequence: The company invests in something that has no value.

Think: A Type II error is when you mistakenly fail to reject a false null hypothesis. In this case, if the null hypothesis is false, that would mean the cars really get a mean mpg > 32.0. If you fail to reject the null hypothesis claim of 32.0, you don't think the additive helps when it really does.

> *Answer:* Type II error: The additive does improve gas mileage, but the company mistakenly thinks it doesn't. Consequence: The company doesn't invest in a potentially lucrative product.

(c) The company decides to work with a level of significance of α = 0.05. What is the probability that a Type I error will be committed?

Think: That's easy. The significance level or α-risk gives the cutoff score for rejecting the null hypothesis, but it's based on the assumption that the null hypothesis is true. This rejecting a true null hypothesis is what we call a Type I error.

> *Answer:* 0.05 (the probability of a Type I error is the same as the level of significance).

TIP

You've already practiced so many hypotheses problems that you should be able to name the parameter and the hypotheses in your sleep!

TIP

Nothing new yet, but you should probably think for a moment what Type I and Type II errors are in general before putting in context.

TIP

Is there some trick here? These questions are not designed to trick you. Some may be just as straightforward as they seem!

(d)   What is the null hypothesis rejection region? In other words, what critical value of the test statistic $\bar{x}$ will lead the company to conclude that there is statistically significant evidence that the new fuel additive increases gas mileage?

Think: You are given the population standard deviation ($\sigma = 2.5$), so you can use a normal distribution for the sampling distribution. The mean is μ and standard deviation is $\frac{\sigma}{\sqrt{n}}$. 0.05 in a tail of the normal curve corresponds to a z-score of 1.645, and the critical value is 1.645 standard deviations from the mean.

Answer: $\sigma_{\bar{x}} = \dfrac{2.5}{\sqrt{30}} = 0.4564$ and $32.0 + 1.645(0.4564) = 32.751$, so any $\bar{x} \geq 32.751$ will lead the company to conclude that there is statistically significant evidence that the new fuel additive increases gas mileage. (Note: with the population standard deviation known, the sampling distribution is roughly normal, thus giving a critical score of 1.645.)

(e)   The R&D statistician plots the following *power curve* showing the "performance" that can be expected from this hypothesis test. That is, for each possible true value of the population parameter, the probability of rejecting the null hypothesis is shown.

Explain why the graph hits the pictured vertical axis above $\mu = 32$ at a y-value of 0.05.

Think: If the true value is 32.0, then it's actually the null hypothesis. However, you just explained in part (c) that the probability of rejecting a true null hypothesis is 0.05!

Answer: H_0: $\mu = 32.0$, so if the true mean was 32, then the probability of rejecting the null hypothesis is simply the level of significance, which is $\alpha = 0.05$.

(f)   What is the meaning of the point (33, 0.7) in terms of a Type II error?

Think: A Type II error is when you mistakenly fail to reject a false null hypothesis. What does that have to do with the given graph? According to the graph, (33, 0.7) signifies that if the true mean is 33, the probability of rejecting it is 0.7. Then the probability of not rejecting it must be 1 – 0.7 = 0.3. Not rejecting it is exactly what the Type II error is all about.

Answer: If the true mean gas mileage using the additive was 33, then the probability of correctly rejecting the null hypothesis is 0.7. So the probability of committing a Type II error, that is, mistakenly failing to reject the false null hypothesis, is 1 – 0.7 = 0.3.

(See page 419 for scoring.)

TIP

Don't get nervous if you haven't yet come across anything new in the Investigative Task. Just keep moving along and applying what you know!

TIP

This is a graph you never heard of! However, all you're being asked to explain is why the graph goes through one particular point.

TIP

You are almost done! Again, this question is something about that new type of graph, but don't quit! There is some claimed relationship between how this graph is defined and the meaning of a Type II error. Look for that relationship!

➡ EXAMPLE 14.2

When you have a sore throat, your doctor has two major options to diagnose strep pharyngitis: a *rapid antigen test* to analyze the bacteria in your throat quickly (10–15 minutes) or a much more accurate *throat culture* that tests to see what grows on a Petri dish (takes 1–2 days).

In a random sample of 500 children brought in to a pediatrician's office with sore throat symptoms, both rapid antigen tests and throat cultures were performed for this study with the following summary counts:

		Patients With or Without Strep Pharyngitis as Confirmed by a Throat Culture	
		Strep Throat	No Strep Bacteria
Results of Rapid Antigen Test	Positive Test	180	15
	Negative Test	20	285

(a) *Sensitivity* is defined as the probability of a positive test given that the subject has the disease. What was the sensitivity of this study?

Think: The phrase "given that" indicates this is a conditional probability. For such probabilities from a two-way table, you should first calculate row and column totals.

	Strep Throat	No Strep Throat	Totals
Positive Test	180	15	195
Negative Test	20	285	305
Totals	200	300	500

You're "given that" the subject has the disease (strep throat in this case). So the first column of numbers is indicated. Note that 200 subjects had strep throat, and 180 of these had a positive test.

Answer:

$$P(\text{positive test} \mid \text{strep}) = \frac{180}{200} = 0.90$$

(b) *Specificity* is defined as the probability of a negative test given that the subject is healthy. What was the specificity of this study?

Think: Here you're given that the subject is healthy; there is "no strep throat." So the second column of numbers is indicated. Note that 300 subjects did not have strep throat, and 285 of these had a negative test.

Answer:

$$P(\text{negative test} \mid \text{healthy}) = \frac{285}{300} = 0.95$$

(c) A valuable tool for assessing the value of such diagnostic tests is the *positive diagnostic likelihood ratio* (LR^+). It gives the ratio of the probability a positive test result will be observed in a diseased person compared to the probability that the same result will be observed in a healthy person.

$$LR^+ = \frac{\text{sensitivity}}{1 - \text{specificity}}$$

Calculate LR^+ in this study. Explain why the larger the value of LR^+ is, the more useful the test.

Think: To calculate the LR^+, it looks like you simply plug in the answers from parts (a) and (b). However, the expository part of the question involves some thought. LR^+ is a fraction. So it is larger when either its numerator is larger or its denominator is smaller. The numerator, sensitivity, is larger when there is a greater probability of a positive test when a child does have strep, and this clearly makes the test more useful. The denominator, 1-specificity, is smaller when specificity is larger and closer to 1 (it's a probability so can't be greater than 1). Thus there is a greater probability of a negative test when a child is healthy, and this too clearly makes the test more useful.

Answer:

$$LR^+ = \frac{0.90}{1-0.95} = 18.0$$

A positive test given a child with strep and a negative test given a healthy child are both desired outcomes, so higher values for both sensitivity and specificity are good. Higher values of sensitivity, the numerator of LR^+, lead to greater values of LR^+. Higher values of specificity give lower values of 1-specificity, the denominator of LR^+, again leading to greater values of LR^+.

(d) Another valuable tool is the *negative diagnostic likelihood ratio* (LR^-). It gives the ratio of the probability a negative test result will be observed in a diseased person compared to the probability that the same result will be observed in a healthy person. Give an expression for LR^- in terms of *sensitivity* and *specificity*.

Answer: Since sensitivity is the probability of a positive test given that the subject has the disease, 1-sensitivity gives the probability of a negative test given that the subject has the disease. The probability that the same result (a negative test) will be observed in a healthy person is the definition of "specificity." The answer is:

$$LR^- = \frac{1\text{-sensitivity}}{\text{specificity}}$$

(e) Suppose in one such sample study, $LR^+ = 14.7$. To determine whether or not this is sufficient evidence that the population LR^+ is below the desired value of 15.0, 100 samples from a population with a known LR^+ of 15.0 are generated and the resulting simulated values of LR^+ are shown in the dotplot:

Positive Diagnostic Likelihood Ratio (LR^+)

TIP

This type of problem has appeared frequently. In other words, you are given a "new" statistic and use simulation and a dotplot to estimate the *P*-value.

Based on this dotplot and the sample $LR^+ = 14.7$, is there evidence that the population LR^+ is below the desired value of 15.0? Explain.

TIP

When the
question says to
use two givens,
in this case the
dotplot and the
sample LR⁺, that's
a very strong hint
on what you must
base your answer!

Think: There are 4 dots out of 100 that correspond to values of 14.7 or less. However, this is really a P-value for the sample LR⁺. With a P-value of 0.04, the LR⁺ value of 14.7 is statistically significant at the 5% significance level. We have enough evidence to reject the null hypothesis of LR⁺ = 15.0.

Answer: The estimated P-value is the proportion of the simulated statistics that are less than or equal to the sample statistic of 14.7. Counting values in the dotplot, 4 out of 100, gives a P-value of 0.04. With this small a P-value, $0.04 < 0.05$, there is sufficient evidence that the population LR⁺ is below the desired value of 15.0.

(See page 419 for scoring.)

➡ EXAMPLE 14.3

A grocery chain executive in charge of picking locations for future developments is interested in developing a function to predict weekly **sales** (in $1000) based on store **size** (in 1000 ft²), median family **income** (in $1000) in the neighborhood, median **family size** in the neighborhood, and number of **competitor** stores in the neighborhood. Data gathered on 15 existing stores yield the following scatterplot matrix:

(a) The scatterplot matrix can be used to determine which variables seem to be the best predictors of sales. Explain which variables seem to be the best predictors of sales and the relationship between sales and those variables.

Think: What is going on? There are lots of scatterplots. In fact, you are given 10 of them with 5 different variables. You're asked about predicting "sales," so sales must be the response variable. "Sales" is to the left of the top row. So the top row of 4 scatterplots must be sales against the other 4 variables. Do any of these 4 scatterplots look like they could be used for prediction, that is, do any of them look like there is a linear relationship? Yes, it occurs in the first and fourth scatterplots along the top! The first seems to indicate some kind of positive association between sales and store size. The fourth seems to indicate some kind of negative association between sales and competitors.

TIP

Wow, this question
starts right off
with a brand-new,
complex picture!
If you have no idea
how to interpret
this, simply move
on. You may still
be able to make
sense out of later
problem parts!

Answer: "Store size" and "number of competitor stores" seem to be the best predictors of "weekly sales." The first scatterplot in the top row indicates a strong positive association between weekly sales and store size. The last scatterplot in the top row indicates a strong negative association between weekly sales and number of competitor stores.

(b) Individual regressions of sales on each of the possible predictors give:

Predictor	Coef	SE Coef	T	P
Constant	-115.51	51.74	-2.23	0.044
St Size	5.8801	0.9808	6.00	0.000

Predictor	Coef	SE Coef	T	P
Constant	66.69	86.15	0.77	0.453
Income	3.859	2.691	1.43	0.175

Predictor	Coef	SE Coef	T	P
Constant	214.19	91.31	2.35	0.036
Fam Size	-11.58	36.71	-0.32	0.758

Predictor	Coef	SE Coef	T	P
Constant	374.64	33.41	11.21	0.000
Comp	-24.777	4.109	-6.03	0.000

TIP

Even though this full table looks unfamiliar, you should recognize each of the 4 parts as a standard linear regression summary.

We will not use any variable whose t-test P-value is over 0.15 and will first use the variable whose t-test P-value is least. Under these guidelines, which, if any, variables will be discarded and which variable will be used first? Explain.

Think: The question is about t-tests and P-values, so this must be about hypothesis tests for regression. You're given a generic computer printout that seems to give separate information about using each of the 4 possible predictors. From working with simpler computer output, remember that the line with the predictor first gives the slope, then the SE of the slope, followed by the t- and P-values from a linear regression test. You're told to throw out the predictor variables with P over 0.15. You see that Income has $P = 0.175$ and Fam Size has $P = 0.758$. (Looking back, this gives you confidence that your answer to part (a) is correct!) You're asked which predictor variable has the smallest P, but both St Size and Comp have $P = 0.000$ rounded to 3 decimals. How can you tell which is smaller? Well, you know the P-values come from the t-values, and you see that Comp has the larger t-value in absolute value (6.03 > 6.00), so it must have the smaller P if you were given more decimal places!

Answer: The variables "median family income" and "median family size" with t-test P-values of 0.175 and 0.758, both over 0.15, should be removed. Although both "store size" and "number of competitor stores" have t-test P-values of 0.000 to three decimal places, "number of competitor stores" will be used first as it has the smaller P-value because it has the larger t-statistic in absolute value, 6.03 versus 6.00.

(c) Multiple regression of sales on the two variables, store size and number of competitors, gives:

Predictor	Coef	SE Coef	T	P
Constant	122.79	94.44	1.30	0.218
St Size	3.342	1.200	2.78	0.017
Comp	-14.213	5.051	-2.81	0.016

S = 37.3598   R-Sq = 84.0%

TIP

It seems like you're simply adding another variable to the standard linear regression analysis—which is not unreasonable!

What sales does this predict for a store of 50,000 ft^2 with 8 competitors?

Think: With a linear regression generic computer printout, you know that the "Coef" column gives the "Constant" and under it the slope or coefficient of the independent variable in the formula for predicting the response variable. Here you have an extra row under the Constant, though, which means two predictor variables, "St Size" and "Comp." The constant in the formula must be 122.79, and the coefficients of "St Size" and "Comp" must be 3.342 and –14.213, respectively.

Answer: 122.79 + 3.342(50) – 14.213(8) = 176.186 thousand or $176,186

NOTE

The coefficient of determination, the proportion of variation in the response variable that can be explained by the regression model, is a curriculum concept often appearing on the exam.

(d)   What proportion of the variation in weekly sales is explained by its linear relationship with store size and number of competitors?

Think: This sounds like the definition of the "coefficient of determination" or R^2. Sure enough, "R-Sq" is in the computer output! So *Answer:*

Answer: 0.84 To be safe, I'll write it out: The proportion of the variation in weekly sales explained by its linear relationship with store size and number of competitors is 0.84.

(e)   Assuming normality of residuals and using the computer output, estimate the probability that the weekly sales will be at least $50,000 over what is predicted by the regression line.

NOTE

You should recognize this as a straightforward normal probability calculation (where the variable is the residuals).

Think: You're being asked about the difference between an actual weekly sale and what is predicted, and this is what you know is a "residual." What is the distribution of the residuals? The question says the distribution is roughly normal. Remember that the sum and thus the mean of the residuals is always 0. By looking at the computer output, you see that the standard deviation of the residuals is $s = 37.3598$. So you can simply use normalcdf on your calculator to find the probability!

Answer: The sum and thus the mean of the residuals is always 0. The standard deviation of the residuals is estimated with $s = 37.3598$. With a z-score of

$$\frac{50 - 0}{37.3598} = 1.338$$

and a roughly normal distribution, we have $P(z > 1.338) \approx 0.090$.

(See page 420 for scoring.)

Directions: Do the following Investigative Tasks on your own, using the methodology illustrated in the previous three examples.

1. A basic problem in ecology is to estimate the number of animals in a wildlife population. Suppose a wildlife management team captures and tags a random sample of 108 springboks in a southwest African wildlife preserve. The springboks are released and given time to mingle with the population. One month later, a second random sample capture is made of 80 springboks, among which 12 are noted to be tagged from the original capture.

    (a) Assuming that the proportion of marked individuals within the second sample is equal to the proportion of marked individuals in the whole population, estimate the total number of springboks in the wildlife preserve.

    (b) A formula for variance of this estimate is $Var(N) = \dfrac{ac(a-b)(c-b)}{b^3}$, where $N =$ estimate of total population size, $a =$ number of animals originally captured and tagged, $b =$ number of tagged animals who are recaptured, and $c =$ number of animals in second sample. What is the standard deviation for the population size estimate above?

    The sampling distribution for estimates of N calculated as above is approximately normal under the condition that both samples are random and both $b \geq 10$ and $c - b \geq 10$.

    (c) Determine a 90% confidence interval for the total population size of springboks in this wildlife preserve.

    The above analysis assumes that there is no change to the population during the investigation (the population is closed).

    (d) Suppose that some of the tagged springboks were killed by hunters or escaped from the preserve before they could rejoin the herd. Would this result in an underestimation or an overestimation of the herd? Explain.

    (e) Suppose that the detection device missed some of the wire tags on the springboks in the second random sample. Would this result in an underestimation or an overestimation of the herd? Explain.

2. When operating at full capacity, a stainless steel bottle filler can fill twelve 64-ounce containers per minute. Once a day, a random set of twelve containers is picked, the containers are weighed, and if $W = \dfrac{\text{Max} + \text{Min}}{2}$ varies significantly from 64 ounces, the machine is stopped and adjustments are performed. To explore this quality control procedure further, the machine is perfectly calibrated. Then 100 random samples of size 12 bottles are taken from the assembly line. For each of these 100 samples, the W measurement is calculated and shown plotted below.

(a) Describe the above sampling distribution.

(b) Assuming a median of 64.005 and quartiles of 63.97 and 64.04, are there any outliers in the above distribution? Explain.

(c) When a sample of size 12 is picked, what are the null and alternative hypotheses being tested?

(d) One day, the random sample is {63.8, 63.84, 63.87, 63.88, 63.91, 63.95, 63.97, 63.97, 63.98, 64.0, 64.02, 64.04}. Based on the dotplot above, is this sufficient evidence at the 5% significance level to conclude that the machine needs adjustment? Justify your answer.

3. In 10 years of measurements, the number of people who died by becoming tangled in their bedsheets (deaths per year) versus per capita consumption of cheese (pounds per year) in the United States is illustrated in the following scatterplot:

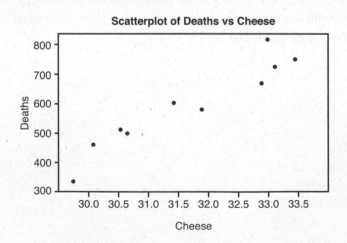

Scatterplot of Deaths vs Cheese

Below is the computer regression output from the 10 data points:

Residuals vs Cheese

Residuals

```
Predictor       Coef    SE Coef        T        P
Constant     -2977.3      427.6    -6.96    0.000
Cheese        113.13      13.56     8.35    0.000

S = 50.0007    R-Sq = 89.7%    R-Sq(adj) = 88.4%
```

(a) Discuss how conditions for inference are met.

(b) During the year when the per capita consumption of cheese was 30.1 pounds, there were 456 deaths from people becoming tangled in their bedsheets. Calculate and interpret the residual for this point.

(c) Determine a 95% confidence interval for the regression slope, and interpret it in context.

(d) What kind of cause-and-effect conclusion is appropriate? Explain.

(e) Using the computer output, give a rough estimate of the probability that the number of deaths in a year is at least 75 over what is predicted by the regression line.

4. Probiotics are microorganisms that are believed to give health benefits, especially with regard to healthy digestion, when consumed. One brand advertises a strength of 30 billion CFU (colony-forming units) per capsule. A company performs a quality check of 100 randomly chosen capsules. The 100 CFU values are arranged in order, the percentile rank of each is noted, and the standard normal value (z-score) corresponding to those percentile ranks is found. (For example, if a value had a percentile rank of 75%, then the corresponding z-score would be invNorm(0.75) = 0.6745.) The resulting plot of z-scores versus CFU values (in billions) is shown below, together with summary statistics:

Probability Plot of CFU
Normal

Mean	32.48
StDev	1.174
N	100
AD	5.672
P-value	< 0.005

(a) What is the *range* of CFU values?

(b) What is the 5th percentile of the data? Explain.

The diagonal (sloping) straight line in the plot above is the graph of a theoretical normal distribution with the same mean and standard deviation of the sample data.

(c) In a normal distribution with the mean and SD of this data, what is the 95th percentile? Explain.

(d) What shape will a histogram of the CFU values have? Explain.

(e) Determine a 95% confidence interval for the mean CFU value (in billions).

5. Not counting the 1919–1921 World Series, which went to the first team to win 5 games, all others go to the first team to win 4 games. Assume that each game is an independent event and the teams are evenly matched.

(a) What is the probability that Team A wins a World Series in 4 straight games? What then is the probability that a series is over in 4 games (Team A or Team B wins the first 4 straight games)?

(b) The probability that a series lasts exactly 5 games is twice the probability that a team wins 3 out of 4 games and then the next game, that is,

$$2\left[\binom{4}{3}(0.5)^3(0.5)\right](0.5) = 0.25$$

Similarly, the probability of a 6-game series is

$$2\left[\binom{5}{3}(0.5)^3(0.5)^2\right](0.5) = 0.3125$$

What is the probability a series lasts 7 games? (Shout out to 2016 Cubs!)

(c) Given a sample of 103 games, with the lengths of the World Series having a binomial distribution with evenly matched teams (as above), what are the expected values for the numbers (4 through 7) of games in a series?

(d) Not counting the 1919–1921 World Series, the counts for lengths of the 103 series from 1909 to 2014 are:

	Number of Games Played in the World Series			
	4	5	6	7
Number of Years	20	22	24	37

Test the null hypothesis that lengths of World Series follow the above-calculated probability distribution.

6. At the end of the school year, a state administers a mathematics exam to a sample of fifth-grade students. Census results show that the state population consists of 50% city, 30% suburb, and 20% rural. The state decides to use a proportionate stratified random sample of size $n = 2000$.

   (a) How many fifth graders from each of the three regions should be included in the sample?

   Partial computer output of the exam scores is as follows:

Variable	Mean	StDev
City	74.3	10.2
Suburb	80.4	9.3
Rural	69.8	12.1

   As an estimator for the mean score of all fifth graders, the state is using the following sample statistic:

   $$\bar{x}_{overall} = 0.5\bar{x}_{city} + 0.3\bar{x}_{suburb} + 0.2\bar{x}_{rural}$$

   (b) Calculate the sample statistic $\bar{x}_{overall}$ for this data.

   (c) Using part (a) and the above computer output, calculate the three standard errors, one for each region, $SE(\bar{x}_{city})$, $SE(\bar{x}_{suburb})$, and $SE(\bar{x}_{rural})$.

   (d) Calculate the standard error, $SE(\bar{x}_{overall})$, of the sampling distribution of $\bar{x}_{overall}$.

   (e) Assuming all conditions for inference are satisfied and using z as an approximation for t (given the very large sample size), determine a 99% confidence interval for the mean score of all fifth graders if they were to take this exam.

7. Sometimes it's possible to fit one type of statistical model to the lower part of the data values and another type to the higher part, separated by a breakpoint, whereby the overall fit is improved. El Niño is associated with a warming in the Pacific Ocean sea surface, which in turn feeds thunderstorms above and creates increased rainfall. The rainfall over the coastal regions of South America associated with strong El Niño events is one of the most dramatic climatic anomalies found anywhere on earth. When El Niño is not present, these localities become semiarid to desert. In contrast, when El Niño is present, the localities become wet and tropical.

   In one specific region over a 10-year study, there were alternating periods of roughly two years each under and not under El Niño effects. The total monthly rainfalls were split evenly with 60 months under an El Niño effect and with 60 months not. Numbering the months 1 through 120, two regressions are run, one of monthly rainfall (in inches) versus months with El Niño present and a second of monthly rainfall (in inches) versus the other months.

60 Months with El Niño present:

```
Predictor          Coef        SE Coef
Constant         186.061         2.741
Month            0.69460        0.04263
```

S = 11.2658   R-Sq = 82.1%   R-Sq(adj) = 81.8%

Variable	Mean	SE Mean	StDev	Minimum	Q1	Median	Q3	Maximum
Rainfall	223.92	3.41	26.38	180.00	200.00	220.00	245.00	275.00

60 Months with El Niño not present:

```
Predictor          Coef        SE Coef
Constant         108.944         2.993
Month            0.13493        0.04004
```

S = 10.5808   R-Sq = 16.4%   R-Sq(adj) = 14.9%

Variable	Mean	SE Mean	StDev	Minimum	Q1	Median	Q3	Maximum
Rainfall	117.92	1.48	11.47	90.00	110.00	120.00	125.00	140.00

(a) What is the mean monthly rainfall during the 10-year period?

(b) What is the median monthly rainfall during the 10-year period?

(c) What is the interquartile range of the 120 monthly rainfalls?

(d) Name a feature that will be apparent in a histogram of monthly rainfall numbers but not in a boxplot.

(e) Explain why a cumulative frequency plot of these rainfall totals would show a long horizontal line.

(f) Assuming all conditions for inference are satisfied, is there statistical evidence of increasing monthly rainfall during the 60 months with El Niño present? Justify your answer, being sure to give the parameter of interest and to state the null and alternative hypotheses.

ANSWERS EXPLAINED

Scoring for the Examples

SCORING FOR EXAMPLE 14.1

Part (a) is essentially correct for correct statements of the parameter, the null hypothesis, and the alternative hypothesis. Part (a) is partially correct for two of these three statements.

Part (b) is essentially correct for correctly explaining Type I and Type II errors in context and for giving reasonable consequences of each. Part (b) is partially correct for two of the four parts.

Parts (c) and (d) together are essentially correct for both giving 0.05 in (c) and calculating 32.751 in (d). They are partially correct for one of these two calculations.

Parts (e) and (f) together are essentially correct for correct explanations of the y-value of 0.05 and of why the Type II error is 0.3. They are partially correct for one of these two.

Count partially correct answers as one-half an essentially correct answer.

4 Complete Answer   Four essentially correct answers.

3 Substantial Answer   Three essentially correct answers.

2 Developing Answer   Two essentially correct answers.

1 Minimal Answer   One essentially correct answer.

Use a holistic approach to decide a score totaling between two numbers.

SCORING FOR EXAMPLE 14.2

Parts (a) and (b) together are essentially correct for correctly calculating 0.90 and 0.95. They are partially correct for one of these results.

Part (c) is essentially correct for a correct calculation of LR^+ together with a clear explanation of why larger values are desired. It is partially correct for one of these two.

Part (d) is essentially correct for a correct expression for LR^- in terms of sensitivity and specificity and is incorrect otherwise.

Part (e) is essentially correct for a correct calculation of the P-value together with a correct conclusion in context with linkage to the P-value. It is partially correct if missing one part such as an incorrect P-value, missing context, or missing linkage in the conclusion.

Count partially correct answers as one-half an essentially correct answer.

4 Complete Answer   Four essentially correct answers.

3 Substantial Answer   Three essentially correct answers.

2 Developing Answer   Two essentially correct answers.

1 Minimal Answer   One essentially correct answer.

Use a holistic approach to decide a score totaling between two numbers.

Part (a) is essentially correct for picking "store size" and "number of competitor stores" as the best predictors and for referring to the two scatterplots together with "strength" and "direction" for each. Part (a) is partially correct for picking these two predictors but giving a weak explanation missing either strength or direction.

Part (b) is essentially correct for picking "median family income" and "median family size" to be removed, for picking "number of competitor stores" to be used first, and for giving correct P-value explanations. Part (b) is partially correct for correctly picking either the two variables to be removed or the variable to be used first together with a correct explanation.

Parts (c) and (d) together are essentially correct for a correction calculation of $176,186 with work and for giving the answer of 0.84. They are partially correct for one of these two parts.

Part (e) is essentially correct for noting that the distribution of residuals is roughly normal with mean 0 and standard deviation 37.3598 and for then using this to calculate the probability correctly. Part (e) is partially correct for correctly noting the distribution of residuals but incorrectly calculating the probability, or for making a calculation based on a normal distribution with mean 0 but using an incorrect standard deviation.

Count partially correct answers as one-half an essentially correct answer.

4	Complete Answer	Four essentially correct answers.
3	Substantial Answer	Three essentially correct answers.
2	Developing Answer	Two essentially correct answers.
1	Minimal Answer	One essentially correct answer.

Use a holistic approach to decide a score totaling between two numbers.

Questions on Topic Fourteen: The Investigative Task

1. (a) **Think:** This sounds like a straightforward algebra problem in which you name a variable, set up an equation where two proportions (two ratios) are set equal, and solve for the unknown. So let N = the whole population size, and note that 108 springboks out of N were tagged, while 12 out of the sample of 80 springboks were seen to be tagged.

   Answer: $\frac{108}{N} = \frac{12}{80}$ gives $N = 720$

   (b) **Think:** That sure is a complex-looking definition of variance, but you're only asked to plug in. And note that the formula is for variance, and you're asked for the standard deviation.

   Answer: $Var(N) = \frac{108 \times 80 \times (108-12)(80-12)}{12^3} = 32{,}640$ and so

   $SD(N) = \sqrt{32{,}640} = 180.67$

(c) **Think:** The sampling distribution is given to be approximately normal, so you can use a z-distribution. Even though this is not one of our standard confidence interval questions, it is still statistical inference, and as always, you need to check conditions and give a conclusion in context, referring to the entire population.

Answer: Check conditions: Both captures are given to be random samples, $b = 12 \geq 10$ and $c - b = 80 - 12 \geq 10$. A 90% z-interval has critical z-scores of ± 1.645, so the confidence interval is $720 \pm 1.645(180.67) = 720 \pm 297.2$. Thus we are 90% confident that the total population of springboks in this wildlife preserve is between 422.8 and 1017.2.

(d) **Think:** This is not difficult, but careful reasoning is required! If some of the marked springboks are lost, then the proportion obtained in the recapture will be smaller than it probably would have been, and if you set $\dfrac{108}{N}$ equal to a smaller number, you'll get a larger N. Now your task is to fully and clearly explain this!

Answer: We are assuming that the proportion of marked individuals within the second sample is equal to the proportion of marked individuals in the whole population. If fewer of the marked springboks are available for recapture, the proportion of marked individuals in the second sample will be smaller than it should be. Therefore, we will think that the number of originally marked individuals are a smaller proportion of the population than it really is. Thus we will think that the population is larger than it really is.

(e) **Think:** Seems This seems to be the same as in (d), that is, fewer correctly marked springboks around.

Answer: Again, if we think that the proportion of marked individuals in the second sample is smaller than it really is, we will think that the number of originally marked individuals is a smaller proportion of the population than it really is. Thus we will think that the population is larger than it really is.

SCORING

Part (a) is essentially correct for the correct proportion and calculation of *N* and is partially correct for the correct value of *N* with no work shown.

Part (b) is essentially correct for the correct calculation with work shown and is partially correct for a calculation of only variance with work shown.

Part (c) is essentially correct for a check of the ≥10 conditions, a correct calculation of the interval, and a correct statement in context. It is partially correct if missing the ≥10 check or if missing a statement in context.

Parts (d) and (e) together are essentially correct for the correct conclusions with reasonable explanations. They are partially correct for one correct conclusion with reasonable explanation or for both correct conclusions but with weak explanations.

Count partially correct answers as one-half an essentially correct answer.

4	**Complete Answer**	Four essentially correct answers.
3	**Substantial Answer**	Three essentially correct answers.
2	**Developing Answer**	Two essentially correct answers.
1	**Minimal Answer**	One essentially correct answer.

Use a holistic approach to decide a score totaling between two numbers.

2. (a) **Think:** A complete answer considers shape, center, spread, and unusual features, and mentions context.

Answer: A complete answer considers shape, center, and spread, and mentions context.

Shape: bimodal and roughly symmetric

Center: The center of the $W = \dfrac{\text{Max} + \text{Min}}{2}$ calculations (where Min and Max are the minimum and maximum weights, respectively, in a set of 12 containers) is around 64 ounces.

Spread: From 63.91 to 64.10 (or range of 0.19 or IQR of 0.07)

(b) **Think:** Outliers are values more than $1.5 \times$ IQR below the first quartile, or $1.5 \times$ IQR above the third quartile. You are given the quartiles and then can look at the dotplot as to whether any values are more than $1.5 \times$ IQR from the quartiles.

Answer: IQR = 64.04 − 63.97 = 0.07 and 1.5(IQR) = 0.105

$Q_1 - 0.105 = 63.865$ and $Q_3 + 0.105 = 64.145$

Since no values are below 63.865 or above 64.145, there are no outliers.

(c) **Think:** This is a two-sided hypothesis test because you are asked whether a value varies significantly from 64, not whether it is larger or smaller than 64.

Answer: H_0: $\dfrac{\text{Min} + \text{Max}}{2} = 64.0$ and H_a: $\dfrac{\text{Min} + \text{Max}}{2} \neq 64.0$

(d) **Think:** Remember that simulation can be used to determine what values of a test statistic are likely to occur by random chance alone, assuming the null hypothesis is true. Then looking at where the test statistic falls, you can estimate a *P*-value. can be estimated.

Answer: For this day's sample, $W = \dfrac{63.8 + 64.04}{2} = 63.92$, which is 0.08 away from 64.0.

There are 4 values (2 above and 2 below) out of 100 this far away or farther from 64.0. With a *P*-value of 0.04, there is sufficient evidence at the 5% significance level (0.04 < 0.05) to reject the null hypothesis. In other words, there is sufficient evidence to conclude that the machine needs adjustment.

3. (a) **Think:** The conditions for regression inference are: (1) the sample must be randomly selected; (2) the scatterplot should be approximately linear; (3) there should be no apparent pattern in the residual plot; and (4) the distribution of the residuals should be approximately normal.

    Answer: First, the scatterplot of Deaths vs. Cheese is roughly linear. Second, the residual plot shows no clear pattern. Third, the histogram of residuals appears roughly bell-shaped (unimodal, symmetric, and without clear skewness or outliers). We must assume a representative sample.

    (b) **Think:** Residual = observed – predicted; you're given the observed, and you can calculate the predicted from the regression line equation. Remember that the interpretation must be in context and discuss direction, not only that the residual is of a certain size.

    Answer: The predicted number of deaths from the regression line is

    $$-2977.3 + 113.13(30.1) = 427.9$$

    So the residual = observed – predicted = 456 – 427.9 = 28.1. The actual number of deaths was 28.1 greater than what was predicted by the regression model for the year when the per capita consumption of cheese was 30.1 pounds.

(c) **Think:** Remember this is a t-distribution with $df = n - 2$, and don't forget that confidence interval questions always involve conclusions in context.

Answer: With $df = 10 - 2 = 8$ and 0.025 in each tail, the critical t-values are ± 2.306.

$b_1 \pm t s_{b_1} = 113.13 \pm 2.306(13.56) = 113.13 \pm 31.27$. We are 95% confident that for each additional pound of cheese consumption per capita, the average increase in deaths per year from bedsheet tangling is between 81.86 and 144.40.

(d) **Think:** Correlation does not imply causation! Most likely the two unrelated data sets follow the same time trend, resulting in correlation, but no connection.

Answer: No cause-and-effect conclusion is appropriate. Evidence of an association is not evidence of a cause-and-effect relationship. In cases like this, most likely the two unrelated data sets follow the same time trend, resulting in correlation but not connection.

(e) **Think:** The sum and thus the mean of the residuals is 0, and the SD is given by "S" in typical computer output. The residuals appear to have a roughly normal distribution, so the asked-for calculation is a straightforward normal probability!

Answer: The sum and thus the mean of the residuals is always 0. The standard deviation of the residuals is estimated with $s = 50.0007$. With a z-score of $\frac{75}{50} = 1.5$ and a roughly normal distribution, we have $P(z > 1.5) \approx 0.067$.

SCORING

Part (a) is essentially correct for noting the three conditions (roughly linear scatterplot, no clear pattern in the residual plot, and roughly normal histogram of residuals). Part (a) is partially correct for correctly noting two of the three conditions.

Part (b) is essentially correct for a correct calculation and a correct interpretation of the residual, including direction. Part (b) is partially correct for one of these two parts correct.

Parts (c) and (d) together are essentially correct for a correct calculation of the confidence interval, a correct interpretation in context, and a correct statement about causation. Parts (c) and (d) together are partially correct for two of the three steps correct.

Part (e) is essentially correct for noting that the distribution of residuals is roughly normal with mean 0 and standard deviation 50.0007 and then using this to calculate the probability correctly. Part (e) is partially correct for correctly noting the distribution of residuals but incorrectly calculating the probability, or for making a calculation based on a normal distribution with mean 0 but using an incorrect standard deviation.

Count partially correct answers as one-half an essentially correct answer.

4   **Complete Answer**   Four essentially correct answers.

3   **Substantial Answer**   Three essentially correct answers.

2   **Developing Answer**   Two essentially correct answers.

1   **Minimal Answer**   One essentially correct answer.

Use a holistic approach to decide a score totaling between two numbers.

4. (a) **Think:** Wow, this is a complex-looking graph right from the start. Take your time and look carefully at what the axes represent: values and their z-scores. You are only asked here for the range of CFU values, and the min and max CFU values can be easily estimated from the graph.

Answer: 33.5 − 28.5 = 5.0 (billion)

(b) **Think:** There are 100 CFU values, so the fifth percentile is simply the value indicated by the fifth dot up from the minimum CFU on the graph.

Answer: Either counting 5 dots (out of 100) or going over to the data plot from a z-score of −1.645, the 5th percentile of the data is seen to be approximately 29.75 (billion).

(c) **Think:** Getting more complicated! You're not looking at the data here, only at that diagonal line. The 95th percentile in a normal distribution corresponds to a z-score of 1.645, which is fortunately marked on the graph.

Answer: Going over to the slant line from a z-score of 1.645, the 95th percentile of a normal distribution with the same mean and standard deviation is seen to be approximately 34.4 (billion).

(d) **Think:** What is the shape of a histogram? In the graph you can see that lots of values are piled up at the upper end between 33 and 34, while at the lower end, the values are spread out from the minimum value toward the median. Remember that skewness is indicated when values are concentrated at one end and spread thinly on the other.

Answer: Note that there are relatively few CFU values between 28 and 32, that these lower values are spread out while the values between 32 and 33 are more concentrated, and that the values between 33 and 34 are very concentrated. Conclude that the distribution is skewed left (skewed toward the lower values).

(e) **Think:** To receive full credit for a confidence interval inference question, you must name the procedure, check conditions, find the interval, and give a conclusion in context about the whole population.

Answer: Procedure: t-interval of a population mean

Checks: Random sample (given), the sample size $n = 100$ is large enough for the CLT to apply, and $n = 100$ is less than 10% of all probiotic capsules of this brand.

Mechanics: With $\bar{x} = 32.48$, $s = 1.174$, and $n = 100$, a 95% t-interval is (32.247, 32.713).

We are 95% confident that the mean CFU value (in billions) of all probiotic capsules of this brand is between 32.247 and 32.713.

TIP

Even if you have no idea how to do the first parts, a later part might be straightforward; in this case the last part is simply a t-interval mean question.

5. (a) **Think:** You're starting out with a straightforward binomial probability.

Answer: The probability that Team A wins four straight games is $(0.5)^4 = 0.0625$. So the probability that the World Series is over in four games is $0.0625 + 0.0625 = 0.125$.

(b) **Think:** You're shown a helpful pattern, so say "thank you" and use it!

Answer: $2\left[\binom{6}{3}(0.5)^3(0.5)^3\right](0.5) = 0.3125$ or $1 - (0.125 + 0.25 + 0.3125) = 0.3125$.

(c) Think: You calculated the relevant probabilities, and the number of games is 103.

Answer: Multiplying each of the probabilities by 103 gives the expected number of occurrences:

	4	5	6	7
Expected Number of Years	12.875	25.75	32.1875	32.1875

(d) **Think:** You're given observations and asked if a certain distribution is followed—a standard χ^2-GOF test! Remember to state the hypotheses, identify the test and check the assumptions, calculate the test statistic χ^2 and the *P*-value, and linking to the *P*-value, give a conclusion in context.

Answer: First, state the hypothesis. H_0: The lengths of World Series follows a distribution of 4, 5, 6, and 7, with probabilities 0.125, 0.25, 0.3125, and 0.3125, respectively. H_a: The probability distribution is different from that of 0.125, 0.25, 0.3125, and 0.3125, respectively.

Second, identify the test and check the assumptions. This is a chi-square test of goodness of fit. We must assume that the 103 series are a representative sample. We note from part (c) that all expected cells are ≥ 5.

Third, calculate the test statistic χ^2 and the P-value. A calculator gives $\chi^2 = 7.291$ and $P = 0.0632$.

Fourth, linking to the P-value, give a conclusion in context. With a P-value this large, $0.0632 > 0.05$, there is not sufficient evidence to reject H_0. In other words, there is not sufficient evidence that the lengths of World Series do not follow a distribution of 4, 5, 6, and 7, with probabilities 0.125, 0.25, 0.3125, and 0.3125, respectively. Alternatively, with a P-value this small, $0.0632 < 0.10$, there is sufficient evidence to reject H_0. In other words, there is sufficient evidence that the lengths of World Series do not follow a distribution of 4, 5, 6, and 7, with probabilities 0.125, 0.25, 0.3125, and 0.3125, respectively.

SCORING

Parts (a) and (b) together are essentially correct for correctly calculating 0.0625, 0.125, and 0.3125. They are partially correct for correctly calculating two of these three probabilities.

Part (c) is essentially correct for correctly calculating expected values using the probabilities calculated in (a) and (b), and incorrect otherwise.

Part (d1) is essentially correct for correctly stating the hypotheses, identifying the test, and noting that the expected cells are all ≥ 5. It is partially correct for two out of three of these statements.

Part (d2) is essentially correct for correctly giving the test statistic χ^2, giving the P-value, and giving a correct conclusion in context. It is partially correct for a mistake in one of these areas.

Count partially correct answers as one-half an essentially correct answer.

4 **Complete Answer**   Four essentially correct answers.

3 **Substantial Answer**   Three essentially correct answers.

2 **Developing Answer**   Two essentially correct answers.

1 **Minimal Answer**   One essentially correct answer.

Use a holistic approach to decide a score totaling between two numbers.

---

6. (a) **Think:** You are asked for a proportionate sample and are given the proportions and the total sample size.

   Answer: City: $(0.50)(2000) = 1000$ students
   Suburbs: $(0.30)(2000) = 600$ students
   Rural: $(0.20)(2000) = 400$ students

   (b) **Think:** You are given a formula and simply asked to plug in!

   Answer: $\bar{x}_{overall} = (0.5)(74.3) + (0.3)(80.4) + (0.2)(69.8) = 75.23$

(c) **Think:** Remember that the standard error of the sampling distribution of $\bar{x}$ is $\frac{s}{\sqrt{n}}$.

Answer: $SE(\bar{x}_{city}) = \dfrac{10.2}{\sqrt{1000}} = 0.3226$

$SE(\bar{x}_{suburb}) = \dfrac{9.3}{\sqrt{600}} = 0.3797$

$SE(\bar{x}_{rural}) = \dfrac{12.1}{\sqrt{400}} = 0.605$

(d) **Think:** Remember that when independent random variables are combined, it's the variances that add!

Answer: $SE(\bar{x}_{overall}) = \sqrt{[(0.5)(0.3226)]^2 + [(0.3)(0.3797)]^2 + [(0.2)(0.605)]^2} = 0.4105$

(e) **Think:** You are given a statement of the procedure and are told to assume all conditions for inference are satisfied. You calculated $\bar{x}_{overall}$ in (b) and $SE(\bar{x}_{overall})$ in (d), so calculating the interval is straightforward. Don't forget a conclusion in context referring to the whole population.

Answer: $75.23 \pm 2.576(0.4105) = 75.23 \pm 1.06$, or $(74.17, 76.29)$. We are 99% confident that if all fifth graders took this mathematics exam, their mean score would be between 74.17 and 76.29.

SCORING

Parts (a) and (b) together are essentially correct for the three correct numbers of students and a correct calculation of the overall sample mean. Parts (a) and (b) together are partially correct for either the correct numbers of students and an incorrect mean calculation or for a correct mean calculation using an incorrect set of numbers of students.

Part (c) is essentially correct for three correct calculations of standard error. It is partially correct for just calculating variances or for dividing by n instead of $\sqrt{n}$. Do not penalize for arithmetic errors if correct formulas are shown.

Part (d) is essentially correct for a correct calculation of $SE(\bar{x})$ and is partially correct if the coefficients (0.5, 0.3, 0.2) are misplaced or missing.

Part (e) is essentially correct for a correct confidence interval based on the answers found in parts (b) and (d) together with a correct statement in context. It is partially correct if the only mistake is either using an incorrect z-score or not putting the statement in context.

Count partially correct answers as one-half an essentially correct answer.

4	**Complete Answer**	Four essentially correct answers.
3	**Substantial Answer**	Three essentially correct answers.
2	**Developing Answer**	Two essentially correct answers.
1	**Minimal Answer**	One essentially correct answer.

Use a holistic approach to decide a score totaling between two numbers.

7. (a) **Think:** There are 60 months averaging 223.92 inches and 60 months averaging 117.92 inches of rainfall.

Answer: The mean rainfall of the 60 months with El Niño not present is 117.92, while the mean rainfall of the 60 months with El Niño present is 223.92. The mean rainfall of all 120 months is $\frac{60(117.92)+60(223.92)}{120} = 170.92$ inches.

(b) **Think:** You need to see that all the rainfall values with El Niño are greater than all the values without El Niño so there is a gap between 140 (max of "without") and 180 (min of "with").

Answer: The rainfall during the 60 months with El Niño not present has a maximum value of 140. The rainfall during the 60 months with El Niño present has a minimum value of 180. The median rainfall of all 120 months is the average of the 60th and 61st values, that is, $\frac{140+180}{2} = 160$ inches.

(c) **Think:** Again, all the rainfall values with El Niño are greater than all the values without El Niño.

Answer: The median of the lower 60 values is the first quartile, Q_1, of the 120 values. The median of the upper 60 values is the third quartile, Q_3, of the 120 values. The interquartile range of the 120 values is IQR $= Q_3 - Q_1 = 220 - 120 = 100$.

(d) **Think:** Histograms can show gaps; these are missed by boxplots, and you have a big gap between 140 and 180.

Answer: A histogram (with reasonable bins) will show two clusters of values with a gap between, while a boxplot will not show this feature.

(e) **Think:** A horizontal line in a cumulative frequency plot represents no values in a specific interval.

Answer: In a cumulative frequency plot, a horizontal line represents that there are no values in a specific interval and there are no rainfall values x where $140 < x < 180$.

(f) **Think:** The hypotheses for regression involve the population slope β; you must calculate the test statistic and P-value, and then linking to the P-value, give a conclusion in context.

Answer: This is a linear regression t-test with $H_0: \beta = 0$ and $H_a: \beta > 0$ where β is the slope of the regression line that relates monthly rainfall to months. It is given that all conditions for inference are satisfied.

$$t = \frac{0.69460 - 0}{0.04263} = 16.29, \text{ and with } df = 60 - 2 = 58, P = 0.000$$

With this small a P-value, $0.000 < 0.05$, there is very strong evidence to reject H_0 and conclude that there is significant evidence of a positive slope. In other words, there is statistical evidence of increasing monthly rainfall during the 60 months with El Niño present.

Parts (a), (b), and (c) together are scored essentially correct for correct calculations of the median, mean, and interquartile range with supporting work. They are partially correct for two of the three correct calculations with supporting work or for all three correct calculations but missing supporting work.

Parts (d) and (e) together are scored essentially correct for pointing out the gap in the histogram and correctly explaining the horizontal line in the cumulative frequency plot. They are partially correct for one of these two components.

Part (f1) is scored essentially correct for naming the test, giving both hypotheses, and defining the parameter. It is partially correct for two of these three components correct.

Part (f2) is scored essentially correct for correctly calculating the test statistic t and the P-value and for giving a correct conclusion in context with linkage to the P-value. It is partially correct if missing one of these components.

Count partially correct answers as one-half an essentially correct answer.

4 **Complete Answer**   Four essentially correct answers.

3 **Substantial Answer**   Three essentially correct answers.

2 **Developing Answer**   Two essentially correct answers.

1 **Minimal Answer**   One essentially correct answer.

Use a holistic approach to decide a score totaling between two numbers.

Statistical Insights into Social Issues

15

→ **DISTRIBUTION OF WEALTH**

→ **INCARCERATION BY AGE AND RACE**

→ **FAMILY INCOME AND SAT SCORES**

→ **REPORTING CONFLICT VIOLENCE**

→ **JURY SELECTION**

→ **ENVIRONMENTAL RACISM**

→ **FLAWED FORENSIC TESTIMONY**

→ **MODERN-DAY SLAVERY**

→ **HATE CRIMES**

→ **BULLYING**

→ **FIREARM SUICIDES**

→ **DANGEROUS LEAD LEVELS**

→ **LAW ENFORCEMENT–ASSISTED DIVERSION**

→ **POLICE PROFILING**

→ **INEQUALITY AND HEALTH**

→ **WHY JAILS ARE FULL**

→ **GLOBAL WARMING**

→ **VACCINES AND AUTISM**

In the spirit of AP Statistics, the following comprehensive review questions aim to give an appreciation of the power of statistics in better understanding some of society's most pressing issues. As you review and apply the statistical tools you have learned, you should also reflect on the content of the examples and the empowerment that statistics gives you to be aware of, to approach, to discuss, and maybe some day to help overcome some of the ills present in the broad social world in which we live. More specifically, statistics can help you to understand the relationships among power, resource inequities, and uneven opportunities among different social groups and to understand explicit discrimination based on race, class, and gender. Of course, in addition to reflecting on these significant issues, don't forget to read carefully what the questions ask, to check assumptions/conditions when appropriate, to show your work, and to communicate your answers clearly!

There is great interest in how statistics can help better society. Statistics can be an incredibly powerful tool both in helping you understand social issues and in helping you effect social change. From disparities in income levels to disparities in the distribution of wealth, from unequal access to education to unequal access to health care, from unfair criminal justice enforcement to unfair housing market discrimination, statistics can be the key for you to become an engaged participant in our democracy. Based on your background, your interests, and your available time, you can decide how far to take the opportunities for deeper discussion raised in the following examples.

➡ EXAMPLE 15.1 _____

A 2016 study found that the richest 62 people in the world have as much wealth as the poorest half of the global population.[1] One way in which the distribution of wealth throughout the world can be analyzed is illustrated below, where GDP refers to Gross Domestic Product and Oceania is a region centered on the islands of the tropical Pacific Ocean.

(a) Fill in the blank cells in the table below.

Region	Population (in millions)	% of World Population	Wealth (GDP) (in billions)	% of World GDP
Africa	1111		2600	
Asia	4299		18,500	
Oceania	38		1800	
Europe	742		24,400	
US/Canada	355		20,300	
Latin America	617		4200	
World Total	7162	100.0	71,800	100.0

(b) Construct a segmented bar chart with two bars from the above table, and comment about what it illustrates.

(c) Consider the following graph:

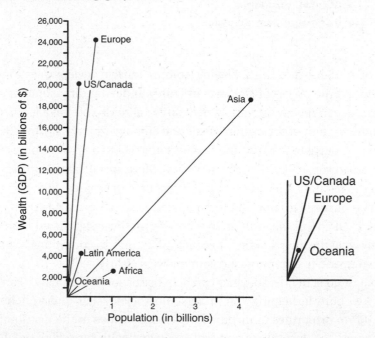

What do the slopes of the six slanted lines represent? Use these slopes to give an appropriate ranking of the six regions.

[1]Oxfam International, using data from Credit Suisse.

(d) The particular quantitative index used in a statistical study can make a marked difference. For example, an alternative to the GDP for measuring the wealth of a nation is the GPI (Genuine Progress Indicator).[2] The GPI takes into account negative factors, such as industrial pollution, depletion of natural resources, and crime rates, as well as positive factors, such as socially productive time uses like volunteerism. Give an example of how changing the quantitative index from GDP to GPI might affect the relative position of US/Canada and Oceania in part (c).

[2]Talberth, Cobb, and Slattery, "Gross Domestic Product, A Tool for Sustainable Development," February 2007.

➡ **EXAMPLE 15.2** _____

The United States has the second highest incarceration rate in the world (second only to the small African country of Seychelles).[3] One study[4] of incarceration rates and ages summarized their findings in the following graph:

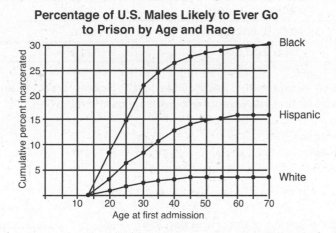

Percentage of U.S. Males Likely to Ever Go to Prison by Age and Race

(a) What is the meaning of the point (25, 15) on the Black graph?

(b) What is the meaning of the horizontal segment on the White graph from ages 60 to 70?

(c) What is the median age at first admission to prison of Hispanics who ever went to prison?

(d) Some states are considering legislation to address the disproportionate suspensions among students of color. How might this affect the above graphs?

[3]"U.S. Prison Boom: Impact on Age of State Prisoners," *Journalist's Resource*, February 2016.
[4]"The Use of Incarceration in the United States," *American Society of Criminology*, November 2000.

➡ EXAMPLE 15.3 _____

SAT used to stand for Scholastic Aptitude Test, but some researchers suggest it also stands for Student Affluence Test.[5] There appears to be an association between mean total SAT scores and family income level. A recent study[6] is summarized in the following table:

2015 College-Bound High School Students

Income Bracket	Total SAT Score
0–20,000	1314
20,000–40,000	1399
40,000–60,000	1458
60,000–80,000	1500
80,000–100,000	1544
100,000–120,000	1581
120,000–140,000	1591
140,000–160,000	1616
160,000–200,000	1636
Over 200,000	1720

Using the median income within each of the first 9 income brackets and a $210,000 income from the last bracket, computer output yields the following:

```
Predictor      Coef       SE Coef        T        P
Constant     1351.37        18.81     71.85    0.000
Income      0.0017915    0.0001565    11.45    0.000

    S = 30.6630 R-Sq = 94.2% R-Sq(adj) = 93.5%
```

(a) What is the equation of the regression line?

(b) Interpret the slope of the regression line in context.

(c) What is the coefficient of determination. Interpret it in context.

(d) What is the residual if a family has $63,500 income and a student has a score of 1450? What does this mean in context?

(e) What is the meaning of $S = 30.6630$?

(f) Is it reasonable to conclude that the best way to raise SAT scores is to raise family incomes? Explain, and if the answer is "no," give a possible confounding variable that could suggest an alternative approach to raising SAT scores.

---

[5] "SAT Scores and Income Inequality: How Wealthier Kids Rank Higher," *Wall Street Journal*, October 2014.
[6] College Board and FairTest (2015 scores).

➡ EXAMPLE 15.4 _____

In analyzing conflict violence, the data are often incomplete, as obtaining data in such situations is very difficult if not impossible. However, accurate data are crucial for systematic studies of the origins of conflict, conflict dynamics, and conflict resolution.

(a) Suppose during one conflict month in Syria, the Syrian Observation for Human Rights (SOHR) reports 3060 deaths, the United Nations (UN) reports 3627 deaths, and 2635 of the deaths are found on *both* lists. Letting D be the actual number of deaths, fill in the following table:

		SOHR		
		Reported Deaths	Unreported Deaths	*Totals*
UN	Reported Deaths			3627
	Unreported Deaths			$D - 3627$
	Totals	3060	$D - 3060$	D

(b) Assuming the two reports above represent independent random samples of all deaths that month, solve for D.

(c) In reality, what are available are only *convenience samples*—perhaps posted videos of public executions, perhaps text messages sent out by protesters, or perhaps victims' testimonies recorded by local reporters. In general, why do convenience samples often lead to misleading conclusions? In our context, how are assumptions made in part (b) above violated?

(d) Another source of *undercoverage bias* found in conflict situations can be called *size bias*. Events involving large numbers of casualties are much more likely to be documented than events involving only one victim. For example, a cluster bombing of a marketplace with many victims will attract media attention where targeted assassinations may be unreported because the victim's family is afraid to make a report or because the body is never found. Explain how this affects the independence assumption made in part (b) above.

(e) Would you expect the answer found for *D* above to be an underestimate or an overestimate? Explain.

_____

➡ EXAMPLE 15.5 _____

In 1986, the Supreme Court ruled that it is unconstitutional to exclude potential jurors because of race.[7] However, many studies since then have shown that racial discrimination in jury selection remains widespread, particularly in murder trials. A university law department plans a study using mock trials to determine if outcomes are different with all-White versus mixed-race juries. The law department will first randomly select two counties from among its state's 67 counties.

(a) What is this sampling method called, and what is an advantage of using this method over a simple random sample?

(b) Give a procedure for randomly selecting the two counties.

(c) Forty-five people will then be randomly selected from among the pool of all eligible White jurors in each of the two counties to obtain a sample of size 90. What is the purpose of randomization in this two-step process?

(d) The proposed study will involve 10 mock cases involving a young African-American male charged with murdering an elderly White female. The sample of 90 white jurors will be randomly assigned to two groups—60 to make up five all-white juries (each of size 12) and 30 to be part of five mixed-race juries (each with 6 Whites and 6 non-Whites). What is the purpose of random assignment here? How might allowing jurors to self-select whether or not to be part of mixed-race juries lead to bias?

[7]Batson v. Kentucky, 476 U.S. 79 (1986)

➡ EXAMPLE 15.6 _____

"Environmental racism" is said to occur when race is associated with the location of hazardous sites. A 2015 California study was designed to assess who was at the greatest risk from oil trains, where a major risk is from derailment explosions. Of the 5.5 million Californians who live within a blast zone defined as within one mile of the railroad tracks, 75% are Latino and only 10% are White. California's population of 38.8 million is 39.0% Latino and 38.9% White.

(a) What is the probability a randomly selected Californian is a Latino living within a blast zone?

(b) What is the probability that a randomly selected Californian Latino lives within a blast zone?

(c) In California, are being Latino and living within a blast zone independent events? Justify your answer mathematically.

(d) In a random sample of 5 Californians, what is the probability that at least 2 of them live in a blast zone?

(e) What is another explanatory variable that could account for the relatively high probability that someone living in a blast zone is non-White?

➡ EXAMPLE 15.7

Although there is no accepted research on how often hair from different people may appear the same, for over two decades the FBI used visual hair comparison matches to help obtain convictions in major criminal cases. Acknowledging possible flawed forensic testimony based on unsound science, the FBI agreed to review over 2500 cases. Flawed forensic testimony was found in 257 of the first 268 cases reviewed.[8] In the following, assume that the sample proportion of cases involving flawed testimony is an accurate measure of the population proportion.

(a) The FBI agreed to review 2500 cases involving possible flawed forensic testimony. Assuming this is a random sample of relevant cases, what is the expected number of these 2500 cases with actual flawed forensic testimony? What is the standard deviation for the expected number of cases with actual flawed forensic testimony?

(b) Among the 257 cases with flawed forensic testimony, 32 resulted in death penalty convictions. Assume this is illustrative of the overall proportion. In examining a random sequence of cases involving flawed testimony, what is the probability that the first case with a resulting death penalty doesn't occur until at least the third case examined?

(c) According to the Innocence Project, the misapplication of forensic science is the second-leading contributor to wrongful conviction, playing a role in 50% of known wrongful convictions, while eyewitness misidentification has played a role in nearly 75% of known wrongful convictions.[9] In wrongful conviction cases, can "the misapplication of forensic science" and "eyewitness misidentification" be mutually exclusive? Can they be independent? Explain.

[8]National Association of Criminal Defense Lawyers and Innocence Project Analysis of FBI and Justice Department Data as of March 2015.
[9]http://www.innocenceproject.org

➡ EXAMPLE 15.8

We often think that slavery ended in the 19th century, but the sad fact is that there are more people in slavery today than at any other time in human history. According to the not-for-profit *Force4Compassion*, an average of 3287 people are sold or kidnapped and forced into slavery every day.[10] Assume this daily total has a roughly normal distribution with a standard deviation of 250 people.

(a) What is the probability that on a randomly selected day, over 3500 people are forced into slavery?

(b) In a random sample of 5 days, what is the probability that on a majority of the days over 3500 people are forced into slavery?

(c) In a random sample of 5 days, what is the probability that the average number of people forced into slavery is over 3500?

(d) Are there any statistics that, if more widely disseminated, might mobilize the world to fight this social blight?

[10]http://www.f-4-c.org/statistics/

EXAMPLE 15.9

When bias-motivated crimes (triggered by prejudice and hate) are committed, whole communities can feel victimized. A study[11] of profiles on America's most popular online hate site revealed that the site members are very young. Consider the following boxplot of self-reported ages of the site members:

Age

(a) Describe the distribution of ages.

(b) Is the quotient $\dfrac{\text{Mean age}}{\text{Median age}}$ greater than 1, less than 1, or approximately 1? Explain.

(c) Internet and social network users tend to be younger than the general population. What does this possibly indicate about the median age of all hate group members? Explain.

(d) According to recent FBI statistics,[12] 18.6% of bias incidents are religious based. In a random sample of 3 bias incidents, what is the probability that a majority are religious based?

[11]"The Data of Hate," *The New York Times*, 7/13/2014.
[12]"Bias Breakdown," *www.fbi.gov*, November 2015.

EXAMPLE 15.10

Schoolyard and online bullying are reaching dangerous levels with violence and depression all too frequent outcomes. In a national survey[13] of 15,686 students in grades six through ten, 19% were involved in moderate to frequent bullying as perpetrators, 17% as victims, and 6% as both vicims and perpetrators.

(a) What proportion of the students were involved in moderate to frequent bullying?

(b) Find a 95% confidence interval for the proportion of students in grades six through ten who are victims of moderate to frequent bullying.

(c) Based on this confidence interval, is there evidence that the proportion of students who are victims of moderate to frequent bullying is different from the 22% claimed in one study?[14] Explain.

(d) School administrators tend to underestimate the rates of bullying because students rarely report it and it often happens when adults aren't around. However, knowing the numbers and types of bullying at all grade levels in your school are critical before a plan for bullying prevention and intervention can be developed. How would you plan a survey at your school to assess the frequency and types of bullying present?

[13]Nansel, T. R., Overpeck, M., Pilla, R. S., Ruan, W. J., Simons-Morton, B., & Scheidt, P. "Bullying Behaviors Among U.S. Youth: Prevalence and Association with Psychosocial Adjustment," *Journal of the American Medical Association*, 285(16), 2094–2100.
[14]National Center for Educational Statistics, 2015.

➡ EXAMPLE 15.11 _____

The Second Amendment to the U.S. Constitution reads, "A well-regulated militia being necessary to the security of a free state, the right of the people to keep and bear arms shall not be infringed." For many years, this amendment drew very little attention, but now its meaning and intention are being fiercely debated. When we think of deaths by firearms, we usually think of murders. However, most firearm-related deaths in the United States are suicides, and most suicides are by guns. According to the CDC,[15] in 2013 there were 21,175 suicides in the United States using firearms.

(a) Assume we have a random sample of 7 years that are considered representative with regard to numbers of firearm suicides:

Yearly firearm suicides

$$\bar{x} = 21{,}128.6,\ s = 779.6$$

Calculate a 95% confidence interval estimate for the mean number of firearm suicides in the United States each year.

(b) According to a major recognized survey,[16] the United States has the highest rate of gun ownership in the world at 88.8 guns per 100 people. The same survey reports Israel at 7.3. Given that these figures report only privately owned guns and exclude weapons that are technically owned by the government, how might the apparent wide discrepancy between the United States and Israel not be as wide as indicated?

(c) When presented with any data set, we should always try to understand what and whose values are implicitly represented. Israel rejects 40% of gun applications and requires yearly renewals for continuing gun ownership. How would this affect or change the conversation from the statistics quoted in part (b)?

[15]http://www.cdc.gov/nchs/fastats/suicide.htm
[16]UNODC (United Nations Office on Drugs and Crime) and the Small Arms Survey (Geneva).

➡ EXAMPLE 15.12 _____

As a cost-cutting measure in April 2014, Flint, Michigan, began temporarily drawing its drinking water from the Flint River. The result was dangerous lead contamination in the water supply. (Lead poisoning especially impacts a child's brain and nervous system.)

(a)   An independent research team[17] tested the Flint water supply for lead. The resulting cumulative frequency plot of the lead levels in ppb (parts per billion) in the water supply of sampled homes is as follows:

What is the shape of the histogram of the lead levels in the water supply of sampled homes? How would you expect the mean and median lead levels to compare to each other? Explain.

(b)   It is recognized that if over 10% of a city's homes have lead levels in their water supply of over 5 ppb, then the city has "a very serious problem with lead."[18] The research team determined that 15 ppb corresponded to the 75th percentile of Flint home water supply lead levels. Does this indicate a very serious problem?

(c)   Suppose the concern is whether or not the mean lead level in the water supply of the homes is greater than 10 ppb. (The EPA regards 15 ppb in the water supply as a very dangerous level.[19]) Describe Type I and Type II errors in context. Which is of greater concern, a Type I or Type II error, and what would be a possible consequence? Explain.

[17]2015 Virginia Tech study.
[18]*http://flintwaterstudy.org*
[19]CDC 24/7: "Saving Lives, Protecting People."

EXAMPLE 15.13

Law Enforcement Assisted Diversion (LEAD) is a prebooking diversion program whereby low-level drug and prostitution offenders are redirected to community-based services instead of jail and prosecution.[20] 318 offenders participated in a pilot program in which they were either assigned to the LEAD program ($n = 203$) or control (i.e., booking as usual; $n = 115$). All 318 offenders were deemed eligible for participation in the LEAD program. Although a nonrandomized design was used, it was felt that both the LEAD and the control groups were representative of the targeted offender population.

(a)   The average yearly criminal and legal system costs of those participating in the LEAD program were $4763. For the control group, the average was $11,695. (Assume standard deviations of $875 and $2125, respectively.) Is this statistically significant evidence that the LEAD program has lower costs than the usual booking program?

(b)   After participation in the program, the LEAD group showed significantly fewer average number of days per year in jail than the control group ($P < 0.05$). Which error, Type I or Type II, might have been committed and what would be a possible consequence?

(c)   Using your answer from part (a), make an argument in favor of the LEAD program to a city council whose members are unfamiliar with statistical arguments.

[20]"(LEAD): Reducing the Role of Criminalization in Local Drug Control," *http://www.drugpolicy.org*

EXAMPLE 15.14

Local activists and community leaders in Berkeley, CA, have long maintained that police "over police" people of color. The claim was that people of color were being stopped at a higher rate than their population would indicate and that the stops were often for no legitimate reasons. In response, the Berkeley City Council directed the Police Department to collect data on the subject of all traffic and pedestrian stops. In September 2015 the Police Department filed a report stating the following:

During the period from January 26 through August 12, 2015, 4658 civilians were stopped, and of that total, 1710 were White, 1423 were African-American, and 543 were Hispanic. Of those stopped, 652 of the Whites, 942 of the African-Americans, and 306 of the Hispanics were released without arrest or citation.

According to 2013 census data, the population of Berkeley is 116,774, which includes 64,412 Whites, 10,076 African-Americans, and 11,600 Hispanics.

(a)   Sketch side-by-side bar charts to illustrate the population percentage distribution and the percentage distribution of who was stopped by race.

(b)   Is there statistically significant evidence that African-Americans are being stopped at a higher rate than their population would indicate? (Assume that we have a representative sample of what has been going on for years.)

(c)   Is there statistically significant evidence of a relationship between race (White, African-American, Hispanic) and disposition (arrest/citation or not)? (Assume that we have a representative sample of what has been going on for years.)

(d)   If it turned out that, for example, people of color drove a different type or age of car than Whites, explain how this could be a confounding variable.

➡ EXAMPLE 15.15 _____

Is there a relationship between inequality and health? The Equality Trust, a nonprofit advocacy group, has created an Index of Health and Social Problems (HSP Index)[21] to combine several troublesome indicators (including lower life expectancy, higher infant mortality, and weak social mobility) into a single variable that can be used to describe the overall "health" of a society. A measure of income inequality is the ratio of the income of the top 20% to the bottom 20% in a country. Computer output of a regression analysis of HSP versus Income ratio in 21 industrialized countries is as follows:

Scatterplot of HSP Index vs Income Ratio

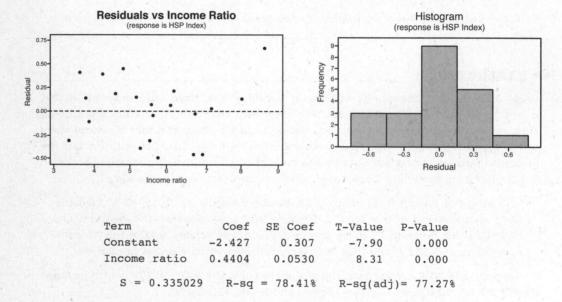

Residuals vs Income Ratio (response is HSP Index)

Histogram (response is HSP Index)

Term	Coef	SE Coef	T-Value	P-Value
Constant	-2.427	0.307	-7.90	0.000
Income ratio	0.4404	0.0530	8.31	0.000

S = 0.335029   R-sq = 78.41%   R-sq(adj)= 77.27%

(a) Find a 95% confidence interval for the slope of the regression line.

(b) Is there sufficient evidence that the higher the income ratio (that is, the higher the income inequality in a country), the greater the HSP Index (that is, the greater the health and social problems of the country)?

(c) Assuming that a lower HSP Index is desirable, what can be concluded about working to lessen the income ratio, that is, to lessen the income inequality in a country?

[21]http://inequality.org/inequality-health/#sthash.H77M9bp9.dpuf

➥ EXAMPLE 15.16

According to the Bureau of Justice Statistics, 1 out of every 35 adults in the United States is under correctional supervision (probation, parole, jail, or prison).[22] According to a 2014 Human Rights Watch report, "tough-on-crime" laws have filled U.S. prisons with mostly nonviolent offenders.[23]

(a) A previous study[24] on the population of local jails showed that 60.6% were people too poor to make bail or pay fines. More specifically, 34.4% were convicted males, 5.0% convicted females, 53.4% unconvicted males, and 7.2% unconvicted females. Suppose a current random sample of local jail populations counts 81 convicted males, 5 convicted females, 110 unconvicted males, and 4 unconvicted females. Is there sufficient evidence that the population distribution has changed from the findings of the previous study?

(b) Advocacy groups are pushing for counties across the country to reform their local criminal justice systems to make them fairer—and cut ever-increasing costs dedicated to running decades-old, overcrowded lockups. Local jails hold nearly 700,000 inmates on any given day, with small counties holding 44% of the total.[25] More data are very much needed. How would you proportionately choose 500 inmates to interview and study?

[22]"Correctional Populations in the United States," Bureau of Justice Statistics, 2013.
[23]"Nation Behind Bars: A Human Rights Solution," Human Rights Watch, May 2014.
[24]Bureau of Justice Statistics, 2012.
[25]Pearson, Jake, "Study: Smaller Counties Driving U.S. Jail Population Growth," December 2015.

➡ EXAMPLE 15.17 _____

Immunizations are a cornerstone of the nation's efforts to protect people from a host of infectious diseases. After a fraudulent, discredited 1998 research paper claiming to show a link between vaccines and ASD (autism spectrum disorder), vaccination rates dropped sharply. In 2015, the United States experienced a large, multistate measles outbreak, mostly with unvaccinated children.

(a) Is there a link between vaccines and autism? A recent major study[26] concluded that between 30% and 36% of all parents have some such concerns. Assuming this is a 95% confidence interval, what was the minimum sample size necessary for such a conclusion?

(b) After looking at total cumulative immunogens from birth to three months, the researchers produced the graph below. Compare the ASD Case distribution and the Control distribution.

(c) After looking at the maximum number of immunogens in one day between birth and two years, the researchers produced the graph below. Compare the ASD Case distribution and the Control distribution.

(d) The study performed a large-scale test with the following:

H_0: There is no association between autism and vaccines
H_a: There is an association between autism and vaccines

Given a very large P-value, what was the proper conclusion?

[26]DeStefano, Price, and Weintraub, "Increasing Exposure to Antibody-Stimulating Proteins and Polysaccharides in Vaccines Is Not Associated with Risk of Autism," 2013.

➡ EXAMPLE 15.18 _____

There is a strong consensus among scientists that global surface temperatures have increased markedly in recent years, and that the trend is caused by emissions of greenhouse gases. However, in the popular media, especially in the United States, there is much debate over whether there is any problem, and even if there is a problem, is there any human responsibility. (China and the United States are the two largest producers of greenhouse gases.)

(a) Given the following graph, describe the relationship between Temperature Anomaly (deviation) and Time.

Global Land–Ocean Temperature Index

(b) In a Pew Global Attitudes Project survey in a random sample of 2029 Indians, 1319 expressed a great deal of worry about global warming. In a random sample of 2180 Chinese, 436 expressed a great deal of worry about global warming. Is the difference statistically significant?

(c) Suppose we wish to conduct a survey among the student body of your high school and calculate a confidence interval for the proportion who worry a great deal about global warming. Comment about bias and direction of bias, from the wording of the following survey:

> Charles, Prince of Wales, has consistently taken a strong stance criticizing both climate change deniers and corporate lobbyists by likening the Earth to a dying patient. Do you have a great deal of worry about global warming?

ANSWERS EXPLAINED

1. (a)

Region	Population (in millions)	% of World Population	Wealth (GDP) (in billions)	% of World GDP
Africa	1111	15.5	2600	3.6
Asia	4299	60.0	18,500	25.8
Oceania	38	0.5	1800	2.5
Europe	742	10.4	24,400	34.0
US/Canada	355	5.0	20,300	28.3
Latin America	617	8.6	4200	5.8
World Total	7162	100.0	71,800	100.0

(b) The segmented bar chart treats each bar as a whole and divides each bar proportionally into segments corresponding to the percentage in each group.

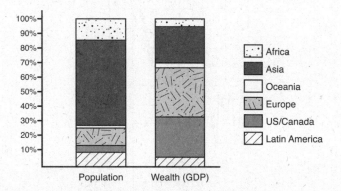

The segmented bar chart shows that both Europe and the US/Canada have a hugely disproportionate share of the wealth of the world when compared to their populations, while both Asia and Africa have a low share of the wealth of the world when compared to their populations.

(c) The slopes give each region's wealth in dollars per person. The US/Canada region has the steepest slope and thus the highest $/person, followed by Oceania, Europe, Latin America, Asia, and Africa.

(d) There are many possible answers. For example, switching to GPI takes into account industrial pollution, which is likely much higher in the US/Canada than in the islands of the South Pacific. With GDP per capita as the index, the US/Canada ranked slightly higher than Oceania. With CPI per capita as the index, the value for Oceania might go up and the value for US/Canada might go down sufficiently for the ranking to be reversed.

2. (a) Fifteen percent of all U.S. Black males are likely to go to prison for the first time by age 25.

(b) No U.S. White males are likely to go to prison for the first time between ages 60 and 70.

(c) Age 30. The graph indicates that 16% of all U.S. Hispanic males are likely to go to prison, and half of them (corresponding to 8% on the vertical axis) will go to prison for the first time by age 30.

(d) Answers should explain how addressing the disproportionate suspensions among students of color should lower the initial slopes of the Black and Hispanic graphs. An answer might also conclude that dismantling the school-to-prison pipeline would move the x-intercept for all 3 curves to the right, that is, raise the ages for first admissions to prison for all young people.

3. (a) Predicted total SAT = 1351.37 + 0.0017915 (Income)

(b) Every $1000 of additional family income is associated with an average of 1.7915 points higher on a student's total SAT score.

(c) The coefficient of determination, $R^2 = 94.2\%$. That is, 94.2% of the variation in total SAT scores is explained by the linear regression model of total SAT on Family Income.

(d) Predicted Total SAT = 1351.37 + 0.0017915(63,500) = 1465.13

Residual = Actual – Predicted = 1450 – 1465.13 = –15.13, that is, the actual total SAT score of this student is 15.13 *less* than what was predicted by the family income.

(e) $S = 30.6630$ is the standard deviation of the residuals. In other words, it is a measure of how the points are spread around the regression line.

(f) No, correlation does not imply causation. A possible confounding variable, for example, could be instructional money spent per student. Leveling per student expenditures across school districts might lead to higher SAT scores in economically disadvantaged areas.

4. (a)

SOHR

UN		Reported Deaths	Unreported Deaths	*Totals*
	Reported Deaths	2635	992	3627
	Unreported Deaths	425	$D - 4052$	$D - 3627$
	Totals	3060	$D - 3060$	D

(b) Relative frequencies should be approximately equal if there is independence. The expected cell formula $\text{Exp} = \dfrac{(\text{Row sum}) \times (\text{Column sum})}{N}$ gives $2635 = \dfrac{(3627)(3060)}{D}$.

(If you look at rows, you get instead $\dfrac{2635}{3627} = \dfrac{425}{D - 3627}$. If you look at columns you get $\dfrac{2635}{3060} = \dfrac{992}{D - 3060}$.) Solving any of these equations gives $D = 4212$.

(c) *Convenience samples* are based on choosing individuals who are easy to reach and tend to produce data highly unrepresentative of the entire population. In this case, only very readily obtainable reports of deaths are included, violating the assumption of a random sample of deaths for that month.

(d) Both samples will tend to report the same large casualty events, thus violating the assumption of independence.

(e) The calculated value of D is almost certainly an underestimate. As shown in parts (c) and (d), the reported deaths tend to be convenience samples of more widely reported deaths and with different samples tending to report the same deaths.

5. (a) This is an example of cluster sampling that allows for a quicker, less expensive method of sampling than an SRS. It assumes that each cluster, in this case each county, is representative of the whole state.

(b) Number the counties 1 through 67. Use a random number generator to pick two unique integers between 1 and 67 (ignoring repeats). The two counties to be used will consist of the two counties with numbers corresponding to the two unique numbers.

(c) The purpose of randomization in picking the 90 jurors is to obtain a sample that is representative of all eligible White jurors in the state so that any conclusions can be generalized to all eligible White jurors in the state.

(d) Random assignment here is used to minimize the effect of possible confounding variables. For example, if self-selection is allowed, perhaps some jurors are uncomfortable with non-Whites and would choose to be on the all-White juries. If these same jurors are more likely to have less understanding of non-White defendants, it might be impossible to say whether or not any conclusions are due solely to whether or not a juror serves on an all-White jury as opposed to a mixed-race jury.

6. (a) $P(\text{Blast zone}) = \dfrac{5.5}{38.8} = 0.14175$ and

$P(\text{Latino} \cap \text{Blast zone}) = P(\text{Blast zone})P(\text{Latino} \mid \text{Blast zone}) = (0.14175)(0.75) = 0.10631$

(b) $P(\text{Blast zone} \mid \text{Latino}) = \dfrac{P(\text{Latino} \cap \text{Blast zone})}{P(\text{Latino})} = \dfrac{0.10631}{0.390} = 0.27259$

(c) Is $P(\text{Blast zone} \mid \text{Latino}) = P(\text{Blast zone})$? No, $0.27259 \neq 0.14175$. So in California, being Latino and living within a blast zone are not independent events.

(d) This is a binomial with $n = 5$ and $p = 0.14175$.
Then $P(X \geq 2) = 1 - [(0.85825)^5 + 5(0.14175)(0.85825)^4] = 0.14979$.
Calculator software gives $1 - \text{binomcdf}(5, 0.14175, 1) = 0.14979$.

(e) Another possible explanatory variable is poverty. It could be that blast zones tend to be associated with low-income housing and that a higher percentage of non-Whites than Whites are low income.

7. (a) This is a binomial distribution with $n = 2500$ and $p = \dfrac{257}{268} = 0.959$.

Thus $E(X) = np = 2500(0.959) = 2397.4$.

The standard deviation $= \sqrt{np(1-p)} = \sqrt{2500(0.959)(0.041)} = 9.915$.

(b) The probability of a case resulting in the death penalty, given it involves flawed forensic testimony, is $\dfrac{32}{257} = 0.1245$. Thus, the probability of the case not resulting in the death penalty is $1 - 0.1245 = 0.8755$. The probabilities of a death penalty occurring on the first or second case examined, respectively, are 0.1245 and

$(0.8755)(0.1245) = 0.1090$. So the probability that the first case with a resulting death penalty doesn't occur until at least the third case examined is

$$1 - (0.1245 + 0.1090) = 0.7665$$

(c) "The misapplication of forensic science" and "eyewitness misidentification" are not mutually exclusive because, if so, their intersection would be the empty set. Then the probability of their union would be $0.50 + 0.75 - 0 = 1.25 > 1$. It is possible that they are independent because, if so, the probability of their intersection would be $(0.50)(0.75)$ and the probability of their union would be

$$0.50 + 0.75 - (0.5)(0.75) = 0.875 \leq 1$$

8. (a) We have a normal distribution with mean 3287 and standard deviation 250.

Then $P(X \geq 3500) = P\left(z \geq \dfrac{3500 - 3287}{250}\right) = 0.1971$.

(b) We have a binomial distribution with $n = 5$ and $p = 0.1971$.

Then $P(X \geq 3) = 1 - P(X \leq 2) = 1 - \text{binomcdf}(5, 0.1971, 2) = 0.05572$.

(c) We have a normal distribution with mean 3287 and standard deviation $\dfrac{250}{\sqrt{5}} = 111.80$.

Then $P(\bar{x} \geq 3,500) = P\left(\dfrac{3,500 - 3,287}{111.80}\right) = 0.0284$.

(d) Possible answers might range from average yearly proceeds (about \$35 billion) to average number of slaves in the world on any given day (21 to 29 million), from numbers of child slaves to numbers of women in sexual slavery.

9. (a) A complete answer addresses center, spread, and shape and mentions context. The median age of hate site members is approximately 23. The ages go from 14 to 65 with a range of $65 - 14 = 51$, and an $\text{IQR} = 30 - 17 = 13$.

There are 6 outliers between 51 and 65. Although it is not usually possible to describe distribution shape from just a boxplot, in this case the distribution of ages does appear to be skewed to the right (to the larger values).

(b) In a skewed right distribution, the mean is usually larger than the median. So we would conclude that the quotient $\dfrac{\text{Mean age}}{\text{Median age}}$ is greater than 1.

(c) If mostly the younger, computer savvy members of the hate groups are the ones who are online, then the median age of all hate group members is probably greater than the median age of 23 of those self-reporting on the site.

(d) This is a binomial with $n = 3$ and $p = 0.186$. Then $P(X \geq 2) = 0.0909$ or

$$3(0.186)^2(0.814) + (0.186)^3 = 0.0909$$

10. (a) $0.19 + 0.17 - 0.06 = 0.30$

(b) We are calculating a one-sample z-interval for a population proportion.

We must assume this was a random sample, $n\hat{p} = (15,686)(0.17) = 2667 \geq 10$, $n\hat{q} = (15,686)(0.83) = 13,019 \geq 10$, and $n = 15,686$ is less than 10% of all students in grades six through ten.

Calculator software gives (0.16415, 0.1759).

We are 95% confident that the proportion of all students in grades six through ten who are victims of moderate to frequent bullying is between 0.16415 and 0.1759.

(c) Because 0.22 is not in the interval of plausible values for the population proportion, there *is* evidence that the proportion of all students who are victims of moderate to frequent bullying is different from the 22% claimed in the one study.

(d) Issues that might be addressed in planning a survey include stratification among the grade levels, anonymity, and timing (early enough in the school year to allow implementation of a plan and again later in the year to assess the effectiveness of any prevention and intervention program).

11. (a) We are calculating a one-sample t-interval for the mean number of firearm suicides in the United States per year. We are given a random sample. The nearly normal condition seems reasonable because the sample data are very roughly unimodal and symmetric. Under these conditions, the mean firearm suicides per year can be modeled by a t-distribution with $n - 1 = 7 - 1 = 6$ degrees of freedom. Calculator software gives the interval (20408,21850).

[We can instead calculate $21,128.6 \pm 2.447 \dfrac{779.6}{\sqrt{7}} = 21,128.6 \pm 721.0$.]

We are 95% confident that the mean number of firearm suicides in the United States each year is between 20,408 and 21,850.

(b) It could well be true (actually, it is true) that most guns in Israel are technically owned by the government (by the IDF—the Israel Defense Forces). The 7.3 guns per 100 people statistic is a significant underestimate of guns in the population.

(c) The conversation would change from number of guns in the population to strictness of gun laws pertaining to gun ownership.

12. (a) A cumulative frequency plot that rises quickly at first and then rises at a much slower rate later corresponds to a histogram with a large area under the curve early and much less area later. Thus it is *skewed right* (skewed to the higher values of lead). In skewed right distributions, we expect the *mean* to be greater than the median.

(b) This absolutely indicates a very serious lead problem in the Flint water supply. The research team found that 25% of the homes had a lead water supply level of at least 15 ppb. 25% > 10% and 15 ppb > 5 ppb, so the percent of homes with at least 5 ppb is larger than 10%.

(c) With $H_0: \mu = 10$ and $H_a: \mu > 10$, a Type I error (mistaken rejection of a true null hypothesis) means that we think the lead level is greater than 10 ppb when, in fact, it is 10 ppb. So we might institute costly and unnecessary cleanup measures. A Type II error (mistaken failure to reject a false null hypothesis) means that the mean lead level is greater than 10 ppb but we don't have enough evidence to think so. So we might fail to institute a necessary cleanup. Clearly, with people's, and especially children's, health at stake, a Type II error is more of a concern.

13. (a) This is a two-sample t-test for means. $H_0: \mu_{LEAD} = \mu_{control}$ and $H_0: \mu_{LEAD} < \mu_{control}$. We are given samples that, by design, are assumed representative and independent.

The sample sizes, 203 and 115, are large enough so that by the CLT, the distribution of sample means is approximately normal and a t-test may be run. The sample sizes are assumed to be less than 10% of the population of all low-level drug and prostitution offenders. Calculator software gives $t = -33.41$ with a P-value of 0.000. With this small a P-value, $0.000 < 0.05$, there is sufficient evidence to reject H_0. In other words, there is sufficient evidence that for those adults eligible for the LEAD program, the mean yearly criminal and legal system costs of those participating in the program are less than that for those not participating.

(b) The null hypothesis was rejected. So the possibility is of a Type I error, that is, of rejecting a true null hypothesis. A Type I error would mean that in reality, after participating in the program, the LEAD group would have the same mean days per year in jail as the control group. However, we think the LEAD group has significantly fewer mean days per year in jail than the control group. A possible consequence is that the LEAD program is expanded, thinking it will reduce the future mean days per year in jail when, in reality, it won't.

(c) "We ran a large sample statistical study and noted that for our sample, the average yearly criminal and legal costs spent on those offenders participating in the LEAD program were significantly lower than the average spent on those not participating in the program. The difference was so great that it is highly unlikely that it could have occurred by chance."

14. (a) $\dfrac{1710}{4658} = 36.7\%$, $\dfrac{1423}{4658} = 30.5\%$, and $\dfrac{543}{4658} = 11.7\%$

$\dfrac{64,412}{116,774} = 55.2\%$, $\dfrac{10,076}{116,774} = 8.6\%$, and $\dfrac{11,600}{116,774} = 9.9\%$

or

(b) A one proportion z-test gives:

$$np_0 = 4658(0.086) = 400.6 \geq 10 \text{ and } n(1 - p_0) = 4658(0.914) = 4257.4 \geq 10$$

A representative (random) sample is assumed. The sample size is $< 10\%$ of the population. H_0: $p = 0.086$ and H_a: $p > 0.086$.

Software such as 1-PropZTest on the TI-84 gives $z = 53.4$ with a P-value of 0.000.

With this low a P-value, $0.000 < 0.001$, there is strong evidence that African-Americans are being stopped at a higher rate than what would be indicated by their population percentage.

(c) A chi-square test of independence gives:

H_0: Race (White, African-American, Hispanic) and disposition (arrest/citation or not) are independent.

H_0: There is a relationship between race (White, African-American, Hispanic) and disposition (arrest/citation or not).

Representative (random sample assumed) and all expected cells as shown below are ≥ 5.

OBS	White	African-American	Hispanic
Citation	1058	481	237
None	652	942	306

EXP	White	African-American	Hispanic
Citation	826.2	687.5	262.3
None	883.8	735.5	280.7

Software (such as χ^2-Test on the TI-84) gives the above expected values and $\chi^2 = 250.61$ with a P-value of 0.000.

With this low a P-value, $0.000 < 0.001$, there is strong evidence of a relationship between race (White, African-American, Hispanic) and disposition (arrest/citation or not).

(d) If police stopped drivers of certain model or condition cars, for example, flashier models or beat-up cars, and if people of color drove these particular cars more than Whites, then the higher rates of stoppage could be due to the model or condition of the car driven rather than to race. The model/condition would be a confounding variable.

15. (a) The confidence interval for the slope of a regression line:

We assume a random sample of countries (or at least a representative sample). The scatterplot looks roughly linear. The residual plot shows no pattern. The distribution of residuals is very roughly normal.

From the printout, the standard error of the slope is $s_{b_1} = 0.0530$. With $df = 21 - 2 = 19$ and 0.025 in each tail, the critical t-values are ± 2.093. The 95% confidence interval of the true slope is $0.4404 \pm 2.093(0.0530) = 0.4404 \pm 0.1109$. We are 95% confident that for every 1-unit increase in the Income ratio, the mean increase in the HSP Index is between 0.3295 and 0.5513.

(b) This is a test of significance for the slope of the regression line. The conditions for inference are the same as in part (a) above.

$H_0: \beta = 0$ and $H_a: \beta > 0$.

The computer printout gives $t = 8.31$ with a P-value of $\frac{1}{2}(0.000) = 0.000$.

With this small a P-value, $0.000 < 0.05$, there is strong evidence to reject H_0. In other words, there is strong evidence that the higher the income ratio (that is, the higher the income inequality in a country), the greater the HSP Index (that is, the greater the health and social problems of the country).

(c) There is strong evidence that income ratios (income inequality) have a positive linear association with HSP Indexes (health and social problems). In fact, 78.41% of the variation in HSP Indexes can be explained by this linear model. However, we cannot conclude causation. We cannot say that if we work to lessen income inequality, we will reduce health and social problems in the country.

16. (a) H_0: The population of local jails is distributed according to 34.4% convicted males, 5.0% convicted females, 53.4% unconvicted males, and 7.2% unconvicted females.

H_a: The population of local jails is not distributed according to 34.4% convicted males, 5.0% convicted females, 53.4% unconvicted males, and 7.2% unconvicted females.

This is a χ^2 goodness of fit test. We are given a random sample. With a sample of size $81 + 5 + 110 + 4 = 200$, all expected cells are at least 5: 34.4% of 200 = 68.8, 5.0% of 200 = 10, 53.4% of 200 = 106.8, and 7.2% of 200 = 14.4.

A calculator gives $\chi^2 = \sum \frac{(\text{obs} - \text{exp})^2}{\text{exp}} = 12.27$. With $df = 4 - 1 = 3$, $P = 0.0065$.

With this small a P-value, $0.0065 < 0.05$, there is sufficient evidence to reject H_0. In other words, there is sufficient evidence that the population distribution of local jails has changed from the previous study's finding of 34.4% convicted males, 5.0% convicted females, 53.4% unconvicted males, and 7.2% unconvicted females.

(b) There are many possible ways of proportionately choosing the 500 inmates to study. We should stratify and randomly pick 44% of 500 = 220 inmates from small county jails and 56% of 500 = 280 inmates from large county jails. For example, we might use cluster sampling to pick 22 small county jails and 28 large county jails randomly. Then in every chosen jail, number the inmates. Using a random integer generator, pick 10 unique numbers in the appropriate range corresponding to the number of inmates in each jail. The inmates with those numbers, together from all the 50 chosen jails, will form the sample to be studied.

17. (a) We have a margin of error of ±0.03. The critical z-scores associated with 95% are ±1.96. Assuming that the proportion of interest was unknown, we use $p = 0.5$ as a "worst case," which leads to $1.96\sqrt{\frac{0.5(1-0.5)}{n}} \leq 0.03$ or $1.96\frac{0.5}{\sqrt{n}} \leq 0.03$, $\sqrt{n} \geq 32.667$, $n \geq 1067.11$. So the sample size should have been at least 1068.

(b) Compare the ASD Case distribution and the Control distribution for total cumulative immunogens from birth to three months. Both distributions are bimodal. Their centers and spreads are roughly the same.

(c) Compare the ASD Case distribution and the Control distribution for the maximum number of immunogens in one day between birth and two years. Both distributions are skewed left. Their centers and spreads are roughly the same.

(d) With a very large P-value, there is little or no evidence to reject the null hypothesis. In other words, there is little or no evidence of an association between autism and vaccines.

18. (a) There is a positive, linear, moderate to strong relationship between Temperature Anomaly and Time.

(b) Use a hypothesis test for a difference between two proportions. There are two independent random samples (given).

$n_{\text{Ind}}\hat{p}_{\text{Ind}} = 1319 \geq 10$, $n_{\text{Ind}}\hat{q}_{\text{Ind}} = 710 \geq 10$

$n_{\text{Chi}}\hat{p}_{\text{Chi}} = 436 \geq 10$, $n_{\text{Chi}}\hat{q}_{\text{Chi}} = 1744 \geq 10$

$H_0\text{: } p_{\text{Ind}} = p_{\text{Chi}}$, $H_a\text{: } p_{\text{Ind}} \neq p_{\text{Chi}}$

$z = 29.59$, $P = 0.000$

With this small a P-value, $0.000 < 0.05$, there is sufficient evidence to reject H_0. In other words, there is sufficient evidence of a difference between the proportion of Indians who express a great deal of worry about global warming and the proportion of Chinese who express a great deal of worry about global warming.

(c) Phrases like "climate change deniers," "corporate lobbyists," and "a dying planet" are clearly nonneutral and leading. The calculated proportion of people who express a great deal of worry about global warming will most likely be greater than the true proportion.

> Remember, this book comes with online access to three additional full-length practice tests. These online tests can be accessed at the following:
>
> *http://bit.ly/Barrons-AP-Statistics*
>
> Be sure to have your copy of *AP Statistics*, 10th edition on hand to complete the registration process,

ANSWER SHEET
Practice Test 1

1. Ⓐ Ⓑ Ⓒ Ⓓ Ⓔ
2. Ⓐ Ⓑ Ⓒ Ⓓ Ⓔ
3. Ⓐ Ⓑ Ⓒ Ⓓ Ⓔ
4. Ⓐ Ⓑ Ⓒ Ⓓ Ⓔ
5. Ⓐ Ⓑ Ⓒ Ⓓ Ⓔ
6. Ⓐ Ⓑ Ⓒ Ⓓ Ⓔ
7. Ⓐ Ⓑ Ⓒ Ⓓ Ⓔ
8. Ⓐ Ⓑ Ⓒ Ⓓ Ⓔ
9. Ⓐ Ⓑ Ⓒ Ⓓ Ⓔ
10. Ⓐ Ⓑ Ⓒ Ⓓ Ⓔ

11. Ⓐ Ⓑ Ⓒ Ⓓ Ⓔ
12. Ⓐ Ⓑ Ⓒ Ⓓ Ⓔ
13. Ⓐ Ⓑ Ⓒ Ⓓ Ⓔ
14. Ⓐ Ⓑ Ⓒ Ⓓ Ⓔ
15. Ⓐ Ⓑ Ⓒ Ⓓ Ⓔ
16. Ⓐ Ⓑ Ⓒ Ⓓ Ⓔ
17. Ⓐ Ⓑ Ⓒ Ⓓ Ⓔ
18. Ⓐ Ⓑ Ⓒ Ⓓ Ⓔ
19. Ⓐ Ⓑ Ⓒ Ⓓ Ⓔ
20. Ⓐ Ⓑ Ⓒ Ⓓ Ⓔ

21. Ⓐ Ⓑ Ⓒ Ⓓ Ⓔ
22. Ⓐ Ⓑ Ⓒ Ⓓ Ⓔ
23. Ⓐ Ⓑ Ⓒ Ⓓ Ⓔ
24. Ⓐ Ⓑ Ⓒ Ⓓ Ⓔ
25. Ⓐ Ⓑ Ⓒ Ⓓ Ⓔ
26. Ⓐ Ⓑ Ⓒ Ⓓ Ⓔ
27. Ⓐ Ⓑ Ⓒ Ⓓ Ⓔ
28. Ⓐ Ⓑ Ⓒ Ⓓ Ⓔ
29. Ⓐ Ⓑ Ⓒ Ⓓ Ⓔ
30. Ⓐ Ⓑ Ⓒ Ⓓ Ⓔ

31. Ⓐ Ⓑ Ⓒ Ⓓ Ⓔ
32. Ⓐ Ⓑ Ⓒ Ⓓ Ⓔ
33. Ⓐ Ⓑ Ⓒ Ⓓ Ⓔ
34. Ⓐ Ⓑ Ⓒ Ⓓ Ⓔ
35. Ⓐ Ⓑ Ⓒ Ⓓ Ⓔ
36. Ⓐ Ⓑ Ⓒ Ⓓ Ⓔ
37. Ⓐ Ⓑ Ⓒ Ⓓ Ⓔ
38. Ⓐ Ⓑ Ⓒ Ⓓ Ⓔ
39. Ⓐ Ⓑ Ⓒ Ⓓ Ⓔ
40. Ⓐ Ⓑ Ⓒ Ⓓ Ⓔ

Practice Test 1

SECTION I

Questions 1–40

Spend 90 minutes on this part of the exam.

> **Directions:** The questions or incomplete statements that follow are each followed by five suggested answers or completions. Choose the response that best answers the question or completes the statement.

1. Following is a histogram of 87 home sale prices (in thousands of dollars) in one community:

   Which of the following could be the median of the sale prices?

   (A) $162,500
   (B) $175,000
   (C) $187,500
   (D) $212,500
   (E) $237,500

2. Which of the following is most useful in establishing cause-and-effect relationships?

   (A) A complete census
   (B) A least squares regression line showing high correlation
   (C) A simple random sample (SRS)
   (D) A well-designed, well-conducted survey incorporating chance to ensure a representative sample
   (E) An experiment

3. The average yearly snowfall in a city is 55 inches. What is the standard deviation if 15% of the years have snowfalls above 60 inches? Assume yearly snowfalls are normally distributed.

(A) $\dfrac{60-55}{0.15}$

(B) $\dfrac{60-55}{0.35}$

(C) $\dfrac{60-55}{0.385}$

(D) $\dfrac{60-55}{0.85}$

(E) $\dfrac{60-55}{1.036}$

4. To determine the average cost of running for a congressional seat, a simple random sample of 50 politicians is chosen and the politicians' records examined. The cost figures show a mean of $125,000 with a standard deviation of $32,000. Which of the following is the best interpretation of a 90% confidence interval estimate for the average cost of running for office?

(A) 90% of politicians running for a congressional seat spend between $117,500 and $132,500.
(B) 90% of politicians running for a congressional seat spend a mean dollar amount that is between $117,500 and $132,500.
(C) We are 90% confident that politicians running for a congressional seat spend between $117,500 and $132,500.
(D) We are 90% confident that politicians running for a congressional seat spend a mean dollar amount between $117,500 and $132,500.
(E) We are 90% confident that in the chosen sample, the mean dollar amount spent running for a congressional seat is between $117,500 and $132,500.

5. In one study on the effect that eating meat products has on weight level, an SRS of 500 subjects who admitted to eating meat at least once a day had their weights compared with those of an independent SRS of 500 people who claimed to be vegetarians. In a second study, an SRS of 500 subjects were served at least one meat meal per day for 6 months, while an independent SRS of 500 others were chosen to receive a strictly vegetarian diet for 6 months, with weights compared after 6 months.

(A) The first study is a controlled experiment, while the second is an observational study.
(B) The first study is an observational study, while the second is a controlled experiment.
(C) Both studies are controlled experiments.
(D) Both studies are observational studies.
(E) Each study is part controlled experiment and part observational study.

6. A plumbing contractor obtains 60% of her boiler circulators from a company whose defect rate is 0.005, and the rest from a company whose defect rate is 0.010. If a circulator is defective, what is the probability that it came from the first company?

(A) $(0.6)(0.005)$

(B) $(0.6)(0.005) + (0.4)(0.010)$

(C) $\dfrac{0.6}{(0.6)(0.005)+(0.4)(0.010)}$

(D) $\dfrac{0.005}{(0.6)(0.005)+(0.4)(0.010)}$

(E) $\dfrac{(0.6)(0.005)}{(0.6)(0.005)+(0.4)(0.010)}$

7. A kidney dialysis center periodically checks a sample of its equipment and performs a major recalibration if readings are sufficiently off target. Similarly, a fabric factory periodically checks the sizes of towels coming off an assembly line and halts production if measurements are sufficiently off target. In both situations, we have the null hypothesis that the equipment is performing satisfactorily. For each situation, which is the *more* serious concern, a Type I or Type II error?

(A) Dialysis center: Type I error, towel manufacturer: Type I error
(B) Dialysis center: Type I error, towel manufacturer: Type II error
(C) Dialysis center: Type II error, towel manufacturer: Type I error
(D) Dialysis center: Type II error, towel manufacturer: Type II error
(E) This is impossible to answer without making an expected value judgment between human life and accurate towel sizes.

8. A coin is weighted so that the probability of heads is 0.75. The coin is tossed 10 times and the number of heads is noted. This procedure is repeated a total of 50 times, and the number of heads is recorded each time. What kind of distribution has been simulated?

(A) The sampling distribution of the sample proportion with $n = 10$ and $p = 0.75$
(B) The sampling distribution of the sample proportion with $n = 50$ and $p = 0.75$
(C) The sampling distribution of the sample mean with $\bar{x} = 10(0.75)$ and
  $\sigma = \sqrt{10(0.75)(0.25)}$
(D) The binomial distribution with $n = 10$ and $p = 0.75$
(E) The binomial distribution with $n = 50$ and $p = 0.75$

9. During the years 1886 through 2000 there were an average of 8.7 tropical cyclones per year, of which an average of 5.1 became hurricanes. Assuming that the probability of any cyclone becoming a hurricane is independent of what happens to any other cyclone, if there are five cyclones in one year, what is the probability that at least three become hurricanes?

(A) 0.313
(B) 0.345
(C) 0.586
(D) 0.658
(E) 0.686

10. Given the probabilities $P(A) = 0.3$ and $P(B) = 0.2$, what is the probability of the union $P(A \cup B)$ if A and B are mutually exclusive? If A and B are independent? If B is a subset of A?

   (A) 0.44, 0.5, 0.2
   (B) .44, 0.5, 0.3
   (C) .5, 0.44, 0.2
   (D) .5, 0.44, 0.3
   (E) 0, 0.5, 0.3

11. A nursery owner claims that a recent drought stunted the growth of 22% of all her evergreens. A botanist tests this claim by examining an SRS of 1100 evergreens and finds that 268 trees in the sample show signs of stunted growth. With $H_0: p = 0.22$ and $H_a: p \neq 0.22$, what is the value of the test statistic?

   (A) $z = \dfrac{0.244 - 0.22}{\sqrt{1100(0.22)(1 - 0.22)}}$

   (B) $z = 2\dfrac{0.244 - 0.22}{\sqrt{1100(0.22)(1 - 0.22)}}$

   (C) $z = \dfrac{0.244 - 0.22}{\sqrt{\dfrac{(0.22)(1 - 0.22)}{1100}}}$

   (D) $z = 2\dfrac{0.244 - 0.22}{\sqrt{\dfrac{(0.22)(1 - 0.22)}{1100}}}$

   (E) $z = \dfrac{0.244 - 0.22}{\sqrt{\dfrac{(0.244)(1 - 0.244)}{1100}}}$

12. A medical research team tests for tumor reduction in a sample of patients using three different dosages of an experimental cancer drug. Which of the following is true?

   (A) There are three explanatory variables and one response variable.
   (B) There is one explanatory variable with three levels of response.
   (C) Tumor reduction is the only explanatory variable, but there are three response variables corresponding to the different dosages.
   (D) There are three levels of a single explanatory variable.
   (E) Each explanatory level has an associated level of response.

13. The graph below shows cumulative proportion plotted against age for a population.

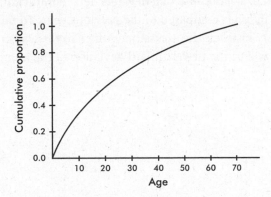

The median is approximately what age?

(A) 17
(B) 30
(C) 35
(D) 40
(E) 45

14. Mr. Bee's statistics class had a standard deviation of 11.2 on a standardized test, while Mr. Em's class had a standard deviation of 5.6 on the same test. Which of the following is the most reasonable conclusion concerning the two classes' performance on the test?

(A) Mr. Bee's class is less heterogeneous than Mr. Em's.
(B) Mr. Em's class is more homogeneous than Mr. Bee's.
(C) Mr. Bee's class performed twice as well as Mr. Em's.
(D) Mr. Em's class did not do as well as Mr. Bee's.
(E) Mr. Bee's class had the higher mean, but this may not be statistically significant.

15. A college admissions officer is interested in comparing the SAT math scores of high school applicants who have and have not taken AP Statistics. She randomly pulls the files of five applicants who took AP Statistics and five applicants who did not, and proceeds to run a t-test to compare the mean SAT math scores of the two groups. Which of the following is a necessary assumption?

(A) The population variances from each group are known.
(B) The population variances from each group are unknown.
(C) The population variances from the two groups are equal.
(D) The population of all SAT scores from each group is roughly normally distributed.
(E) The samples must be independent simple random samples, and for each sample np and $n(1 - p)$ must both be at least 10.

16. The appraised values of houses in a city have a mean of $125,000 with a standard deviation of $23,000. Because of a new teachers' contract, the school district needs an extra 10% in funds compared to the previous year. To raise this additional money, the city instructs the assessment office to raise all appraised house values by $5,000. What will be the new standard deviation of the appraised values of houses in the city?

(A) $23,000
(B) $25,300
(C) $28,000
(D) $30,300
(E) $30,800

17. A psychologist hypothesizes that scores on an aptitude test are normally distributed with a mean of 70 and a standard deviation of 10. In a random sample of size 100, scores are distributed as in the table below. What is the χ^2 statistic for a goodness-of-fit test?

Score:	below 60	60–70	70–80	above 80
Number of people:	10	40	35	15

(A) $\dfrac{(10-16)^2}{10} + \dfrac{(40-34)^2}{40} + \dfrac{(35-34)^2}{35} + \dfrac{(15-16)^2}{15}$

(B) $\dfrac{(10-16)^2}{16} + \dfrac{(40-34)^2}{34} + \dfrac{(35-34)^2}{34} + \dfrac{(15-16)^2}{16}$

(C) $\dfrac{(10-25)^2}{25} + \dfrac{(40-25)^2}{25} + \dfrac{(35-25)^2}{25} + \dfrac{(15-25)^2}{25}$

(D) $\dfrac{(10-25)^2}{10} + \dfrac{(40-25)^2}{40} + \dfrac{(35-25)^2}{35} + \dfrac{(15-25)^2}{15}$

(E) $\dfrac{(10-25)^2}{16} + \dfrac{(40-25)^2}{34} + \dfrac{(35-25)^2}{34} + \dfrac{(15-25)^2}{16}$

18. A talk show host recently reported that in response to his on-air question, 82% of the more than 2500 e-mail messages received through his publicized address supported the death penalty for anyone convicted of selling drugs to children. What does this show?

(A) The survey is meaningless because of voluntary response bias.
(B) No meaningful conclusion is possible without knowing something more about the characteristics of his listeners.
(C) The survey would have been more meaningful if he had picked a random sample of the 2500 listeners who responded.
(D) The survey would have been more meaningful if he had used a control group.
(E) This was a legitimate sample, randomly drawn from his listeners, and of sufficient size to be able to conclude that most of his listeners support the death penalty for such a crime.

19. Define a new measurement as the difference between the 60th and 40th percentile scores in a population. This measurement will give information concerning

  (A) central tendency.
  (B) variability.
  (C) symmetry.
  (D) skewness.
  (E) clustering.

20. Suppose X and Y are random variables with $E(X) = 37$, var$(X) = 5$, $E(Y) = 62$, and var$(Y) = 12$. What are the expected value and variance of the random variable $X + Y$?

  (A) $E(X + Y) = 99$, var$(X + Y) = 8.5$
  (B) $E(X + Y) = 99$, var$(X + Y) = 13$
  (C) $E(X + Y) = 99$, var$(X + Y) = 17$
  (D) $E(X + Y) = 49.5$, var$(X + Y) = 17$
  (E) There is insufficient information to answer this question.

21. Which of the following can affect the value of the correlation r?

  (A) A change in measurement units
  (B) A change in which variable is called x and which is called y
  (C) Adding the same constant to all values of the x-variable
  (D) All of the above can affect the r value.
  (E) None of the above can affect the r value.

22. The American Medical Association (AMA) wishes to determine the percentage of obstetricians who are considering leaving the profession because of the rapidly increasing number of lawsuits against obstetricians. The AMA would like an answer to within ±3% at the 95% confidence level. Which of the following should be used to find the sample size (n) needed?

  (A) $1.645\sqrt{\dfrac{0.5}{n}} \leq 0.03$

  (B) $1.645\sqrt{\dfrac{0.5}{n}} \leq 0.06$

  (C) $1.645\sqrt{\dfrac{(0.5)(0.5)}{n}} \leq 0.06$

  (D) $1.96\sqrt{\dfrac{0.5}{n}} \leq 0.03$

  (E) $1.96\sqrt{\dfrac{(0.5)(0.5)}{n}} \leq 0.03$

23. Suppose you did 10 independent tests of the form $H_0: \mu = 25$ versus $H_a: \mu < 25$, each at the $\alpha = 0.05$ significance level. What is the probability of committing a Type I error and incorrectly rejecting a true H_0 with at least one of the 10 tests?

   (A) 0.05
   (B) 0.40
   (C) 0.50
   (D) 0.60
   (E) 0.95

24. Which of the following sets has the smallest standard deviation? Which has the largest?

   I. 1, 2, 3, 4, 5, 6, 7
   II. 1, 1, 1, 4, 7, 7, 7
   III. 1, 4, 4, 4, 4, 4, 7

   (A) I, II
   (B) II, III
   (C) III, I
   (D) II, I
   (E) III, II

25. A researcher plans a study to examine long-term confidence in the U.S. economy among the adult population. She obtains a simple random sample of 30 adults as they leave a Wall Street office building one weekday afternoon. All but two of the adults agree to participate in the survey. Which of the following conclusions is correct?

   (A) Proper use of chance as evidenced by the simple random sample makes this a well-designed survey.
   (B) The high response rate makes this a well-designed survey.
   (C) Selection bias makes this a poorly designed survey.
   (D) A voluntary response study like this gives too much emphasis to persons with strong opinions.
   (E) Lack of anonymity makes this a poorly designed survey.

26. Which among the following would result in the narrowest confidence interval?

   (A) Small sample size and 95% confidence
   (B) Small sample size and 99% confidence
   (C) Large sample size and 95% confidence
   (D) Large sample size and 99% confidence
   (E) This cannot be answered without knowing an appropriate standard deviation.

27. Consider the following three scatterplots:

What is the relationship among r_1, r_2, and r_3, the correlations associated with the first, second, and third scatterplots, respectively?

(A) $r_1 < r_2 < r_3$
(B) $r_1 < r_3 < r_2$
(C) $r_2 < r_3 < r_1$
(D) $r_3 < r_1 < r_2$
(E) $r_3 < r_2 < r_1$

28. There are two games involving flipping a fair coin. In the first game you win a prize if you can throw between 45% and 55% heads. In the second game you win if you can throw more than 80% heads. For each game would you rather flip the coin 30 times or 300 times?

(A) 30 times for each game
(B) 300 times for each game
(C) 30 times for the first game and 300 times for the second
(D) 300 times for the first game and 30 times for the second
(E) The outcomes of the games do not depend on the number of flips.

29. City planners are trying to decide among various parking plan options ranging from more on-street spaces to multilevel facilities to spread-out small lots. Before making a decision, they wish to test the downtown merchants' claim that shoppers park for an average of only 47 minutes in the downtown area. The planners have decided to tabulate parking durations for 225 shoppers and to reject the merchants' claim if the sample mean exceeds 50 minutes. If the merchants' claim is wrong and the true mean is 51 minutes, what is the probability that the random sample will lead to a mistaken failure to reject the merchants' claim? Assume that the standard deviation in parking durations is 27 minutes.

(A) $P\left(z < \dfrac{50 - 47}{27/\sqrt{225}}\right)$

(B) $P\left(z > \dfrac{50 - 47}{27/\sqrt{225}}\right)$

(C) $P\left(z < \dfrac{50 - 51}{27/\sqrt{225}}\right)$

(D) $P\left(z > \dfrac{50 - 51}{27/\sqrt{225}}\right)$

(E) $P\left(z > \dfrac{51 - 47}{27/\sqrt{225}}\right)$

30. Consider the following parallel boxplots indicating the starting salaries (in thousands of dollars) for blue collar and white collar workers at a particular production plant:

Which of the following is a correct conclusion?

(A) The ranges of the distributions are the same.
(B) In each distribution the mean is equal to the median.
(C) Each distribution is symmetric.
(D) Each distribution is roughly normal.
(E) The distributions are outliers of each other.

31. The mean Law School Aptitude Test (LSAT) score for applicants to a particular law school is 650 with a standard deviation of 45. Suppose that only applicants with scores above 700 are considered. What percentage of the applicants considered have scores below 740? (Assume the scores are normally distributed.)

(A) 13.3%
(B) 17.1%
(C) 82.9%
(D) 86.7%
(E) 97.7%

32. If all the other variables remain constant, which of the following will increase the power of a hypothesis test?

  I. Increasing the sample size.
 II. Increasing the significance level.
III. Increasing the probability of a Type II error.

(A) I only
(B) II only
(C) III only
(D) I and II only
(E) All are true.

33. A researcher planning a survey of school principals in a particular state has lists of the school principals employed in each of the 125 school districts. The procedure is to obtain a random sample of principals from each of the districts rather than grouping all the lists together and obtaining a sample from the entire group. Which of the following is a correct conclusion?

   (A) This is a simple random sample obtained in an easier and less costly manner than procedures involving sampling from the entire population of principals.
   (B) This is a cluster sample in which the population was divided into heterogeneous groups called clusters.
   (C) This is an example of systematic sampling, which gives a reasonable sample as long as the original order of the list is not related to the variables under consideration.
   (D) This is an example of proportional sampling based on sizes of the school districts.
   (E) This is a stratified sample, which may give comparative information that a simple random sample wouldn't give.

34. A simple random sample is defined by

   (A) the method of selection.
   (B) examination of the outcome.
   (C) both of the above.
   (D) how representative the sample is of the population.
   (E) the size of the sample versus the size of the population.

35. Changing from a 90% confidence interval estimate for a population proportion to a 99% confidence interval estimate, with all other things being equal,

   (A) increases the interval size by 9%.
   (B) decreases the interval size by 9%.
   (C) increases the interval size by 57%.
   (D) decreases the interval size by 57%.
   (E) This question cannot be answered without knowing the sample size.

36. Which of the following is a true statement about hypothesis testing?

   (A) If there is sufficient evidence to reject a null hypothesis at the 10% level, then there is sufficient evidence to reject it at the 5% level.
   (B) Whether to use a one- or a two-sided test is typically decided after the data are gathered.
   (C) If a hypothesis test is conducted at the 1% level, there is a 1% chance of rejecting the null hypothesis.
   (D) The probability of a Type I error plus the probability of a Type II error always equal 1.
   (E) The power of a test concerns its ability to detect an alternative hypothesis.

37. A population is normally distributed with mean 25. Consider all samples of size 10. The variable $\dfrac{\bar{x} - 25}{\frac{s}{\sqrt{10}}}$

(A) has a normal distribution.
(B) has a t-distribution with $df = 10$.
(C) has a t-distribution with $df = 9$.
(D) has neither a normal distribution or a t-distribution.
(E) has either a normal or a t-distribution depending on the characteristics of the population standard deviation.

38. A correlation of 0.6 indicates that the percentage of variation in y that is explained by the variation in x is how many times the percentage indicated by a correlation of 0.3?

(A) 2
(B) 3
(C) 4
(D) 6
(E) There is insufficient information to answer this question.

39. As reported on CNN, in a May 1999 national poll 43% of high school students expressed fear about going to school. Which of the following best describes what is meant by the poll having a margin of error of 5%?

(A) It is likely that the true proportion of high school students afraid to go to school is between 38% and 48%.
(B) Five percent of the students refused to participate in the poll.
(C) Between 38% and 48% of those surveyed expressed fear about going to school.
(D) There is a 0.05 probability that the 43% result is in error.
(E) If similar size polls were repeatedly taken, they would be wrong about 5% of the time.

40. In a high school of 1650 students, 132 have personal investments in the stock market. To estimate the total stock investment by students in this school, two plans are proposed. Plan I would sample 30 students at random, find a confidence interval estimate of their average investment, and then multiply both ends of this interval by 1650 to get an interval estimate of the total investment. Plan II would sample 30 students at random from among the 132 who have investments in the market, find a confidence interval estimate of their average investment, and then multiply both ends of this interval by 132 to get an interval estimate of the total investment. Which is the better plan for estimating the total stock market investment by students in this school?

(A) Plan I
(B) Plan II
(C) Both plans use random samples and so will produce equivalent results.
(D) Neither plan will give an accurate estimate.
(E) The resulting data must be seen to evaluate which is the better plan.

STOP

If there is still time remaining, you may review your answers.

SECTION II

Part A

QUESTIONS 1–5

Spend about 65 minutes on this part of the exam.
Percentage of Section II grade—75

> **Directions:** You must show all work and indicate the methods you use. You will be graded on the correctness of your methods and on the accuracy of your results and explanations.

1. Cumulative frequency graphs of the heights (in inches) of athletes playing college baseball (A), basketball (B), and football (C) are given below:

Write a few sentences comparing the distributions of heights of athletes playing the three college sports.

2. When students do not understand an author's vocabulary, there is no way they can fully understand the text; thus, learning vocabulary is recognized as one of the most important skills in all subject areas. A school system is planning a study to see which of two programs is more effective in ninth grade. Program A involves teaching students to apply morphemic analysis, the process of deriving a word's meaning by analyzing its meaningful parts, such as roots, prefixes, and suffixes. Program B involves teaching students to apply contextual analysis, the process of inferring the meaning of an unfamiliar word by examining the surrounding text.

   (a) Explain the purpose of incorporating a control group in this experiment.

   (b) To comply with informed consent laws, parents will be asked their consent for their children to be randomly assigned to be taught in one of the new programs or by the local standard method. Students of parents who do not return the consent forms will continue to be taught by the local standard method. Explain why these students should not be considered part of the control group.

(c) Given 90 students randomly selected from ninth grade students with parental consent, explain how to assign them to the three groups, Program A, Program B, and Control, for a completely randomized design.

3. An experiment is designed to test a home insulation product. Two side-by-side identical houses are chosen, and in one the insulation product is installed (blown into the attic). Both homes use natural gas for heating, and the thermostat in each house is fixed at 68 degrees Fahrenheit. On 30 randomly chosen winter days, the quantity of gas (in hundreds of cubic feet) used to heat each house is measured. The table below summarizes the data.

	Number	Mean (in 100 cu ft)	Standard Deviation (in 100 cu ft)
Without insulation product	30	7.112	1.579
With insulation product	30	5.555	1.198
Daily differences	30	1.557	1.331

(a) Is there evidence that the insulation product does result in a lower mean quantity of gas being used?

Two least square regression calculations are performed with the following computer output, where Temp is the average outside temperature, and WithOut and With are the quantities of gas (in hundreds of cubic feet) used to heat the two homes on a day with that outside temperature:

```
Regression Analysis: WithOut versus Temp
Predictor      Coef    SE Coef         T          P
Constant    10.0228     0.4009     25.00      0.000
Temp       -0.14926    0.01879     -7.94      0.000
S = 0.890921   R-Sq = 69.3%   R-Sq(adj) = 68.2%


Regression Analysis: With versus Temp
Predictor      Coef    SE Coef         T          P
Constant     7.5850     0.3531     21.48      0.000
Temp       -0.10408    0.01655     -6.29      0.000
S = 0.784653   R-Sq = 58.5%   R-Sq(adj) = 57.1%
```

(b) Compare the slopes of the regression lines in context of this problem.

(c) Compare the y-intercepts of the regression lines in context of this problem.

4. A new anti-spam software program is field tested on 1000 e-mails and the results are summarized in the following table (positive test = program labels e-mail as spam):

	Spam	Legitimate	
Positive test	205	90	295
Negative test	45	660	705
	250	750	1000

Using the above empirical results, determine the following probabilities.

(a) (i) What is the *predictive value* of the test? That is, what is the probability that an e-mail is spam and the test is positive?

 (ii) What is the *false-positive* rate? That is, what is the probability of testing positive given that the e-mail is legitimate?

(b) (i) What is the *sensitivity* of the test? That is, what is the probability of testing positive given that the e-mail is spam?

 (ii) What is the *specificity* of the test? That is, what is the probability of testing negative given that the e-mail is legitimate?

(c) (i) Given a random sample of five legitimate e-mails, what is the probability that the program labels at least one as spam?

 (ii) Given a random sample of five spam e-mails, what is the probability that the program correctly labels at least three as spam?

(d) (i) If 35% of the incoming e-mails from one source are spam, what is the probability that an e-mail from that source will be labeled as spam?

 (ii) If an e-mail from this source is labeled as spam, what is the probability it really is spam?

5. Cumulative exposure to nitrogen dioxide (NO_2) is a major risk factor for lung disease in tunnel construction workers. In a 2004 study, researchers compared cumulative exposure to NO_2 for a random sample of drill and blast workers and for an independent random sample of concrete workers. Summary statistics are shown below.

Tunnel worker activity	Sample size	Mean cumulative NO_2 exposure (ppm*/yr)	Standard deviation in cumulative NO_2 exposure (ppm*/yr)
Drill and blast	115	4.1	1.8
Concrete	69	4.8	2.4

*Parts per million

(a) Find a 95% confidence interval estimate for the difference in mean cumulative NO_2 exposure for drill and blast workers and concrete workers.

(b) Using this confidence interval, is there evidence of a difference in mean cumulative NO_2 exposure for drill and blast workers and concrete workers? Explain.

(c) It turns out that both data sets were somewhat skewed. Does this invalidate your analysis? Explain.

SECTION II

Part B

QUESTION 6

Spend about 25 minutes on this part of the exam.

Percentage of Section II grade—25

6. Do strong magnetic fields have an effect on the early development of mice? A study is performed on a random sample of 120 one-month-old mice. Each is subjected to a strong magnetic field for one week and no magnetic field for one week, the order being decided randomly for each mouse. The weight gain in grams during each week for each mouse is recorded. The weight gains with no magnetic field are approximately normally distributed with a mean of 25.1 grams and a standard deviation of 3.1 grams, while the weight gains in the magnetic field are approximately normally distributed with a mean of 17.3 grams and a standard deviation of 5.7 grams. Of the 120 mice, 101 gained more weight with no magnetic field.

   (a) Calculate and interpret a 95% confidence interval for the proportion of mice that gain more weight with no magnetic field.

   (b) Suppose two of these mice were chosen at random. What is the probability that the first mouse gained more weight with no magnetic field than the second mouse gained in the magnetic field?

   (c) Based on your answers to (a) and (b), is there evidence that for mice, weight gain with no magnetic field and in a magnetic field are not independent?

   (d) In a test for variances on whether or not the variances of the two sets of weight gains are significantly different, $H_0 : \sigma_X^2 = \sigma_Y^2$, $H_a : \sigma_X^2 \neq \sigma_Y^2$, with $\alpha = 0.05$, for samples of this size, the critical cut-off scores for $\dfrac{S_Y^2}{S_X^2}$ are 0.698 (left tail of 0.025) and 1.43 (right tail of 0.025). What is the resulting conclusion about variances?

STOP

If there is still time remaining, you may review your answers.

Answer Key

SECTION I

1. **D**	9. **D**	17. **B**	25. **C**	33. **E**
2. **E**	10. **D**	18. **A**	26. **C**	34. **A**
3. **E**	11. **C**	19. **B**	27. **D**	35. **C**
4. **D**	12. **D**	20. **E**	28. **D**	36. **E**
5. **B**	13. **A**	21. **E**	29. **C**	37. **C**
6. **E**	14. **B**	22. **E**	30. **A**	38. **C**
7. **C**	15. **D**	23. **B**	31. **C**	39. **A**
8. **D**	16. **A**	24. **E**	32. **D**	40. **B**

Answers Explained

SECTION I

1. **(D)** With 87 home sale prices, the median is between the 43rd and 44th values when placed in either ascending or descending order. There appear to be around 40 values between $225,000 and $250,000 and 25 more values between $200,000 and $225,000. So both the 43rd and 44th values are in the 200 to 225 interval in the histogram.

2. **(E)** Regression lines show association, not causation. Surveys suggest relationships, which experiments can help to show to be cause and effect.

3. **(E)** The critical z-score is 1.036. Thus $60 - 55 = 1.036\sigma$ and $\sigma = \dfrac{60-55}{1.036}$.

4. **(D)** We are 90% confident that the population mean is within the interval calculated using the data from the sample.

5. **(B)** The first study was observational because the subjects were not chosen for treatment.

6. **(E)**

$$P(\text{def}) = P(\text{1st} \cap \text{def}) + P(\text{2nd} \cap \text{def})$$
$$= (0.6)(0.005) + (0.4)(0.010)$$
$$P\left(\text{1st} \,|\, \text{def}\right) = \frac{P(\text{1st} \cap \text{def})}{P(\text{def})} = \frac{(0.6)(0.005)}{(0.6)(0.005)+(0.4)(0.010)}$$

7. **(C)** At the dialysis center the more serious concern would be a Type II error, which is that the equipment is not performing correctly, yet the check does not pick this up; while at the towel manufacturing plant the more serious concern would be a Type I error, which is that the equipment is performing correctly, yet the check causes a production halt.

8. **(D)** There are two possible outcomes (heads and tails), with the probability of heads always 0.75 (independent of what happened on the previous toss), and we are interested in the number of heads in 10 tosses. Thus, this is a binomial model with $n = 10$ and $p = 0.75$. Repeating this over and over (in this case 50 times) simulates the resulting binomial distribution.

9. **(D)** In a binomial distribution with probability of success $p = \dfrac{5.1}{8.7} = 0.586$, the probability of at least 3 successes is

$$\binom{5}{3}(0.586)^3(0.414)^2 + \binom{5}{4}(0.586)^4(0.414) + (0.586)^5 = 0.658$$

or 1-binomcdf(5, 0.586, 2) = 0.658.

10. **(D)** If A and B are mutually exclusive, then $P(A \cap B) = 0$, and so $P(A \cup B) = 0.3 + 0.2 - 0 = 0.5$. If A and B are independent, then $P(A \cap B) = P(A)P(B)$, and so $P(A \cup B) = 0.3 + 0.2 - (0.3)(0.2) = 0.44$. If B is a subset of A, then $A \cup B = A$, and so $P(A \cup B) = P(A) = 0.3$.

11. **(C)** The standard deviation of the test statistic is $\sigma_{\hat{p}} = \sqrt{\dfrac{p(1-p)}{n}}$. Since this is a two-sided test, the P-value will be twice the tail probability of the test statistic; however, the test statistic itself is not doubled.

12. **(D)** Dosage is the only explanatory variable, and it is being tested at three levels. Tumor reduction is the single response variable.

13. **(A)** The median corresponds to a cumulative proportion of 0.5.

14. **(B)** Standard deviation is a measure of variability. The less variability, the more homogeneity.

15. **(D)** Since the sample sizes are small, the samples must come from normally distributed populations. While the samples should be independent simple random samples, np and $n(1 - p)$ refer to conditions for tests involving sample proportions, not means.

16. **(A)** Adding the same constant to all values in a set will increase the mean by that constant, but will leave the standard deviation unchanged.

17. **(B)** With about 68% of the values within 1 standard deviation of the mean, the expected numbers for a normal distribution are as follows:

16	34	34	16

$$\text{Thus } \chi^2 = \sum \frac{(obs - exp)^2}{exp}$$

$$= \frac{(10 - 16)^2}{16} + \frac{(40 - 34)^2}{34} + \frac{(35 - 34)^2}{34} + \frac{(15 - 16)^2}{16}$$

18. **(A)** This is a good example of voluntary response bias, which often overrepresents strong or negative opinions. The people who chose to respond were very possibly the parents of children facing drug problems or people who had had bad experiences with drugs being sold in their neighborhoods. There is very little chance that the 2500 respondents were representative of the population. Knowing more about his listeners or taking a sample of the sample would not have helped.

19. **(B)** The range (difference between largest and smallest values), the interquartile range $(Q_3 - Q_1)$, and this difference between the 60th and 40th percentile scores all are measures of variability, or how spread out is the population or a subset of the population.

20. **(E)** Without independence we cannot determine $\operatorname{var}(X+Y)$ from the information given.

21. **(E)** The correlation coefficient r is not affected by changes in units, by which variable is called x or y, or by adding or multiplying all the values of a variable by the same constant.

22. **(E)** 1.96 is the critical z-score for a 95% confidence interval. With unknown p, we use 0.5 in our calculation of n.

23. **(B)** $P(\text{at least one Type I error}) = 1 - P(\text{no Type I errors}) = 1 - (0.95)^{10} = 0.40$

24. **(E)** Note that all three sets have the same mean and the same range. The third set has most of its values concentrated right at the mean, while the second set has most of its values concentrated far from the mean.

25. **(C)** People coming out of a Wall Street office building are a very unrepresentative sample of the adult population, especially given the question under consideration. Using chance and obtaining a high response rate will not change the selection bias and make this into a well-designed survey. This is a convenience sample, not a voluntary response sample.

26. **(C)** Larger samples (so $\dfrac{\sigma}{\sqrt{n}}$ is smaller) and less confidence (so the critical z or t is smaller) both result in smaller intervals.

27. **(D)** The third scatterplot shows perfect negative association, so $r_3 = -1$. The first scatterplot shows strong, but not perfect, negative correlation, so $-1 < r_1 < 0$. The second scatterplot shows no correlation, so $r_2 = 0$.

28. **(D)** The probability of throwing heads is 0.5. By the law of large numbers, the more times you flip the coin, the more the relative frequency tends to become closer to this probability. With fewer tosses there is more chance for wide swings in the relative frequency.

29. **(C)** $\sigma_{\bar{x}} = \dfrac{\sigma}{\sqrt{n}} = \dfrac{27}{\sqrt{225}}$. If the true mean parking duration is 51 minutes, the normal curve should be centered at 51. The critical value of 50 has a z-score of $\dfrac{50-51}{27/\sqrt{225}}$.

30. **(A)** The overall lengths (between tips of whiskers) are the same, and so the ranges are the same. Just because the min and max are equidistant from the median, and Q_1 and Q_3 are equidistant from the median, does not imply that a distribution is symmetric or that the mean and median are equal. And even if a distribution is symmetric, this does not imply that it is roughly normal. Particular values, not distributions, may be outliers.

31. **(C)** Critical z-scores are $\dfrac{700-650}{45} = 1.11$ and $\dfrac{740-650}{45} = 2$ with right tail probabilities of 0.1335 and 0.0228, respectively. The percentage below 740 given that the scores are above 700 is $\dfrac{0.1335-0.0228}{0.1335} = 82.9\%$. Or $\dfrac{\text{normalcdf}(700,740,650,45)}{\text{normalcdf}(700,1000000,650,45)} = 0.82928$.

32. **(D)** There is a different probability of Type II error for each possible correct value of the population parameter, and 1 minus this probability is the power of the test against the associated correct value. Increasing the sample size will make the standard deviations smaller and thus can decrease the probabilities of Type I and Type II errors, and

so increase the power. Increasing the significance level α (making it easier to reject the null) will lower the probability of a Type II error and thus also increase the power.

33. **(E)** This is not a simple random sample because all possible sets of the required size do not have the same chance of being picked. For example, a set of principals all from just half the school districts has no chance of being picked to be the sample. This is not a cluster sample in that there is no reason to believe that each school district resembles the population as a whole, and furthermore, there was no random sample taken of the school districts. This is not systematic sampling as the districts were not put in some order with every nth district chosen. Stratified samples are often easier and less costly to obtain and also make comparative data available. In this case responses can be compared among various districts.

34. **(A)** A simple random sample can be any size and may or may not be representative of the population. It is a method of selection in which every possible sample of the desired size has an equal chance of being selected.

35. **(C)** The critical z-scores go from ±1.645 to ±2.576, resulting in an increase in the interval size: $\frac{2.576}{1.645} = 1.57$ or an increase of 57%.

36. **(E)** If the P-value is less than 0.10, it does not follow that it is less than 0.05. Decisions such as whether a test should be one- or two-sided are made before the data are gathered. If $\alpha = 0.01$, there is a 1% chance of rejecting the null hypothesis *if* the null hypothesis is true. There is one probability of a Type I error, the significance level, while there is a different probability of a Type II error associated with each possible correct alternative, so the sum does not equal 1.

37. **(C)** $\dfrac{\bar{x} - \mu}{\frac{s}{\sqrt{n}}}$ has a t-distribution with $df = n - 1$.

38. **(C)** A correlation of 0.6 explains $(0.6)^2$ or 36% of the variation in y, while a correlation of 0.3 explains only $(0.3)^2$ or 9% of the variation in y.

39. **(A)** Using a measurement from a sample, we are never able to say *exactly* what a population proportion is; rather we always say we have a certain *confidence* that the population proportion lies in a particular *interval*. In this case that interval is 43% ± 5% or between 38% and 48%.

40. **(B)** With Plan I the expected number of students with stock investments is only 2.4 out of 30. Plan II allows an estimate to be made using a full 30 investors.

SECTION II

1. A complete answer compares shape, center, and spread, and mentions context.

   Shape: The baseball players, (A), for which the cumulative frequency plot rises steeply at first, include more shorter players, and thus the distribution is skewed to the right (toward the greater heights). The football players, (C), for which the cumulative frequency plot rises slowly at first, and then steeply toward the end, include more taller players, and thus the distribution is skewed to the left (toward the lower heights). The basketball players, (B), for which the cumulative frequency plot rises slowly at each end, and steeply in the middle, have a more bell-shaped distribution of heights.

   Center: The medians correspond to relative frequencies of 0.5. Reading across from 0.5 and then down to the *x*-axis shows the median heights to be about 63.5 inches for baseball players, about 72.5 inches for basketball players, and about 79 inches for football players. Thus, the center of the baseball height distribution is the least, and the center of the football height distribution is the greatest.

   Spread: The range of the football players is the smallest, $80 - 65 = 15$ inches, then comes the range of the baseball players, $80 - 60 = 20$ inches, and finally the range of the basketball players is the greatest, $85 - 60 = 25$ inches.

SCORING

The discussion of shape is essentially correct for correctly identifying which distribution is skewed left, skewed right, and more bell-shaped, and for giving a correct justification based on the cumulative frequency plots. The discussion of shape is partially correct for correctly identifying which distribution is skewed left, skewed right, and more bell-shaped without giving a good explanation.

The discussion of center is essentially correct for correctly noting that the baseball players have the lowest median height and the football players have the greatest median height, and giving some numerical justification. The discussion of center is partially correct for correctly noting that the baseball players have the lowest median height and the football players have the greatest median height but without giving a good explanation.

The discussion of spread is essentially correct for correctly noting that the football players have the smallest range for their heights and the basketball players have the greatest range for their heights, and giving some numerical justification. The discussion of spread is partially correct for correctly noting that the football players have the smallest range for their heights and the basketball players have the greatest range for their heights but without giving a good explanation.

4 Complete Answer   All three parts essentially correct.

3 Substantial Answer   Two parts essentially correct and one part partially correct.

2 Developing Answer   Two parts essentially correct OR one part essentially correct and one or two parts partially correct OR all three parts partially correct.

1 Minimal Answer   One part essentially correct OR two parts partially correct.

Lower a 4 to a 3, or a 3 to a 2, if context is never mentioned.

2. (a) A control group would allow the school system to compare the effectiveness of each of the new programs to the local standard method currently being used.

   (b) Parents who fail to return the consent form are a special category who may well place less priority on education. The effect of using their children may distort results, since their children could only be placed in the control group.

   (c) Assign each student a unique number 01–90. Using a random number table or a random number generator, pick numbers between 01 and 90, throwing out repeats. The students corresponding to the first 30 such numbers picked will be assigned to Program A, the next 30 picked to Program B, and the remaining to the control group.

SCORING

Part (a) is essentially correct if the purpose is given for using a control group in this study. Part (a) is partially correct if a correct explanation for the use of a control group is given, but not in context of this study.

Part (b) is essentially correct for a clear explanation in context. Part (b) is partially correct if the explanation is weak.

Part (c) is essentially correct if randomization is used correctly and the method is clear. Part (c) is partially correct if randomization is used but the method is not clearly explained.

4 Complete Answer   All three parts essentially correct.

3 Substantial Answer   Two parts essentially correct and one part partially correct.

2 Developing Answer   Two parts essentially correct OR one part essentially correct and one or two parts partially correct OR all three parts partially correct.

1 Minimal Answer   One part essentially correct OR two parts partially correct.

---

3. (a) This is a paired data test, not a two-sample test. There are four parts to a complete solution.

   Part 1: Must state a correct pair of hypotheses and define the parameter of interest.

   H_0: $\mu_d = 0$ and H_a: $\mu_d > 0$, where μ_d is the mean daily difference in quantity of gas used (in 100 cu ft) between the house without insulation and the house with insulation.

   Part 2: Must name the test and check the conditions.

   This is a paired t-test, that is, a single sample hypothesis test on the set of differences. Conditions: Random sample (given), $n = 30$ is less than 10% of all possible winter days, and $n = 30$ is sufficiently large for the CLT to apply.

   Part 3: Must find the test statistic t and the P-value.

   Calculator software (such as T-Test on the TI-84 or 1-Sample tTest on the Casio Prizm) gives $t = 6.4072$ and $P = 0.0000$.

Part 4: Must state the conclusion in context with linkage to the *P*-value.

With this small a *P*-value, $0.0000 < 0.05$, there is sufficient evidence to reject H_0, that is, there is sufficient evidence that the mean daily gas usage in a home without the insulation product is greater than the mean daily gas usage in a home with the insulation product.

(b) In the house without the insulation product an estimate for the average *increase* in quantity of gas used when the outside temperature goes *down* 1 degree Fahrenheit is 0.15 hundred cubic feet, while in the house with the insulation product an estimate for the average *increase* in quantity of gas used when the outside temperature goes *down* 1 degree Fahrenheit is 0.10 hundred cubic feet. So when the temperature goes down, the increase in gas used is less in the house with the insulation product than in the house without the insulation product.

(c) In the house without the insulation product an estimate for the average quantity of gas used when the outside temperature is 0 degrees Fahrenheit is 10.02 hundred cubic feet, while in the house with the insulation product an estimate for the average quantity of gas used when the outside temperature is 0 degrees Fahrenheit is 7.59 hundred cubic feet. So at an outside temperature of 0 degrees Fahrenheit, the average quantity of gas used is less in the house with the insulation product than in the house without the insulation product.

SCORING

In Part (a), Parts 1–2 are essentially correct for a correct statement of the hypotheses (in terms of *population* means), together with naming the test and checking the conditions, and partially correct for one of Part 1 or Part 2 completely correct.

Parts 3–4 are essentially correct for a correct calculation of both the test statistic *t* and the *P*-value, together with a correct conclusion in context linked to the *P*-value, and partially correct for one of Part 3 or Part 4 completely correct.

Part (b) is essentially correct for giving the correct values for the slopes, in context of the problem, and making a direct comparison. Part (b) is partially correct for giving the correct slopes but missing the context or the direct comparison.

Part (c) is essentially correct for giving the correct values for the *y*-intercepts, in context of the problem, and making a direct comparison. Part (c) is partially correct for giving the correct *y*-intercepts but missing the context or the direct comparison.

Count essentially correct answers as one point and partially correct answers as one-half point.

4	**Complete Answer**	Four points
3	**Substantial Answer**	Three points
2	**Developing Answer**	Two points
1	**Minimal Answer**	One point

Use a holistic approach to decide a score totaling between two numbers.

4. (a) (i) $P(spam \cap positive) = \dfrac{205}{1000} = 0.205$

(ii) $P(positive \mid legitimate) = \dfrac{90}{750} = 0.12$

(b) (i) $P(positive \mid spam) = \dfrac{205}{250} = 0.82$

(ii) $P(negative \mid legitimate) = \dfrac{660}{750} = 0.88$

(c) (i) $1 - (0.88)^5 = 0.4723$

(ii) $10(0.82)^3(0.18)^2 + 5(0.82)^4(0.18) + (0.82)^5 = 0.9563$
Or $1 - \text{binomedf}(5, 0.82, 3) = 1 - 0.0437 = 0.9563$

(d) (i) $P(positive) = P(positive \cap spam) + P(positive \cap legitimate) =$
$P(spam)\,P(positive \mid spam) + P(legitimate)\,P(positive \mid legitimate) =$
$(0.35)(0.82) + (0.65)(0.12) = 0.365$

(ii) $P(spam \mid positive) = \dfrac{P(spam \cap positive)}{P(positive)} = \dfrac{(0.35)(0.82)}{0.365} = 0.7863$

SCORING

There are two probabilities to calculate in each Part (a)–(d). Each Part is essentially correct for both probabilities correctly calculated and partially correct for one probability correctly calculated. In Parts that use results from previous parts, full credit is given for correctly using the results of the earlier Part, whether that earlier calculation was correct or not. For credit for Part (d), a correct methodology must also be shown.

Count partially correct answers as one-half an essentially correct answer.

4   Complete Answer   Four essentially correct answers.
3   Substantial Answer   Three essentially correct answers.
2   Developing Answer   Two essentially correct answers.
1   Minimal Answer   One essentially correct answer.

Use a holistic approach to decide a score totaling between two numbers.

5. (a) Identify the confidence interval and check conditions: two-sample t-interval for $\mu_{db} - \mu_c$, the difference in mean cumulative NO_2 exposure for drill and blast workers and concrete workers. We are given that we have independent random samples, 115 and 69 are less than 10% of all "drill and blast" and "concrete" workers, respectively, and we note that the sample sizes are large ($n_{db} = 115 \geq 30$ and $n_c = 69 \geq 30$).

Calculate the confidence interval:
[2-SampTInt on the TI-84 and 2-Sample tInterval on the Casio Prizm both give (−1.362, −0.0381).]

(Or with $df = \min(115-1, 69-1) = 68$, $(4.1 - 4.8) \pm 1.995 \sqrt{\dfrac{(1.8)^2}{115} + \dfrac{(2.4)^2}{69}} = -0.7 \pm 0.667$.)

Interpretation in context: We are 95% confident that the true difference in the mean cumulative NO_2 exposure for all drill and blast workers and all concrete workers is between −1.367 and −0.033 ppm/yr.

(b) Zero is not in the above 95% confidence interval, so at the $\alpha = 0.05$ significance level, there is evidence to reject $H_0: \mu_{db} - \mu_c = 0$ in favor of $H_a: \mu_{db} - \mu_c \neq 0$. That is, there is evidence of a difference in mean cumulative NO_2 exposure for drill and blast workers and for concrete workers.

(c) With sample sizes this large, the central limit theorem applies and our analysis is valid.

SCORING

Part (a) has two components. The first component, identifying the confidence interval and checking conditions, is essentially correct for naming the confidence interval procedure, noting independent random samples, and noting the large sample sizes. This component is partially correct for correctly noting two of the three points.

The second component of Part (a) is essentially correct for correct mechanics in calculating the confidence interval and for a correct (based on the shown mechanics) interpretation in context, and is partially correct for one of these two features.

Part (b) is essentially correct for noting that zero is not in the interval so the observed difference is significant and stating this in context of the problem. Part (b) is partially correct for noting that zero is not in the interval so the observed difference is significant, but failing to put this conclusion in context of the problem.

Part (c) is essentially correct for relating the central limit theorem (CLT) to the samples being large. Part (c) is partially correct for referring to the CLT or to the large samples, but not linking the two.

Count partially correct answers as one-half an essentially correct answer.

4	**Complete Answer**	Four essentially correct answers.
3	**Substantial Answer**	Three essentially correct answers.
2	**Developing Answer**	Two essentially correct answers.
1	**Minimal Answer**	One essentially correct answer.

Use a holistic approach to decide a score totaling between two numbers.

6. (a) Identify the confidence interval by name or formula.

95% confidence interval for the population proportion

$$\text{or } \hat{p} \pm 1.96\sqrt{\frac{\hat{p}(1-\hat{p})}{n}}$$

Check the conditions.

Random sample (given), $n = 120$ is less than 10% of all one-month-old mice, $n\hat{p} = 101 \geq 10$, and $n(1 - \hat{p}) = 19 \geq 10$.

Calculate the confidence interval.

Calculator software (such as 1-PropZInt on the TI-84 or 1-PropZInterval on the Casio Prizm) gives (0.77635, 0.90698).

Interpret the confidence interval in context.

We are 95% confident that between 77.6% and 90.7% of all one-month-old mice will gain more weight with no magnetic field.

(b) The distribution of the set of all possible differences of weight gains without and with magnetic fields is approximately normal, with mean $25.1 - 17.3 = 7.8$ and standard deviation $\sqrt{(3.1)^2 + (5.7)^2} = 6.49$. Then the probability that the first mouse gained more weight than the second equals

$$P\left(z > \frac{0 - 7.8}{6.49}\right) = P(z > -1.202) = 0.885$$

(c) From the answer to (b), if weight gains without or with the magnetic field are independent, we would expect approximately 88.5% of mice will gain more weight with no magnetic field. From the answer to (a) we estimate the percent of mice who gain more weight with no magnetic field to be between 77.6% and 90.7%. Since 88.5% is in this interval, there is no evidence to suggest that weight gain with no magnetic field and in a magnetic field are not independent.

(d) Calculating $\dfrac{S_Y^2}{S_X^2} = \dfrac{(3.1)^2}{(5.7)^2} = 0.296$ or $\dfrac{(5.7)^2}{(3.1)^2} = 3.38$, the test statistic is outside the critical cutoff scores of 0.698 and 1.43. Thus there *is* evidence that the variances of the distributions of weight gains without and with magnetic fields are different.

SCORING

Part (a) has four parts: 1) identifying the confidence interval; 2) checking assumptions; 3) calculating the confidence interval; and 4) interpreting the confidence interval in context. Part (a) is essentially correct if three or four of these parts are correct and partially correct if one or two of these parts are correct.

Part (b) is essentially correct for a complete answer, and partially correct for finding the mean and SD of the set of differences but incorrectly finding the probability, OR making a mistake in finding the mean or standard deviation but using these correctly in finding the probability.

Part (c) is essentially correct for a complete answer and is partially correct if the conclusion is right but the links to (a) and (b) are weak. Part (c) is incorrect if the conclusion is correct but there is no explanation.

Part (d) is essentially correct for a correct calculation of the quotient of sample variances together with a correct conclusion in context, and is partially correct for one of these two parts correct.

Count partially correct answers as one-half an essentially correct answer.

4	**Complete Answer**	Four essentially correct answers.
3	**Substantial Answer**	Three essentially correct answers.
2	**Developing Answer**	Two essentially correct answers.
1	**Minimal Answer**	One essentially correct answer.

Use a holistic approach to decide a score totaling between two numbers.

ANSWER SHEET
Practice Test 2

1. (A) (B) (C) (D) (E)   11. (A) (B) (C) (D) (E)   21. (A) (B) (C) (D) (E)   31. (A) (B) (C) (D) (E)

2. (A) (B) (C) (D) (E)   12. (A) (B) (C) (D) (E)   22. (A) (B) (C) (D) (E)   32. (A) (B) (C) (D) (E)

3. (A) (B) (C) (D) (E)   13. (A) (B) (C) (D) (E)   23. (A) (B) (C) (D) (E)   33. (A) (B) (C) (D) (E)

4. (A) (B) (C) (D) (E)   14. (A) (B) (C) (D) (E)   24. (A) (B) (C) (D) (E)   34. (A) (B) (C) (D) (E)

5. (A) (B) (C) (D) (E)   15. (A) (B) (C) (D) (E)   25. (A) (B) (C) (D) (E)   35. (A) (B) (C) (D) (E)

6. (A) (B) (C) (D) (E)   16. (A) (B) (C) (D) (E)   26. (A) (B) (C) (D) (E)   36. (A) (B) (C) (D) (E)

7. (A) (B) (C) (D) (E)   17. (A) (B) (C) (D) (E)   27. (A) (B) (C) (D) (E)   37. (A) (B) (C) (D) (E)

8. (A) (B) (C) (D) (E)   18. (A) (B) (C) (D) (E)   28. (A) (B) (C) (D) (E)   38. (A) (B) (C) (D) (E)

9. (A) (B) (C) (D) (E)   19. (A) (B) (C) (D) (E)   29. (A) (B) (C) (D) (E)   39. (A) (B) (C) (D) (E)

10. (A) (B) (C) (D) (E)   20. (A) (B) (C) (D) (E)   30. (A) (B) (C) (D) (E)   40. (A) (B) (C) (D) (E)

Practice Test 2

━━━

SECTION I

Questions 1–40

Spend 90 minutes on this part of the exam.

> **Directions:** The questions or incomplete statements that follow are each followed by five suggested answers or completions. Choose the response that best answers the question or completes the statement.

1. Suppose that the regression line for a set of data, $\hat{y} = mx + 3$, passes through the point (2, 7). If $\bar{x}$ and $\bar{y}$ are the sample means of the x- and y-values, respectively, then $\bar{y} =$

   (A) $\bar{x}$.
   (B) $\bar{x} - 2$.
   (C) $\bar{x} + 3$.
   (D) $2\bar{x} + 3$.
   (E) $3.5\bar{x} + 3$.

2. A study is made to determine whether more hours of academic studying leads to higher point scoring by basketball players. In surveying 50 basketball players, it is noted that the 25 who claim to study the most hours have a higher point average than the 25 who study less. Based on this study, the coach begins requiring the players to spend more time studying. Which of the following is a correct statement?

   (A) While this study may have its faults, it still does prove causation.
   (B) There could well be a confounding variable responsible for the seeming relationship.
   (C) While this is a controlled experiment, the conclusion of the coach is not justified.
   (D) To get the athletes to study more, it would be more meaningful to have them put in more practice time on the court to boost their point averages, as higher point averages seem to be associated with more study time.
   (E) No proper conclusion is possible without somehow introducing *blinding*.

3. The longevity of people living in a certain locality has a standard deviation of 14 years. What is the mean longevity if 30% of the people live longer than 75 years? Assume a roughly normal distribution for life spans.

(A) 75 − 0.30(14)
(B) 75 + 0.30(14)
(C) 75 − 0.4756(14)
(D) 75 + 0.5244(14)
(E) 75 − 0.5244(14)

4. Which of the following is a correct statement about correlation?

(A) If the slope of the regression line is exactly 1, then the correlation is exactly 1.
(B) If the correlation is 0, then the slope of the regression line is undefined.
(C) Switching which variable is called x and which is called y changes the sign of the correlation.
(D) The correlation r is equal to the slope of the regression line when z-scores for the y-variable are plotted against z-scores for the x-variable.
(E) Changes in the measurement units of the variables may change the correlation.

5. Which of the following are affected by outliers?

  I. Mean
 II. Median
III. Standard deviation
 IV. Range
  V. Interquartile range

(A) I, III, and V only
(B) II and IV only
(C) I and V only
(D) III and IV only
(E) I, III, and IV only

6. An engineer wishes to determine the quantity of heat being generated by a particular electronic component. She knows that the standard deviation is 2.4 and wants to be 99% sure of knowing the mean quantity to within ±0.6. Which of the following should be used to find the sample size (n) needed?

(A) $1.645\sqrt{\dfrac{2.4}{n}} \le 0.6$

(B) $1.96\left(\dfrac{2.4}{\sqrt{n}}\right) \le 0.6$

(C) $2.576\left(\dfrac{2.4}{\sqrt{n}}\right) \le 0.6$

(D) $2.326\sqrt{\dfrac{2.4}{n}} \le 0.6$

(E) $2.576\sqrt{\dfrac{2.4}{n}} \le 0.6$

7. A company that produces facial tissues continually monitors tissue strength. If the mean strength from sample data drops below a specified level, the production process is halted and the machinery inspected. Which of the following would result from a Type I error?

(A) Halting the production process when sufficient customer complaints are received.
(B) Halting the production process when the tissue strength is below specifications.
(C) Halting the production process when the tissue strength is within specifications.
(D) Allowing the production process to continue when the tissue strength is below specifications.
(E) Allowing the production process to continue when the tissue strength is within specifications.

8. Two possible wordings for a questionnaire on a proposed school budget increase are as follows:

 I. This school district has one of the highest per student expenditure rates in the state. This has resulted in low failure rates, high standardized test scores, and most students going on to good colleges and universities. Do you support the proposed school budget increase?

 II. This school district has one of the highest per student expenditure rates in the state. This has resulted in high property taxes, with many people on fixed incomes having to give up their homes because they cannot pay the school tax. Do you support the proposed school budget increase?

One of these questions showed that 58% of the population favor the proposed school budget increase, while the other question showed that only 13% of the population support the proposed increase. Which produced which result and why?

(A) The first showed 58% and the second 13% because of the lack of randomization as evidenced by the wording of the questions.
(B) The first showed 13% and the second 58% because of a placebo effect due to the wording of the questions.
(C) The first showed 58% and the second 13% because of the lack of a control group.
(D) The first showed 13% and the second 58% because of response bias due to the wording of the questions.
(E) The first showed 58% and the second 13% because of response bias due to the wording of the questions.

9. Suppose that for a certain Caribbean island in any 3-year period the probability of a major hurricane is 0.25, the probability of water damage is 0.44, and the probability of both a hurricane and water damage is 0.22. What is the probability of water damage given that there is a hurricane?

(A) 0.25 + 0.44 – 0.22

(B) $\dfrac{0.22}{0.44}$

(C) 0.25 + 0.44

(D) $\dfrac{0.22}{0.25}$

(E) 0.25 + 0.44 + 0.22

10. A union spokesperson is trying to encourage a college faculty to join the union. She would like to argue that faculty salaries are not truly based on years of service as most faculty believe. She gathers data and notes the following scatterplot of salary versus years of service.

Which of the following most correctly interprets the overall scatterplot?

(A) The faculty member with the fewest years of service makes the lowest salary, and the faculty member with the most service makes the highest salary.

(B) A faculty member with more service than another has the greater salary than the other.

(C) There is a strong positive correlation with little deviation.

(D) There is no clear relationship between salary and years of service.

(E) While there is a strong positive correlation, there is a distinct deviation from the overall pattern for faculty with fewer than ten years of service.

11. Two random samples of students are chosen, one from those taking an AP Statistics class and one from those not. The following back-to-back stemplots compare the GPAs.

AP Statistics		No AP Statistics	
	1	89	Key: 1	8 means 1.8
97653	2	015688	
98775332110	3	133344777888	
1100	4		

Which of the following is true about the ranges and standard deviations?

(A) The first set has both a greater range and a greater standard deviation.
(B) The first set has a greater range, while the second has a greater standard deviation.
(C) The first set has a greater standard deviation, while the second has a greater range.
(D) The second set has both a greater range and a greater standard deviation.
(E) The two sets have equal ranges and equal standard deviations.

12. In a group of 10 scores, the largest score is increased by 40 points. What will happen to the mean?

(A) It will remain the same.
(B) It will increase by 4 points.
(C) It will increase by 10 points.
(D) It will increase by 40 points.
(E) There is not sufficient information to answer this question.

13. Suppose X and Y are random variables with $\mu_x = 32$, $\sigma_x = 5$, $\mu_y = 44$, and $\sigma_y = 12$. Given that X and Y are independent, what are the mean and standard deviation of the random variable $X + Y$?

(A) $\mu_{x+y} = 76$, $\sigma_{x+y} = 8.5$
(B) $\mu_{x+y} = 76$, $\sigma_{x+y} = 13$
(C) $\mu_{x+y} = 76$, $\sigma_{x+y} = 17$
(D) $\mu_{x+y} = 38$, $\sigma_{x+y} = 17$
(E) There is insufficient information to answer this question.

14. Suppose you toss a fair die three times and it comes up an even number each time. Which of the following is a true statement?

   (A) By the law of large numbers, the next toss is more likely to be an odd number than another even number.

   (B) Based on the properties of conditional probability the next toss is more likely to be an even number given that three in a row have been even.

   (C) Dice actually do have memories, and thus the number that comes up on the next toss will be influenced by the previous tosses.

   (D) The law of large numbers tells how many tosses will be necessary before the percentages of evens and odds are again in balance.

   (E) The probability that the next toss will again be even is 0.5.

15. A pharmaceutical company is interested in the association between advertising expenditures and sales for various over-the-counter products. A sales associate collects data on nine products, looking at sales (in $1000) versus advertising expenditures (in $1000). The results of the regression analysis are shown below.

Dependent variable: Sales

Source	df	Sum of Squares	Mean Square	F-ratio
Regression	1	9576.1	9576.1	1118.45
Residual	7	59.9	8.6	

Variable	Coefficient	SE Coef	t-ratio	P
Constant	123.800	1.798	68.84	0.000
Advertising	12.633	0.378	33.44	0.000

R—Sq = 99.4%   R—Sq (adj) = 99.3%

s = 2.926 with 9—2 = 7 degrees of freedom

Which of the following gives a 90% confidence interval for the slope of the regression line?

   (A) $12.633 \pm 1.415(0.378)$
   (B) $12.633 \pm 1.895(0.378)$
   (C) $123.800 \pm 1.414(1.798)$
   (D) $123.800 \pm 1.895(1.798)$
   (E) $123.800 \pm 1.645(1.798/\sqrt{9})$

16. Suppose you wish to compare the AP Statistics exam results for the male and female students taking AP Statistics at your high school. Which is the most appropriate technique for gathering the needed data?

   (A) Census
   (B) Sample survey
   (C) Experiment
   (D) Observational study
   (E) None of these is appropriate.

17. Jonathan obtained a score of 80 on a statistics exam, placing him at the 90th percentile. Suppose five points are added to everyone's score. Jonathan's new score will be at the

  (A) 80th percentile.
  (B) 85th percentile.
  (C) 90th percentile.
  (D) 95th percentile.
  (E) There is not sufficient information to answer this quesiton.

18. To study the effect of music on piecework output at a clothing manufacturer, two experimental treatments are planned: day-long classical music for one group versus day-long light rock music for another. Which one of the following groups would serve best as a control for this study?

  (A) A third group for which no music is played
  (B) A third group that randomly hears either classical or light rock music each day
  (C) A third group that hears day-long R & B music
  (D) A third group that hears classical music every morning and light rock every afternoon
  (E) A third group in which each worker has earphones and chooses his or her own favorite music

19. Suppose H_0: $p = 0.6$, and the power of the test for H_a: $p = 0.7$ is 0.8. Which of the following is a valid conclusion?

  (A) The probability of committing a Type I error is 0.1.
  (B) If H_a is true, the probability of failing to reject H_0 is 0.2.
  (C) The probability of committing a Type II error is 0.3.
  (D) All of the above are valid conclusions.
  (E) None of the above are valid conclusions.

20. Following is a histogram of the numbers of ties owned by bank executives.

Which of the following is a correct statement ?

  (A) The median number of ties is five.
  (B) More than four executives own over eight ties each.
  (C) An executive is equally likely to own fewer than five ties or more than seven ties.
  (D) One tie is a reasonable estimate for the standard deviation.
  (E) Removing all the executives with three, nine, and ten ties may change the median.

21. Which of the following is a binomial random variable?

(A) The number of tosses before a "5" appears when tossing a fair die.
(B) The number of points a hockey team receives in 10 games, where two points are awarded for wins, one point for ties, and no points for losses.
(C) The number of hearts out of five cards randomly drawn from a deck of 52 cards, without replacement.
(D) The number of motorists not wearing seat belts in a random sample of five drivers.
(E) None of the above.

22. Company I manufactures demolition fuses that burn an average of 50 minutes with a standard deviation of 10 minutes, while company II advertises fuses that burn an average of 55 minutes with a standard deviation of 5 minutes. Which company's fuse is more likely to last at least 1 hour? Assume roughly normal distributions of fuse times.

(A) Company I's, because of its greater standard deviation
(B) Company II's, because of its greater mean
(C) For both companies, the probability that a fuse will last at least 1 hour is 0.159
(D) For both companies, the probability that a fuse will last at least 1 hour is 0.841
(E) The problem cannot be solved from the information given.

23. Which of the following is *not* important in the design of experiments?

(A) Control of confounding variables
(B) Randomization in assigning subjects to different treatments
(C) Use of a confounding variable to control the placebo effect
(D) Replication of the experiment using sufficient numbers of subjects
(E) All of the above are important in the design of experiments.

24. The travel miles claimed in weekly expense reports of the sales personnel at a corporation are summarized in the following boxplot.

Miles claimed

Which of the following is the most reasonable conclusion?

(A) The mean and median numbers of travel miles are roughly equal.
(B) The mean number of travel miles is greater than the median number.
(C) Most of the claimed numbers of travel miles are in the [0, 200] interval.
(D) Most of the claimed numbers of travel miles are in the [200, 240] interval.
(E) The left and right whiskers contain the same number of values from the set of personnel travel mile claims.

25. For the regression line, which of the following statements about residuals is true?

   (A) Influential scores have large residuals.
   (B) If the linear model is good, the number of positive residuals will be the same as the number of negative residuals.
   (C) The mean of the residuals is always zero.
   (D) If the correlation is 0, there will be a distinct pattern in the residual plot.
   (E) If the correlation is 1, there will not be a distinct pattern in the residual plot.

26. Four pairs of data are used in determining a regression line $\hat{y} = 3x + 4$. If the four values of the independent variable are 32, 24, 29, and 27, respectively, what is the mean of the four values of the dependent variable?

   (A) 68
   (B) 84
   (C) 88
   (D) 100
   (E) The mean cannot be determined from the given information.

27. According to one poll, 12% of the public favor legalizing all drugs. In a simple random sample of six people, what is the probability that at least one person favors legalization?

   (A) $6(0.12)(0.88)^5$
   (B) $(0.88)^6$
   (C) $1 - (0.88)^6$
   (D) $1 - 6(0.12)(0.88)^5$
   (E) $6(0.12)(0.88)^5 + (0.88)^6$

28. Sampling error occurs

   (A) when interviewers make mistakes resulting in bias.
   (B) because a sample statistic is used to estimate a population parameter.
   (C) when interviewers use judgment instead of random choice in picking the sample.
   (D) when samples are too small.
   (E) in all of the above cases.

29. A telephone executive instructs an associate to contact 104 customers using their service to obtain their opinions in regard to an idea for a new pricing package. The associate notes the number of customers whose names begin with *A* and uses a random number table to pick four of these names. She then proceeds to use the same procedure for each letter of the alphabet and combines the 4 × 26 = 104 results into a group to be contacted. Which of the following is a correct conclusion?

(A) Her procedure makes use of chance.
(B) Her procedure results in a simple random sample.
(C) Each customer has an equal probability of being included in the survey.
(D) Her procedure introduces bias through *sampling error*.
(E) With this small a sample size, it is better to let the surveyor pick representative customers to be surveyed based on as many features such as gender, political preference, income level, race, age, and so on, as are in the company's data banks.

30. The graph below shows cumulative proportions plotted against GPAs for high school seniors.

What is the approximate interquartile range?

(A) 0.85
(B) 2.25
(C) 2.7
(D) 2.75
(E) 3.1

31. PCB contamination of a river by a manufacturer is being measured by amounts of the pollutant found in fish. A company scientist claims that the fish contain only 5 parts per million, but an investigator believes the figure is higher. The investigator catches six fish that show the following amounts of PCB (in parts per million): 6.8, 5.6, 5.2, 4.7, 6.3, and 5.4. In performing a hypothesis test with H_0: $\mu = 5$ and H_a: $\mu > 5$, what is the test statistic?

(A) $t = \dfrac{5.67 - 5}{0.763}$

(B) $t = \dfrac{5.67 - 5}{\sqrt{0.763/5}}$

(C) $t = \dfrac{5.67 - 5}{\sqrt{0.763/6}}$

(D) $t = \dfrac{5.67 - 5}{0.763/\sqrt{5}}$

(E) $t = \dfrac{5.67 - 5}{0.763/\sqrt{6}}$

32. The distribution of weights of 16-ounce bags of a particular brand of potato chips is approximately normal with a standard deviation of 0.28 ounce. How does the weight of a bag at the 40th percentile compare with the mean weight?

(A) 0.40 ounce above the mean
(B) 0.25 ounce above the mean
(C) 0.07 ounce above the mean
(D) 0.07 ounce below the mean
(E) 0.25 ounce below the mean

33. In general, how does tripling the sample size change the confidence interval width?

(A) It triples the interval width.
(B) It divides the interval width by 3.
(C) It multiples the interval width by 1.732.
(D) It divides the interval width by 1.732.
(E) This question cannot be answered without knowing the sample size.

34. Which of the following statements is *false*?

(A) Like the normal distribution, the t-distributions are symmetric.
(B) The t-distributions are lower at the mean and higher at the tails, and so are more spread out than the normal distribution.
(C) The greater the df, the closer the t-distributions are to the normal distribution.
(D) The smaller the df, the better the 68–95–99.7 Rule works for t-models.
(E) The areas under all t-distribution curves are 1.

35. A study on school budget approval among people with different party affiliations resulted in the following segmented bar chart:

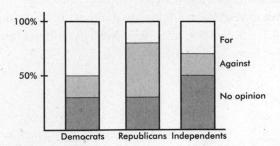

Which of the following is greatest?

(A) Number of Democrats who are for the proposed budget
(B) Number of Republicans who are against the budget
(C) Number of Independents who have no opinion on the budget
(D) The above are all equal.
(E) The answer is impossible to determine without additional information.

36. The sampling distribution of the sample mean is close to the normal distribution

(A) only if both the original population has a normal distribution and n is large.
(B) if the standard deviation of the original population is known.
(C) if n is large, no matter what the distribution of the original population.
(D) no matter what the value of n or what the distribution of the original population.
(E) only if the original population is not badly skewed and does not have outliers.

37. What is the probability of a Type II error when a hypothesis test is being conducted at the 10% significance level ($\alpha = 0.10$)?

(A) 0.05
(B) 0.10
(C) 0.90
(D) 0.95
(E) There is insufficient information to answer this question.

38.

Above is the dotplot for a set of numbers. One element is labeled X. Which of the following is a correct statement?

(A) X has the largest z-score, in absolute value, of any element in the set.
(B) A modified boxplot will plot an outlier like X as an isolated point.
(C) A stemplot will show X isolated from two clusters.
(D) Because of X, the mean and median are different.
(E) The IQR is exactly half the range.

39. A 2018 survey of 500 households concluded that 82% of the population uses grocery coupons. Which of the following best describes what is meant by the poll having a margin of error of 3%?

(A) Three percent of those surveyed refused to participate in the poll.
(B) It would not be unexpected for 3% of the population to begin using coupons or stop using coupons.
(C) Between 395 and 425 of the 500 households surveyed responded that they used grocery coupons.
(D) If a similar survey of 500 households were taken weekly, a 3% change in each week's results would not be unexpected.
(E) It is likely that between 79% and 85% of the population use grocery coupons.

40.

	College Plans	
	Public	Private
Taking AP Statistics:	18	27
Not taking AP Statistics:	26	40

The above two-way table summarizes the results of a survey of a random sample of high school seniors conducted to determine if there is a relationship between whether or not a student is taking AP Statistics and whether he or she plans to attend a public or a private college after graduation. Which of the following is the most reasonable conclusion about the relationship between taking AP Statistics and the type of college a student plans to attend?

(A) There appears to be no association since the proportion of AP Statistics students planning to attend public schools is almost identical to the proportion of students not taking AP Statistics who plan to attend public schools.
(B) There appears to be an association since the proportion of AP Statistics students planning to attend public schools is almost identical to the proportion of students not taking AP Statistics who plan to attend public schools.
(C) There appears to be an association since more students plan to attend private than public schools.
(D) There appears to be an association since fewer students are taking AP Statistics than are not taking AP Statistics.
(E) These data do not address the question of association.

STOP

If there is still time remaining, you may review your answers.

SECTION II

Part A

QUESTIONS 1–5

Spend about 65 minutes on this part of the exam.

Percentage of Section II grade—75

> **Directions:** You must show all work and indicate the methods you use. You will be graded on the correctness of your methods and on the accuracy of your results and explanations.

1. Ten volunteer male subjects are to be used for an experiment to study four drugs (aloe, camphor, eucalyptus oil, and benzocaine) and a placebo with regard to itching relief. Itching is induced on a forearm with an itch stimulus (cowage), a drug is topically administered, and the duration of itching is recorded.

   (a) If the experiment is to be done in one sitting, how would you assign treatments for a completely randomized design?

   (b) If 5 days are set aside for the experiment, with one sitting a day, how would you assign treatments for a randomized block design where the subjects are the blocks?

   (c) What limits are there on generalization of any results?

2. A comprehensive study of more than 3,000 baseball games results in the graph below showing relative frequencies of runs scored by home teams.

   (a) Calculate the mean and the median.

   (b) Between the mean and the median, is the one that is greater what was to be expected? Explain.

   (c) What is the probability that in 4 randomly selected games, the home team is shut out (scores no runs) in at least one of the games?

   (d) If X is the random variable for the runs scored per game by home teams, its standard deviation is 2.578. Suppose 200 games are selected at random, and $\bar{x}$, the mean number of runs scored by the home teams, is calculated. Describe the sampling distribution of $\bar{x}$.

3. A study is performed to explore the relationship, if any, between 24-hour urinary metabolite 3-methoxy-4-hydroxyphenylglycol (MHPG) levels and depression in bipolar patients. The MHPG level is measured in micrograms per 24 hours while the manic-depression (MD) scale used goes from 0 (manic delirium), through 5 (euthymic), up to 10 (depressive stupor). A partial computer printout of regression analysis with MHPG as the independent variable follows:

```
Average MHPG = 1243.1 with SD = 384.9
Average MD = 5.4 with SD = 2.875
95% confidence interval for slope b₁ = (−0.0096, −0.0016)
```

(a) Calculate the slope of the regression line and interpret it in the context of this problem.

(b) Find the equation of the regression line.

(c) Calculate and interpret the value of r^2 in the context of this problem.

(d) What does the correlation say about causation in the context of this problem?

4. An experiment is run to test whether daily stimulation of specific reflexes in young infants will lead to earlier walking. Twenty infants were recruited through a pediatrician's service, and were randomly split into two groups of ten. One group received the daily stimulation while the other was considered a control group. The ages (in months) at which the infants first walked alone were recorded.

Ages in months at which first steps alone were taken

With stimulation: 10, 12, 11, 10.5, 11, 11.5, 11.5, 11, 12, 11.5
Control: 10, 13, 12, 11, 11.5, 11.5, 12.5, 12, 11.5, 11

Is there statistical evidence that infants walk earlier with daily stimulation of specific reflexes?

5. A city sponsors two charity runs during the year, one 5 miles and the other 10 kilometers. Both runs attract many thousands of participants. A statistician is interested in comparing the types of runners attracted to each race. She obtains a random sample of 100 runners from each race, calculates five-number summaries of the times, and displays the results in the parallel boxplots below.

(a) Compare the distributions of times above.

The statistician notes that 5 miles equals 8 kilometers and so decides to multiply the times from the 5-mile event by $\frac{10}{8}$ and then compare box plots. This display is below.

(b) How do the distributions now compare?

(c) Given that the 10-kilometer race was a longer distance than the 5-mile race, was the change from the first set of boxplots to the second set as expected? Explain.

(d) Based on the boxplots, would you expect the difference in mean times in the first set of parallel boxplots to be less than, greater than, or about the same as the difference in mean times in the second set of parallel boxplots? Explain.

SECTION II

Part B

QUESTION 6

Spend about 25 minutes on this part of the exam.
Percentage of Section II grade—25

6. A random sample of 250 SAT mathematics scores tabulates as follows:

Score range	200–300	300–400	400–500	500–600	600–700	700–800
Number of scores	7	39	74	82	36	12

(a) Find a 95% confidence interval for the proportion of scores over 500.

(b) Test the null hypothesis that the data follow a *normal distribution* with $\mu = 500$ and $\sigma = 100$.

(c) Assume the above data comes from a normally distributed population with unknown μ and σ. In this case, the sampling distribution of $\dfrac{(n-1)s^2}{\sigma^2}$ follows a chi-square distribution. If computer output of the above data yields $n = 250$, $\bar{x} = 504.8$, and $s = 114.4$, and if in this problem the χ^2-values for tails of 0.025 and 0.975 are 294.6 and 207.2, respectively, find a 95% confidence interval for the standard deviation σ.

STOP

If there is still time remaining, you may review your answers.

PRACTICE TEST 2

Answer Key

SECTION I

1. **D**	9. **D**	17. **C**	25. **C**	33. **D**
2. **B**	10. **E**	18. **A**	26. **C**	34. **D**
3. **E**	11. **D**	19. **B**	27. **C**	35. **E**
4. **D**	12. **B**	20. **C**	28. **B**	36. **C**
5. **E**	13. **B**	21. **D**	29. **A**	37. **E**
6. **C**	14. **E**	22. **C**	30. **A**	38. **C**
7. **C**	15. **B**	23. **C**	31. **E**	39. **E**
8. **E**	16. **A**	24. **D**	32. **D**	40. **A**

Answers Explained

SECTION I

1. **(D)** Since (2, 7) is on the line $y = mx + 3$, we have $7 = 2m + 3$ and $m = 2$. Thus the regression line is $y = 2x + 3$. The point $(\bar{x}, \bar{y})$ is always on the regression line, and so we have $\bar{y} = 2\bar{x} + 3$.

2. **(B)** It could well be that conscientious students are the same ones who both study and do well on the basketball court. If students could be randomly assigned to study or not study, the results would be more meaningful. Of course, ethical considerations might make it impossible to isolate the confounding variable in this way.

3. **(E)** invNorm(.7) = 0.5244 so $75 - \mu = 0.5244(14)$ and $\mu = 75 - 0.5244(14)$.

4. **(D)** The slope of the regression line and the correlation are related by $b_1 = r\dfrac{s_y}{s_x}$. When using z-scores, the standard deviations s_x and s_y are 1. If $r = 0$, then $b_1 = 0$. Switching which variable is x and which is y, or changing units, will not change the correlation.

5. **(E)** The median and interquartile range are specifically used when outliers are suspected of unduly influencing the mean, range, or standard deviation.

6. **(C)** 2.576 is the critical z-score for a 99% confidence interval, and $\sigma_{\bar{x}} = \dfrac{\sigma}{\sqrt{n}} = \dfrac{2.4}{\sqrt{n}}$.

7. **(C)** This is a hypothesis test with H_0: tissue strength is within specifications, and H_a: tissue strength is below specifications. A Type I error is committed when a true null hypothesis is mistakenly rejected.

8. **(E)** The wording of questions can lead to response bias. The neutral way of asking this question would simply have been, "Do you support the proposed school budget increase?"

9. **(D)**

$$P(\text{water} \mid \text{hurricane}) = \frac{P(\text{water} \cap \text{hurricane})}{P(\text{hurricane})} = \frac{0.22}{0.25}$$

10. **(E)** While it is important to look for basic patterns, it is also important to look for deviations from these patterns. In this case, there is an overall positive correlation; however, those faculty with under ten years of service show little relationship between years of service and salary. While (A) is a true statement, it does not give an overall interpretation of the scatterplot.

11. **(D)** The second set has a greater range, $3.8 - 1.8 = 2.0$ as compared to $4.1 - 2.3 = 1.8$, and with its skewness it also has a greater standard deviation.

12. **(B)** With $n = 10$, increasing Σx by 40 increases $\dfrac{\Sigma x}{n}$ by $\dfrac{40}{10} = 4$.

13. **(B)** Means can always be added, and in this example, because of independence, variances can also be added. Thus the new variance is $5^2 + 12^2 = 169$, and the new standard deviation is $\sqrt{169} = 13$.

14. **(E)** Dice have no memory, so the probability that the next toss will be an even number is 0.5 and the probability that it will be an odd number is 0.5. The law of large numbers says that as the number of tosses becomes larger, the proportion of even numbers tends to become closer to 0.5.

15. **(B)** The sample slope is 12.633, the standard error of the slope is 0.378, and the critical t-scores for 90% confidence with $df = 7$ are ± 1.895.

16. **(A)** Either directly or anonymously, you should be able to obtain the test results for *every* student.

17. **(C)** Percentile ranking is a measure of relative position. Adding five points to everyone's score will not change the relative positions.

18. **(A)** The control group should have experiences identical to those of the experimental groups except for the treatment under examination. They should not be given a new treatment.

19. **(B)** If H_a is true, the probability of failing to reject H_0 and thus committing a Type II error is 1 minus the power, that is, $1 - 0.8 = 0.2$.

20. **(C)** Five does not split the area in half, so 5 is not the median. Histograms such as these show relative frequencies, not actual frequencies. The area from 1.5 to 4.5 is the same as that between 7.5 and 10.5, each being about 25% of the total. Given the spread, 1 is too small an estimate of the standard deviation. The area above 3 looks to be the same as the area above 9 and 10, so the median won't change.

21. **(D)** There must be a fixed number of trials, which rules out (A); only two possible outcomes, which rules out (B); and a constant probability of success on any trial, which rules out (C).

22. **(C)** In both cases 1 hour is one standard deviation from the mean with a right tail probability of 0.1587.

23. **(C)** Control, randomization, and replication are all important aspects of well-designed experiments. We try to control confounding variables, not to use them to control something else.

24. **(D)** The data are strongly skewed to the left, indicating that the mean is less than the median. The median appears to be roughly 215, indicating that the interval [200, 240] probably has more than 50% of the values. While in a standard boxplot each whisker contains 25% of the values, this is a modified boxplot showing four outliers, and so the left whisker has four fewer values than the right whisker.

25. **(C)** For the regression line, the sum and thus the mean of the residuals are always zero. An influential score may have a small residual but still have a great effect on the regression line. If the correlation is 1, all the residuals would be 0, resulting in a very distinct pattern.

26. **(C)** $\bar{x} = \dfrac{32 + 24 + 29 + 27}{4} = 28$. Since $(\bar{x}, \bar{y})$ is a point on the regression line, $\bar{y} = 3(28) + 4 = 88$.

27. **(C)** $P(\text{at least } 1) = 1 - P(\text{none}) = 1 - (0.88)^6$

28. **(B)** Different samples give different sample statistics, all of which are estimates for the same population parameter, and so error, called *sampling error*, is naturally present.

29. **(A)** While the associate does use chance, each customer would have the same chance of being selected only if the same number of customers had names starting with each letter of the alphabet. This selection does not result in a simple random sample because each possible set of 104 customers does not have the same chance of being picked as part of the sample. For example, a group of customers whose names all start with *A* will not be chosen. Sampling error, the natural variation inherent in a survey, is always present and is not a source of bias. Letting the surveyor have free choice in selecting the sample, rather than incorporating chance in the selection process, is a recipe for disaster!

30. **(A)** Corresponding to cumulative proportions of 0.25 and 0.75 are $Q_1 = 2.25$ and $Q_3 = 3.1$, respectively, and so the interquartile range is $3.1 - 2.25 = 0.85$.

31. **(E)** The standard deviation of the test statistic is $\sigma_{\bar{x}} = \dfrac{\sigma}{\sqrt{n}} \approx \dfrac{s}{\sqrt{n}}$.

32. **(D)** From a table or a calculator (for example `invNorm` on the TI-84), the 40th percentile corresponds to a z-score of -0.2533, and $-0.2533(0.28) = -0.0709$.

33. **(D)** Increasing the sample size by a multiple of d divides the interval width by $\sqrt{d}$.

34. **(D)** The t-distributions are symmetric; however, they are lower at the mean and higher at the tails and so are more spread out than the normal distribution. The greater the *df*, the closer the t-distributions are to the normal distribution. The 68–95–99.7 Rule applies to the z-distribution and will work for t-models with very large *df*. All probability density curves have an area of 1 below them.

35. **(E)** The given bar chart shows percentages, not actual numbers.

36. **(C)** This follows from the central limit theorem.

37. **(E)** There is a different Type II error for each possible correct value for the population parameter.

38. **(C)** X is close to the mean and so will have a z-score close to 0. Modified boxplots show only outliers that are far from the mean. X and the two clusters are clearly visible in a stemplot of these data. In symmetric distributions the mean and median are equal. The IQR here is close to the range.

39. **(E)** Using a measurement from a sample, we are never able to say *exactly* what a population proportion is; rather we always say we have a certain *confidence* that the population proportion lies in a particular *interval*. In this case that interval is 82% ± 3% or between 79% and 85%.

40. **(A)** Whether or not students are taking AP Statistics seems to have no relationship to which type of school they are planning to go to. Chi-square is close to 0.

SECTION II

1. (a) Number the volunteers 1 through 10. Use a random number generator to pick numbers between 1 and 10, throwing out repeats. The volunteers corresponding to the first two numbers chosen will receive aloe, the next two will receive camphor, the next two eucalyptus oil, the next two benzocaine, and the remaining two a placebo.

   (b) Each volunteer (the volunteers are "blocks") should receive all five treatments, one a day, with the time-order randomized. For example, label aloe 1, camphor 2, eucalyptus oil 3, benzocaine 4, and the placebo 5. Then for each volunteer use a random number generator to pick numbers between 1 and 5, throwing away repeats. The order picked gives the day on which each volunteer receives each treatment.

   (c) Results cannot be generalized to women.

<table>
<tr><td colspan="2" align="center">**SCORING**</td></tr>
<tr><td colspan="2">

Part (a) is essentially correct for giving a procedure which randomly assigns 2 volunteers to each of the five treatments. Part (a) is partially correct for giving a procedure which randomly assigns one of the treatments for each volunteer, but may not result in *two* volunteers receiving each treatment.

Part (b) is essentially correct for giving a procedure which assigns a random order for each of the volunteers to have all 5 treatments. Part (b) is partially correct for having each volunteer take all five treatments, one a day, but not clearly randomizing the time-order.

Part (c) is essentially correct for stating that the results cannot be generalized to women and is incorrect otherwise.

</td></tr>
<tr><td>**4 Complete Answer**</td><td>All three parts essentially correct.</td></tr>
<tr><td>**3 Substantial Answer**</td><td>Two parts essentially correct and one part partially correct.</td></tr>
<tr><td>**2 Developing Answer**</td><td>Two parts essentially correct OR one part essentially correct and one or two parts partially correct OR all three parts partially correct.</td></tr>
<tr><td>**1 Minimal Answer**</td><td>One part essentially correct OR two parts partially correct.</td></tr>
</table>

2. (a) $\bar{x} = \sum xp(x) = 0(0.05) + 1(0.10) + 2(0.13) + 3(0.15) + 4(0.14) + 5(0.12) + 6(0.10) + 7(0.08)$
$+ 8(0.06) + 9(0.04) + 10(0.03) = 4.27$.

With $(0.05 + 0.10 + 0.13 + 0.15) = 0.43$ below 4 runs, and $(0.12 + 0.10 + 0.08 + 0.06 + 0.04 + 0.03) = 0.43$ above 4 runs, the median must be 4.

(b) The mean is greater than the median, as was to be expected because the distribution is skewed to the right.

(c) This is a binomial with $n = 4$ and $p = 0.05$.
P(at least one shutout in 4 games) $= 1 - P$(no shutouts in the 4 games)
$= 1 - (1 - 0.05)^4 = 1 - (0.95)^4 = 0.1855$

(d) The distribution of $\bar{x}$ is approximately normal with mean $\mu_{\bar{x}} = 4.27$ (from above)
and standard deviation $\sigma_{\bar{x}} = \dfrac{2.578}{\sqrt{200}} = 0.1823$.

SCORING

Part (a) is essentially correct for correctly calculating both the mean and median. Part (a) is partially correct for correctly calculating one of these two measures.

Part (b) is essentially correct for noting that the mean is greater than the median and relating this to the skew.

Part (c) is essentially correct for recognizing this as a binomial probability calculation and making the correct calculation. Part (c) is partially correct for recognizing this as a binomial probability calculation, but with an error such as $1 - (0.05)^4$ or $4(0.05)(0.95)^3$.

Part (d) is essentially correct for "approximately normal," $\mu_{\bar{x}} = 4.27$, and $\sigma_{\bar{x}} = 0.1823$, and partially correct for two of these three answers.

Count partially correct answers as one-half an essentially correct answer.

4	**Complete Answer**	Four essentially correct answers.
3	**Substantial Answer**	Three essentially correct answers.
2	**Developing Answer**	Two essentially correct answers.
1	**Minimal Answer**	One essentially correct answer.

Use a holistic approach to decide a score totaling between two numbers.

---

3. (a) The slope b_1 is in the center of the confidence interval, so $b_1 = \dfrac{-0.0096 - 0.0016}{2} = -0.0056$. In context, 0.0056 *estimates* the *average* decrease in the manic-depressive scale score for each 1-microgram increase in the level of urinary MHPG. (Thus high levels of MHPG are associated with increased mania, and conversely, low levels of MHPG are associated with increased depression.)

(b) Recalling that the regression line goes through the point $(\bar{x}, \bar{y})$ or using the AP exam formula page, $b_o = \bar{y} - b_1\bar{x} = 5.4 - (-0.0056)(1243.1) = 12.36$, and thus the equation of the regression line is $\widehat{MD} = 12.36 - 0.0056(MHPG)$ (or $\hat{y} = 12.36 - 0.0056x$, where x is the level of urinary MHPG in micrograms per 24 hours, and y is the score on a 0–10 manic-depressive scale).

(c) Recalling that on the regression line, each one SD increase in the independent variable corresponds to an increase of r SD in the dependent variable, or using the AP exam formula page, $b_1 = r\dfrac{S_y}{S_x}$, so

$$r = b_1 \frac{S_x}{S_y} = -0.0056\left(\frac{384.9}{2.875}\right) = -0.750$$

and $r^2 = 56.2\%$. Thus, 56.2% of the variation in the manic-depression scale level is explained by the regression line with urinary MHPG levels as the explanatory variable.

(d) Correlation never proves causation. It could be that depression causes biochemical changes leading to low levels of urinary MHPG, or it could be that low levels of urinary MHPG cause depression, or it could be that some other variable (a confounding variable) simultaneously affects both urinary MHPG levels and depression.

SCORING

Part (a) is essentially correct if the slope is correctly calculated and correctly interpreted in context. Part (a) is partially correct if the slope is not correctly calculated but a correct interpretation is given using the incorrect value for the slope.

Part (b) is essentially correct if the regression equation is correctly calculated (using the slope found in part (a)) and it is clear what the variables stand for. Part (b) is partially correct if the correct equation is found (using the slope found in part (a)) but it is unclear what the variables stand for.

Part (c) is essentially correct if the coefficient of determination, r^2, is correctly calculated and correctly interpreted in context. Part (c) is partially correct if r^2 is not correctly calculated but a correct interpretation is given using the incorrect value for r^2.

Part (d) is essentially correct for noting that correlation never proves causation, and referring to context. Part (d) is partially correct for a correct statement about correlation and causation, but with no reference to context.

Count partially correct answers as one-half an essentially correct answer.

4	**Complete Answer**	Four essentially correct answers.
3	**Substantial Answer**	Three essentially correct answers.
2	**Developing Answer**	Two essentially correct answers.
1	**Minimal Answer**	One essentially correct answer.

Use a holistic approach to decide a score totaling between two numbers.

4. There are four parts to this solution.

(a) State the hypotheses and define the parameters of interest.

$$H_0: \mu_1 - \mu_2 = 0 \text{ and } H_a: \mu_1 - \mu_2 < 0$$

where μ_1 = mean number of months at which first steps alone are taken for infants receiving daily stimulation, and μ_2 = mean number of months at which first steps are taken for infants not receiving daily simulation. [Other notations are also possible, for example, $H_0: \mu_1 = \mu_2$ and $H_a: \mu_1 < \mu_2$.] If notation is clear, for example, $H_0: \mu_{\text{stimulation}} = \mu_{\text{control}}$ and $H_a: \mu_{\text{stimulation}} < \mu_{\text{control}}$, then this identifies the variables, and a separate explanation is unnecessary.

(b) Identify the test by name or formula and check the assumptions.

Two-sample t-test   OR $t = \dfrac{\bar{x}_1 - \bar{x}_2}{\sqrt{\dfrac{s_1^2}{n_1} + \dfrac{s_2^2}{n_2}}}$

Checks: We are given that the infants were randomly assigned to the two groups, the two groups are independent, and dotplots of the two groups show no outliers and are roughly bell-shaped.

Age (mo)
(with stimulation)

Age (mo)
(control)

(c) Mechanics: Putting the data into two Lists, calculator software (such as 2-SampTTest on the TI-84 or 2-Sample tTest on the Casio Prizm) gives $t = -1.2$ and $P = 0.12345$.

(d) State the conclusion in context with linkage to the P-value.

With this large a P-value, $0.12345 > 0.05$, there is not sufficient evidence to reject H_0, that is, there is not sufficient evidence that infants walk earlier with daily stimulation of specific reflexes.

Part (a) is essentially correct for stating the hypotheses and identifying the variables. Part (a) is partially correct for correct hypotheses, but missing identification of the variables.

Part (b) is essentially correct for identifying the test and checking the assumptions. Part (b) is partially correct for only one of these two elements.

Part (c) is essentially correct for a correct conclusion, in context, linked to the P-value. Part (c) is partially correct for a correct conclusion but missing either context or linkage.

Part (d) is essentially correct for a conclusion in context with linkage to the P-value. Part (d) is partially correct for only one of these two elements.

Count partially correct answers as one-half an essentially correct answer.

4   Complete Answer   Four essentially correct answers.
3   Substantial Answer   Three essentially correct answers.
2   Developing Answer   Two essentially correct answers.
1   Minimal Answer   One essentially correct answer.

Use a holistic approach to decide a score totaling between two numbers.

5. (a) A complete answer compares shape, center, and spread, and mentions context.
   Shape: Although shape usually cannot be deduced from a boxplot, the 10-kilometer distribution appears roughly symmetric, while the 5-mile distribution appears skewed right.
   Center: The median of the 10-kilometer distribution is 4 minutes greater than the median of the 5-mile distribution.
   Spread: The ranges of the two distributions are equal, both about 7 minutes.

   (b) *Shape:* Again, the 10-kilometer distribution appears roughly symmetric, while the 5-mile distribution still appears skewed right.
   Center: Now the median of the 10-kilometer distribution is about 3.5 minutes less than the median of the 5-mile distribution.
   Spread: Now the range of the 10-kilometer distribution is less than the range of the 5-mile distribution.

   (c) One possible answer is that this was not as expected with the following explanation: One would expect that for 5 miles, the shorter run, the speeds would be faster, so adjusting the 5-mile run times would result in faster times (less minutes) than the 10-kilometer times, but this was not the case. [Perhaps slower, less serious runners participate in the shorter race, so when their times are adjusted, the minutes are greater than the times for the 10-kilometer run.]

(d) In each set of parallel boxplots, the 10-kilometer distribution appears roughly symmetric, so that the mean will be about the same as the median, while the 5-mile distribution is skewed right so that the mean will in all likelihood be greater than the median. In the first set of boxplots (where the 5-mile median < 10-kilometer median) this will result in the means being closer, while in the second set of boxplots (where the 5-mile median > 10-kilometer median) this will result in the means being further apart. Thus we would expect the difference in mean times in the first set of parallel boxplots to be less than the difference in mean times in the second set of parallel boxplots.

SCORING

Part (a) is essentially correct for correctly comparing shape, center, and spread. Part (a) is partially correct for correctly comparing two of the three features. Lower an E to a P or a P to an I, if context is never mentioned.

Part (b) is essentially correct for correctly comparing shape, center, and spread. Part (b) is partially correct for correctly comparing two of the three features.

Part (c) is essentially correct for a reasonable statement about the change between the two sets of parallel boxplots together with a correct explanation to go along with the statement. Part (c) is partially correct if the explanation is weak.

Part (d) is essentially correct for a correct prediction about the means together with a reasonable justification based on symmetry of the 10-kilometer distribution and skewness of the 5-mile distribution. Part (d) is partially correct if the justification is weak.

Count partially correct answers as one-half an essentially correct answer.

4	**Complete Answer**	Four essentially correct answers.
3	**Substantial Answer**	Three essentially correct answers.
2	**Developing Answer**	Two essentially correct answers.
1	**Minimal Answer**	One essentially correct answer.

Use a holistic approach to decide a score totaling between two numbers.

6. (a) Identify the confidence interval by name or formula.

$$z\text{-interval of a population proportion or } \hat{p} \pm z^* \sqrt{\frac{\hat{p}(1-\hat{p})}{n}}$$

Conditions: Random sample (given), $n = 250$ is less than 10% of all SAT math scores, $n\hat{p} = 82 + 36 + 12 = 130 \geq 10$, and $n(1 - \hat{p}) = 250 - 130 = 120 \geq 10$.

Mechanics: Calculator software (such as 1-PropZInt on the TI-84 or 1-Prop zInterval on the Casio Prizm) gives (0.45807, 0.58193).

Interpret the confidence interval in context.

We are 95% confident that between 45.8% and 58.2% of SAT mathematics scores are over 500.

(b) State the hypotheses.

H_0: The distribution of SAT mathematics scores is normal with $\mu = 500$ and $\sigma = 100$.

H_a: The distribution of SAT mathematics scores is not normal with $\mu = 500$ and $\sigma = 100$.

Identify the test by name or formula and check the assumptions.

Chi-square goodness-of-fit test

$$\text{or} \qquad \chi^2 = \sum \frac{(observed - expected)^2}{expected}$$

Check conditions

We have a random sample.

A normal distribution has 34.13% on each side of the mean and within one SD, 13.59% on each side between one and two SDs from the mean, and 2.28% on each side more than two SDs from the mean. This gives expected cell counts of $(0.0228)250 = 5.7$, $(0.1359)250 = 34.0$, and $(0.3413)250 = 85.3$, each of which is at least 5.

Mechanics: Put {7, 39, 74, 82, 36, 12} in one List and {5.7, 34.0, 85.3, 85.3, 34.0, 5.7} in a second List. With $df = 6 - 1 = 5$, calculator software (such as χ^2GOF-Test on the TI-84) gives $\chi^2 = 9.7372$ and $P = 0.0830$.

State the conclusion in context with linkage to the P-value.

With this large a P-value, $0.0830 > 0.05$, there is not sufficient evidence to reject H_0, that is, there is not sufficient evidence that the distribution of SAT math scores is not normal.

(c) $207.2 < \dfrac{(250-1)(114.4)^2}{\sigma^2} < 294.6$

$207.2 < \dfrac{3,259,000}{\sigma^2} < 294.6$

$\dfrac{1}{207.2} > \dfrac{\sigma^2}{3,259,000} > \dfrac{1}{294.6}$

$15,729 > \sigma^2 > 11,062$

$125.4 > \sigma > 105.2$

Thus we are 95% confident that the standard deviation for the distribution of SAT mathematics scores is between 105.2 and 125.4.

Part (a) has four parts: 1) identifying the confidence interval; 2) checking assumptions; 3) calculating the confidence interval; and 4) interpreting the confidence interval in context. Part (a) is essentially correct if three or four of these parts are correct and partially correct if one or two of these parts are correct.

Part (b) has four parts: 1) stating the hypotheses; 2) identifying the test and checking assumptions; 3) calculating the test statistic and the P-value; and 4) giving a conclusion in context with linkage to the P-value. Part (b) is essentially correct if three or four of these parts are correct and partially correct if one or two of these parts are correct.

Part (c) is essentially correct for calculating the confidence interval and interpreting it in context. Part (c) is partially correct for one of these two parts correct.

4 Complete Answer   All three parts essentially correct.

3 Substantial Answer   Two parts essentially correct and one part partially correct.

2 Developing Answer   Two parts essentially correct OR one part essentially correct and one or two parts partially correct OR all three parts partially correct.

1 Minimal Answer   One part essentially correct OR two parts partially correct.

ANSWER SHEET
Practice Test 3

1. Ⓐ Ⓑ Ⓒ Ⓓ Ⓔ   11. Ⓐ Ⓑ Ⓒ Ⓓ Ⓔ   21. Ⓐ Ⓑ Ⓒ Ⓓ Ⓔ   31. Ⓐ Ⓑ Ⓒ Ⓓ Ⓔ

2. Ⓐ Ⓑ Ⓒ Ⓓ Ⓔ   12. Ⓐ Ⓑ Ⓒ Ⓓ Ⓔ   22. Ⓐ Ⓑ Ⓒ Ⓓ Ⓔ   32. Ⓐ Ⓑ Ⓒ Ⓓ Ⓔ

3. Ⓐ Ⓑ Ⓒ Ⓓ Ⓔ   13. Ⓐ Ⓑ Ⓒ Ⓓ Ⓔ   23. Ⓐ Ⓑ Ⓒ Ⓓ Ⓔ   33. Ⓐ Ⓑ Ⓒ Ⓓ Ⓔ

4. Ⓐ Ⓑ Ⓒ Ⓓ Ⓔ   14. Ⓐ Ⓑ Ⓒ Ⓓ Ⓔ   24. Ⓐ Ⓑ Ⓒ Ⓓ Ⓔ   34. Ⓐ Ⓑ Ⓒ Ⓓ Ⓔ

5. Ⓐ Ⓑ Ⓒ Ⓓ Ⓔ   15. Ⓐ Ⓑ Ⓒ Ⓓ Ⓔ   25. Ⓐ Ⓑ Ⓒ Ⓓ Ⓔ   35. Ⓐ Ⓑ Ⓒ Ⓓ Ⓔ

6. Ⓐ Ⓑ Ⓒ Ⓓ Ⓔ   16. Ⓐ Ⓑ Ⓒ Ⓓ Ⓔ   26. Ⓐ Ⓑ Ⓒ Ⓓ Ⓔ   36. Ⓐ Ⓑ Ⓒ Ⓓ Ⓔ

7. Ⓐ Ⓑ Ⓒ Ⓓ Ⓔ   17. Ⓐ Ⓑ Ⓒ Ⓓ Ⓔ   27. Ⓐ Ⓑ Ⓒ Ⓓ Ⓔ   37. Ⓐ Ⓑ Ⓒ Ⓓ Ⓔ

8. Ⓐ Ⓑ Ⓒ Ⓓ Ⓔ   18. Ⓐ Ⓑ Ⓒ Ⓓ Ⓔ   28. Ⓐ Ⓑ Ⓒ Ⓓ Ⓔ   38. Ⓐ Ⓑ Ⓒ Ⓓ Ⓔ

9. Ⓐ Ⓑ Ⓒ Ⓓ Ⓔ   19. Ⓐ Ⓑ Ⓒ Ⓓ Ⓔ   29. Ⓐ Ⓑ Ⓒ Ⓓ Ⓔ   39. Ⓐ Ⓑ Ⓒ Ⓓ Ⓔ

10. Ⓐ Ⓑ Ⓒ Ⓓ Ⓔ   20. Ⓐ Ⓑ Ⓒ Ⓓ Ⓔ   30. Ⓐ Ⓑ Ⓒ Ⓓ Ⓔ   40. Ⓐ Ⓑ Ⓒ Ⓓ Ⓔ

Practice Test 3

SECTION I

Questions 1–40

Spend 90 minutes on this part of the exam.

> **Directions:** The questions or incomplete statements that follow are each followed by five suggested answers or completions. Choose the response that best answers the question or completes the statement.

1. Which of the following is a true statement?

   (A) While properly designed experiments can strongly suggest cause-and-effect relationships, a complete census is the only way of establishing such a relationship.

   (B) If properly designed, observational studies can establish cause-and-effect relationships just as strongly as properly designed experiments.

   (C) Controlled experiments are often undertaken later to establish cause-and-effect relationships first suggested by observational studies.

   (D) A useful approach to overcome bias in observational studies is to increase the sample size.

   (E) In an experiment, the control group is a self-selected group who choose not to receive a designated treatment.

2. Two classes take the same exam. Suppose a certain score is at the 40th percentile for the first class and at the 80th percentile for the second class. Which of the following is the most reasonable conclusion?

   (A) Students in the first class generally scored higher than students in the second class.

   (B) Students in the second class generally scored higher than students in the first class.

   (C) A score at the 20th percentile for the first class is at the 40th percentile for the second class.

   (D) A score at the 50th percentile for the first class is at the 90th percentile for the second class.

   (E) One of the classes has twice the number of students as the other.

3. In an experiment, the control group should receive

(A) treatment opposite that given the experimental group.
(B) the same treatment given the experimental group without knowing they are receiving the treatment.
(C) a procedure identical to that given the experimental group except for receiving the treatment under examination.
(D) a procedure identical to that given the experimental group except for a random decision on receiving the treatment under examination.
(E) none of the procedures given the experimental group.

4. In a random sample of Toyota car owners, 83 out of 112 said they were satisfied with the Toyota front-wheel drive, while in a similar survey of Subaru owners, 76 out of 81 said they were satisfied with the Subaru four-wheel drive. A 90% confidence interval estimate for the difference in proportions between Toyota and Subaru car owners who are satisfied with their drive systems is reported to be -0.197 ± 0.081. Which is a proper conclusion?

(A) The interval is invalid because probabilities cannot be negative.
(B) The interval is invalid because it does not contain zero.
(C) Subaru owners are approximately 19.7% more satisfied with their drive systems than are Toyota owners.
(D) 90% of Subaru owners are approximately 19.7% more satisfied with their drive systems than are Toyota owners.
(E) We are 90% confident that the difference in proportions between Toyota and Subaru car owners who are satisfied with their drive systems is between -0.278 and -0.116.

5. In a study on the effect of music on worker productivity, employees were told that a different genre of background music would be played each day and the corresponding production outputs noted. Every change in music resulted in an increase in production. This is an example of

(A) the effect of a treatment unit.
(B) the placebo effect.
(C) the control group effect.
(D) sampling error.
(E) voluntary response bias.

6. A computer manufacturer sets up three locations to provide technical support for its customers. Logs are kept noting whether or not calls about problems are solved successfully. Data from a sample of 1000 calls are summarized in the following table:

	Location			Total
	1	2	3	
Problem solved	325	225	150	700
Problem not solved	125	100	75	300
Total	450	325	225	1000

Assuming there is no association between location and whether or not a problem is resolved successfully, what is the expected number of successful calls (problem solved) from location 1?

(A) $\dfrac{(325)(450)}{700}$

(B) $\dfrac{(325)(700)}{450}$

(C) $\dfrac{(325)(450)}{1000}$

(D) $\dfrac{(325)(700)}{1000}$

(E) $\dfrac{(450)(700)}{1000}$

7. Which of the following statements about the correlation coefficient is true?

(A) The correlation coefficient and the slope of the regression line may have opposite signs.
(B) A correlation of 1 indicates a perfect cause-and-effect relationship between the variables.
(C) Correlations of +0.87 and –0.87 indicate the same degree of clustering around the regression line.
(D) Correlation applies equally well to quantitative and categorical data.
(E) A correlation of 0 shows little or no association between two variables.

8. Suppose X and Y are random variables with $E(X) = 780$, $\text{var}(X) = 75$, $E(Y) = 430$, and $\text{var}(Y) = 25$. Given that X and Y are independent, what is the variance of the random variable $X - Y$?

(A) $75 - 25$
(B) $75 + 25$
(C) $\sqrt{75 - 25}$
(D) $\sqrt{75 + 25}$
(E) $\sqrt{75} - \sqrt{25}$

9. What is a sample?

   (A) A measurable characteristic of a population
   (B) A set of individuals having a characteristic in common
   (C) A value calculated from raw data
   (D) A subset of a population
   (E) None of the above

10. A histogram of the cholesterol levels of all employees at a large law firm is as follows:

   Which of the following is the best estimate of the standard deviation of this distribution?

   (A) $\dfrac{(230-170)}{6} = 10$
   (B) 15
   (C) $\dfrac{200}{6} = 33.3$
   (D) $230 - 170 = 60$
   (E) $245 - 155 = 90$

11. A soft drink dispenser can be adjusted to deliver any fixed number of ounces. If the machine is operating with a standard deviation in delivery equal to 0.3 ounce, what should be the mean setting so that a 12-ounce cup will overflow less than 1% of the time? Assume a normal distribution for ounces delivered.

   (A) $12 - 0.99(0.3)$ ounces
   (B) $12 - 2.326(0.3)$ ounces
   (C) $12 - 2.576(0.3)$ ounces
   (D) $12 + 2.326(0.3)$ ounces
   (E) $12 + 2.576(0.3)$ ounces

12. An insurance company wishes to study the number of years drivers in a large city go between automobile accidents. They plan to obtain and analyze the data from a sample of drivers. Which of the following is a true statement?

   (A) A reasonable time-and-cost-saving procedure would be to use systematic sampling on an available list of all AAA (Automobile Association of America) members in the city.
   (B) A reasonable time-and-cost-saving procedure would be to randomly choose families and include all drivers in each of these families in the sample.
   (C) To determine the mean number of years between accidents, randomness in choosing a sample of drivers is not important as long as the sample size is very large.
   (D) The larger a simple random sample, the more likely its standard deviation will be close to the population standard deviation divided by the square root of the sample size.
   (E) None of the above are true statements.

13. The probability that a person will show a certain gene-transmitted trait is 0.8 if the father shows the trait and 0.06 if the father doesn't show the trait. Suppose that the children in a certain community come from families in 25% of which the father shows the trait. Given that a child shows the trait, what is the probability that her father shows the trait?

   (A) $(0.25)(0.8)$
   (B) $(0.25)(0.8) + (0.75)(0.06)$
   (C) $\dfrac{0.25}{(0.25)(0.8)+(0.75)(0.06)}$
   (D) $\dfrac{0.8}{(0.25)(0.8)+(0.75)(0.06)}$
   (E) $\dfrac{(0.25)(0.8)}{(0.25)(0.8)+(0.75)(0.06)}$

14. Given an experiment with $H_0 : \mu = 10$, $H_a : \mu > 10$, and a possible correct value of 11, which of the following increases as n increases?

   I. The probability of a Type I error.
   II. The probability of a Type II error.
   III. The power of the test.

   (A) I only
   (B) II only
   (C) III only
   (D) II and III
   (E) None will increase.

15. If all the values of a data set are the same, all of the following must equal zero except for which one?

    (A) Mean
    (B) Standard deviation
    (C) Variance
    (D) Range
    (E) Interquartile range

16. To determine the average number of minutes it takes to manufacture one unit of a new product, an assembly line manager tracks a random sample of 15 units and records the number of minutes it takes to make each unit. The assembly times are assumed to have a roughly normal distribution. If the mean and standard deviation of the sample are 3.92 and 0.45 minutes respectively, which of the following gives a 90% confidence interval for the mean assembly time, in minutes, for units of the new product?

    (A) $3.92 \pm 1.645 \dfrac{0.45}{\sqrt{14}}$

    (B) $3.92 \pm 1.753 \dfrac{0.45}{\sqrt{14}}$

    (C) $3.92 \pm 1.761 \dfrac{0.45}{\sqrt{14}}$

    (D) $3.92 \pm 1.753 \dfrac{0.45}{\sqrt{15}}$

    (E) $3.92 \pm 1.761 \dfrac{0.45}{\sqrt{15}}$

17. A company has 1000 employees evenly distributed throughout five assembly plants. A sample of 30 employees is to be chosen as follows. Each of the five managers will be asked to place the 200 time cards of their respective employees in a bag, shake them up, and randomly draw out six names. The six names from each plant will be put together to make up the sample. Will this method result in a simple random sample of the 1000 employees?

    (A) Yes, because every employee has the same chance of being selected.
    (B) Yes, because every plant is equally represented.
    (C) Yes, because this is an example of stratified sampling, which is a special case of simple random sampling.
    (D) No, because the plants are not chosen randomly.
    (E) No, because not every group of 30 employees has the same chance of being selected.

18. Given that $P(E) = 0.32$, $P(F) = 0.15$, and $P(E \cap F) = 0.048$, which of the following is a correct conclusion?

(A) The events E and F are both independent and mutually exclusive.

(B) The events E and F are neither independent nor mutually exclusive.

(C) The events E and F are mutually exclusive but not independent.

(D) The events E and F are independent but not mutually exclusive.

(E) The events E and F are independent, but there is insufficient information to determine whether or not they are mutually exclusive.

19. The number of leasable square feet of office space available in a city on any given day has a roughly normal distribution with mean 640,000 square feet and standard deviation 18,000 square feet. What is the interquartile range for this distribution?

(A) 652,000 – 628,000

(B) 658,000 – 622,000

(C) 667,000 – 613,000

(D) 676,000 – 604,000

(E) 694,000 – 586,000

20. Consider the following back-to-back stemplot:

Left	Stem	Right
	0	348
	1	01256
843	2	29
65210	3	2557
92	4	
7552	5	6
	6	1458
6	7	09
8541	8	
90	9	

Key: 2|5|6 represents a value of 52 on the left and 56 on the right.

Which of the following is a correct statement?

(A) The distributions have the same mean.

(B) The distributions have the same median.

(C) The interquartile range of the distribution to the left is 20 greater than the interquartile range of the distribution to the right.

(D) The distributions have the same variance.

(E) None of the above is correct.

21. Which of the following is a correct statement?

    (A) A study results in a 99% confidence interval estimate of (34.2, 67.3). This means that in about 99% of all samples selected by this method, the sample means will fall between 34.2 and 67.3.
    (B) A high confidence level may be obtained no matter what the sample size.
    (C) The central limit theorem is most useful when drawing samples from normally distributed populations.
    (D) The sampling distribution for a mean has standard deviation $\frac{\sigma}{\sqrt{n}}$ only when n is sufficiently large (typically one uses $n \geq 30$).
    (E) The center of any confidence interval is the population parameter.

22. The binomial distribution is an appropriate model for which of the following?

    (A) The number of minutes in an hour for which the Dow-Jones average is above its beginning average for the day.
    (B) The number of cities among the 10 largest in New York State for which the weather is cloudy for most of a given day.
    (C) The number of drivers wearing seat belts if 10 consecutive drivers are stopped at a police roadblock.
    (D) The number of A's a student receives in his/her five college classes.
    (E) None of the above.

23. Suppose two events, E and F, have nonzero probabilities p and q, respectively. Which of the following is impossible?

    (A) $p + q > 1$
    (B) $p - q < 0$
    (C) $p / q > 1$
    (D) E and F are neither independent nor mutually exclusive.
    (E) E and F are both independent and mutually exclusive.

24. An inspection procedure at a manufacturing plant involves picking four items at random and accepting the whole lot if at least three of the four items are in perfect condition. If in reality 90% of the whole lot are perfect, what is the probability that the lot will be accepted?

    (A) $(0.9)^4$
    (B) $1 - (0.9)^4$
    (C) $4(0.9)^3(0.1)$
    (D) $0.1 - 4(0.9)^3(0.1)$
    (E) $4(0.9)^3(0.1) + (0.9)^4$

25. A town has one high school, which buses students from urban, suburban, and rural communities. Which of the following samples is recommended in studying attitudes toward tracking of students in honors, regular, and below-grade classes?

(A) Convenience sample
(B) Simple random sample (SRS)
(C) Stratified sample
(D) Systematic sample
(E) Voluntary response sample

26. Suppose there is a correlation of $r = 0.9$ between number of hours per day students study and GPAs. Which of the following is a reasonable conclusion?

(A) 90% of students who study receive high grades.
(B) 90% of students who receive high grades study a lot.
(C) 90% of the variation in GPAs can be explained by variation in number of study hours per day.
(D) 10% of the variation in GPAs cannot be explained by variation in number of study hours per day.
(E) 81% of the variation in GPAs can be explained by variation in number of study hours per day.

27. To determine the average number of children living in single-family homes, a researcher picks a simple random sample of 50 such homes. However, even after one follow-up visit the interviewer is unable to make contact with anyone in 8 of these homes. Concerned about nonresponse bias, the researcher picks another simple random sample and instructs the interviewer to keep trying until contact is made with someone in a total of 50 homes. The average number of children is determined to be 1.73. Is this estimate probably too low or too high?

(A) Too low, because of undercoverage bias.
(B) Too low, because convenience samples overestimate average results.
(C) Too high, because of undercoverage bias.
(D) Too high, because convenience samples overestimate average results.
(E) Too high, because voluntary response samples overestimate average results.

28. The graph below shows cumulative proportions plotted against land values (in dollars per acre) for farms on sale in a rural community.

What is the median land value?

(A) $2000
(B) $2250
(C) $2500
(D) $2750
(E) $3000

29. An experiment is to be conducted to determine whether taking fish oil capsules or garlic capsules has more of an effect on cholesterol levels. In past studies it was noted that daily exercise intensity (low, moderate, high) is associated with cholesterol level, but average sleep length (< 5, 5 – 8, > 8 hours) is not associated with cholesterol level. This experiment should be done

(A) by blocking on exercise intensity
(B) by blocking on sleep length
(C) by blocking on cholesterol level
(D) by blocking on capsule type
(E) without blocking

30. A confidence interval estimate is determined from the monthly grocery expenditures in a random sample of n families. Which of the following will result in a smaller margin of error?

 I. A smaller confidence level
 II. A smaller sample standard deviation
 III. A smaller sample size

(A) II only
(B) I and II only
(C) I and III only
(D) II and III only
(E) I, II, and III

31. A medical research team claims that high vitamin C intake increases endurance. In particular, 1000 milligrams of vitamin C per day for a month should add an average of 4.3 minutes to the length of maximum physical effort that can be tolerated. Army training officers believe the claim is exaggerated and plan a test on an SRS of 400 soldiers in which they will reject the medical team's claim if the sample mean is less than 4.0 minutes. Suppose the standard deviation of added minutes is 3.2. If the true mean increase is only 4.2 minutes, what is the probability that the officers will fail to reject the false claim of 4.3 minutes?

(A) $P\left(z < \dfrac{4.0 - 4.3}{3.2/\sqrt{400}}\right)$

(B) $P\left(z > \dfrac{4.0 - 4.3}{3.2/\sqrt{400}}\right)$

(C) $P\left(z < \dfrac{4.3 - 4.2}{3.2/\sqrt{400}}\right)$

(D) $P\left(z > \dfrac{4.3 - 4.2}{3.2/\sqrt{400}}\right)$

(E) $P\left(z > \dfrac{4.0 - 4.2}{3.2/\sqrt{400}}\right)$

32. Consider the two sets $X = \{10, 30, 45, 50, 55, 70, 90\}$ and $Y = \{10, 30, 35, 50, 65, 70, 90\}$. Which of the following is false?

(A) The sets have identical medians.
(B) The sets have identical means.
(C) The sets have identical ranges.
(D) The sets have identical boxplots.
(E) None of the above are false.

33. The weight of an aspirin tablet is 300 milligrams according to the bottle label. An FDA investigator weighs a simple random sample of seven tablets, obtains weights of 299, 300, 305, 302, 299, 301, and 303, and runs a hypothesis test of the manufacturer's claim. Which of the following gives the P-value of this test?

(A) $P(t > 1.54)$ with $df = 6$
(B) $2P(t > 1.54)$ with $df = 6$
(C) $P(t > 1.54)$ with $df = 7$
(D) $2P(t > 1.54)$ with $df = 7$
(E) $0.5P(t > 1.54)$ with $df = 7$

34. A teacher believes that giving her students a practice quiz every week will motivate them to study harder, leading to a greater overall understanding of the course material. She tries this technique for a year, and everyone in the class achieves a grade of at least C. Is this an experiment or an observational study?

    (A) An experiment, but with no reasonable conclusion possible about cause and effect
    (B) An experiment, thus making cause and effect a reasonable conclusion
    (C) An observational study, because there was no use of a control group
    (D) An observational study, but a poorly designed one because randomization was not used
    (E) An observational study, and thus a reasonable conclusion of association but not of cause and effect

35. Which of the following is *not* true with regard to contingency tables for chi-square tests for independence?

    (A) Both variables are categorical.
    (B) Observed frequencies should be whole numbers.
    (C) Expected frequencies should be whole numbers.
    (D) Expected frequencies in each cell should be at least 5, and to achieve this, one sometimes combines categories for one or the other or both of the variables.
    (E) The expected frequency for any cell can be found by multiplying the row total by the column total and dividing by the sample size.

36. Which of the following is a correct statement?

    (A) The probability of a Type II error does not depend on the probability of a Type I error.
    (B) In conducting a hypothesis test, it is possible to simultaneously make both a Type I and a Type II error.
    (C) A Type II error will result if one incorrectly assumes the data are normally distributed.
    (D) In medical disease testing with the null hypothesis that the patient is healthy, a Type I error is associated with a *false negative*; that is, the test incorrectly indicates that the patient is disease free.
    (E) When you choose a significance level α, you're setting the probability of a Type I error to exactly α.

37.

Above is a scatterplot with one point labeled *X*. Suppose you find the least squares regression line. Which of the following is a correct statement?

(A) *X* has the largest residual, in absolute value, of any point on the scatterplot.
(B) *X* is an influential point.
(C) The residual plot will show a curved pattern.
(D) The association between the *x* and *y* variables is very weak.
(E) If the point *X* were removed, the correlation would be 1.

38. A banking corporation advertises that 90% of the loan applications it receives are approved within 24 hours. In a random sample of 50 applications, what is the expected number of loan applications that will be turned down?

(A) 50(0.90)
(B) 50(0.10)
(C) 50(0.90)(0.10)
(D) $\sqrt{50(0.90)(0.10)}$
(E) $\sqrt{\dfrac{(0.90)(0.10)}{50}}$

39. The parallel boxplots below show monthly rainfall summaries for Liberia, West Africa.

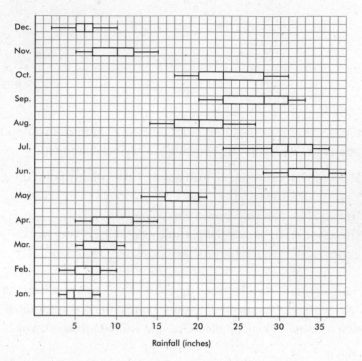

Rainfall (inches)

Which of the following months has the least variability as measured by *interquartile range*?

(A) January
(B) February
(C) March
(D) May
(E) December

40. In comparing the life expectancies of two models of refrigerators, the average years before complete breakdown of 10 model A refrigerators is compared with that of 15 model B refrigerators. The 90% confidence interval estimate of the difference is (6, 12). Which of the following is the most reasonable conclusion?

(A) The mean life expectancy of one model is twice that of the other.
(B) The mean life expectancy of one model is 6 years, while the mean life expectancy of the other is 12 years.
(C) The probability that the life expectancies are different is 0.90.
(D) The probability that the difference in life expectancies is greater than 6 years is 0.90.
(E) We should be 90% confident that the difference in life expectancies is between 6 and 12 years.

STOP

If there is still time remaining, you may review your answers.

SECTION II

Part A

QUESTIONS 1–5

Spend about 65 minutes on this part of the exam.
Percentage of Section II grade—75

> **Directions:** You must show all work and indicate the methods you use. You will be graded on the correctness of your methods and on the accuracy of your results and explanations.

1. The Information Technology Services division at a university is considering installing a new spam filter software product on all campus computers to combat unwanted advertising and spyware. A sample of 60 campus computers was randomly divided into two groups of 30 computers each. One group of 30 was considered to be a control group, while each computer in the other group had the spam filter software installed. During a two-week period each computer user was instructed to keep track of the number of unwanted spam e-mails received. The back-to-back stemplot below shows the distribution of such e-mails received for the control and treatment groups.

```
              Control        Treatment
            9 8 7 5    0   2 2 3 5 5 6 6 6 7 8 8 9 9 9
  9 8 8 7 5 5 5 4 2 2 0    1   0 0 1 1 2 2 5 7 9
      8 7 5 4 4 3 2 1 1 0    2   0 2 4 6
              4 3 2 0 0    3   2 5
                           4   1
```

Key: 0|3|2 represents 30 and 32 spam e-mails for the control and treatment groups, respectively

(a) Compare the distribution of spam e-mails from the control and treatment groups.

(b) The standard deviation of the numbers of e-mails in the control group is 8.1. How does this value summarize variability for the control group data?

(c) A researcher in Information Technology Services calculates a 95% confidence interval for the difference in mean number of spam e-mails received between the control group and the treatment group with the new software and obtains (1.5, 10.9). Assuming all conditions for a two-sample t-interval are met, comment on whether or not there is evidence of a difference in the means for the number of spam e-mails received during a two-week period by computers with and without the software.

(d) The computer users on campus fall into four groups: administrators, staff, faculty, and students. Explain why a researcher might decide to use blocking in setting up this experiment.

2. A game contestant flips three fair coins and receives a score equal to the absolute value of the difference between the number of heads and number of tails showing.

(a) Construct the probability distribution table for the possible scores in this game.

(b) Calculate the expected value of the score for a player.

(c) What is the probability that if a player plays this game three times the total score will be exactly 3?

(d) Suppose a player wins a major prize if he or she can average a score of at least 2. Given the choice, should he or she try for this average by playing 10 times or by playing 15 times? Explain.

3. A high school has a room set aside after school for students to play games. The attendance data are summarized in the following table.

Week	1	2	3	4	5	6	7	8
Attendance	73	78	84	88	29	35	39	44

A student plugs these data into his calculator and comes up with:

$$\widehat{Attendance} = 91.1 - 7.19\,(Week) \text{ with } r = -0.726$$

(a) Interpret the slope of the regression line in context. Does this seem to adequately explain the data? Why or why not?

Another student points out that during the first 4 weeks the school allowed computer games, but then the policy changed.

(b) Give a more appropriate regression analysis modeling attendance in terms of week. Include correlation and interpretation of slope in your analysis.

(c) Explain the above results in terms of a scatterplot.

4. State investigators believe that a particular auto repair facility is fraudulently charging customers for repairs they don't need. As part of their investigation they pick a random sample of ten damaged cars, do their own cost estimate for repair work, and then send the cars to the facility under suspicion for an estimate. The data obtained are shown in the table below.

Car

	1	2	3	4	5	6	7	8	9	10
Investigator Estimate ($)	2585	3040	560	8250	3800	1575	3590	2830	6550	1820
Facility Estimate ($)	2250	3600	800	9100	4675	1920	3710	4050	6300	2200

Is the mean estimate of the facility under suspicion significantly greater than the mean estimate by the investigators? Justify your answer.

5. Five new estimators are being evaluated with regard to quality control in manufacturing professional baseballs of a given weight. Each estimator is tested every day for a month on samples of sizes $n = 10$, $n = 20$, and $n = 40$. The baseballs actually produced that month had a consistent mean weight of 146 grams. The distributions given by each estimator are as follows:

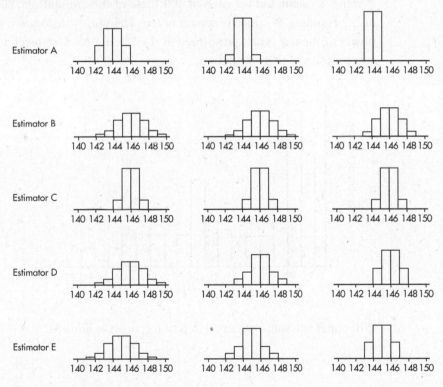

(a) Which of the above appear to be unbiased estimators of the population parameter? Explain.

(b) Which of the above exhibits the lowest variability for $n = 40$? Explain.

(c) Which of the above is the best estimator if the selected estimator will eventually be used with a sample of size $n = 100$? Explain.

SECTION II

Part B

QUESTION 6

Spend about 25 minutes on this part of the exam.
Percentage of Section II grade—25

6. A state congressman will support a proposed heavy fine for driving while texting if there is significant evidence that over 20% of drivers text while driving. An aide to the congressman conducts a small preliminary study, using a hidden highway camera, that counts 49 out of a random sample of 200 drivers texting.

(a) At the 5% significance level, is there sufficient evidence for the congressman to support the proposed fine?

(b) Based on the conclusion in part (a), which type of error, Type I or Type II, is possible? What would be the consequence in context of this study?

(c) Because of the closeness of the preliminary study, the congressman asks about an increase in power of the study if the sample size is increased. The aide decides to investigate this by performing a simulation. For this simulation, the aide assumes that 22% of drivers text and assumes a sample size of 300. The *P*-value is calculated for each of 100 trials of the simulation. The histogram of the resulting *P*-values is shown below. Using this histogram, estimate the power of the test when performed at the 5% significance level and given the assumptions of this simulation.

(d) What does the value obtained in part (c) mean in context?

If there is still time remaining, you may review your answers.

534 AP STATISTICS

Answer Key

SECTION I

1. **C**	9. **D**	17. **E**	25. **C**	33. **B**
2. **A**	10. **B**	18. **D**	26. **E**	34. **A**
3. **C**	11. **B**	19. **A**	27. **C**	35. **C**
4. **E**	12. **E**	20. **D**	28. **A**	36. **E**
5. **B**	13. **E**	21. **B**	29. **A**	37. **C**
6. **E**	14. **C**	22. **E**	30. **B**	38. **B**
7. **C**	15. **A**	23. **E**	31. **E**	39. **E**
8. **B**	16. **E**	24. **E**	32. **E**	40. **E**

Answers Explained

SECTION I

1. **(C)** A complete census can give much information about a population, but it doesn't necessarily establish a cause-and-effect relationship among seemingly related population parameters. While the results of well-designed observational studies might suggest relationships, it is difficult to conclude that there is cause and effect without running a well-designed experiment. If bias is present, increasing the sample size simply magnifies the bias. The control group is selected by the researchers making use of chance procedures.

2. **(A)** In the first class only 40% of the students scored below the given score, while in the second class 80% scored below the same score.

3. **(C)** The control group should have experiences identical to those of the experimental groups except for the treatment under examination. They should not be given a new treatment.

4. **(E)** The negative sign comes about because we are dealing with the difference of proportions. The confidence interval estimate means that we have a certain *confidence* that the difference in population proportions lies in a particular *interval*.

5. **(B)** The desire of the workers for the study to be successful led to a placebo effect. In this example, the result was increased production no matter what the change in music.

6. **(E)** The proportion of successful calls (problem solved) is $\frac{700}{1000}$, so $\frac{700}{1000}$ (450) is the expected number of calls from location 1 that are successful. Alternatively, the proportion of calls from location 1 is $\frac{450}{1000}$, so $\frac{450}{1000}$ (700) gives the expected number of successful calls from location 1. $\left(\text{Or, Expected} = \frac{\text{(column total)(row total)}}{\text{(grand total)}} = \frac{(450)(700)}{1000} \right)$

7. **(C)** The slope and the correlation always have the same sign. Correlation shows association, not causation. Correlation does not apply to categorical data. Correlation measures linear association, so even with a correlation of 0, there may be very strong nonlinear association.

8. **(B)** If two random variables are independent, the mean of the difference of the two random variables is equal to the difference of the two individual means; however, the

variance of the difference of the two random variables is equal to the *sum* of the two individual variances.

9. **(D)** A sample is simply a subset of a population.

10. **(B)** The markings, spaced 15 apart, clearly look like the standard deviation spacings associated with a normal curve.

11. **(B)** With a right tail having probability 0.01, the critical z-score is 2.326. Thus $\mu + 2.326(0.3) = 12$, giving $\mu = 12 - 2.326(0.3)$.

12. **(E)** There is no reason to think that AAA members are representative of the city's drivers. Family members may have similar driving habits and the independence condition would be violated. Random selection is important regardless of the sample size. The larger a random sample, the closer its standard deviation will be to the population standard deviation.

13. **(E)**

$$P\left(\begin{array}{c}\text{child}\\\text{shows}\end{array}\right) = P\left(\begin{array}{c}\text{father}\\\text{shows}\end{array} \cap \begin{array}{c}\text{child}\\\text{shows}\end{array}\right) + P\left(\begin{array}{c}\text{father}\\\text{doesn't}\end{array} \cap \begin{array}{c}\text{child}\\\text{shows}\end{array}\right)$$

$$= (0.25)(0.8) + (0.75)(0.06)$$

$$P\left(\begin{array}{c}\text{father}\\\text{shows}\end{array} \middle| \begin{array}{c}\text{child}\\\text{shows}\end{array}\right) = \dfrac{P\left(\begin{array}{c}\text{father}\\\text{shows}\end{array} \cap \begin{array}{c}\text{child}\\\text{shows}\end{array}\right)}{P\left(\begin{array}{c}\text{child}\\\text{shows}\end{array}\right)} = \dfrac{(0.25)(0.8)}{(0.25)(0.8) + (0.75)(0.06)}$$

14. **(C)** As n increases the probabilities of Type I and Type II errors both decrease.

15. **(A)** The mean equals the common value of all the data elements. The other terms all measure variability, which is zero when all the data elements are equal.

16. **(E)** $SE(\bar{x}) = \dfrac{s}{\sqrt{n}} = \dfrac{0.45}{\sqrt{15}}$ and with $df = 15 - 1 = 14$, the critical t-scores are ±1.761.

17. **(E)** In a simple random sample, every possible group of the given size has to be equally likely to be selected, and this is not true here. For example, with this procedure it is impossible for the employees in the final sample to all be from a single plant. This method is an example of stratified sampling, but stratified sampling does not result in simple random samples.

18. **(D)** $(0.32)(0.15) = 0.048$ so $P(E \cap F) = P(E)P(F)$ and thus E and F are independent. $P(E \cap F) \neq 0$, so E and F are not mutually exclusive.

19. **(A)** The quartiles Q_1 and Q_3 have z-scores of ± 0.67, so

$$Q_1 = 640{,}000 - (0.67)18{,}000 \approx 628{,}000,$$

while $Q_3 = 640{,}000 + (0.67)18{,}000 \approx 652{,}000$. The interquartile range is the difference $Q_3 - Q_1$.

20. **(D)** One set is a shift of 20 units from the other, so they have different means and medians, but they have identical shapes and thus the same variability including IQR, standard deviation, and variance.

21. **(B)** A 99% confidence interval estimate means that in about 99% of all samples selected by this method, the population mean will be included in the confidence interval. The wider the confidence interval, the higher the confidence level, so one may obtain as high a confidence level as one wishes if a very wide confidence interval is accepted. The central limit theorem applies to any population, no matter if it is normally distributed or not. The sampling distribution for a mean always has standard deviation $\frac{\sigma}{\sqrt{n}}$; large enough sample size n refers to the closer the distribution will be to a normal distribution. The center of a confidence interval is the sample statistic, not the population parameter.

22. **(E)** In none of these are the trials independent. For example, as each consecutive person is stopped at a roadblock, the probability the next person has a seat belt on will quickly increase; if a student has one A, the probability is increased that he or she has another A.

23. **(E)** Independence implies $P(E \cap F) = P(E)P(F)$, while mutually exclusive implies $P(E \cap F) = 0$. So to be both independent and mutually exclusive in this example would mean $pq = 0$, which is impossible because p and q are given to be nonzero.

24. **(E)** In a binomial with $n = 4$ and $p = 0.9$:

$$P(\text{at least 3 successes}) = P(\text{exactly 3 successes}) + P(\text{exactly 4 successes})$$
$$= 4(0.9)^3(0.1) + (0.9)^4$$

25. **(C)** In stratified sampling the population is divided into representative groups, and random samples of persons from each group are chosen. In this case it might well be important to have sufficient numbers from each group and to be able to consider separately the responses from each of the three groups—urban, suburban, and rural.

26. **(E)** r^2, the coefficient of determination, indicates the percentage of variation in y that is explained by variation in x.

27. **(C)** It is most likely that the homes at which the interviewer had difficulty finding someone home were homes with fewer children living in them. Replacing these homes with other randomly picked homes will most likely replace homes with fewer children with homes with more children.

28. **(A)** The median corresponds to the 0.5 cumulative proportion.

29. **(A)** Blocking divides the subjects into groups of similar individuals, in this case individuals with similar exercise habits, and runs the experiment on each separate group. This controls the known effect of variation in exercise level on cholesterol level.

30. **(B)** The margin of error varies directly with the critical z-value and directly with the standard deviation of the sample, but inversely with the square root of the sample size.

31. **(E)** $\sigma_{\bar{x}} = \dfrac{3.2}{\sqrt{400}}$. With a true mean increase of 4.2, the z-score for 4.0 is

$$\frac{4.0 - 4.2}{3.2 / \sqrt{400}}$$

and the officers fail to reject the claim if the sample mean has z-score greater than this.

32. **(E)** Both have 50 for their means and medians, both have a range of $90 - 10 = 80$, and both have identical boxplots, with first quartile 30 and third quartile 70.

33. **(B)** Since we are not told that the investigator suspects that the average weight is over 300 mg or is under 300 mg, and since a tablet containing too little or too much of a drug clearly should be brought to the manufacturer's attention, this is a two-sided test. Thus, the P-value is twice the tail probability obtained (using the t-distribution with $df = n - 1 = 6$).

34. **(A)** This study was an experiment because a treatment (weekly quizzes) was imposed on the subjects. However, it was a poorly designed experiment with no use of randomization and no control over confounding variables. An example of an experiment in which cause and effect might have been possible would have been if half the class was randomly assigned to take weekly quizzes, while the other half did not take the quizzes.

35. **(C)** The expected frequencies, as calculated by the rule in (E), may not be whole numbers.

36. **(E)** The probabilities of Type I and Type II error are related; for example, lowering the Type I error increases the probability of a Type II error. A Type I error can be made only if the null hypothesis is true, while a Type II error can be made only if the null hypothesis is false. In medical testing, with the usual null hypothesis that the patient is healthy, a Type I error is that a healthy patient is diagnosed with a disease, that is, a *false positive*. We reject H_0 when the P-value falls below α, and when H_0 is true this rejection will happen precisely with probability α.

37. **(C)** X is probably very close to the least squares regression line and so has a small residual. Removing X will change the regression line very little if at all, and so it is not an influential point. The association between the x and y variables is very strong, just not linear. Correlation measures the strength of a *linear* relationship, which is very weak regardless of whether the point X is present or not.

38. **(B)** The probability of an application being turned down is $1 - 0.90 = 0.10$, and the expected value of a binomial with $n = 50$ and $p = 0.10$ is $np = 50(0.10)$.

39. **(E)** A boxplot gives a five-number summary: smallest value, 25th percentile (Q_1), median, 75th percentile (Q_3), and largest value. The interquartile range is given by $Q_3 - Q_1$, or the total length of the two "boxes" minus the "whiskers."

40. **(E)** Using a measurement from a sample, we are never able to say *exactly* what a population mean is; rather we always say we have a certain *confidence* that the population mean lies in a particular *interval*.

SECTION II

1. (a) A complete answer compares shape, center, and spread, and mentions content.
   Shape. The control group distribution is somewhat bell-shaped and symmetric, while the treatment group distribution is somewhat skewed right.
   Center. The center of the control group distribution is around 20, which is greater than the center of the treatment group distribution, which is somewhere around 10 to 12.
   Spread. The spread of the control group distribution, 5 to 34, is less than the spread of the treatment group distribution, which is 2 to 41.

   (b) For computers in the control group (no spam software), the number of spam e-mails received varies an "average" amount of 8.1 from the mean number of spam e-mails received in the control group.

   (c) Since the 95% confidence interval for the difference does not contain zero, the researcher can conclude the observed difference in mean numbers of spam e-mails received between the control group and the treatment group that received spam software is significant.

   (d) It may well be that the four groups—administrators, staff, faculty, and students—are each exposed to different kinds of spam e-mail risks, and possibly the software will be more or less of a help to each group. In that case, the researcher should in effect run four separate experiments on the homogeneous groups, called blocks. Even though results from the different blocks will be combined at the end, some conclusions may be more specific.

SCORING

Part (a) is essentially correct for correctly comparing shape, center, and spread. Part (a) is partially correct for correctly comparing two of the three features. Lower an E to a P or a P to an I, if context is never mentioned.

Part (b) is essentially correct if standard deviation is explained correctly in context of this problem. Part (b) is partially correct if there is a correct explanation of SD but no reference to context.

Part (c) is essentially correct for noting that zero is not in the interval so the observed difference is significant, and stating this in context of the problem. Part (c) is partially correct for noting that zero is not in the interval so the observed difference is significant, but failing to put this conclusion in context of the problem.

Part (d) is essentially correct if the purpose of blocking is correctly explained in context of this problem. Part (d) is partially correct if the general purpose of blocking is correctly explained but not in context of this problem.

Count partially correct answers as one-half an essentially correct answer.

4	**Complete Answer**	Four essentially correct answers.
3	**Substantial Answer**	Three essentially correct answers.
2	**Developing Answer**	Two essentially correct answers.
1	**Minimal Answer**	One essentially correct answer.

Use a holistic approach to decide a score totaling between two numbers.

2. (a) Listing the eight possibilities: {HHH, HHT, HTH, HTT, THH, THT, TTH, TTT} clearly shows that the only possibilities for the absolute value of the difference are 1 with probability 6/8 = 0.75 and 3 with probability 2/8 = 0.25. The table is:

Absolute difference	Probability
1	0.75
3	0.25

(b) $E = \Sigma\, xP(x) = 1(0.75) + 3(0.25) = 1.5$

(c) Since the only possible scores for each game are 1 and 3, the only way to have a total score of 3 in three games is to score 1 in each game. The probability of this is $(0.75)^3 = 0.421875$ [or $\left(\frac{3}{4}\right)^3 = \frac{27}{64}$].

(d) The more times the game is played, the closer the average score will be to the expected value of 1.5. The player does not want to average close to 1.5, so should prefer playing 10 times rather than 15 times.

SCORING

Part (a) is essentially correct for the correct probability distribution table. Part (a) is partially correct for one minor error.

Part (b) is essentially correct for the correct calculation of expected value based on the answer given in Part (a). Part (b) is partially correct for the correct formula for expected value but an incorrect calculation based on the answer given in Part (a).

Part (c) is essentially correct for the correct probability calculation with some indication of where the answer is coming from. Part (c) is partially correct for the correct probability with no work shown.

Part (d) is essentially correct for choosing 10 and giving a clear explanation. Part (d) is partially correct for choosing 10 and giving a weak explanation. Part (d) is incorrect for choosing 10 with no explanation or with an incorrect explanation.

Count partially correct answers as one-half an essentially correct answer.

4	**Complete Answer**	Four essentially correct answers.
3	**Substantial Answer**	Three essentially correct answers.
2	**Developing Answer**	Two essentially correct answers.
1	**Minimal Answer**	One essentially correct answer.

Use a holistic approach to decide a score totaling between two numbers.

3. (a) The slope of –7.19 says that the attendance *dropped* an average of 7.19 students per week. No, this does not seem to adequately explain the data, because the attendance *increased* every week during the first 4 weeks, and again *increased* every week during the final 4 weeks.

(b) Modeling the first 4 weeks (computer games allowed) gives $\widehat{Attendance} = 68 + 5.1(Week)$ with $r = 0.997$, and modeling the final 4 weeks (no computer games) gives $\widehat{Attendance} = 4.9 + 4.9(Week)$ with $r = 0.997$. These models give an average increase in attendance each week of 5.1 and 4.9, respectively, as well as much higher correlations.

(c)

The scatterplot clearly shows the linearity in the first 4 weeks and in the final 4 weeks, and also the nonlinearity of the full set of data.

SCORING

Part (a) is essentially correct if the slope is correctly interpreted in context and a reasonable explanation is given as to why this slope does not explain the data. Part (a) is partially correct for one of these two components correct.

Part (b) is essentially correct for two separate regression models, correct interpretations of the slopes, and noting the increased correlation. Part (b) is partially correct if one of these components is missing.

Part (c) is essentially correct if the scatterplot is correctly drawn and the strong linearity in the first four weeks and separately in the final four weeks is noted. Part (c) is partially correct for a correctly drawn scatterplot but no observations on linearity.

4 Complete Answer   All three parts essentially correct.

3 Substantial Answer   Two parts essentially correct and one part partially correct.

2 Developing Answer   Two parts essentially correct OR one part essentially correct and one or two parts partially correct OR all three parts partially correct.

1 Minimal Answer   One part essentially correct OR two parts partially correct.

4. This is a paired data test, not a two-sample test, with four parts to a complete solution.

Part 1: Must state a correct pair of hypotheses and defiine the parameter of interest.

Either $H_0: \mu_d = 0$ and $H_a: \mu_d < 0$ where μ_d is the mean difference between what an investigator would estimate and what the facility would estimate.

$H_0: \mu_1 - \mu_2 = 0$ and $H_a: \mu_1 - \mu_2 < 0$ where μ_1 is the mean estimate of the investigator and μ_2 is the mean estimate of the facility.

Part 2: Must name the test and check the conditions.

This is a paired t-test, that is, a single sample hypothesis test on the set of differences.

Conditions:

Random sample (given) and it is reasonable to assume that the 10 data pairs are independent of each other. Normality of the population distribution of differences should be checked graphically on the sample data using a histogram, or a dotplot, or a normal probability plot:

The histogram and dotplot show rough symmetry and no extreme skewness or outliers, while the normal probability plot is roughly linear.

Part 3: Must find the test statistic t and the P-value.

Putting the *differences* {335, −560, ..., −380} in a List, calculator software (such as T-Test on the TI-84 or 1-Sample tTest on the Casio Prizm) gives $t = -2.564$ and $P = 0.0152$.

Part 4: Linking to the P-value, give a correct conclusion in context.

With this small a P-value, $0.0152 < 0.05$, there is sufficient evidence to reject H_0. That is, there is sufficient evidence that the mean estimate of the facility under suspicion is greater than the mean estimate by investigators.

Part 1 is essentially correct for a correct statement of the hypotheses (in terms of a population mean). Part 1 is partially correct for a correct statement of hypotheses but missing identification of the variable.

Part 2 is essentially correct if the test is correctly identified by name or formula and a graphical check of the normality condition is given. Part 2 is partially correct for only one of these two elements.

Part 3 is essentially correct for a correct calculation of both the test statistic t and the P-value. Part 3 is partially correct for only one of these two elements.

Part 4 is essentially correct for a correct conclusion in context, linked to the P-value. Part 4 is partially correct for a correct conclusion but missing either context or linkage.

Count partially correct answers as one-half an essentially correct answer.

4	**Complete Answer**	All four essentially correct answers.
3	**Substantial Answer**	Three essentially correct answers.
2	**Developing Answer**	Two essentially correct answers.
1	**Minimal Answer**	One essentially correct answer.

Use a holistic approach to decide a score totaling between two numbers.

5. (a) A statistic used to estimate a population parameter is unbiased if the mean of the sampling distribution of the statistic is equal to the true value of the parameter being estimated. Estimators B, C, and D appear to have means equal to the population mean of 146.

(b) For $n = 40$, estimator A exhibits the lowest variability, with a range of only 2 grams compared to the other ranges of 6 grams, 4 grams, 4 grams, and 4 grams.

(c) The estimator should have a distribution centered at 146, thus eliminating A and E. As n increases, D shows tighter clustering around 146 than does B. Finally, while C looks better than D for $n = 40$, the estimator will be used with $n = 100$, and the D distribution is clearly converging as the sample size increases while the C distribution remains the same. Choose D.

Part (a) is essentially correct for a correct answer with a good explanation of what unbiased means, and is partially correct for a correct answer with a weak explanation. Part (a) is incorrect for a correct answer with no explanation or with an incorrect explanation.

Part (b) is essentially correct for a correct answer together with some numerical justification, and is partially correct for a correct answer with a weak explanation. Part (b) is incorrect for a correct answer with no explanation or with an incorrect explanation.

Part (c) is essentially correct for a correct answer with a good explanation, and is partially correct for a correct answer with a weak explanation. Part (c) is incorrect for a correct answer with no explanation or with an incorrect explanation.

4	**Complete Answer**	All three parts essentially correct.
3	**Substantial Answer**	Two parts essentially correct and one part partially correct.
2	**Developing Answer**	Two parts essentially correct OR one part essentially correct and one or two parts partially correct OR all three parts partially correct.
1	**Minimal Answer**	One part essentially correct OR two parts partially correct.

6. (a) Procedure: One-sample z-test of a population proportion.

Hypotheses: H_0: $p = 0.20$ and H_a: $p > 0.20$ where p = proportion of all drivers who text while driving.

Conditions: Both $np = (200)(0.20) = 40 \geq 10$ and $n(1 - p) = (200)(0.80) = 160 \geq 10$

Random sample (given).

$n = 200$ is less than 10% of all drivers.

Mechanics: $z = 1.591$ with a P-value of 0.0558.

Conclusion: With this large a P-value, $0.0558 > 0.05$, there is not sufficient evidence to reject the null hypothesis. In other words, there is not sufficient evidence for the congressman to support the proposed fine.

(b) We did not reject the null hypothesis and so possibly committed a Type II error, that is, a mistaken failure to reject a false null hypothesis. The consequence would be that the congressman would fail to support the proposed fine even though over 20% of drivers text and the congressman intended to support the proposed fine under this circumstance.

(c) To estimate the power of the test, we must estimate the probability that the simulated P-value is less than the significance level of 0.05. We note that the frequency from 0.0 to 0.05 is 18. Since there were 100 simulations, our answer is $\frac{18}{100} = 0.18$.

(d) Under the assumption that 22% of drivers text and a sample size of 300, the probability of rejecting the null hypothesis (which says that $p = 0.20$) is estimated to be 0.18.

SCORING

Part (a1) is essentially correct for naming the procedure, checking the conditions, and stating the hypotheses. It is partially correct for having correct answers to two of these three parts.

Part (a2) is essentially correct for correct mechanics (z and P), and for giving a correct conclusion in context with linkage to the P-value. It is partially correct for a correct conclusion based on incorrect mechanics or for correct mechanics and a conclusion missing either linkage or context.

Part (b) is essentially correct for correctly picking a Type II error and giving a reasonable consequence. It is partially correct for picking a Type II error without a reasonable consequence or for picking a Type I error and giving a reasonable consequence of a Type I error.

Parts (c) and (d) together are essentially correct for a correct calculation of power with work shown and a correct explanation in context of the value obtained. They are partially correct for having one of these two parts correct.

Count partially correct answers as one-half an essentially correct answer.

4 Complete Answer   Four essentially correct answers.

3 Substantial Answer   Three essentially correct answers.

2 Developing Answer   Two essentially correct answers.

1 Minimal Answer   One essentially correct answer.

Use a holistic approach to decide a score totaling between two numbers.

ANSWER SHEET
Practice Test 4

1. Ⓐ Ⓑ Ⓒ Ⓓ Ⓔ   11. Ⓐ Ⓑ Ⓒ Ⓓ Ⓔ   21. Ⓐ Ⓑ Ⓒ Ⓓ Ⓔ   31. Ⓐ Ⓑ Ⓒ Ⓓ Ⓔ

2. Ⓐ Ⓑ Ⓒ Ⓓ Ⓔ   12. Ⓐ Ⓑ Ⓒ Ⓓ Ⓔ   22. Ⓐ Ⓑ Ⓒ Ⓓ Ⓔ   32. Ⓐ Ⓑ Ⓒ Ⓓ Ⓔ

3. Ⓐ Ⓑ Ⓒ Ⓓ Ⓔ   13. Ⓐ Ⓑ Ⓒ Ⓓ Ⓔ   23. Ⓐ Ⓑ Ⓒ Ⓓ Ⓔ   33. Ⓐ Ⓑ Ⓒ Ⓓ Ⓔ

4. Ⓐ Ⓑ Ⓒ Ⓓ Ⓔ   14. Ⓐ Ⓑ Ⓒ Ⓓ Ⓔ   24. Ⓐ Ⓑ Ⓒ Ⓓ Ⓔ   34. Ⓐ Ⓑ Ⓒ Ⓓ Ⓔ

5. Ⓐ Ⓑ Ⓒ Ⓓ Ⓔ   15. Ⓐ Ⓑ Ⓒ Ⓓ Ⓔ   25. Ⓐ Ⓑ Ⓒ Ⓓ Ⓔ   35. Ⓐ Ⓑ Ⓒ Ⓓ Ⓔ

6. Ⓐ Ⓑ Ⓒ Ⓓ Ⓔ   16. Ⓐ Ⓑ Ⓒ Ⓓ Ⓔ   26. Ⓐ Ⓑ Ⓒ Ⓓ Ⓔ   36. Ⓐ Ⓑ Ⓒ Ⓓ Ⓔ

7. Ⓐ Ⓑ Ⓒ Ⓓ Ⓔ   17. Ⓐ Ⓑ Ⓒ Ⓓ Ⓔ   27. Ⓐ Ⓑ Ⓒ Ⓓ Ⓔ   37. Ⓐ Ⓑ Ⓒ Ⓓ Ⓔ

8. Ⓐ Ⓑ Ⓒ Ⓓ Ⓔ   18. Ⓐ Ⓑ Ⓒ Ⓓ Ⓔ   28. Ⓐ Ⓑ Ⓒ Ⓓ Ⓔ   38. Ⓐ Ⓑ Ⓒ Ⓓ Ⓔ

9. Ⓐ Ⓑ Ⓒ Ⓓ Ⓔ   19. Ⓐ Ⓑ Ⓒ Ⓓ Ⓔ   29. Ⓐ Ⓑ Ⓒ Ⓓ Ⓔ   39. Ⓐ Ⓑ Ⓒ Ⓓ Ⓔ

10. Ⓐ Ⓑ Ⓒ Ⓓ Ⓔ   20. Ⓐ Ⓑ Ⓒ Ⓓ Ⓔ   30. Ⓐ Ⓑ Ⓒ Ⓓ Ⓔ   40. Ⓐ Ⓑ Ⓒ Ⓓ Ⓔ

Practice Test 4

Spend 90 minutes on this part of the exam.

> **Directions:** The questions or incomplete statements that follow are each followed by five suggested answers or completions. Choose the response that best answers the question or completes the statement.

1. A company wishes to determine the relationship between the number of days spent training employees and their performances on a job aptitude test. Collected data result in a least squares regression line, $\hat{y} = 12.1 + 6.2x$, where x is the number of training days and $\hat{y}$ is the predicted score on the aptitude test. Which of the following statements best interprets the slope and y-intercept of the regression line?

   (A) The base score on the test is 12.1, and for every day of training one would expect, on average, an increase of 6.2 on the aptitude test.
   (B) The base score on the test is 6.2, and for every day of training one would expect, on average, an increase of 12.1 on the aptitude test.
   (C) The mean number of training days is 12.1, and for every additional 6.2 days of training one would expect, on average, an increase of one unit on the aptitude test.
   (D) The mean number of training days is 6.2, and for every additional 12.1 days of training one would expect, on average, an increase of one unit on the aptitude test.
   (E) The mean number of training days is 12.1, and for every day of training one would expect, on average, an increase of 6.2 on the aptitude test.

2. To survey the opinions of the students at your high school, a researcher plans to select every twenty-fifth student entering the school in the morning. Assuming there are no absences, will this result in a simple random sample of students attending your school?

(A) Yes, because every student has the same chance of being selected.
(B) Yes, but only if there is a single entrance to the school.
(C) Yes, because the 24 out of every 25 students who are not selected will form a control group.
(D) Yes, because this is an example of systematic sampling, which is a special case of simple random sampling.
(E) No, because not every sample of the intended size has an equal chance of being selected.

3. Consider a hypothesis test with $H_0 : \mu = 70$ and $H_a : \mu < 70$. Which of the following choices of significance level and sample size results in the greatest power of the test when $\mu = 65$?

(A) $\alpha = 0.05$, $n = 15$
(B) $\alpha = 0.01$, $n = 15$
(C) $\alpha = 0.05$, $n = 30$
(D) $\alpha = 0.01$, $n = 30$
(E) There is no way of answering without knowing the strength of the given power.

4. Suppose that 60% of a particular electronic part last over 3 years, while 70% last less than 6 years. Assuming a normal distribution, what are the mean and standard deviation with regard to length of life of these parts?

(A) $\mu = 3.677$, $\sigma = 3.561$
(B) $\mu = 3.977$, $\sigma = 3.861$
(C) $\mu = 4.177$, $\sigma = 3.561$
(D) $\mu = 4.377$, $\sigma = 3.261$
(E) The mean and standard deviation cannot be computed from the information given.

5. The label on a package of cords claims that the breaking strength of a cord is 3.5 pounds, but a hardware store owner believes the real value is less. She plans to test 36 such cords; if their mean breaking strength is less than 3.25 pounds, she will reject the claim on the label. If the standard deviation for the breaking strengths of all such cords is 0.9 pounds, what is the probability of mistakenly rejecting a true claim?

(A) $P\left(z < \dfrac{3.25 - 3.5}{0.9}\right)$

(B) $P\left(z > \dfrac{3.25 - 3.5}{0.9}\right)$

(C) $P\left(z < \dfrac{3.25 - 3.5}{0.9/\sqrt{36}}\right)$

(D) $2P\left(z < \dfrac{3.25 - 3.5}{0.9/\sqrt{36}}\right)$

(E) $P\left(z > \dfrac{3.25 - 3.5}{0.9/\sqrt{36}}\right)$

6. The graph below shows cumulative proportions plotted against numbers of employees working in mid-sized retail establishments.

What is the approximate interquartile range?

(A) 18
(B) 35
(C) 57
(D) 68
(E) 75

7. What is a placebo?

(A) A method of selection
(B) An experimental treatment
(C) A control treatment
(D) A parameter
(E) A statistic

8. To study the effect of alcohol on reaction time, subjects were randomly selected and given three beers to consume. Their reaction time to a simple stimulus was measured before and after drinking the alcohol. Which of the following is a correct statement?

(A) This study was an observational study.
(B) Lack of blocking makes this a poorly designed study.
(C) The placebo effect is irrelevant in this type of study.
(D) This study was an experiment with no controls.
(E) This study was an experiment in which the subjects were used as their own controls.

9. Suppose that the regression line for a set of data, $\hat{y} = 7x + b$, passes through the point $(-2, 4)$. If $\bar{x}$ and $\bar{y}$ are the sample means of the x- and y-values, respectively, then $\bar{y} =$

(A) $\bar{x}$.
(B) $\bar{x} + 2$.
(C) $\bar{x} - 4$.
(D) $7\bar{x}$.
(E) $7\bar{x} + 18$.

10. Which of the following is a false statement about simple random samples?

(A) A sample must be reasonably large to be properly considered a simple random sample.
(B) Inspection of a sample will give no indication of whether or not it is a simple random sample.
(C) Attributes of a simple random sample may be very different from attributes of the population.
(D) Every element of the population has an equal chance of being picked.
(E) Every sample of the desired size has an equal chance of being picked.

11. A local school has seven math teachers and seven English teachers. When comparing their mean salaries, which of the following is most appropriate?

(A) A two-sample z-test of population means
(B) A two-sample t-test of population means
(C) A one-sample z-test on a set of differences
(D) A one-sample t-test on a set of differences
(E) None of the above are appropriate.

12. Following is a histogram of ages of people applying for a particular high school teaching position.

Which of the following is a correct statement?

(A)   The median age is between 24 and 25.
(B)   The mean age is between 22 and 23.
(C)   The mean age is greater than the median age.
(D)   More applicants are under 23 years of age than are over 23.
(E)   There are a total of 10 applicants.

13. To conduct a survey of which long distance carriers are used in a particular locality, a researcher opens a telephone book to a random page, closes his eyes, puts his finger down on the page, and then calls the next 75 names. Which of the following is a correct statement?

(A)   The procedure results in a simple random sample.
(B)   While the survey design does incorporate chance, the procedure could easily result in selection bias.
(C)   This is an example of cluster sampling with 26 clusters.
(D)   This is an example of stratified sampling with 26 strata.
(E)   Given that the researcher truly keeps his eyes closed, this is a good example of blinding.

14. Which of the following is *not* true about *t*-distributions?

(A)   There are different *t*-distributions for different values of *df* (degrees of freedom).
(B)   *t*-distributions are bell-shaped and symmetric.
(C)   *t*-distributions always have mean 0 and standard deviation 1.
(D)   *t*-distributions are more spread out than the normal distribution.
(E)   The larger the *df* value, the closer the distribution is to the normal distribution.

15. Suppose the probability that a person picked at random has lung cancer is 0.035 and the probability that the person both has lung cancer and is a heavy smoker is 0.014. Given that someone picked at random has lung cancer, what is the probability that the person is a heavy smoker?

(A) 0.035 − 0.014

(B) 0.035 + 0.014

(C) 0.035 + 0.014 − (0.035)(0.014)

(D) $\dfrac{0.035}{1-0.014}$

(E) $\dfrac{0.014}{0.035}$

16. Taxicabs in a metropolitan area are driven an average of 75,000 miles per year with a standard deviation of 12,000 miles. What is the probability that a randomly selected cab has been driven less than 100,000 miles if it is known that it has been driven over 80,000 miles? Assume a roughly normal distribution of miles per year among cabs.

(A) 0.06

(B) 0.34

(C) 0.66

(D) 0.68

(E) 0.94

17. A plant manager wishes to determine the difference in number of accidents per day between two departments. She wants to be 90% certain of the difference in daily averages to within 0.25 accidents per day. Assume standard deviations of 0.8 and 0.5 accidents per day in the two departments, respectively. Which of the following should be used to determine how many days' (*n*) records should be examined?

(A) $1.645\dfrac{\sqrt{0.8^2+0.5^2}}{\sqrt{n}} \le 0.25$

(B) $1.645\dfrac{0.8^2+0.5^2}{\sqrt{n}} \le 0.25$

(C) $1.645\sqrt{\dfrac{0.8+0.5}{n}} \le 0.25$

(D) $1.645\dfrac{\sqrt{0.8^2+0.5^2}}{n} \le 0.25$

(E) $1.645\dfrac{\sqrt{0.8+0.5}}{n} \le 0.25$

18. Suppose the correlation between two variables is $r = 0.19$. What is the new correlation if 0.23 is added to all values of the x-variable, every value of the y-variable is doubled, and the two variables are interchanged?

(A) 0.19
(B) 0.42
(C) 0.84
(D) −0.19
(E) −0.84

19. Two commercial flights per day are made from a small county airport. The airport manager tabulates the number of on-time departures for a sample of 200 days.

Number of on-time departures	0	1	2
Observed number of days	10	80	110

What is the χ^2 statistic for a goodness-of-fit test that the distribution is binomial with probability equal to 0.8 that a flight leaves on time?

(A) $\dfrac{(10-8)^2}{8} + \dfrac{(80-64)^2}{64} + \dfrac{(110-128)^2}{128}$

(B) $\dfrac{(10-8)^2}{10} + \dfrac{(80-64)^2}{80} + \dfrac{(110-128)^2}{110}$

(C) $\dfrac{(10-10)^2}{10} + \dfrac{(80-30)^2}{30} + \dfrac{(110-160)^2}{160}$

(D) $\dfrac{(10-10)^2}{10} + \dfrac{(80-30)^2}{80} + \dfrac{(110-160)^2}{110}$

(E) $\dfrac{(10-66)^2}{10} + \dfrac{(80-67)^2}{80} + \dfrac{(110-67)^2}{110}$

20. A company has a choice of three investment schemes. Option I gives a sure $25,000 return on investment. Option II gives a 50% chance of returning $50,000 and a 50% chance of returning $10,000. Option III gives a 5% chance of returning $100,000 and a 95% chance of returning nothing. Which option should the company choose?

(A) Option II if it wants to maximize expected return
(B) Option I if it needs at least $20,000 to pay off an overdue loan
(C) Option III if it needs at least $80,000 to pay off an overdue loan
(D) All of the above answers are correct.
(E) Because of chance, it really doesn't matter which option it chooses.

21. Suppose $P(X) = 0.35$ and $P(Y) = 0.40$. If $P(X|Y) = 0.28$, what is $P(Y|X)$?

(A) $\dfrac{(0.28)(0.35)}{0.40}$

(B) $\dfrac{(0.28)(0.40)}{0.35}$

(C) $\dfrac{(0.35)(0.40)}{0.28}$

(D) $\dfrac{0.28}{0.40}$

(E) $\dfrac{0.28}{0.35}$

22. To test whether extensive exercise lowers the resting heart rate, a study is performed by randomly selecting half of a group of volunteers to exercise 1 hour each morning, while the rest are instructed to perform no exercise. Is this study an experiment or an observational study?

(A) An experiment with a control group and blinding
(B) An experiment with blocking
(C) An observational study with comparison and randomization
(D) An observational study with little if any bias
(E) None of the above

23. The waiting times for a new roller coaster ride are normally distributed with a mean of 35 minutes and a standard deviation of 10 minutes. If there are 150,000 riders the first summer, which of the following is the shortest time interval associated with 100,000 riders?

(A) 0 to 31.7 minutes
(B) 31.7 to 39.3 minutes
(C) 25.3 to 44.7 minutes
(D) 25.3 to 35 minutes
(E) 39.3 to 95 minutes

24. A medical researcher, studying the effective durations of three over-the-counter pain relievers, obtains the following boxplots from three equal-sized groups of patients, one group using each of the pain relievers.

Which of the following is a correct statement with regard to comparing the effective durations in minutes of the three pain relievers?

(A) All three have the same interquartile range.
(B) More patients had over 210 minutes of pain relief in the (I) group than in either of the other two groups.
(C) More patients had over 240 minutes of pain relief in the (I) group than in either of the other two groups.
(D) More patients had less than 120 minutes of pain relief in the (III) group than in either of the other two groups.
(E) The durations of pain relief in the (II) group form a roughly normal distribution.

25. For their first exam, students in an AP Statistics class studied an average of 4 hours with a standard deviation of 1 hour. Almost everyone did poorly on the exam, and so for the second exam every student studied 10 hours. What is the correlation between the numbers of hours students studied for each exam?

(A) −1
(B) 0
(C) 0.4
(D) 1
(E) There is not sufficient information to answer this question.

26. Suppose that 54% of the graduates from your high school go on to 4-year colleges, 20% go on to 2-year colleges, 19% find employment, and the remaining 7% search for a job. If a randomly selected student is not going on to a 2-year college, what is the probability she will be going on to a 4-year college?

(A) 0.460
(B) 0.540
(C) 0.630
(D) 0.675
(E) 0.730

27. We are interested in the proportion p of people who are unemployed in a large city. Eight percent of a simple random sample of 500 people are unemployed. What is the midpoint for a 95% confidence interval estimate of p?

(A) 0.012
(B) 0.025
(C) 0.475
(D) p
(E) None of the above.

28. Random samples of size n are drawn from a population. The mean of each sample is calculated, and the standard deviation of this set of sample means is found. Then the procedure is repeated, this time with samples of size $4n$. How does the standard deviation of the second group compare with the standard deviation of the first group?

(A) It will be the same.
(B) It will be twice as large.
(C) It will be four times as large.
(D) It will be half as large.
(E) It will be one-quarter as large.

29. Leech therapy is used in traditional medicine for treating localized pain. In a double blind experiment on 50 patients with osteoarthritis of the knee, half are randomly selected to receive injections of leech saliva while the rest receive a placebo. Pain levels 7 days later among those receiving the saliva show a mean of 19.5, while pain levels among those receiving the placebo show a mean of 25.6 (higher numbers indicate more pain). Partial calculator output is shown below.

```
        2-SampTTest
   μ1<μ2
   t=-3.939503313
   df=43.43159286
   x̄1=19.5
   x̄2=25.6
   Sx1=4.5
   Sx2=6.3
   n1=25
   n2=25
```

Which of the following is a correct conclusion?

(A) After 7 days, the mean pain level with the leech treatment is significantly lower than the mean pain level with the placebo at the 0.01 significance level.

(B) After 7 days, the mean pain level with the leech treatment is significantly lower than the mean pain level with the placebo at the 0.05 significance level, but not at the 0.01 level.

(C) After 7 days, the mean pain level with the leech treatment is significantly lower than the mean pain level with the placebo at the 0.10 significance level, but not at the 0.05 level.

(D) After 7 days, the mean pain level with the leech treatment is not significantly lower than the mean pain level with the placebo at the 0.10 significance level.

(E) The proper test should be a one-sample t-test on a set of differences.

30. A survey was conducted to determine the percentage of parents who would support raising the legal driving age to 18. The results were stated as 67% with a margin of error of ±3%. What is meant by ±3%?

(A) Three percent of the population were not surveyed.

(B) In the sample, the percentage of parents who would support raising the driving age is between 64% and 70%.

(C) The percentage of the entire population of parents who would support raising the driving age is between 64% and 70%.

(D) It is unlikely that the given sample proportion result could be obtained unless the true percentage was between 64% and 70%.

(E) Between 64% and 70% of the population were surveyed.

31. It is estimated that 30% of all cars parked in a metered lot outside City Hall receive tickets for meter violations. In a random sample of 5 cars parked in this lot, what is the probability that at least one receives a parking ticket?

(A) $1 - (0.3)^5$
(B) $1 - (0.7)^5$
(C) $5(0.3)(0.7)^4$
(D) $5(0.3)^4 (0.7)$
(E) $5(0.3)^4(0.7) + 10(0.3)^3(0.7)^2 + 10(0.3)^2(0.7)^3 + 5(0.3)(0.7)^4 + (0.7)^5$

32. Data are collected on income levels x versus number of bank accounts y. Summary calculations give $\bar{x} = 32,000$; $s_x = 11,500$; $\bar{y} = 1.7$; $s_y = 0.4$, and $r = 0.42$. What is the slope of the least squares regression line of number of bank accounts on income level?

(A) $\dfrac{(0.42)(0.4)}{11,500}$

(B) $\dfrac{(0.42)(11,500)}{0.4}$

(C) $\dfrac{(1.7)(11,500)}{0.42}$

(D) $\dfrac{(1.7)(0.42)}{11,500}$

(E) $\dfrac{(0.42)(11,500)}{1.7}$

33. Given that the sample has a standard deviation of zero, which of the following is a true statement?

(A) The standard deviation of the population is also zero.
(B) The sample mean and sample median are equal.
(C) The sample may have outliers.
(D) The population has a symmetric distribution.
(E) All samples from the same population will also have a standard deviation of zero.

34. In one study half of a class were instructed to watch exactly 1 hour of television per day, the other half were told to watch 5 hours per day, and then their class grades were compared. In a second study students in a class responded to a questionnaire asking about their television usage and their class grades.

(A) The first study was an experiment without a control group, while the second was an observational study.
(B) The first study was an observational study, while the second was a controlled experiment.
(C) Both studies were controlled experiments.
(D) Both studies were observational studies.
(E) Each study was part controlled experiment and part observational study.

35. All of the following statements are true for all discrete random variables except for which one?

(A) The possible outcomes must all be numerical.

(B) The possible outcomes must be mutually exclusive.

(C) The mean (expected value) always equals the sum of the products obtained by multiplying each value by its corresponding probability.

(D) The standard deviation of a random variable can never be negative.

(E) Approximately 95% of the outcomes will be within two standard deviations of the mean.

36. In leaving for school on an overcast April morning you make a judgment on the null hypothesis: The weather will remain dry. What would the results be of Type I and Type II errors?

(A) Type I error: get drenched
    Type II error: needlessly carry around an umbrella

(B) Type I error: needlessly carry around an umbrella
    Type II error: get drenched

(C) Type I error: carry an umbrella, and it rains
    Type II error: carry no umbrella, but weather remains dry

(D) Type I error: get drenched
    Type II error: carry no umbrella, but weather remains dry

(E) Type I error: get drenched
    Type II error: carry an umbrella, and it rains

37. The mean thrust of a certain model jet engine is 9500 pounds. Concerned that a production process change might have lowered the thrust, an inspector tests a sample of units, calculating a mean of 9350 pounds with a z-score of -2.46 and a P-value of 0.0069. Which of the following is the most reasonable conclusion?

(A) 99.31% of the engines produced under the new process will have a thrust under 9350 pounds.

(B) 99.31% of the engines produced under the new process will have a thrust under 9500 pounds.

(C) 0.69% of the time an engine produced under the new process will have a thrust over 9500 pounds.

(D) There is evidence to conclude that the new process is producing engines with a mean thrust under 9350 pounds.

(E) There is evidence to conclude that the new process is producing engines with a mean thrust under 9500 pounds.

38. In a group of 10 third graders, the mean height is 50 inches with a median of 47 inches, while in a group of 12 fourth graders, the mean height is 54 inches with a median of 49 inches. What is the median height of the combined group?

(A) 48 inches

(B) 52 inches

(C) $\frac{10(47)+12(49)}{22}$ inches

(D) $\frac{10(50)+12(54)}{22}$ inches

(E) The median of the combined group cannot be determined from the given information.

39. A reading specialist in a large public school system believes that the more time students spend reading, the better they will do in school. She plans a middle school experiment in which an SRS of 30 eighth graders will be assigned four extra hours of reading per week, an SRS of 30 seventh graders will be assigned two extra hours of reading per week, and an SRS of 30 sixth graders with no extra assigned reading will be a control group. After one school year, the mean GPAs from each group will be compared. Is this a good experimental design?

(A) Yes.

(B) No, because while this design may point out an association between reading and GPA, it cannot establish a cause-and-effect relationship.

(C) No, because without blinding, there is a strong chance of a placebo effect.

(D) No, because any conclusion would be flawed because of blocking bias.

(E) No, because grade level is a confounding variable.

40. A study at 35 large city high schools gives the following back-to-back stemplot of the percentages of students who say they have tried alcohol.

School Year 2014–15		School Year 2017–18
	0	
2	1	
9	2	
6 6 3	3	1 8
4 3 1 1	4	0 2 9
9 9 8 6 5 3 2 2 0	5	3 3 4 6
9 8 7 6 6 5 1 1	6	1 2 2 2 7
5 4 3 2 2	7	0 1 3 3 5 5 6 7 8 9
5 4 0	8	2 3 4 5 8 8 9
0	9	0 1 1 2

Key: 1|4|0 represents 41% for the 2014–15 school year and 40% for the 2017–18 school year

Which of the following does *not* follow from the above data?

(A) In general, the percentage of students trying alcohol seems to have increased from 2014–15 to 2017–18.

(B) The median alcohol percentage among the 35 schools increased from 2014–15 to 2017–18.

(C) The spread between the lowest and highest alcohol percentages decreased from 2014–15 to 2017–18.

(D) For both school years in most of the 35 schools, most of the students said they had tried alcohol.

(E) The percentage of students trying alcohol increased in each of the schools between 2014–15 and 2017–18.

STOP

If there is still time remaining, you may review your answers.

SECTION II

Part A

QUESTIONS 1–5

Spend about 65 minutes on this part of the exam.
Percentage of Section II grade—75

> **Directions:** You must show all work and indicate the methods you use. You will be graded on the correctness of your methods and on the accuracy of your results and explanations.

1. A guidance counselor at a large college is interested in interviewing students about their college experience. After talking with a local AP Statistics teacher, the guidance counselor plans the following three-stage sampling procedure.

   (1) The counselor will obtain separate lists of the four groups of students on the college campus: freshmen, sophomores, juniors, and seniors.

   (2) Each of the four groups takes a number of large classes together. The counselor will use a random number generator to pick three of these classes for each of the four groups.

   (3) In each of the chosen classes, the counselor will pick every fifth student entering the classroom.

   (a) The first stage above represents what kind of sampling procedure? Give an advantage in using it in this context.

   (b) The second stage above represents what kind of sampling procedure? Give an advantage in using it in this context.

   (c) The third stage above represents what kind of sampling procedure? Give a *disadvantage* in using it in this context.

2. (a) Suppose that nationwide, 69% of all registered voters would answer "Yes" to the question "Do you consider yourself highly focused on this year's presidential election?" Which of the following is more likely: an SRS of 50 registered voters having over 75% answer "Yes," or an SRS of 100 registered voters having over 75% answer "Yes" to the given question? Explain.

   (b) A particular company with 95 employees wishes to survey their employees with regard to interest in the presidential election. They pick an SRS of 30 employees and ask "Do you consider yourself highly focused on this year's presidential election?" Suppose that in fact 60% of all 95 employees would have answered "Yes." Explain why it is not reasonable to say that the distribution for the count in the sample who say "Yes" is a binomial with $n = 30$ and $p = 0.6$.

(c) Suppose that nationwide, 78% of all registered voters would answer "Yes" to the question "Do you consider yourself highly focused on this year's presidential election?" You plan to interview an SRS of 20 registered voters. Explain why it is not reasonable to say that the distribution for the proportion in the sample who say "Yes" is approximately a normal distribution.

3. An instructor takes an anonymous survey and notes exam score, hours studied, and gender for the first exam in a large college statistics class ($n = 250$). A resulting regression model is:

$$\widehat{Score} = 50.90 + 9.45(Hours) + 4.40(Gender)$$

where *Gender* takes the value 0 for males and 1 for females.

(a) Provide an interpretation in context for each of the three numbers appearing in the above model formula.

(b) Sketch the separate prediction lines for males and females resulting from using 0 or 1 in the above model.

Looking at the data separately by gender results in the following two regression models:

For males: $\widehat{Score} = 51.8 + 8.5(Hours)$
For females: $\widehat{Score} = 54.4 + 10.4(Hours)$

(c) What comparative information do the slope coefficients from these two models give that does not show in the original model?

(d) Sketch the prediction lines given by these last two models.

4. A laboratory is testing the concentration level in milligrams per milliliter for the active ingredient found in a pharmaceutical product. In a random sample of five vials of the product, the concentrations were measured at 2.46, 2.57, 2.70, 2.64, and 2.54 mg/ml.

(a) Determine a 95% confidence interval for the mean concentration level in milligrams per milliliter for the active ingredient found in this pharmaceutical product.

(b) Explain in words what effect an increase in confidence level would have on the width of the confidence interval.

(c) Suppose a concentration above 2.70 milligrams per milliliter is considered dangerous. What conclusion is justified by your answers to (a) and (b)?

5. A point is said to have *high leverage* if it is an outlier in the *x*-direction, and a point is said to be *influential* if its removal sharply changes the regression line.

(a) In the scatterplots below, compare points *A* and *B* with regard to having high leverage and with regard to being influential.

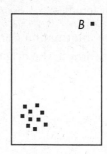

(b) In the scatterplots below, compare points C and D with regard to residuals and influence on β_1, the slope of the regression line.

(c) In the scatterplot below, compare the effect of removing point E or point F to that of removing both E and F.

(d) In the scatterplot below, compare the effect of removing point G or point H to that of removing both G and H.

SECTION II

Part B

QUESTION 6

Spend about 25 minutes on this part of the exam.
Percentage of Section II grade—25

6. Can the number of juvenile arrests for possession of marijuana in any given year be predicted by noting the number of honey-producing bee colonies that year? In a representative sample of 20 years, number of arrests versus number of colonies (in 1000s) shows the following scatterplot and regression output:

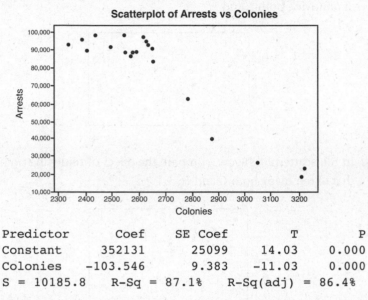

```
Predictor        Coef     SE Coef         T          P
Constant       352131       25099     14.03      0.000
Colonies     -103.546       9.383    -11.03      0.000
S = 10185.8     R-Sq = 87.1%     R-Sq(adj) = 86.4%
```

(a) What is the value of the correlation r, and what is a proper cause-and-effect conclusion?

(b) During the year when there were 2622 thousand honey-producing bee colonies, there were 95,962 juvenile arrests for possession of marijuana. What is the residual for this point? Interpret it in context.

(c) Given that the mean and standard deviation for the colonies variable are $\bar{x} = 2663.8$ and $s_x = 249.0$, what are the mean and standard deviation for the arrests variable?

(d) Assuming all conditions for inference are met, determine a 95% confidence interval for the regression slope. Interpret the confidence interval in context.

(e) Assuming all conditions for inference are met and using a t-distribution with $df = n - 2$, find the 90% confidence interval for the y-intercept. Interpret this in context.

(f) Comment on any concern about extrapolation in the answer to part (e).

If there is still time remaining, you may review your answers.

Answer Key

SECTION I

1. **A**	9. **E**	17. **A**	25. **B**	33. **B**
2. **E**	10. **A**	18. **A**	26. **D**	34. **A**
3. **C**	11. **E**	19. **A**	27. **E**	35. **E**
4. **B**	12. **C**	20. **D**	28. **D**	36. **B**
5. **C**	13. **B**	21. **B**	29. **A**	37. **E**
6. **A**	14. **C**	22. **E**	30. **D**	38. **E**
7. **C**	15. **E**	23. **C**	31. **B**	39. **E**
8. **E**	16. **E**	24. **B**	32. **A**	40. **E**

Answers Explained

SECTION I

1. **(A)** The slope, 6.2, gives the predicted average increase in the y-variable for each unit increase in the x-variable.

2. **(E)** For a simple random sample, every possible group of the given size has to be equally likely to be selected, and this is not true here. For example, with this procedure it will be impossible for all the early arrivals to be together in the final sample. This procedure is an example of systematic sampling, but systematic sampling does not result in simple random samples.

3. **(C)** Power $= 1 - \beta$, and β is smallest when α is more and n is more.

4. **(B)** The critical z-scores for 60% to the right and 70% to the left are -0.253 and 0.524, respectively. Then $\{\mu - 0.253\sigma = 3,\ \mu + 0.524\sigma = 6\}$ gives $\mu = 3.977$ and $\sigma = 3.861$.

5. **(C)** We have H_0: $\mu = 3.5$ and H_a: $\mu < 3.5$. Then $\sigma_{\bar{x}} = \dfrac{0.9}{\sqrt{36}}$ and the z-score of 3.25 is

$$\frac{3.25 - 3.5}{\frac{0.9}{\sqrt{36}}}.$$

6. **(A)** The cumulative proportions of 0.25 and 0.75 correspond to $Q_1 = 57$ and $Q_3 = 75$, respectively, and so the interquartile range is $75 - 57 = 18$.

7. **(C)** A placebo is a control treatment in which members of the control group do not realize whether or not they are receiving the experimental treatment.

8. **(E)** In experiments on people, subjects can be used as their own controls, with responses noted before and after the treatment. However, with such designs there is always the danger of a placebo effect. In this case, subjects might well have slower reaction times after drinking the alcohol because they think they should. Thus the design of choice would involve a separate control group to use for comparison. Blocking is not necessary for a well-designed experiment, and there is no indication that it would be useful here.

9. **(E)** Since $(-2, 4)$ is on the line $\hat{y} = 7x + b$, we have $4 = -14 + b$ and $b = 18$. Thus the regression line is $\hat{y} = 7x + 18$. The point $(\bar{x}, \bar{y})$ is always on the regression line, and so we have $\bar{y} = 7\bar{x} + 18$.

10. **(A)** A simple random sample can be any size.

11. **(E)** With such small populations, censuses instead of samples are used, and there is no resulting probability statement about the difference.

12. **(C)** Half the area is on either side of 23, so 23 is the median. The distribution is skewed to the right, and so the mean is greater than the median. With half the area to each side of 23, half the applicants' ages are to each side of 23. Histograms such as this show relative frequencies, not actual frequencies.

13. **(B)** While the procedure does use some element of chance, all possible groups of size 75 do not have the same chance of being picked, so the result is not a simple random sample. There is a real chance of selection bias. For example, a number of relatives with the same name and all using the same long-distance carrier might be selected.

14. **(C)** While t-distributions do have mean 0, their standard deviations are greater than 1.

15. **(E)**
$$P(\text{smoker}\,|\,\text{cancer}) = \frac{P(\text{smoker} \cap \text{cancer})}{P(\text{cancer})} = \frac{0.014}{0.035}$$

16. **(E)** The critical z-scores are $\dfrac{80{,}000 - 75{,}000}{12{,}000} = 0.42$ and $\dfrac{100{,}000 - 75{,}000}{12{,}000} = 2.08$, with corresponding right tail probabilities of 0.3372 and 0.0188. The probability of being less than 100,000 given that the mileage is over 80,000 is $\dfrac{0.3372 - 0.0188}{0.3372} = 0.94$.
$$\left(\text{Or } \frac{\text{normalcdf}(80000,100000,75000,12000)}{\text{normalcdf}(80000,1E99,75000,12000)} = 0.9450\right)$$

17. **(A)** $\sigma_{\bar{x}_1 - \bar{x}_2} = \sqrt{\sigma_{\bar{x}_1}^2 + \sigma_{\bar{x}_2}^2} = \sqrt{\left(\dfrac{\sigma_1}{\sqrt{n}}\right)^2 + \left(\dfrac{\sigma_2}{\sqrt{n}}\right)^2} = \sqrt{\dfrac{0.8^2}{n} + \dfrac{0.5^2}{n}} = \dfrac{\sqrt{0.8^2 + 0.5^2}}{\sqrt{n}}$

18. **(A)** The correlation coefficient is not changed by adding the same number to every value of one of the variables, by multiplying every value of one of the variables by the same positive number, or by interchanging the x- and y-variables.

19. **(A)** The binomial distribution with $n = 2$ and $p = 0.8$ is $P(0) = (0.2)^2 = 0.04$, $P(1) = 2(0.2)(0.8) = 0.32$, and $P(2) = (0.8)^2 = 0.64$, resulting in expected numbers of $0.04(200) = 8$, $0.32(200) = 64$, and $0.64(200) = 128$. Thus,

$$\chi^2 = \sum \frac{(obs - exp)^2}{exp} = \frac{(10 - 8)^2}{8} + \frac{(80 - 64)^2}{64} + \frac{(110 - 128)^2}{128}$$

20. **(D)** Option II gives the highest expected return: $(50{,}000)(0.5) + (10{,}000)(0.5) = 30{,}000$, which is greater than 25,000 and is also greater than $(100{,}000)(0.05) = 5000$. Option I guarantees that the $20,000 loan will be paid off. Option III provides the only chance of paying off the $80,000 loan. The moral is that the highest expected value is not automatically the "best" answer.

21. **(B)** $P(X \cap Y) = P(X\,|\,Y)\,P(Y) = (0.28)(0.40)$.

Then $P(Y\,|\,X) = \dfrac{P(X \cap Y)}{P(X)} = \dfrac{(0.28)(0.40)}{0.35}$.

22. **(E)** This study is an experiment because a treatment (extensive exercise) is imposed. There is no blinding because subjects clearly know whether or not they are exercising. There is no blocking because subjects are not divided into blocks before random assignment to treatments. For example, blocking would have been used if subjects had been separated by gender or age before random assignment to exercise or not.

23. **(C)** From the shape of the normal curve, the answer is in the middle. The middle two-thirds is between z-scores of ± 0.97, and $35 \pm 0.97(10)$ gives $(25.3, 44.7)$.

24. **(B)** The interquartile range is the length of the box, so they are not all equal. More than 25% of the patients in the A group had over 210 minutes of pain relief, which is not the case for the other two groups. There is no way to positively conclude a normal distribution from a boxplot.

25. **(B)** A scatterplot would be horizontal; the correlation is zero.

26. **(D)** $\dfrac{0.54}{0.54+0.19+0.07} = 0.675$

27. **(E)** The midpoint of the confidence interval is 0.08.

28. **(D)** $\sigma_{\bar{x}} = \dfrac{\sigma}{\sqrt{n}}$, so the standard deviation of sample means is inversely related to the square root of the sample size. Thus, increasing the sample size by a multiple of d^2 divides the standard deviation of the set of sample means by d.

29. **(A)** With $df = 43.43$ and $t = -3.94$, tcdf gives that the P-value is $0.000146 < 0.01$.

30. **(D)** While the sample proportion is between 64% and 70% (more specifically, it is 67%), this is not the meaning of $\pm 3\%$. While the percentage of the entire population is likely to be between 64% and 70%, this is not known for certain.

31. **(B)** In this binomial situation, the probability that a car does not receive a ticket is $1 - 3 = 0.7$, the probability that none of the five cars receives a ticket is $(0.7)^5$, and thus the probability that at least one receives a ticket is $1 - (0.7)^5$.

32. **(A)** $b_1 = r\dfrac{s_y}{s_x} = 0.42\dfrac{0.4}{11,500} = \dfrac{(0.42)(0.4)}{11,500}$

33. **(B)** If the standard deviation of a set is zero, all the values in the set are equal. The mean and median would both equal this common value and so would equal each other. If all the values are equal, there are no outliers. Just because the sample happens to have one common value, there is no reason for this to be true for the whole population. Statistics from one sample can be different from statistics from any other sample.

34. **(A)** The first study is an experiment with two treatment groups and no control group. The second study is observational; the researcher did not randomly divide the subjects into groups and have each group watch a designated number of hours of television per night.

35. **(E)** This refers only to very particular random variables, for example, random variables whose values are the numbers of successes in a binomial probability distribution with large n.

PRACTICE TEST 4   **571**

36. **(B)** A Type I error means that the null hypothesis is correct (the weather will remain dry), but you reject it (thus you needlessly carry around an umbrella). A Type II error means that the null hypothesis is wrong (it will rain), but you fail to reject it (thus you get drenched).

37. **(E)** If the sample statistic is far enough away from the claimed population parameter, we say that there is sufficient evidence to reject the null hypothesis. In this case the null hypothesis is that $\mu = 9500$. The P-value is the probability of obtaining a sample statistic as extreme or more extreme than the one obtained if the null hypothesis is assumed to be true. The smaller the P-value, the more significant the difference between the null hypothesis and the sample results. With $P = 0.0069$, there is strong evidence to reject H_0.

38. **(E)** There are $10 + 12 = 22$ students in the combined group. In ascending order, where are the two middle scores? At least 5 third graders and 6 fourth graders have heights less than or equal to 49 inches, so at most 11 students have heights greater than or equal to 49 and thus the median is less than or equal to 49. At least 5 third graders and 6 fourth graders have heights greater than or equal to 47 inches, so at most 11 students have heights less than or equal to 47 and thus the median is greater than or equal to 47. All that can be said about the median of the combined group is that it is between 47 and 49 inches.

39. **(E)** Good experimental design aims to give each group the same experiences except for the treatment under consideration, Thus, all three SRSs should be picked from the same grade level.

40. **(E)** The stemplot does not indicate what happened for any individual school.

SECTION II

1. (a) This is an example of *stratified sampling*, where the groups (freshmen, sophomores, juniors, and seniors) are strata. The advantage is that the counselor will ensure that the final sample will represent all four groups of students, each of which might have different views of their college experience.

   (b) This an example of *cluster sampling*, where each class (cluster) resembles the overall group from which it is selected. The advantage is that using these clusters is much more practical than trying to sample from among the large groups (freshman, sophomores, juniors, and seniors).

   (c) This is an example of *systematic sampling*, which is quicker and easier than many other procedures. A possible disadvantage occurs if ordering is related to the variable under consideration. For example, in this study, if the order students enter the classroom is related to their views of their college experience, the counselor could end up with a nonrepresentative sample.

SCORING

Part (a) is essentially correct for identifying the procedure and giving a correct advantage in context. Part (a) is partially correct for one of these two elements.

Part (b) is essentially correct for identifying the procedure and giving a correct advantage in context. Part (b) is partially correct for one of these two elements.

Part (c) is essentially correct for identifying the procedure and giving a correct disadvantage in context. Part (c) is partially correct for one of these two elements.

4	**Complete Answer**	All three parts essentially correct.
3	**Substantial Answer**	Two parts essentially correct and one part partially correct.
2	**Developing Answer**	Two parts essentially correct OR one part essentially correct and one or two parts partially correct OR all three parts partially correct.
1	**Minimal Answer**	One part essentially correct OR two parts partially correct.

---

2. (a) The SRS with $n = 50$ is more likely to have a sample proportion greater than 75%. In each case, the sampling distribution of $\hat{p}$ is approximately normal with a mean of 0.69 and a standard deviation of $\sigma_{\hat{p}} = \sqrt{\dfrac{pq}{n}} = \sqrt{\dfrac{(0.69)(0.31)}{n}} = \dfrac{0.462}{\sqrt{n}}$. Thus, the sampling distribution with $n = 50$ will have more variability than the sampling distribution with $n = 100$. Thus the tail area ($\hat{p} > 0.75$) will be larger for $n = 50$.

   (b) The size of the sample, 30, is much too large compared to the size of the population, 95. With a sample size this close to the population size, the necessary assumption of independence does not follow. A commonly accepted condition is that the sample should be no more than 10% of the population.

(c) A normal distribution may be used to approximate a binomial distribution only if the sample size n is not too small. The commonly used condition check is that both np and nq are at least 10. In this case $nq = (20)(1 - 0.78) = 4.4$.

3. (a) The value 50.90 estimates the average score of males who spend 0 hours studying. The value 9.45 estimates the average increase in score for each additional hour study time. For any fixed number of hours study time, the value 4.40 estimates the average number of points that females score higher than males.

(b)

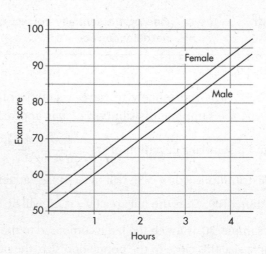

(c) For each additional hour of study time, the scores of males increase an average of 8.5 while those of females increase an average of 10.4. Additional hours of study time appear to benefit females more than males, something that does not show in the original model.

(d)

4. (a) Name the procedure: This is a one-sample *t*-interval for the mean.

Check the conditions: We are given that this is a random sample, $n = 5$ is less than 10% of all vials of this product, and a dotplot

makes the nearly normal condition reasonable.

Mechanics: Putting the data in a List, calculator software (such as TInterval on the TI-84 or 1-Sample tInterval on the Casio Prizm) gives (2.4674, 2.6966).

Conclusion in context: We are 95% confident that the mean concentration level for the active ingredient found in this pharmaceutical product is between 2.467 and 2.697 milligrams per milliliter.

(b) Raising the confidence level would increase the width of the confidence interval.

(c) Since the whole confidence interval (2.467, 2.697) is below the critical 2.7, at the 95% confidence level the mean concentration is at a safe level. However, with 2.697 so close to 2.7, based on the statement in (c), if the confidence level is raised we are no longer confident that the mean concentration is at a safe level.

SCORING

Part (a) has two components. The first component is essentially correct for correctly naming the procedure and checking the conditions, and is partially correct for one of these two. The second component is essentially correct for correct mechanics and a correct conclusion in context, and is partially correct for one of these two.

Part (b) is either essentially correct or incorrect.

Part (c) is essentially correct for a correct conclusion in context for both the 95% interval and for if the confidence is raised, both conclusions referencing the answers from (a) and (b). Part (c) is partially correct for a correct conclusion in context for either of the confidence intervals or for both but with a weak explanation.

Count partially correct answers as one-half an essentially correct answer.

4	**Complete Answer**	Four essentially correct answers.
3	**Substantial Answer**	Three essentially correct answers.
2	**Developing Answer**	Two essentially correct answers.
1	**Minimal Answer**	One essentially correct answer.

Use a holistic approach to decide a score totaling between two numbers.

5. (a) Points A and B both have high leverage; that is, both their x-coordinates are outliers in the x-direction. However, point B is influential (its removal sharply changes the regression line), while point A is not influential (it lies directly on the regression line, so its removal will not change the line).

    (b) Point C lies off the regression line so its residual is much greater than that of point D, whose residual is 0 (point D lies on the regression line). However, the removal of point C will very minimally affect the slope of the regression line, if at all, while the removal of point D dramatically affects the slope of the regression line.

    (c) Removal of either point E or point F minimally affects the regression line, while removal of both has a dramatic effect.

    (d) Removing either point G or point H will definitely affect the regression line (pulling the line toward the remaining of the two points), while removing both will have little, if any, effect on the line.

SCORING

Each of parts (a), (b), (c), and (d) has two components and is scored essentially correct for both components correct and partially correct for one component correct.

Give 1 point for each essentially correct part and $\frac{1}{2}$ point for each partially correct part.

Count partially correct answers as one-half an essentially correct answer.

4	**Complete Answer**	4 points
3	**Substantial Answer**	3 points
2	**Developing Answer**	2 points
1	**Minimal Answer**	1 point

Use a holistic approach to decide a score totaling between two numbers.

---

6. (a) $-\sqrt{0.871} = -0.933$ where the sign is negative because the correlation always has the same sign as the slope, which is negative here. No cause-and-effect conclusion is appropriate. Correlation and association do not imply causation. (In cases like this, most likely the two unrelated data sets follow the same time trend, resulting in significant correlation but no connection.)

    (b) $352{,}131 - 103.546(2622) = 80{,}633$ so the residual $= 95{,}962 - 80{,}633 = 15{,}329$. The residual value means that the number of juvenile arrests for marijuana possession is 15,329 more than would be predicted by the number of honey-producing bee colonies.

    (c) The point $(\bar{x}, \bar{y})$ is on the regression line and $b_1 = r\dfrac{s_y}{s_x}$. So

    $\bar{y} = 352{,}131 - 103.546(2663.8) = 76{,}305$ and

    $s_y = \dfrac{(-103.546)(249.0)}{(-0.933)} = 27{,}634.$

(d) The standard error of the slope is given to be 9.383. With $df = 20 - 2 = 18$ and 95% confidence, the critical t-scores are ±2.101. The confidence interval is $-103.546 \pm 2.101(9.383) = -103.546 \pm 19.714$. We are 95% confident that for each additional 1000 honey-producing bee colonies, the mean number of juvenile arrests for marijuana possession will go down between approximately 83.8 and 123.3 arrests.

(e) With $df = 20 - 2 = 18$ and 90% confidence, the critical t-scores are ±1.734. The confidence interval is $352,131 \pm (1.734)(25,099) = 352,131 \pm 43,521.7$. We are 90% confident that if there were no honey-producing bee colonies, the mean number of juvenile arrests for marijuana possession would be between approximately 308,609.3 and 395,652.7 arrests.

(f) The intercept is so far from the given values of data that it is risky to assume that the linear association still applies. So the confidence interval may be meaningless.

SCORING

Parts (a) and (b) together are essentially correct for correctly calculating both the correlation and the residual, correctly noting that no causation is appropriate, and correctly interpreting the residual value. They are partially correct if two or three of these four calculations/statements are correct.

Part (c) is essentially correct for correctly calculating both the mean and the standard deviation (using the r from part (a)). It is partially correct if one of the two are correct.

Part (d) is essentially correct for a correct calculation of the confidence interval and a correct conclusion statement in context. It is partially correct for a single error such as an incorrect t-value or missing context in the conclusion.

Parts (e) and (f) together are essentially correct for a correct calculation of the confidence interval, a correct conclusion statement in context, and a correct statement about extrapolation. They are partially correct if two of these three calculations/statements are correct.

Count partially correct answers as one-half an essentially correct answer.

4	Complete Answer	Four essentially correct answers.
3	Substantial Answer	Three essentially correct answers.
2	Developing Answer	Two essentially correct answers.
1	Minimal Answer	One essentially correct answer.

Use a holistic approach to decide a score totaling between two numbers.

ANSWER SHEET
Practice Test 5

1. Ⓐ Ⓑ Ⓒ Ⓓ Ⓔ   11. Ⓐ Ⓑ Ⓒ Ⓓ Ⓔ   21. Ⓐ Ⓑ Ⓒ Ⓓ Ⓔ   31. Ⓐ Ⓑ Ⓒ Ⓓ Ⓔ
2. Ⓐ Ⓑ Ⓒ Ⓓ Ⓔ   12. Ⓐ Ⓑ Ⓒ Ⓓ Ⓔ   22. Ⓐ Ⓑ Ⓒ Ⓓ Ⓔ   32. Ⓐ Ⓑ Ⓒ Ⓓ Ⓔ
3. Ⓐ Ⓑ Ⓒ Ⓓ Ⓔ   13. Ⓐ Ⓑ Ⓒ Ⓓ Ⓔ   23. Ⓐ Ⓑ Ⓒ Ⓓ Ⓔ   33. Ⓐ Ⓑ Ⓒ Ⓓ Ⓔ
4. Ⓐ Ⓑ Ⓒ Ⓓ Ⓔ   14. Ⓐ Ⓑ Ⓒ Ⓓ Ⓔ   24. Ⓐ Ⓑ Ⓒ Ⓓ Ⓔ   34. Ⓐ Ⓑ Ⓒ Ⓓ Ⓔ
5. Ⓐ Ⓑ Ⓒ Ⓓ Ⓔ   15. Ⓐ Ⓑ Ⓒ Ⓓ Ⓔ   25. Ⓐ Ⓑ Ⓒ Ⓓ Ⓔ   35. Ⓐ Ⓑ Ⓒ Ⓓ Ⓔ
6. Ⓐ Ⓑ Ⓒ Ⓓ Ⓔ   16. Ⓐ Ⓑ Ⓒ Ⓓ Ⓔ   26. Ⓐ Ⓑ Ⓒ Ⓓ Ⓔ   36. Ⓐ Ⓑ Ⓒ Ⓓ Ⓔ
7. Ⓐ Ⓑ Ⓒ Ⓓ Ⓔ   17. Ⓐ Ⓑ Ⓒ Ⓓ Ⓔ   27. Ⓐ Ⓑ Ⓒ Ⓓ Ⓔ   37. Ⓐ Ⓑ Ⓒ Ⓓ Ⓔ
8. Ⓐ Ⓑ Ⓒ Ⓓ Ⓔ   18. Ⓐ Ⓑ Ⓒ Ⓓ Ⓔ   28. Ⓐ Ⓑ Ⓒ Ⓓ Ⓔ   38. Ⓐ Ⓑ Ⓒ Ⓓ Ⓔ
9. Ⓐ Ⓑ Ⓒ Ⓓ Ⓔ   19. Ⓐ Ⓑ Ⓒ Ⓓ Ⓔ   29. Ⓐ Ⓑ Ⓒ Ⓓ Ⓔ   39. Ⓐ Ⓑ Ⓒ Ⓓ Ⓔ
10. Ⓐ Ⓑ Ⓒ Ⓓ Ⓔ   20. Ⓐ Ⓑ Ⓒ Ⓓ Ⓔ   30. Ⓐ Ⓑ Ⓒ Ⓓ Ⓔ   40. Ⓐ Ⓑ Ⓒ Ⓓ Ⓔ

Practice Test 5

SECTION I

Questions 1–40

Spend 90 minutes on this part of the exam.

> **Directions:** The questions or incomplete statements that follow are each followed by five suggested answers or completions. Choose the response that best answers the question or completes the statement.

1. The mean and standard deviation of the population $\{1, 5, 8, 11, 15\}$ are $\mu = 8$ and $\sigma = 4.8$, respectively. Let S be the set of the 125 *ordered* triples (repeats allowed) of elements of the original population. Which of the following is a correct statement about the mean $\mu_{\bar{x}}$ and standard deviation $\sigma_{\bar{x}}$ of the means of the triples in S?

   (A) $\mu_{\bar{x}} = 8$, $\sigma_{\bar{x}} = 4.8$
   (B) $\mu_{\bar{x}} = 8$, $\sigma_{\bar{x}} < 4.8$
   (C) $\mu_{\bar{x}} = 8$, $\sigma_{\bar{x}} > 4.8$
   (D) $\mu_{\bar{x}} < 8$, $\sigma_{\bar{x}} = 4.8$
   (E) $\mu_{\bar{x}} > 8$, $\sigma_{\bar{x}} > 4.8$

2. Consider the following studies being run by three different AP Statistics instructors.

   I. One rewards students every day with lollipops for relaxation, encouragement, and motivation to learn the material.
   II. One promises that all students will receive A's as long as they give their best efforts to learn the material.
   III. One is available every day after school and on weekends so that students with questions can come in and learn the material.

   (A) None of these studies use randomization.
   (B) None of these studies use control groups.
   (C) None of these studies use blinding.
   (D) Important information can be found from all these studies, but none can establish causal relationships.
   (E) All of the above.

3. A survey to measure job satisfaction of high school mathematics teachers was taken in 2013 and repeated 5 years later in 2018. Each year a random sample of 50 teachers rated their job satisfaction on a 1-to-100 scale with higher numbers indicating greater satisfaction. The results are given in the following back-to-back stemplot.

2013		2018
	0	
98775	1	
98530	2	1
65210	3	3589
96430	4	01122233455667889999
87421	5	035667899
99877555322100	6	1344789
976442	7	22689
7511	8	138
0	9	7

Key: $0|2|1$ represents a rating of 20 in 2013 and a rating of 21 in 2018.

What is the trend from 2013 to 2018 with regard to the standard deviation and range of the two samples?

(A) Both the standard deviation and range increased.
(B) The standard deviation increased, while the range decreased.
(C) The range increased, while the standard deviation decreased.
(D) Both the standard deviation and range decreased.
(E) Both the standard deviation and range remained unchanged.

4. The number of days it takes to build a new house has a variance of 386. A sample of 40 new homes shows an average building time of 83 days. With what confidence can we assert that the average building time for a new house is between 80 and 90 days?

(A) 15.4%
(B) 17.8%
(C) 20.0%
(D) 38.8%
(E) 82.1%

5. A shipment of resistors have an average resistance of 200 ohms with a standard deviation of 5 ohms, and the resistances are roughly normally distributed. Suppose a randomly chosen resistor has a resistance under 194 ohms. What is the probability that its resistance is greater than 188 ohms?

(A) 0.07
(B) 0.12
(C) 0.50
(D) 0.93
(E) 0.97

6. Suppose 4% of the population have a certain disease. A laboratory blood test gives a positive reading for 95% of people who have the disease and for 5% of people who do not have the disease. If a person tests positive, what is the probability the person has the disease?

   (A) 0.038
   (B) 0.086
   (C) 0.442
   (D) 0.558
   (E) 0.950

7. For which of the following is it appropriate to use a census?

   (A) A 95% confidence interval of mean height of teachers in a small town.
   (B) A 95% confidence interval of the proportion of students in a small town who are taking some AP class.
   (C) A two-tailed hypothesis test where the null hypothesis was that the mean expenditure on entertainment by male students at a high school is the same as that of female students.
   (D) None of the above.
   (E) All of the above.

8. On the same test, Mary and Pam scored at the 64th and 56th percentiles, respectively. Which of the following is a true statement?

   (A) Mary scored eight more points than Pam.
   (B) Mary's score is 8% higher than Pam's.
   (C) Eight percent of those who took the test scored between Pam and Mary.
   (D) Thirty-six people scored higher than both Mary and Pam.
   (E) None of the above.

9. Which of the following is a true statement?

   (A) While observational studies gather information on an already existing condition, they still often involve intentionally forcing some treatment to note the response.
   (B) In an experiment, researchers decide on the treatment but typically allow the subjects to self-select into the control group.
   (C) If properly designed, either observational studies or controlled experiments can easily be used to establish cause and effect.
   (D) Wording to disguise hidden interests in observational studies is the same idea as blinding in experimental design.
   (E) Stratifying in sampling is the same idea as blocking for experiments.

10. The random variable describing the number of minutes high school students spend in front of a computer daily has a mean of 200 minutes. Samples of two different sizes result in sampling distributions with the two graphs below.

Which of the following is a true statement?

(A) Based on these graphs, no comparison between the two sample sizes is possible.

(B) More generally, sample sizes have no effect on sampling distributions.

(C) The sample size in A is the same as the sample size in B.

(D) The sample size in A is less than the sample size in B.

(E) The sample size in A is greater than the sample size in B.

11. To determine the mean cost of groceries in a certain city, an identical grocery basket of food is purchased at each store in a random sample of ten stores. If the average cost is $47.52 with a standard deviation of $1.59, find a 98% confidence interval estimate for the cost of these groceries in the city.

(A) $47.52 \pm 2.33\sqrt{1.59}$

(B) $47.52 \pm 2.33\left(\dfrac{1.59}{\sqrt{10}}\right)$

(C) $47.52 \pm 2.33\left(\sqrt{\dfrac{1.59}{10}}\right)$

(D) $47.52 \pm 2.821\left(\dfrac{1.59}{\sqrt{10}}\right)$

(E) $47.52 \pm 2.821\sqrt{\dfrac{1.59}{10}}$

12. A set consists of four numbers. The largest value is 200, and the range is 50. Which of the following statements is true?

(A) The mean is less than 185.

(B) The mean is greater than 165.

(C) The median is less than 195.

(D) The median is greater than 155.

(E) The median is the mean of the second and third numbers if the set is arranged in ascending order.

13. A telephone survey of 400 registered voters showed that 256 had not yet made up their minds 1 month before the election. How sure can we be that between 60% and 68% of the electorate were still undecided at that time?

   (A) 2.4%
   (B) 8.0%
   (C) 64.0%
   (D) 90.4%
   (E) 95.3%

14. Suppose we have a random variable X where the probability associated with the value k is $\binom{15}{k}(0.29)^k(0.71)^{15-k}$ for $k = 0, \ldots, 15$.

   What is the mean of X?

   (A) 0.29
   (B) 0.71
   (C) 4.35
   (D) 10.65
   (E) None of the above

15. The financial aid office at a state university conducts a study to determine the total student costs per semester. All students are charged $7500 for tuition. The mean cost for books is $350 with a standard deviation of $65. The mean outlay for room and board is $4800 with a standard deviation of $380. The mean personal expenditure is $1075 with a standard deviation of $125. Assuming independence among categories, what is the standard deviation of the total student costs?

   (A) $24
   (B) $91
   (C) $190
   (D) $405
   (E) $570

16. Suppose X and Y are random variables with $E(X) = 312$, $\text{var}(X) = 6$, $E(X) = 307$, and $\text{var}(Y) = 8$. What are the expected value and variance of the random variable $X + Y$?

   (A) $E(X + Y) = 619$, $\text{var}(X + Y) = 7$
   (B) $E(X + Y) = 619$, $\text{var}(X + Y) = 10$
   (C) $E(X + Y) = 619$, $\text{var}(X + Y) = 14$
   (D) $E(X + Y) = 309.5$, $\text{var}(X + Y) = 14$
   (E) There is insufficient information to answer this question.

17. In sample surveys, what is meant by *bias*?

    (A) A systematic error in a sampling method that leads to an unrepresentative sample.

    (B) Prejudice, as in ethnic and gender-related studies.

    (C) Natural variability seen between samples.

    (D) Tendency for some distributions to be skewed.

    (E) Tendency for some distributions to vary from normality.

18. The following histogram gives the shoe sizes of people in an elementary school building one morning.

Which of the following is a true statement?

    (A) The distribution of shoe sizes is bimodal.

    (B) The median shoe size is $7\frac{1}{2}$.

    (C) The mean shoe size is probably less than the median shoe size.

    (D) The five-number summary is: $5\frac{1}{2}, 6\frac{1}{2}, 7\frac{1}{2}, 8\frac{1}{2}, 9\frac{1}{2}$.

    (E) Only 10% of the people had size 5 shoes.

19. When comparing the standard normal (z) distribution to the t-distribution with $df = 30$, which of (A)–(D), if any, are false?

    (A) Both are symmetric.

    (B) Both are bell-shaped.

    (C) Both have center 0.

    (D) Both have standard deviation 1.

    (E) All the above are true statements.

20. Given a probability of 0.65 that interest rates will jump this year, and a probability of 0.72 that if interest rates jump the stock market will decline, what is the probability that interest rates will jump and the stock market will decline?

    (A) $0.72 + 0.65 - (0.72)(0.65)$

    (B) $(0.72)(0.65)$

    (C) $1 - (0.72)(0.65)$

    (D) $\dfrac{0.65}{0.72}$

    (E) $1 - \dfrac{0.65}{0.72}$

21. Sampling error is

   (A) the mean of a sample statistic.
   (B) the standard deviation of a sample statistic.
   (C) the standard error of a sample statistic.
   (D) the result of bias.
   (E) the difference between a population parameter and an estimate of that parameter.

22. Suppose that the weights of trucks traveling on the interstate highway system are roughly normally distributed. If 70% of the trucks weigh more than 12,000 pounds and 80% weigh more than 10,000 pounds, what are the mean and standard deviation for the weights of trucks traveling on the interstate system?

   (A) $\mu = 14{,}900$; $\sigma = 6100$
   (B) $\mu = 15{,}100$; $\sigma = 6200$
   (C) $\mu = 15{,}300$; $\sigma = 6300$
   (D) $\mu = 15{,}500$; $\sigma = 6400$
   (E) The mean and standard deviation cannot be computed from the information given.

23. If the correlation coefficient $r = 0.78$, what percentage of variation in y is explained by the linear regression model?

   (A) 22%
   (B) 39%
   (C) 44%
   (D) 61%
   (E) 78%

24. Consider the following scatterplot:

   Which of the following is the best estimate of the correlation between x and y?

   (A) −0.95
   (B) −0.15
   (C) 0
   (D) 0.15
   (E) 0.95

25. For one NBA playoff game the actual percentage of the television viewing public who watched the game was 24%. If you had taken a survey of 50 television viewers that night and constructed a confidence interval estimate of the percentage watching the game, which of the following would have been true?

    I. The center of the interval would have been 24%.
   II. The interval would have contained 24%.
  III. A 99% confidence interval estimate would have contained 24%.

  (A) I and II only
  (B) I and III only
  (C) II and III only
  (D) All are true.
  (E) None is true.

26. Which of the following is a true statement?

  (A) In a well-designed, well-conducted sample survey, sampling error is effectively eliminated.
  (B) In a well-designed observational study, responses are influenced through an orderly, carefully planned procedure during the collection of data.
  (C) In a well-designed experiment, the treatments are carefully planned to result in responses that are as similar as possible.
  (D) In a well-designed experiment, double-blinding is a useful matched pairs design.
  (E) None of the above is a true statement.

27. Consider the following scatterplot showing the relationship between caffeine intake and job performance.

Which of the following is a reasonable conclusion?

  (A) Low caffeine intake is associated with low job performance.
  (B) Low caffeine intake is associated with high job performance.
  (C) High caffeine intake is associated with low job performance.
  (D) High caffeine intake is associated with high job performance.
  (E) Job performance cannot be predicted from caffeine intake.

28. An author of a new book claims that anyone following his suggested diet program will lose an average of 2.8 pounds per week. A researcher believes that the true figure will be lower and plans a test involving a random sample of 36 overweight people. She will reject the author's claim if the mean weight loss in the volunteer group is less than 2.5 pounds per week. Assume that the standard deviation among individuals is 1.2 pounds per week. If the true mean value is 2.4 pounds per week, what is the probability that the researcher will mistakenly fail to reject the author's false claim of 2.8 pounds?

(A) $P\left(z > \dfrac{2.5 - 2.4}{1.2/\sqrt{36}}\right)$

(B) $P\left(z < \dfrac{2.5 - 2.4}{1.2/\sqrt{36}}\right)$

(C) $P\left(z < \dfrac{2.8 - 2.5}{1.2/\sqrt{36}}\right)$

(D) $P\left(z > \dfrac{2.8 - 2.4}{1.2/\sqrt{36}}\right)$

(E) $P\left(z < \dfrac{2.8 - 2.4}{1.2/\sqrt{36}}\right)$

29. Which of the following is the central limit theorem?

(A) No matter how the population is distributed, as the sample size increases, the mean of the sample means becomes closer to the mean of the population.

(B) No matter how the population is distributed, as the sample size increases, the standard deviation of the sample means becomes closer to the standard deviation of the population divided by the square root of the sample size.

(C) If the population is normally distributed, then as the sample size increases, the sampling distribution of the sample mean becomes closer to a normal distribution.

(D) All of the above together make up the central limit theorem.

(E) The central limit theorem refers to something else.

30. What is a sampling distribution?

(A) A distribution of all the statistics that can be found in a given sample

(B) A histogram, or other such visual representation, showing the distribution of a sample

(C) A normal distribution of some statistic

(D) A distribution of all the values taken by a statistic from all possible samples of a given size

(E) All of the above

31. A judge chosen at random reaches a just decision roughly 80% of the time. What is the probability that in randomly chosen cases at least two out of three judges reach a just decision?

(A) $3(0.8)^2(0.2)$
(B) $1 - 3(0.8)^2(0.2)$
(C) $(0.8)^3$
(D) $1 - (0.8)^3$
(E) $3(0.8)^2(0.2) + (0.8)^3$

32. Miles per gallon versus speed (miles per hour) for a new model automobile is fitted with a least squares regression line. Following is computer output of the statistical analysis of the data.

```
Dependent variable: Miles per gallon

Source          df        Sum of Squares   Mean Square   F-ratio
Regression      1             199.34         199.34        3.79
Residual        6             315.54           5.59

Variable    Coefficient      SE Coef        t-ratio         P
Constant      38.929          5.651           6.89        0.000
Speed        -0.2179          0.112          -1.95        0.099

R-Sq = 38.7%      R-Sq(adj) = 28.5%
s = 7.252 with 8 - 2 = 6 degrees of freedom
```

Which of the following gives a 99% confidence interval for the slope of the regression line of miles per gallon versus speed for the new model automobile?

(A) $-0.2179 \pm 3.707(0.112)$

(B) $-0.2179 \pm 3.143(0.112/\sqrt{8})$

(C) $-0.2179 \pm 3.707(0.112/\sqrt{8})$

(D) $38.929 \pm 3.143(5.651/\sqrt{8})$

(E) $38.929 \pm 3.707(5.651/\sqrt{8})$

33. What fault do all these sampling designs have in common?

    I. The Parent-Teacher Association (PTA), concerned about rising teenage pregnancy rates at a high school, randomly picks a sample of high school students and interviews them concerning unprotected sex they have engaged in during the past year.

    II. A radio talk show host asks people to phone in their views on whether the United States should keep troops in Afghanistan indefinitely to fight terrorist groups.

    III. The *Ladies Home Journal* plans to predict the winner of a national election based on a survey of its readers.

(A) All the designs make improper use of stratification.
(B) All the designs have errors that can lead to strong bias.
(C) All the designs confuse association with cause and effect.
(D) All the designs suffer from sampling error.
(E) None of the designs makes use of chance in selecting a sample.

34. Hospital administrators wish to determine the average length of stay for all surgical patients. A statistician determines that for a 95% confidence level estimate of the average length of stay to within ±0.50 days, 100 surgical patients' records would have to be examined. How many records should be looked at for a 95% confidence level estimate to within ±0.25 days?

(A) $\dfrac{100}{4}$

(B) $\dfrac{100}{2}$

(C) 2(100)

(D) 4(100)

(E) There is not enough information given to determine the necessary sample size.

35. A chess master wins 80% of her games, loses 5%, and draws the rest. If she receives 1 point for a win, $\frac{1}{2}$ point for a draw, and no points for a loss, what is true about the sampling distribution X of the points scored in two independent games?

(A) X takes on the values 0, 1, and 2 with respective probabilities 0.10, 0.26, and 0.64.

(B) X takes on the values 0, $\frac{1}{2}$, 1, 1$\frac{1}{2}$, and 2 with respective probabilities 0.0025, 0.015, 0.1025, 0.24, and 0.64.

(C) X takes on values according to a binomial distribution with $n = 2$ and $p = 0.8$.

(D) X takes on values according to a binomial distribution with mean

    $1(0.8) + \frac{1}{2}(0.15) + 0(0.05)$.

(E) X takes on values according to a distribution with mean (2)(0.8) and standard deviation $\sqrt{2(0.8)(0.2)}$.

36. Which of the following is a true statement?

   (A) The P-value is a conditional probability.
   (B) The P-value is usually chosen before an experiment is conducted.
   (C) The P-value is based on a specific test statistic and thus should not be used in a two-sided test.
   (D) P-values are more appropriately used with t-distributions than with z-distributions.
   (E) If the P-value is less than the level of significance, then the null hypothesis is proved false.

37. An assembly line machine is supposed to turn out ball bearings with a diameter of 1.25 centimeters. Each morning the first 30 bearings produced are pulled and measured. If their mean diameter is under 1.23 centimeters or over 1.27 centimeters, the machinery is stopped and an engineer is called to make adjustments before production is resumed. The quality control procedure may be viewed as a hypothesis test with the null hypothesis H_0: $\mu = 1.25$ and the alternative hypothesis H_a: $\mu \neq 1.25$. The engineer is asked to make adjustments when the null hypothesis is rejected. In test terminology, what would a Type II error result in?

   (A) A warranted halt in production to adjust the machinery
   (B) An unnecessary stoppage of the production process
   (C) Continued production of wrong size ball bearings
   (D) Continued production of proper size ball bearings
   (E) Continued production of ball bearings that randomly are the right or wrong size

38. Both over-the-counter niacin and the prescription drug Lipitor are known to lower blood cholesterol levels. In one double-blind study Lipitor outperformed niacin. The 95% confidence interval estimate of the difference in mean cholesterol level lowering was (18, 41). Which of the following is a reasonable conclusion?

   (A) Niacin lowers cholesterol an average of 18 points, while Lipitor lowers cholesterol an average of 41 points.
   (B) There is a 0.95 probability that Lipitor will outperform niacin in lowering the cholesterol level of any given individual.
   (C) There is a 0.95 probability that Lipitor will outperform niacin by at least 23 points in lowering the cholesterol level of any given individual.
   (D) We should be 95% confident that Lipitor will outperform niacin as a cholesterol-lowering drug.
   (E) None of the above.

39. The following parallel boxplots show the average daily hours of bright sunshine in Liberia, West Africa:

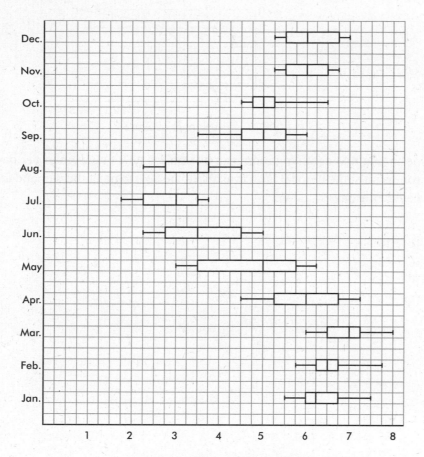

Average daily hours of bright sunshine

For how many months is the median below 4 hours?

(A) One
(B) Two
(C) Three
(D) Four
(E) Five

40. Following is a cumulative probability graph for the number of births per day in a city hospital.

Number of births

Assuming that a birthing room can be used by only one woman per day, how many rooms must the hospital have available to be able to meet the demand at least 90 percent of the days?

(A) 5
(B) 10
(C) 15
(D) 20
(E) 25

STOP

If there is still time remaining, you may review your answers.

SECTION II

Part A
QUESTIONS 1–5

Spend about 65 minutes on this part of the exam.
Percentage of Section II grade—75

> **Directions:** You must show all work and indicate the methods you use. You will be graded on the correctness of your methods and on the accuracy of your results and explanations.

1. An experiment is being planned to study urban land use practices aimed at reviving and sustaining native bird populations. Vegetation types A and B are to be compared, and eight test sites are available. After planting, volunteer skilled birdwatchers will collect data on the abundance of bird species making each of the two habitat types their home. The east side of the city borders a river, while the south side borders an industrial park.

    (a) Suppose the decision is made to block using the scheme below (one block is white, one gray). How would you use randomization, and what is the purpose of the randomization?

    (b) Comment on the strength and weakness of the above scheme as compared to the following blocking scheme (one block is white, one gray).

2. A world organization report gives the following percentages for primary-school-age children enrolled in school in the 17 countries of each of two geographic regions.

Region A: 36, 45, 52, 56, 56, 58, 60, 63, 65, 66, 69, 71, 72, 74, 77, 82, 92
Region B: 35, 37, 41, 43, 43, 48, 50, 54, 65, 71, 78, 82, 83, 87, 89, 91, 92

(a) Draw a back-to-back stemplot of this data.

(b) The report describes both regions as having the same median percentage (for primary-school-age children enrolled in school) among their 17 countries and approximately the same range. What about the distributions is missed by the report?

(c) If the organization has education funds to help only one region, give an argument for which should be helped.

(d) A researcher plans to run a two-sample t-test to study the difference in means between the percentages from each region. Comment on his plan.

3. Data were collected from a random sample of 100 student athletes at a large state university. The plot below shows grade point average (GPA) versus standard normal value (z-score) corresponding to the percentile of each GPA (when arranged in order). Also shown is the *normal line,* that is, a line passing through expected values for a normal distribution with the mean and SD of the given data.

(a) What is the shape of this distribution? Explain.

(b) What is the 95th percentile of the data? Explain.

(c) In a normal distribution with the mean and SD of this data, what is the 95th percentile? Explain.

4. Although blood type frequencies in the United States are in the ratios of 9:8:2:1 for types O, A, B, and AB, respectively, local differences are often found depending upon a variety of demographic characteristics. Two researchers are independently assigned to determine if patients at a particular large city general hospital exhibit blood types supporting the above model. The table below gives the data results from what each researcher claims to be random samples of 500 patient lab results.

	O	A	B	AB
Researcher 1	253	194	38	15
Researcher 2	230	196	48	26

(a) Do the data reported by Researcher 1 support the 9:8:2:1 model for blood types of patients at the particular hospital? Justify your answer.

(b) The editorial board of a medical publication rejects the findings of Researcher 2, claiming that his data are suspicious in that they are too good to be true. Give a statistical justification for the board's decision.

5. In a random sample of automobiles the highway mileage (in mpg) and the engine size (in liters) are measured, and the following computer output for regression is obtained:

```
mpg: Mean = 28.5882 and SD = 5.88493
Mean liter = 2.52353 and SD = 0.898978
Dependent variable is: mpg
R-sq = 49.8%   R-sq(adj) = 46.5%
Variable     Coef       s.e.      t         p
Constant     40.2462    3.197     12.6      0.0001
liter        −4.61969   1.198     −3.86     0.0015
s = 4.306
```

Assume all conditions for regression inference are met.

(a) One of the points on the regression line corresponds to an auto with an engine size of 1.5 liters and highway mileage of 35 miles per gallon. What is the residual of this point? Interpret this residual in context.

Suppose another auto with an engine size of 5.0 liters and highway mileage of 17 mpg is added to the data set.

(b) Explain whether the new slope will be greater than, less than, or about the same as the slope given by the output above.

(c) Explain whether the new correlation will be greater than, less than, or about the same in absolute value as the correlation given by the output above.

SECTION II

Part B

QUESTION 6

Spend about 25 minutes on this part of the exam.
Percentage of Section II grade—25

6. An advertising firm wants to estimate the proportion of residents in a county who have a favorable view of a particular product. Census results show that the county population is 55% city, 30% suburb, and 15% rural.

(a) What is the advantage of using a stratified sample in this context?

The firm decides to use a proportionate stratified random sample of size $n = 1200$. So they obtain random samples of sizes 660, 360, and 180, respectively, from the city, suburb, and rural regions.

The sample results are:

City: 66 out of 660 have a favorable view of the product.

Suburb: 144 out of 360 have a favorable view of the product.

Rural: 90 out of 180 have a favorable view of the product.

(b) Calculate the sample statistics $\hat{p}_{city}$, $\hat{p}_{suburb}$, and $\hat{p}_{rural}$ and the standard errors $SE(\hat{p}_{city})$, $SE(\hat{p}_{suburb})$, and $SE(\hat{p}_{rural})$.

As an estimator for the proportion of all county residents with a favorable view of the product, the firm is using the following sample statistic:

$$\hat{p}_{county} = 0.55\,\hat{p}_{city} + 0.30\,\hat{p}_{suburb} + 0.15\,\hat{p}_{rural}$$

(c) Calculate $\hat{p}_{county}$ and $SE(\hat{p}_{county})$.

(d) Suppose there had been a simple random sample of $n = 1200$ with $x = 66 + 144 + 90 = 300$ favorable views. What then would have been $\hat{p}$ and $SE(\hat{p})$? What advantage to using stratified sampling is shown?

STOP

If there is still time remaining, you may review your answers.

Answer Key

SECTION I

1. **B**	9. **E**	17. **A**	25. **E**	33. **B**
2. **E**	10. **E**	18. **C**	26. **E**	34. **D**
3. **C**	11. **D**	19. **D**	27. **E**	35. **B**
4. **E**	12. **E**	20. **B**	28. **A**	36. **A**
5. **D**	13. **D**	21. **E**	29. **E**	37. **C**
6. **C**	14. **C**	22. **C**	30. **D**	38. **E**
7. **D**	15. **D**	23. **D**	31. **E**	39. **C**
8. **C**	16. **E**	24. **B**	32. **A**	40. **D**

Answers Explained

SECTION I

1. **(B)** $\mu_{\bar{x}} = \mu = 8$, and $\sigma_{\bar{x}} = \frac{\sigma}{\sqrt{n}} = \frac{4.8}{\sqrt{3}} < 4.8$

2. **(E)** None of the studies have any controls such as randomization, a control group, or blinding, and so while they may give valuable information, they cannot establish cause and effect.

3. **(C)** The range increased from $90 - 15 = 75$ to $97 - 21 = 76$, while the standard deviation decreased (note how the values are bunched together more closely in 2018).

4. **(E)** $\sigma_{\bar{x}} = \frac{\sqrt{386}}{\sqrt{40}} = 3.106$. normalcdf(80, 90, 83, 3.106) = 0.8208.

5. **(D)** The critical z-scores are $\frac{188 - 200}{5} = -2.4$ and $\frac{194 - 200}{5} = -1.2$, with corresponding left tail probabilities of 0.0082 and 0.1151, respectively. The probability of being greater than 188 given that it is less than 194 is $\frac{0.1151 - 0.0082}{0.1151} = 0.93$. $\left(\text{Or } \frac{\text{normalcdf}(188,194,200,5)}{\text{normalcdf}(0,194,200,5)} = 0.92876\right)$

6. **(C)**

$$P(\text{pos test}) = P(\text{disease} \cap \text{pos}) + P(\text{healthy} \cap \text{pos})$$
$$= (0.04)(0.95) + (0.96)(0.05)$$
$$= 0.038 + 0.048 = 0.086$$
$$P(\text{disease} \mid \text{pos test}) = \frac{0.038}{0.086} = 0.442$$

7. **(D)** Given a census, the population parameter is known, and there is no need to use the techniques of inference.

8. **(C)** Sixty-four percent of the students scored below Mary and 56% scored below Pam, and so 8% must have scored between them.

9. **(E)** Intentionally forcing some treatment to note the response is associated with controlled experiments, not with observational studies. In experiments, the researchers decide how people are placed in different groups; self-selection is associated with observational studies. Results of observational studies may suggest cause-and-effect relationships; however, controlled studies are used to establish such relationships.

10. **(E)** With $\sigma_{\bar{x}} = \dfrac{\sigma}{\sqrt{n}}$, the greater the sample size n, the smaller the standard deviation $\sigma_{\bar{x}}$.

11. **(D)** $df = 9$, $\sigma_{\bar{x}} = \dfrac{s}{\sqrt{n}}$, and invT(.01, 9) = –2.821.

12. **(E)** The set could be {150, 150, 150, 200} with mean 162.5 and median 150. It might also be {150, 200, 200, 200} with mean 187.5 and median 200. The median of a set of four elements is the mean of the two middle elements.

13. **(D)** $\hat{p} = \dfrac{256}{400} = 0.64$, and $\sigma_{\hat{p}} = \sqrt{\dfrac{(0.64)(0.36)}{400}} = 0.024$. normalcdf(.6, .68, .64, .024) = 0.9044.

14. **(C)** This is a binomial with $n = 15$ and $p = 0.29$, and so the mean is $np = 15(0.29) = 4.35$.

15. **(D)** With independence, variances add, so $\sqrt{65^2 + 380^2 + 125^2} = 405$.

16. **(E)** Without independence we cannot determine var($X + Y$) from the information given.

17. **(A)** Bias is the tendency to favor the selection of certain members of a population. It has nothing to do with the shape of distributions. The natural variability between samples is called *sampling error.*

18. **(C)** There are not two distinct peaks, so the distribution is not bimodal. Relative frequency is given by relative area. The distribution is skewed to the left, and so the mean is probably less than the median.

19. **(D)** While the standard deviation of the z-distribution is 1, the standard deviation of the t-distribution is greater than 1.

20. **(B)**
$$P\left(\begin{array}{c}\text{market}\\\text{decline}\end{array} \cap \begin{array}{c}\text{interest}\\\text{jumps}\end{array}\right) = P\left(\begin{array}{c}\text{market}\\\text{decline}\end{array} \middle| \begin{array}{c}\text{interest}\\\text{jumps}\end{array}\right) P\left(\begin{array}{c}\text{interest}\\\text{jumps}\end{array}\right)$$
$$= (0.72)(0.65)$$

21. **(E)** Different samples give different sample statistics, all of which are estimates for the same population parameter, and so error, called *sampling error,* is naturally present.

22. **(C)** invNorm gives that the critical z-scores for 70% to the right and 80% to the right are –0.5244 and –0.8416, respectively. Then {$\mu - 0.5244\sigma = 12{,}000$, $\mu - 0.8416\sigma = 10{,}000$} gives $\mu = 15{,}306$ and $\sigma = 6305$.

23. **(D)** The percentage of the variation in y explained by the variation in x is given by the coefficient of determination r^2. In this example, $(0.78)^2 = 0.61$.

24. **(B)** There is a weak negative correlation, and so –0.15 is the only reasonable possibility among the choices given.

25. **(E)** There is no guarantee that 24 is anywhere near the interval, and so none of the statements is true.

26. **(E)** Sampling error relates to natural variation between samples, and it can never be eliminated. In good observational studies, responses are not influenced during the collection of data. In good experiments, treatments are compared as to the differences in responses.

27. **(E)** The scatterplot suggests a zero correlation.

28. **(A)** $\sigma_{\overline{x}} = \dfrac{1.2}{\sqrt{36}}$. The z-score of 2.5 is $\dfrac{2.5 - 2.4}{1.2 / \sqrt{36}}$, and to the right is the probability of failing to reject the false claim.

29. **(E)** The central limit theorem says that no matter how the original population is distributed, as the sample size increases, the sampling distribution of the sample mean becomes closer to a normal distribution.

30. **(D)** A sampling distribution is the distribution of all the values taken by a statistic, such as sample mean or sample proportion, from all possible samples of a given size.

31. **(E)** $B(n = 3, p = 0.8)$. $P(\text{at least } 2) = P(\text{exactly } 2) + P(\text{exactly } 3) = 3(0.8)^2(0.2) + (0.8)^3$

32. **(A)** The standard error of the slope is given to be 0.112. The critical t-scores for 99% confidence with $df = 6$ are ±3.707.

33. **(B)** The PTA survey has strong response bias in that students may not give truthful responses to a parent or teacher about their engaging in unprotected sex. The talk show survey results in a voluntary response sample, which typically gives too much emphasis to persons with strong opinions. The *Ladies Home Journal* survey has strong selection bias; that is, people who read the *Journal* are not representative of the general population.

34. **(D)** To divide the interval width by d, the sample size must be increased by a multiple of d^2.

35. **(B)** $P(0 \text{ pts}) = (0.05)^2$, $P(\frac{1}{2} \text{ pt}) = 2(0.05)(0.15)$, $P(1 \text{ pt}) = (0.15)^2 + 2(0.05)(0.8)$, $P(1\frac{1}{2} \text{ pts}) = 2(0.15)(0.8)$, and $P(2 \text{ pts}) = (0.8)^2$.

36. **(A)** The P-value is the probability of obtaining a result as extreme or more extreme as the one seen *given that* the null hypothesis is true; thus, it is a conditional probability. The P-value depends on the sample chosen. The P-value in a two-sided test is calculated by doubling the indicated tail probability. P-values are not restricted to use with any particular distribution. With a small P-value, there is evidence to reject the null hypothesis, but we're not *proving* anything.

37. **(C)** A Type II error is a mistaken failure to reject a false null hypothesis or, in this case, a failure to realize that the machinery is turning out wrong size ball bearings.

38. **(E)** Using a measurement from a sample, we are never able to say *exactly* what a population mean is; rather we always say we have a certain *confidence* that the population mean lies in a particular *interval*. In this case we are 95% confident the mean number of points by which Lipitor outperforms niacin is between 18 and 41.

39. **(C)** The median number of hours is less than four for June, July, and August.

40. **(D)** A horizontal line drawn at the 0.9 probability level corresponds to roughly 19 rooms.

SECTION II

1. (a) In each block, two of the sites will be randomly assigned to receive Type A vegetation, while the remaining two sites in the block will receive Type B vegetation. Randomization of vegetation type to the sites within each block should reduce bias due to any confounding variables. In particular, the randomization in blocks in the first scheme should even out the effect of the distance vegetation in blocks from the industrial park.

(b) The first scheme creates homogeneous blocks with respect to distance from the river, while the second scheme creates homogeneous blocks with respect to the industrial park. Randomization of vegetation types to sites within blocks in the first scheme should even out effects of distance from the industrial park, while randomization of vegetation types to sites within blocks in the second scheme should even out effects of distance from the river.

SCORING

It is important to explain why randomization is important within the context of this problem, so use of terms like bias and confounding must be in context. One must explain the importance of homogeneous experimental units (sites, not vegetation types) within blocks.

4 Complete Answer   Correct explanation both of use of blocking and of use of randomization in context of this problem.

3 Substantial Answer   Correct explanation of use of either blocking or randomization in context and a weak explanation of the other.

2 Developing Answer   Correct explanation of use of either blocking or randomization in context OR weak explanations of both.

1 Minimal Answer   Weak explanation of either blocking or randomization in context.

---

2. (a)

```
            Region A          Region B
                      6 │ 3 │ 5 7            Key:  6│3│5 means
                      5 │ 4 │ 1 3 3 8              36% in Region A
                8 6 6 2 │ 5 │ 0 4                  and 35% in
            9 6 5 3 0 │ 6 │ 5                      Region B
                7 4 2 1 │ 7 │ 1 8
                      2 │ 8 │ 2 3 7 9
                      2 │ 9 │ 1 2
```

(b) While the percentages in both regions have roughly symmetric distributions, the percentages from Region A form a distinctly unimodal pattern, while those from Region B are distinctly bimodal. That is, in Region B the countries tended to show either a very low or a very high percentage, while in Region A most countries showed a percentage near the middle one.

(c) Either an argument can be made for Region A because Region B has so many countries with high percentages (for primary-school-age children enrolled in school), or an argument can be made for Region B because Region B has so many countries with low percentages.

(d) Data are already given on *all* the countries in the two regions. In doing inference, one uses sample statistics to estimate population parameters. If the data are actually the whole population, there is no point of a *t*-test.

SCORING

Part (a) is essentially correct for a correct stemplot (numbers in each row do not have to be in order) with labeling as to which side refers to which region. Part (a) is partially correct for a correct stemplot missing the labeling.

Part (b) is essentially correct for noting that the percentages from Region A form a distinctly unimodal pattern, while those from Region B are distinctly bimodal. Part (b) is partially correct for noting either the unimodal pattern from the Region A data or the bimodal pattern from the Region B data.

Part (c) is essentially correct for choosing either region and giving a reasonable argument for the choice. Part (c) is partially correct if the argument is weak. Part (c) is incorrect for no argument or an incorrect argument for the choice given.

Part (d) is essentially correct for explaining that inference is not proper when the data are the whole population. Part (d) is partially correct if the explanation is correct but weak.

Count partially correct answers as one-half an essentially correct answer.

4	**Complete Answer**	Four essentially correct answers.
3	**Substantial Answer**	Three essentially correct answers.
2	**Developing Answer**	Two essentially correct answers.
1	**Minimal Answer**	One essentially correct answer.

Use a holistic approach to decide a score totaling between two numbers.

3. (a) Note that there are a great number of values between 2.0 and 2.4, fewer values between 2.4 and 2.8, on down to the fewest values being between 3.6 and 4.0. Thus, the distribution is skewed right (skewed toward the higher values).

(b) Note that the greatest five values, out of the 100 values, are above a point roughly at 3.4 or 3.5, so the 95th percentile is roughly 3.4 or 3.5.

(c) In a normal distribution, the 95th percentile has a z-score of 1.645. Reading up from the x-axis to the normal line and across to the y-axis gives a GPA of approximately 3.2.

SCORING

Part (a) is essentially correct if the shape is correctly identified as skewed right, and a correct explanation based on the given normal probability plot is given. Part (a) is partially correct for a correct identification with a weak explanation. Part (a) is incorrect if a correct identification is given with no explanation.

Part (b) is essentially correct for correctly noting 3.4 or 3.5 as the 95th percentile with a correct explanation. Part (b) is partially correct for giving 3.4 or 3.5 with a weak explanation or no explanation.

Part (c) is essentially correct for noting approximately 3.2 as the GPA and stating the method clearly. Part (c) is partially correct for giving 3.2 with a weak explanation or no explanation.

4	**Complete Answer**	All three parts essentially correct.
3	**Substantial Answer**	Two parts essentially correct and one part partially correct.
2	**Developing Answer**	Two parts essentially correct OR one part essentially correct and one or two parts partially correct OR all three parts partially correct.
1	**Minimal Answer**	One part essentially correct OR two parts partially correct.

---

4. (a) There are four elements to this solution.

State the hypotheses:

H_0: The distribution of blood types among patients at this hospital is in the ratios of 9:8:2:1 for types O, A, B, and AB, respectively.

H_a: The distribution of blood types among patients at this hospital is not in the ratios of 9:8:2:1 for types O, A, B, and AB, respectively.

Identify the test by name or formula and check the assumptions:

Chi-square goodness-of-fit test:

$$\chi^2 = \sum \frac{(\text{observed} - \text{expected})^2}{\text{expected}}$$

Check the assumptions:

1. The researcher claims that the data are from a random sample of patient records.

2. The ratios 9:8:2:1 give expected cell frequencies of $\frac{9}{20}(500) = 225$, $\frac{8}{20}(500) = 200$, $\frac{2}{20}(500) = 50$, $\frac{1}{20}(500) = 25$, each of which is at least 5.

Demonstrate correct mechanics:

Use χ^2GOF-Test or calculate:

$$\chi^2 = \frac{(253-225)^2}{225} + \frac{(194-200)^2}{200} + \frac{(38-50)^2}{50} + \frac{(15-25)^2}{25} = 10.544$$

With $df = 4 - 1 = 3$, the P-value is 0.0145.

State the conclusion in context with linkage to the P-value:

With this small a P-value, $0.0145 < 0.05$, there is sufficient evidence to reject H_0, that is, there is sufficient evidence that the distribution of blood types among patients at this hospital is *not* in the ratios of 9:8:2:1 for types O, A, B, and AB, respectively.

(b) In this case the mechanics give:

$$\chi^2 = \frac{(230-225)^2}{225} + \frac{(196-200)^2}{200} + \frac{(48-50)^2}{50} + \frac{(26-25)^2}{25} = 0.311$$

With $df = 4 - 1 = 3$, the probability that χ^2 is *less than or equal to* 0.311 is 0.0421. With such a small probability, there is evidence that the data are too good to be true; that is, there is evidence that Researcher 2 made up the data to fit the 9:8:2:1 model.

5. (a) The regression line is $\widehat{mpg} = 40.2462 - 4.61969 \, liter$.

   Then $40.2462 - 4.61969(1.5) \approx 33.32$.

   Residual = actual − predicted = 35 − 33.32 = 1.68 mpg

   The actual mpg for this auto is 1.68 mpg greater than what was predicted by the regression model.

   (b) The predicted value for mpg is $40.2462 - 4.61969(5.0) \approx 17.15$, which is very close to the actual value of 17. The added point is consistent with the linear pattern given by the computer output, so the new slope should be about the same as the old slope.

   (c) Since the new point fits the old linear pattern so well, and has an x-value (*liter*) much greater than $\bar{x} = 2.52$, the absolute value of the correlation will increase.

SCORING

Part (a) is essentially correct for a correct calculation of the residual, and a correct interpretation in context. Part (a) is partially correct for an error in calculating the residual but a correct interpretation using this error, or a correct residual but an error in interpretation.

Part (b) is essentially correct for showing that the actual value is very close to the predicted value and concluding that the new slope should be about the same. Part (b) is partially correct if the predicted value is miscalculated but a proper conclusion based on the miscalculation is reached.

Part (c) is essentially correct for concluding that the absolute value of the correlation will increase because the new point fits the old linear pattern so well. Part (c) is partially correct for concluding that the absolute value of the correlation will increase but giving a weak explanation.

4 Complete Answer — All three parts essentially correct.

3 Substantial Answer — Two parts essentially correct and one part partially correct.

2 Developing Answer — Two parts essentially correct OR one part essentially correct and one or two parts partially correct OR all three parts partially correct.

1 Minimal Answer — One part essentially correct OR two parts partially correct.

6. (a) Stratified sampling ensures that all strata (city, suburb, and rural regions) are represented in the final sample so that the true diversity of the population is reflected. In addition, stratified sampling results in data that can be used to make comparisons among the different strata of the population; that is, comparisons can be made among the proportions of residents in city, suburb, and rural regions that have favorable views of the product.

(b) $\hat{p}_{city} = \frac{66}{660} = 0.1$, $\hat{p}_{suburb} = \frac{144}{360} = 0.4$, and $\hat{p}_{rural} = \frac{90}{180} = 0.5$.

$$SE(\hat{p}_{city}) = \sqrt{\frac{(0.1)(0.9)}{660}} = 0.0117, \quad SE(\hat{p}_{suburb}) = \sqrt{\frac{(0.4)(0.6)}{360}} = 0.0258, \text{ and}$$

$$SE(\hat{p}_{rural}) = \sqrt{\frac{(0.5)(0.5)}{180}} = 0.0373.$$

(c) $\hat{p}_{county} = (0.55)(0.1) + (0.30)(0.4) + (0.15)(0.5) = 0.25$.

$$SE(\hat{p}_{county}) = \sqrt{(0.55\times0.0117)^2 + (0.30\times0.0258)^2 + (0.15\times0.0373)^2} = 0.0115.$$

(d) $\hat{p} = \frac{300}{1200} = 0.25$ and $SE(\hat{p}) = \sqrt{\frac{(0.25)(0.75)}{1200}} = 0.0125$. At least in this example, the sampling distribution of $\hat{p}$ has a smaller standard error with stratified sampling than with a simple random sample of the same size and same overall proportion.

SCORING

Part (a) is essentially correct for referring in context either to ensuring population representativeness or to making possible comparisons among population subgroups. It is partially correct to reference one of these advantages without mentioning context.

Part (b) is essentially correct for correct calculations of all three sample proportion statistics and all three standard errors of the sample proportions. It is partially correct for correct calculations of all three sample proportion statistics but not of the standard errors.

Part (c) is essentially correct for a correct calculation of both $\hat{p}_{county}$ and $SE(\hat{p}_{county})$. It is partially correct for just a correct calculation of $\hat{p}_{county}$.

Part (d) is essentially correct for correct calculations of $\hat{p}$ and $SE(\hat{p})$ and for a correct statement about a reduced standard error. It is partially correct for two of these three parts.

Count partially correct answers as one-half an essentially correct answer.

4   **Complete Answer**   Four essentially correct answers.

3   **Substantial Answer**   Three essentially correct answers.

2   **Developing Answer**   Two essentially correct answers.

1   **Minimal Answer**   One essentially correct answer.

Use a holistic approach to decide a score totaling between two numbers.

Relating Multiple-Choice Problems to Review Book Topics

Topic 1: Graphical Displays (pages 39–64)

Topic 2: Summarizing Distributions (pages 65–98)

Topic 3: Exploring Bivariate Data (pages 99–136)

Topic 4: Exploring Categorical Data: Two-Way Tables (pages 137–152)

Topic 5: Planning and Conducting Surveys (pages 153–170)

Topic 6: Planning and Conducting Experiments (pages 171–190)

Topic 7: Probability as Relative Frequency (pages 191–222)

Topic 8: Combining and Transforming Random Variables (pages 223–234)

Topic 9: The Normal Distribution (pages 235–256)

Topic 10: Sampling Distributions (pages 257–288)

Topic 11: Confidence Intervals (pages 289–336)

Topic 12: Tests of Significance—Proportions and Means (pages 337–372)

Topic 13: Tests of Significance—Chi-Square and Slope of Least Squares Line (pages 373–404)

The table below gives the above Topic corresponding to each MC question.

Question	Test 1	Test 2	Test 3	Test 4	Test 5
1	1, 2	3	6	3	10
2	5, 6	5, 6	2	5	6
3	9	9	6	12	2
4	11	3	11	9	11
5	6	2	6	12	7, 9
6	7	11	13	1, 2	7
7	12	12	3	6	11, 12
8	7, 10	5	8	6	2
9	7	7	5	3	6
10	7	3	9	5	10
11	12	1, 2	9	5, 12	11
12	6	2	5	1, 2	2
13	1, 2	8	7	5	11
14	2	7	12	10	7
15	12	11	2	7	8
16	2	5, 6	11	7, 9	8
17	9, 13	2	5	11	5
18	5	6	7	3	1, 2
19	2	12	9	13	9, 10
20	8	1, 2	2	7	7
21	3	7	10, 11	7	5
22	11	9	7	6	9
23	12	6	7	9	3
24	2	2	7	2	3
25	5	3	5	3	11
26	11	3	3	7	5, 6
27	3	7	5	11	3
28	7	5	1, 2	10	12
29	12	5	6	12	10
30	2	1, 2	11	11	10
31	7, 9	12	12	7	7
32	12	9	2	3	11
33	5	11	12	2	5
34	5	10	6	6	11
35	11	2	13	7	7
36	12	10	12	12	12
37	10	12	3	12	12
38	3	1, 2	7	2	11
39	11	11	2	6	2
40	11	13	11	2	1

Appendix

CHECKING ASSUMPTIONS FOR INFERENCE

Conditions for constructing confidence intervals and performing hypothesis tests

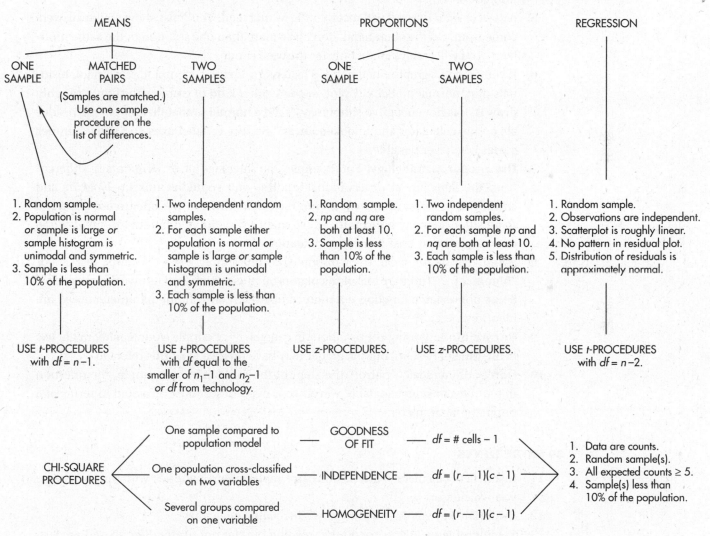

MEANS

ONE SAMPLE | MATCHED PAIRS | TWO SAMPLES

(Samples are matched.) Use one sample procedure on the list of differences.

1. Random sample.
2. Population is normal *or* sample is large *or* sample histogram is unimodal and symmetric.
3. Sample is less than 10% of the population.

USE *t*-PROCEDURES with $df = n-1$.

1. Two independent random samples.
2. For each sample either population is normal *or* sample is large *or* sample histogram is unimodal and symmetric.
3. Each sample is less than 10% of the population.

USE *t*-PROCEDURES with *df* equal to the smaller of n_1-1 and n_2-1 *or df* from technology.

PROPORTIONS

ONE SAMPLE | TWO SAMPLES

1. Random sample.
2. *np* and *nq* are both at least 10.
3. Sample is less than 10% of the population.

USE *z*-PROCEDURES.

1. Two independent random samples.
2. For each sample *np* and *nq* are both at least 10.
3. Each sample is less than 10% of the population.

USE *z*-PROCEDURES.

REGRESSION

1. Random sample.
2. Observations are independent.
3. Scatterplot is roughly linear.
4. No pattern in residual plot.
5. Distribution of residuals is approximately normal.

USE *t*-PROCEDURES with $df = n-2$.

CHI-SQUARE PROCEDURES

One sample compared to population model — GOODNESS OF FIT — $df = \text{\# cells} - 1$

One population cross-classified on two variables — INDEPENDENCE — $df = (r-1)(c-1)$

Several groups compared on one variable — HOMOGENEITY — $df = (r-1)(c-1)$

1. Data are counts.
2. Random sample(s).
3. All expected counts ≥ 5.
4. Sample(s) less than 10% of the population.

40 AP EXAM HINTS

THE TOP TEN

1. Graders want to give you credit—help them! Make them understand *what* you are doing, *why* you are doing it, and *how* you are doing it. Don't make the reader guess at what you are doing. **Communication** is just as important as statistical knowledge!

2. For both multiple-choice and free-response questions, read the question *carefully!* Be sure you understand **exactly what you are being asked to do or find or explain**.

3. *Check assumptions.* **Don't just state them!** Be sure the assumptions to be checked are stated correctly. Verifying assumptions and conditions means more than simply listing them with little check marks—you must show work or give some reason to confirm verification.

4. Learn and practice how to read *generic computer output.* And in answering questions using computer output, realize that you usually will **not** need to use all the information given.

5. *Naked* or *bald answers* will receive little or **no** credit! You must show where answers come from. On the other hand, don't give more than one solution to the same problem—you will receive credit only for the weaker one.

6. If you refer to a graph, whether it is a histogram, boxplot, stemplot, scatterplot, residuals plot, normal probability plot, or some other kind of graph, you should **roughly draw it**. It is not enough to simply say, "I did a normal probability plot of the residuals on my calculator and it looked linear." Be sure to **label axes** and **show number scales** whenever possible.

7. **Use proper terminology!** For example, the language of experiments is different from the language of observational studies—you shouldn't mix up *blocking* and *stratification.* Know what *confounding* means and when it is proper to use this term. *Replication* on many subjects to reduce chance variation is different from *replication* of an experiment itself to achieve validation.

8. **Avoid** "calculator speak"! For example, do not simply write "2-SampTTest …" or "binomcdf…" There are lots of calculators out there, each with its own abbreviations. Some abbreviated function notation can be referenced IF the parameters used are identified.

9. **Be careful** about using abbreviations in general. For example, your teacher might use LOBF (line of best fit), but the grader may have no idea what this refers to.

10. **Don't** automatically "parrot" the stem of the problem! For example, the question might refer to sample data; however, your hypotheses must be stated in terms of a population parameter.

30 MORE HINTS

11. Insert new calculator batteries or carry replacement batteries with you or charge your calculator.

12. Underline key words and phrases while reading questions.

13. Some problems look scary on first reading but are not overly difficult and are surprisingly straightforward if you approach them systematically. And some questions might take you beyond the scope of the AP curriculum; however, remember that they will be phrased in such ways that you should be able to answer them based on what you have learned in your AP Stat class.

14. Read through all six free-response questions, unerlining key points. Go back and do those questions you think are easiest (this will usually, but not always, include Problem 1). Then tackle the heavily weighted Problem 6 for a while. Finally, try the remaining problems. If you have a few minutes at the end to try some more on Problem 6, that's great!

15. Answers do not have to be in paragraph form. Symbols and algebra are fine. Just be sure that your method, reasoning, and calculations will be clear to the reader, and that the explanations and conclusions are given *in context of the problem*.

16. If you show calculations carefully, a wrong answer due to a computational error might still result in full credit.

17. If you can't solve part of a problem, but that solution is necessary to proceed, make up a reasonable answer to the first part and use this to proceed to the remaining parts of the problem.

18. When substituting values into a formula, you do not have to write down the actual formula.

19. Conclusions should be stated in proper English. For example, don't use double negatives.

20. Read carefully and recognize that sometimes very different tests are required in different parts of the same problem.

21. Realize that there may be several reasonable approaches to a given problem. In such cases pick either the one you feel most comfortable with or the one you feel will require the least time.

22. Realize that there may not be one clear, correct answer. Some questions are designed to give you an opportunity to creatively synthesize a relationship among the problem's statistical components.

23. Punching a long list of data into the calculator doesn't show statistical knowledge. Be sure it's necessary!

24. Remember that uniform and symmetric are not the same, and that not all symmetric unimodal distributions are normal.

25. When making predictions and interpreting slopes and intercepts, don't become confused between your original data and your regression model.

26. If the slope is close to ±1, it does *not* follow that the correlation is strong.

27. When describing residual graphs, "randomly scattered" does *not* mean the same as "half below and half above." You should comment on whether or not there are nonlinear patterns, and increasing or decreasing spread, and on whether the residuals are small or large compared to the associated y-values.

28. Simply using a calculator to find a regression line is not enough; you must understand it (for example, be able to interpret the slope and intercepts in context).

29. With regard to hypothesis test problems, (1) the hypotheses must be stated in terms of population parameters with all variables clearly defined; (2) the test must be specified by name or formula, and you must show how the test assumptions are met; (3) the test statistic and associated P-value should be calculated; (4) the P-value must be *linked* to a decision; and (5) a conclusion must be stated *in the context* of the problem.

30. For any inference problem, be sure to ask yourself whether the variable is categorical (usually leading to proportions) or quantitative (usually leading to means), whether you are working with raw data or summary statistics, whether there is a single popu-

lation of interest or two populations being compared, and in the case of comparison whether there are independent samples or a paired comparison.

31. A simple random sample (SRS) and a random assignment of treatments to subjects both have to do with randomness, but they are not the same. Understand the difference!

32. Simply saying to "randomly assign" subjects to treatment groups is usually an incomplete response. You need to explain how to make the assignments, for example, using a random number table or through generating random numbers on a calculator.

33. Blinding and placebos in experiments are important but are not always feasible. You can still have "experiments" without these.

34. The distribution of a particular sample (data set) is not the same as the sampling distribution of a statistic. Understand the difference!

35. Just because a sample is large does not imply that the distribution of the sample will be close to a normal distribution.

36. Distinguish between z, t, χ^2, b statistics. They are not interchangeable.

37. If asked why one thing is better than another, in general you should describe both an advantage of one and a disadvantage of the other.

38. Don't use technical terms in designing a survey or experiment unless specifically asked to do so. In general, clear, plain English is safer!

39. When comparing distributions, make sure to make actual comparisons and not a "laundry list."

40. Always give your answers *in context*.

AP SCORING GUIDE

The Multiple-Choice and Free-Response sections are weighted equally.

There is no penalty for guessing in the Multiple-Choice section.

The Investigative Task counts for 25% of the Free-Response section. Each of the Free-Response questions has a possible four points.

To find your score use the following guide:

Multiple-Choice section (40 questions)

Number correct $\times$ 1.25 = _____

Free-Response section (5 Open-Ended questions plus an Investigative Task)

Question 1 _____ $\times$ 1.875 = _____
out of 4

Question 2 _____ $\times$ 1.875 = _____
out of 4

Question 3 _____ $\times$ 1.875 = _____
out of 4

Question 4 _____ $\times$ 1.875 = _____
out of 4

Question 5 _____ $\times$ 1.875 = _____
out of 4

Question 6 _____ $\times$ 3.125 = _____
out of 4

Total points from Multiple-Choice and Free-Response sections = _____

Conversion chart based on a recent AP exam

Total points	AP Score
70–100	5
57–69	4
44–56	3
33–43	2
0–32	1

In the past, roughly 10% of students scored 5, 20% scored 4, 25% scored 3, 20% scored 2, and 25% scored 1 on the AP Statistics exam. Colleges generally require a score of at least 3 for a student to receive college credit.

BASIC USES OF THE TI-83/TI-84

There are many more useful features than introduced below—see the guidebook that comes with the calculator. For reference, following is a listing of the basic uses with which all students should be familiar. However, always remember that the calculator is only a tool, that it will find minimal use in the multiple-choice section, and that "calculator talk" (calculator syntax) should NOT be used in the free-response section.

Plotting statistical data:

> **STAT PLOT** allows one to show scatterplots, histograms, modified boxplots, and regular boxplots of data stored in lists. Note the use of **TRACE** with the various plots.

Numerical statistical data:

> **1-Var Stats** gives the mean, standard deviation, and 5-number summary of a list of data.

Binomial probabilities:

> **binompdf** (n, p, x) gives the probability of exactly x successes in n trials where p is the probability of success on a single trial.
>
> **binomcdf** (n, p, x) gives the cumulative probability of x or fewer successes in n trials where p is the probability of success on a single trial.

Geometric probabilities:

> **geometpdf** (p, x) gives the probability that the first success occurs on the x-th trial, where p is the probability of success on a single trial.
>
> **geometcdf** (p, x) gives the cumulative probability that the first success occurs on or before the x-th trial, where p is the probability of success on a single trial.

The normal distribution:

> **normalcdf** (lowerbound, upperbound, μ, σ) gives the probability that a score is between the two bounds for the designated mean μ and standard deviation σ. The defaults are $\mu = 0$ and $\sigma = 1$.
>
> **InvNorm** (area, μ, σ) gives the score associated with an area (probability) to the left of the score for the designated mean μ and standard deviation σ. The defaults are $\mu = 0$ and $\sigma = 1$.

The t-distribution:

> **tcdf** (lowerbound, upperbound, df) gives the probability a score is between the two bounds for the specified df (degrees of freedom).
>
> **invT**(area, df) gives the t-score associated with an area (probability) to the left of the score under the student t-probability function for the specified df (degrees of freedom). [Note: this is available on the new operating system for the TI-84+.]

The chi-square distribution:

> χ^2**cdf** (lowerbound, upperbound, df) gives the probability a score is between the two bounds for the specified df (degrees of freedom).

Linear regression and correlation:

> **LinReg (ax + b)** fits the equation $y = ax + b$ to the data in lists L1 and L2 using a least-squares fit. When **DiagnosticOn** is set, the values for r^2 and r are also displayed.

Confidence intervals:

For proportions—

1-PropZInt gives a confidence interval for a proportion of successes.

2-PropZInt gives a confidence interval for the difference between the proportion of successes in two populations.

For means—

TInterval gives a confidence interval for a population mean (use the t-distribution because population variances are never really known).

2-SampTInt gives a confidence interval for the difference between two population means.

Hypothesis tests:

For proportions—

1-PropZTest

2-PropZTest compares the proportion of successes from two populations (making use of the pooled sample proportion).

For means—

T-Test

2-SampTTest

For chi-square test for association—

χ^2**-Test** gives the χ^2-value and P-value for the null hypothesis H_0: no association between row and column variables, and the alternative hypothesis H_a: the variables are related. The observed counts must first be entered into a matrix.

χ^2**GOF-Test** is a chi-square goodness-of-fit test to confirm whether sample data conforms to a specified distribution. [Note: this is available on the new operating system for the TI-84+.]

For linear regression—

LinRegTTest calculates a linear regression and performs a t-test on the null hypothesis H_0: $\beta = 0$ (H_0: $\rho = 0$). The regression equation is stored in **RegEQ** (under **VARS Statistics EQ**) and the list of residuals is stored in **RESID** (under **LIST NAMES**).

Catalog help:

To activate Catalog Help, press APPS, choose CtlgHelp, and press ENTER. Then, for example, if you press 2nd, DISTR, arrow down to normalcdf, and press +, you are prompted to insert (lowerbound, upperbound, [μ, σ]), that is, to insert the bounds and, optionally, the mean and SD.

A 90-day free emulator trial of the TI-84 calculator is available at:

http://education.ti.com/go/84smartview/v4

BASIC USES OF THE CASIO PRIZM

Casio calculators have a module completely devoted to Statistics. You can find this module on the Main Menu and can access it with number 2 on the key pad.

Plotting statistical data:

> **GRAPH** allows one to show bar charts, pie charts, scatterplots, histograms, and boxplots of data in any of 26 stored lists. Note the use of **TRACE** with the various plots. With quantitative graphs, you can select **1-VAR** and access summary statistics. For scatterplots, selecting **DefG** will calculate predictions and **CALC** will find all manner of regressions.

Numerical statistical data:

> **CALC 1-VAR** gives the mean, standard deviation, mode, and 5-number summary of a list of data.

Probability Distributions are accessed in the DIST menu. Here are some examples.

Binomial probabilities:

> **Bpd** gives the probability of exactly x successes in n (Numtrial) where p is the probability of success on a single trial.

> **Bcd** gives the cumulative probability of receiving between a Lower and an Upper number of successes, where p is the probability of success on a single trial.

Geometric probabilities:

> **Gpd** gives the probability that the first success occurs on the x-th trial, where p is the probability of success on a single trial.

> **Gcd** gives the cumulative probability of receiving your first success between a Lower and an Upper number of successes, where p is the probability of success on a single trial.

The normal distribution:

> **Ncd** gives the probability that a score is between a Lower value and an Upper value with a given σ and μ. When using z-scores, the defaults of $\sigma = 1$ and $\mu = 0$ should be used.

> **InvN** gives the requested area that is either on the **Left** of the normal curve, on the **Right** of the normal curve, or in the **Central** area. When using z-scores, the defaults of $\sigma = 1$ and $\mu = 0$ should be used.

The t-distribution:

> **tcdf** gives the probability a score is between the Lower and Upper bounds for the specified df (degrees of freedom).

> **Invt** gives the t-score associated with an area (probability) to the left of the score under the student t-probability function for the specified df (degrees of freedom).

The chi-square distribution:

> **Ccd** gives the probability a score is between the Lower and Upper bounds for the specified df (degrees of freedom).

Linear regression and correlation:

From the ⟨CALC⟩ menu, use the ⟨ X ⟩ menu to fit the least-squares equation $\hat{y} = a + bx$ for data in any two lists. This command can also be accessed when viewing a scatterplot. Use the ⟨SETUP⟩ command to choose where the residuals are stored.

Confidence Intervals are available in the ⟨INTR⟩ menu:

For proportions—

Z-1-PROP gives a confidence interval for a proportion of successes.

Z-2-PROP gives a confidence interval for the difference between the proportion of successes in two populations.

For means—

t-1-SAMPLE gives a confidence interval for a population mean.

t-2-SAMPLE gives a confidence interval for the difference between two population means.

Hypothesis Tests are available in the ⟨TEST⟩ menu:

For proportions—

Z-1-PROP is for a one-sample proportion test.

Z-2-PROP compares the proportion of successes from two populations (making use of the pooled sample proportion).

For means—

t-1-SAMPLE is for a one-sample mean test.

t-2-SAMPLE compares the difference between two population means.

For chi-square—

CHI-GOF is a chi square goodness-of-fit test to confirm whether sample data conform to a specified distribution.

CHI-2WAY gives the χ^2-value and P-value for the null hypothesis H_o: no association between row and column variables, and the alternative hypothesis H_a: the variables are related. Data are entered in a matrix by pressing **F2** in the ⟨2WAY⟩ menu.

Test or Interval for the slope of a regression equation:

t-REG in the ⟨INTR⟩ menu will calculate a confidence interval for slope.

t-REG in the ⟨TEST⟩ calculates a linear regression and performs a t-test on the null hypothesis H_0: $\beta = 0$ (H_0: $\rho = 0$). The regression equation can be saved into any **Y** variable of your choice with the **COPY** command, and the residuals will be placed in the **List** selected in the ⟨SETUP⟩ menu.

A 90-day free emulator trial of the Casio calculator is available at:

https://edu.casio.com/freetrial/en/freetrial_form.php

BASIC USES OF THE HP PRIME

The statistical apps:

Press ⬛APPS INFO and tap to open the app you want (Statistics 1Var, Statistics 2Var, or Inference). Press ⬛NUM SETUP for Numeric view, ⬛SYMB SETUP for Symbolic view, and ⬛PLOT SETUP for Plot view (or press ⬛VIEW COPY and select **Autoscale**).

Statistics 1Var App	
Numeric view	• ⬛STATS Enter data in list D1 or enter data in D1 and frequencies in D2 • Tap to calculate summary statistics (5-number summary and so on)
Symbolic view	• Set analysis H1 to use D1 as data or D1 as data and D2 as frequencies • Select a plot type (Histogram, Box and Whisker, Normal Probability, Dot, Stem and Leaf, Pie, and so on)
Plot view	• Press ⬛VIEW COPY and select **Autoscale** • View the graphical representation of your data; tap to move the cursor, pinch to zoom, drag to pan

Statistics 2Var App	
Numeric view	• Enter independent data in list C1 and dependent data in C2 • Tap ⬛STATS to calculate summary statistics (r, R^2, and so on)
Symbolic view	• Set analysis S1 to use C1 as independent data and C2 as dependent data • Select a fit type (Linear, Exponential, Logarithmic, and so on)
Plot view	• Press ⬛VIEW COPY and select **Autoscale** • View the scatter plot and fit; pinch to zoom, drag to pan • Tap ⬛FIT to toggle the fit on and off; the fit equation is saved to Symbolic view

Inference App	
Symbolic view	• Select a method (Hypothesis Test, Confidence Interval, χ^2 Test, Regression) • Select a type; options vary by Method; for example, if the Method is χ^2 Test, then the types are Goodness-of-fit and 2-way test • Select an alternative hypothesis for hypothesis tests or linear t-test
Numeric view	• Enter your sample statistics or import them from the Statistics 1Var or Statistics 2Var apps • Tap ⬛CALC to see the computed results in a table
Plot view	• View a graphical representation of your test results • For the linear t-test, use the U and D direction keys to cycle through the scatterplot, the residual plot, a histogram of the residuals, and the normal probability plot of the residuals

The probability density functions:

Press ⟨MEM B⟩ to open the Toolbox menu. Tap ⟨MATH⟩ to open the numerical Math menu. The 5th option is **Probability**. Under this option are Density, Cumulative Density, and Inverse Cumulative Density categories for the Normal, Student's T, Chi-Square, Binomial, Geometric, and other probability density functions.

NORMAL_CDF(μ, σ, x) returns the lower-tail probability of the normal probability density function for the value x, given the mean (μ) and standard deviation (σ) of a normal distribution. NORMAL_CDF(μ, σ, x_1, x_2) returns the area under the normal probability density function between the two x-values.

NORMALD_ICDF(μ, σ, p) returns the cumulative normal distribution x-value associated with the lower-tail probability p, given the mean μ, and standard deviation σ, of a normal distribution.

If μ and σ are omitted for the three commands above, results for the standard normal distribution will be returned.

STUDENT_CDF(n, x) returns the lower-tail probability of the Student's t probability density function at x, given n degrees of freedom. STUDENT_CDF(n, x_1, x_2) returns the area under the Student's t probability density function between the two x-values.

STUDENT_ICDF(n, p) returns the value x such that the Student's t lower-tail probability of x, with n degrees of freedom, is p.

The commands CHISQUARE_CDF and CHISQUARE_ICDF work in a manner similar to the above.

BINOMIAL(n, p, k) returns the probability of k successes out of n trials, each with a probability of success p.

BINOMIAL_CDF(n, p, k) returns the probability of k or fewer successes out of n trials, each with a probability of success p. BINOMIAL_CDF(n, p, k_1, k_2) returns the probability of at least k_1 but no more than k_2 successes out of n trials, each with a probability of success p.

BINOMIAL_ICDF(n, p, q) returns the number of successes, k, out of n trials, each with a probability of p, such that the probability of k or fewer successes is q.

The commands GEOMETRIC_CDF and GEOMTRIC_ICDF work in a manner similar to the above.

Help

Help on these and other apps, commands, and functions is always available. Press ⟨MEM B⟩ to open the Toolbox menus, navigate to any command and press ⟨HELP USER⟩ to see help for that command. Tap ⟨EXAMPL⟩ and choose an example to paste it to the command line; press ⟨ENTER⟩ to see the result.

A 90-day free emulator trial of the HP Prime calculator is available at:

www.hp.com/go/primeapptrial

BASIC USES OF THE TI-NSPIRE

The TI-Nspire has additional statistical features, more memory, and greater resolution than the TI-84+.

Entering data:

Add a **Lists & Spreadsheet** page. *Name* the list, and enter the data in the spreadsheet column. Data can then be graphed and/or used to create confidence intervals or run hypothesis tests.

Plotting statistical data:

Add a **Data & Statistics** page. At the bottom center of the screen, **Click to add variable** and select the list *name* to graph. The default graph is a dotplot. Use **Menu → Plot Type** to change to a Box Plot, Histogram, or Normal Probability Plot. If graphing bivariate data, click to add a variable and select second *name* of list to graph. Use Menu → Analyze → Regression → Show Linear(a+bx) to plot the regression line over the graph.

Probability functions:

All probability functions can be found in either of two places: under

Menu → Statistics → Distributions or Menu → Probability → Distributions.

Normal distribution:

Normal Cdf (Lower Bound, Upper Bound, μ, σ) gives the probability that a score is between the two bounds for the designated mean and standard deviation.

Inverse Normal(Area, μ, σ) gives the value associated with an area (probability) to the left of the score for the designated mean and standard deviation.

***t*-distribution:**

t Cdf (Lower Bound, Upper Bound, df) gives the probability a value is between the two bounds for the specified degrees of freedom.

Inverse t (Area, df) gives the *t*-score associated with an area (probability) to the left of the score for the designated degrees of freedom.

Chi-square distribution:

χ^2 **Cdf** (Lower Bound, Upper Bound, df) gives the probability a score is between the two bounds for the specified degrees of freedom.

Binomial probabilities:

Binomial Pdf (*n*, *p*, *x*) gives the probability of exactly *x* successes in *n* trials where *p* is the probability of success on a single trial.

Binomial Cdf (*n*, *p*, Lower Bound, Upper Bound) gives the cumulative probability of getting between a Lower and an Upper number of successes in *n* trials where *p* is the probability of success on a single trial.

Confidence Intervals:

This can be accessed from either a Lists & Spreadsheet page (if using data) or a Calculator page (if using summary statistics or proportions).

Menu → Statistics → Confidence Intervals → desired interval

For proportions:

1-Prop z Interval (*x, n, C-level*) *x* has to be an integer value.

2-Prop z Interval (x1, n1, x2, n2, C-level) *x*1 and *x*2 have to be integer values.

For means:

t Interval can use data (list *name* from List & Spreadsheet) or summary statistics ($\bar{x}$, s, n) along with the level of confidence.

2-Sample t Interval can use data (list *names* from List & Spreadsheet) or two sample means, sample standard deviations, and sample sizes, along with the level of confidence.

For linear regression:

Linear Reg t Intervals (x list, y list, C-level) gives a confidence interval for a slope.

Hypothesis tests:

This can be accessed from either a Lists & Spreadsheet page (if using data) or a Calculator page (if using summary statistics or proportions).

Menu → Statistics → Stat Tests → desired test

For proportions:

1-Prop z Test(Po, *x*, *n*, alt. hyp) *x* has to be an integer value.

2-Prop z Test(*x1, n1, x2, n2*, alt. hyp) *x*1 and *x*2 have to be integer values.

For means:

t Test can use data or summary statistics.

2-Sample t Test can use data or summary statistics.

Chi-squared tests:

χ^2 **GOF Test(observed list, expected list, df)** gives the χ^2 statistic and the *P*-value.
χ^2 **2-way Test(name of matrix)** for tests of association or homogeneity. It runs the hypothesis test and finds the test statistic and *p*-value. The observed counts must first be entered into a matrix.

Linear regression:

Linear Reg t Test(x list, y list, alt hyp) calculates the regression line and performs a *t*-test on the null hypothesis H_0: $\beta = 0$. Output gives both *s* and SE Slope, and it gives a list of residuals.

A 30-day free emulator trial of the TI-NSpire calculator is available at:

https://education.ti.com/en/us/software/details/en/
4DCD444963B04332A10DCF8CC455EF43/ti-nspire_pc_trial

FORMULAS

The following are in the form and notation as recommended by the College Board.

Descriptive Statistics

$$\bar{x} = \frac{\sum x_i}{n}$$

$$s = \sqrt{\frac{1}{n-1}\sum(x_i - \bar{x})^2}$$

$$s_p = \sqrt{\frac{(n_1 - 1)s_1^2 + (n_2 - 1)s_2^2}{(n_1 - 1) + (n_2 - 1)}}$$

$$\hat{y} = b_0 + b_1 x$$

$$b_1 = \frac{\sum(x_i - \bar{x})(y_i - \bar{y})}{\sum(x_i - \bar{x})^2}$$

$$b_0 = \bar{y} - b_1\bar{x}$$

$$r = \frac{1}{n-1} = \sum\left(\frac{x_i - \bar{x}}{s_x}\right)\left(\frac{y_i - \bar{y}}{s_y}\right)$$

$$b_1 = r\frac{s_y}{s_x}$$

$$s_{b1} = \frac{\sqrt{\dfrac{\sum(y_i - \hat{y}_i)^2}{n-2}}}{\sqrt{\sum(x_i - \bar{x})^2}}$$

Probability

$$P(A \cup B) = P(A) + P(B) - P(A \cap B)$$

$$P(A|B) = \frac{P(A \cap B)}{P(B)}$$

$$E(X) = \mu_x = \sum x_i p_i$$

$$\mathrm{var}(X) = \sigma_x^2 = \sum(x_i - \mu_x)^2 p_i$$

If X has a binomial distribution with parameters n and p, then:

$$P = (X = k) = \binom{n}{k}p^k(1-p)^{(n-k)}$$

$$\mu_x = np$$

$$\sigma_x = \sqrt{np(1-p)}$$

$$\mu_{\hat{p}} = p$$

$$\sigma_{\hat{p}} = \sqrt{\frac{p(1-p)}{n}}$$

If X has a normal distribution with mean μ and standard deviation σ, then:

$$\mu_{\bar{x}} = \mu$$

$$\sigma_{\bar{x}} = \frac{\sigma}{\sqrt{n}}$$

INFERENTIAL STATISTICS

$$\text{Standardized statistic: } \frac{\text{Estimate} - \text{parameter}}{\text{Standard deviation of the estimate}}$$

Confidence interval:

Estimate ± (critical value) · (standard deviation of the estimate)

SINGLE SAMPLE

Statistic	Standard Deviation
Mean	$\dfrac{\sigma}{\sqrt{n}}$
Proportion	$\sqrt{\dfrac{p(1-p)}{n}}$

TWO SAMPLE

Statistic	Standard Deviation
Difference of means (unequal variances)	$\sqrt{\dfrac{\sigma_1^2}{n_1} + \dfrac{\sigma_2^2}{n_2}}$
Difference of means (equal variances)	$\sigma \sqrt{\dfrac{1}{n_1} + \dfrac{1}{n_2}}$
Difference of proportions (unequal variances)	$\sqrt{\dfrac{p_1(1-p_1)}{n_1} + \dfrac{p_2(1-p_2)}{n_2}}$
Difference of proportions (equal variances)	$\sqrt{p(1-p)} \sqrt{\dfrac{1}{n_1} + \dfrac{1}{n_2}}$

$$\text{Chi-square statistic} = \sum \frac{(\text{observed} - \text{expected})^2}{\text{expected}}$$

GRAPHICAL DISPLAYS

	Right-skewed	**Symmetric**	**Left-skewed**

Dotplots

Stemplots

Right-skewed	Symmetric	Left-skewed
3 | 0000000002222222244444466666688888	3 | 0246688	3 | 02468
4 | 000022244466688	4 | 000222444466666888888	4 | 00224466888
5 | 002244668	5 | 00000022222244444666688	5 | 000222444666688888
6 | 02468	6 | 000224468	6 | 000002222224444444666668888
7 | 0	7 | 0	7 | 0000

Histograms

Cumulative Frequency Plots

Boxplots

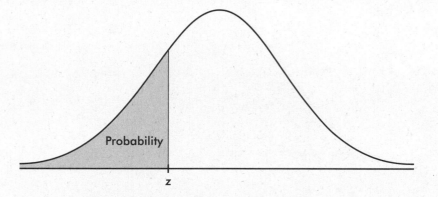

Probability

z

TABLE A: STANDARD NORMAL PROBABILITIES

z	.00	.01	.02	.03	.04	.05	.06	.07	.08	.09
−3.4	.0003	.0003	.0003	.0003	.0003	.0003	.0003	.0003	.0003	.0002
−3.3	.0005	.0005	.0005	.0004	.0004	.0004	.0004	.0004	.0004	.0003
−3.2	.0007	.0007	.0006	.0006	.0006	.0006	.0006	.0005	.0005	.0005
−3.1	.0010	.0009	.0009	.0009	.0008	.0008	.0008	.0008	.0007	.0007
−3.0	.0013	.0013	.0013	.0012	.0012	.0011	.0011	.0011	.0010	.0010
−2.9	.0019	.0018	.0018	.0017	.0016	.0016	.0015	.0015	.0014	.0014
−2.8	.0026	.0025	.0024	.0023	.0023	.0022	.0021	.0021	.0020	.0019
−2.7	.0035	.0034	.0033	.0032	.0031	.0030	.0029	.0028	.0027	.0026
−2.6	.0047	.0045	.0044	.0043	.0041	.0040	.0039	.0038	.0037	.0036
−2.5	.0062	.0060	.0059	.0057	.0055	.0054	.0052	.0051	.0049	.0048
−2.4	.0082	.0080	.0078	.0075	.0073	.0071	.0069	.0068	.0066	.0064
−2.3	.0107	.0104	.0102	.0099	.0096	.0094	.0091	.0089	.0087	.0084
−2.2	.0139	.0136	.0132	.0129	.0125	.0122	.0119	.0116	.0113	.0110
−2.1	.0179	.0174	.0170	.0166	.0162	.0158	.0154	.0150	.0146	.0143
−2.0	.0228	.0222	.0217	.0212	.0207	.0202	.0197	.0192	.0188	.0183
−1.9	.0287	.0281	.0274	.0268	.0262	.0256	.0250	.0244	.0239	.0233
−1.8	.0359	.0351	.0344	.0336	.0329	.0322	.0314	.0307	.0301	.0294
−1.7	.0446	.0436	.0427	.0418	.0409	.0401	.0392	.0384	.0375	.0367
−1.6	.0548	.0537	.0526	.0516	.0505	.0495	.0485	.0475	.0465	.0455
−1.5	.0668	.0655	.0643	.0630	.0618	.0606	.0594	.0582	.0571	.0559
−1.4	.0808	.0793	.0778	.0764	.0749	.0735	.0721	.0708	.0694	.0681
−1.3	.0968	.0951	.0934	.0918	.0901	.0885	.0869	.0853	.0838	.0823
−1.2	.1151	.1131	.1112	.1093	.1075	.1056	.1038	.1020	.1003	.0985
−1.1	.1357	.1335	.1314	.1292	.1271	.1251	.1230	.1210	.1190	.1170
−1.0	.1587	.1562	.1539	.1515	.1492	.1469	.1446	.1423	.1401	.1379
−0.9	.1841	.1814	.1788	.1762	.1736	.1711	.1685	.1660	.1635	.1611
−0.8	.2119	.2090	.2061	.2033	.2005	.1977	.1949	.1922	.1894	.1867
−0.7	.2420	.2389	.2358	.2327	.2296	.2266	.2236	.2206	.2177	.2148
−0.6	.2743	.2709	.2676	.2643	.2611	.2578	.2546	.2514	.2483	.2451
−0.5	.3085	.3050	.3015	.2981	.2946	.2912	.2877	.2843	.2810	.2776
−0.4	.3446	.3409	.3372	.3336	.3300	.3264	.3228	.3192	.3156	.3121
−0.3	.3821	.3783	.3745	.3707	.3669	.3632	.3594	.3557	.3520	.3483
−0.2	.4207	.4168	.4129	.4090	.4052	.4013	.3974	.3936	.3897	.3859
−0.1	.4602	.4562	.4522	.4483	.4443	.4404	.4364	.4325	.4286	.4247
−0.0	.5000	.4960	.4920	.4880	.4840	.4801	.4761	.4721	.4681	.4641

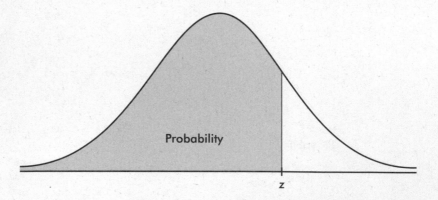

Probability

z

TABLE A: STANDARD NORMAL PROBABILITIES (CONTINUED)

z	.00	.01	.02	.03	.04	.05	.06	.07	.08	.09
0.0	.5000	.5040	.5080	.5120	.5160	.5199	.5239	.5279	.5319	.5359
0.1	.5398	.5438	.5478	.5517	.5557	.5596	.5636	.5675	.5714	.5753
0.2	.5793	.5832	.5871	.5910	.5948	.5987	.6026	.6064	.6103	.6141
0.3	.6179	.6217	.6255	.6293	.6331	.6368	.6406	.6643	.6480	.6517
0.4	.6554	.6591	.6628	.6664	.6700	.6736	.6772	.6808	.6844	.6879
0.5	.6915	.6950	.6985	.7019	.7054	.7088	.7123	.7157	.7190	.7224
0.6	.7257	.7291	.7324	.7357	.7389	.7422	.7454	.7486	.7517	.7549
0.7	.7580	.7611	.7642	.7673	.7704	.7734	.7764	.7794	.7823	.7852
0.8	.7881	.7910	.7939	.7967	.7995	.8023	.8051	.8078	.8106	.8133
0.9	.8159	.8186	.8212	.8238	.8264	.8289	.8315	.8340	.8365	.8389
1.0	.8413	.8438	.8461	.8485	.8508	.8531	.8554	.8577	.8599	.8621
1.1	.8643	.8665	.8686	.8708	.8729	.8749	.8770	.8790	.8810	.8830
1.2	.8849	.8869	.8888	.8907	.8925	.8944	.8962	.8980	.8997	.9015
1.3	.9032	.9049	.9066	.9082	.9099	.9115	.9131	.9147	.9162	.9177
1.4	.9192	.9207	.9222	.9236	.9251	.9265	.9279	.9292	.9306	.9319
1.5	.9332	.9345	.9357	.9370	.9382	.9394	.9406	.9418	.9429	.9441
1.6	.9452	.9463	.9474	.9484	.9495	.9505	.9515	.9525	.9535	.9545
1.7	.9554	.9564	.9573	.9582	.9591	.9599	.9608	.9616	.9625	.9633
1.8	.9641	.9649	.9656	.9664	.9671	.9678	.9686	.9693	.9699	.9706
1.9	.9713	.9719	.9726	.9732	.9738	.9744	.9750	.9756	.9761	.9767
2.0	.9772	.9778	.9783	.9788	.9793	.9798	.9803	.9808	.9812	.9817
2.1	.9821	.9826	.9830	.9834	.9838	.9842	.9846	.9850	.9854	.9857
2.2	.9861	.9864	.9868	.9871	.9875	.9878	.9881	.9884	.9887	.9890
2.3	.9893	.9896	.9898	.9901	.9904	.9906	.9909	.9911	.9913	.9916
2.4	.9918	.9920	.9922	.9925	.9927	.9929	.9931	.9932	.9934	.9936
2.5	.9938	.9940	.9941	.9943	.9945	.9946	.9948	.9949	.9951	.9952
2.6	.9953	.9955	.9956	.9957	.9959	.9960	.9961	.9962	.9963	.9964
2.7	.9965	.9966	.9967	.9968	.9969	.9970	.9971	.9972	.9973	.9974
2.8	.9974	.9975	.9976	.9977	.9977	.9978	.9979	.9979	.9980	.9981
2.9	.9981	.9982	.9982	.9983	.9984	.9984	.9985	.9985	.9986	.9986
3.0	.9987	.9987	.9987	.9988	.9988	.9989	.9989	.9989	.9990	.9990
3.1	.9990	.9991	.9991	.9991	.9992	.9992	.9992	.9992	.9993	.9993
3.2	.9993	.9993	.9994	.9994	.9994	.9994	.9994	.9995	.9995	.9995
3.3	.9995	.9995	.9995	.9996	.9996	.9996	.9996	.9996	.9996	.9997
3.4	.9997	.9997	.9997	.9997	.9997	.9997	.9997	.9997	.9997	.9998

Probability

t

TABLE B: *t*-DISTRIBUTION CRITICAL VALUES

						Tail probability *p*						
df	.25	.20	.15	.10	.05	.025	.02	.01	.005	.0025	.001	.0005
1	1.000	1.376	1.963	3.078	6.314	12.71	15.89	31.82	63.66	127.3	318.3	636.6
2	.816	1.061	1.386	1.886	2.920	4.303	4.849	6.965	9.925	14.09	22.33	31.60
3	.765	.978	1.250	1.638	2.353	3.182	3.482	4.541	5.841	7.453	10.21	12.92
4	.741	.941	1.190	1.533	2.132	2.776	2.999	3.747	4.604	5.598	7.173	8.610
5	.727	.920	1.156	1.476	2.015	2.571	2.757	3.365	4.032	4.773	5.893	6.869
6	.718	.906	1.134	1.440	1.943	2.447	2.612	3.143	3.707	4.317	5.208	5.959
7	.711	.896	1.119	1.415	1.895	2.365	2.517	2.998	3.499	4.029	4.785	5.408
8	.706	.889	1.108	1.397	1.860	2.306	2.449	2.896	3.355	3.833	4.501	5.041
9	.703	.883	1.100	1.383	1.833	2.262	2.398	2.821	3.250	3.690	4.297	4.781
10	.700	.879	1.093	1.372	1.812	2.228	2.359	2.764	3.169	3.581	4.144	4.587
11	.697	.876	1.088	1.363	1.796	2.201	2.328	2.718	3.106	3.497	4.025	4.437
12	.695	.873	1.083	1.356	1.782	2.179	2.303	2.681	3.055	3.428	3.930	4.318
13	.694	.870	1.079	1.350	1.771	2.160	2.282	2.650	3.012	3.372	3.852	4.221
14	.692	.868	1.076	1.345	1.761	2.145	2.264	2.624	2.977	3.326	3.787	4.140
15	.691	.866	1.074	1.341	1.753	2.131	2.249	2.602	2.947	3.286	3.733	4.073
16	.690	.865	1.071	1.337	1.746	2.120	2.235	2.583	2.921	3.252	3.686	4.015
17	.689	.863	1.069	1.333	1.740	2.110	2.224	2.567	2.898	3.222	3.646	3.965
18	.688	.862	1.067	1.330	1.734	2.101	2.214	2.552	2.878	3.197	3.611	3.922
19	.688	.861	1.066	1.328	1.729	2.093	2.205	2.539	2.861	3.174	3.579	3.883
20	.687	.860	1.064	1.325	1.725	2.086	2.197	2.528	2.845	3.153	3.552	3.850
21	.686	.859	1.063	1.323	1.721	2.080	2.189	2.518	2.831	3.135	3.527	3.819
22	.686	.858	1.061	1.321	1.717	2.074	2.183	2.508	2.819	3.119	3.505	3.792
23	.685	.858	1.060	1.319	1.714	2.069	2.177	2.500	2.807	3.104	3.485	3.768
24	.685	.857	1.059	1.318	1.711	2.064	2.172	2.492	2.797	3.091	3.467	3.745
25	.684	.856	1.058	1.316	1.708	2.060	2.167	2.485	2.787	3.078	3.450	3.725
26	.684	.856	1.058	1.315	1.706	2.056	2.162	2.479	2.779	3.067	3.435	3.707
27	.684	.855	1.057	1.314	1.703	2.052	2.158	2.473	2.771	3.057	3.421	3.690
28	.683	.855	1.056	1.313	1.701	2.048	2.154	2.467	2.763	3.047	3.408	3.674
29	.683	.854	1.055	1.311	1.699	2.045	2.150	2.462	2.756	3.038	3.396	3.659
30	.683	.854	1.055	1.310	1.697	2.042	2.147	2.457	2.750	3.030	3.385	3.646
40	.681	.851	1.050	1.303	1.684	2.021	2.123	2.423	2.704	2.971	3.307	3.551
50	.679	.849	1.047	1.299	1.676	2.009	2.109	2.403	2.678	2.937	3.261	3.496
60	.679	.848	1.045	1.296	1.671	2.000	2.099	2.390	2.660	2.915	3.232	3.460
80	.678	.846	1.043	1.292	1.664	1.990	2.088	2.374	2.639	2.887	3.195	3.416
100	.677	.845	1.042	1.290	1.660	1.984	2.081	2.364	2.626	2.871	3.174	3.390
1000	.675	.842	1.037	1.282	1.646	1.962	2.056	2.330	2.581	2.813	3.098	3.300
∞	.674	.841	1.036	1.282	1.645	1.960	2.054	2.326	2.576	2.807	3.091	3.291
	50%	60%	70%	80%	90%	95%	96%	98%	99%	99.5%	99.8%	99.9%

Confidence level *C*

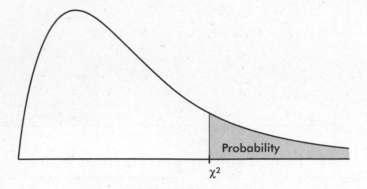

Probability

χ^2

TABLE C: χ^2 CRITICAL VALUES

df	\multicolumn{11}{c}{Tail probability p}
	.25	.20	.15	.10	.05	.025	.02	.01	.005	.0025	.001
1	1.32	1.64	2.07	2.71	3.84	5.02	5.41	6.63	7.88	9.14	10.83
2	2.77	3.22	3.79	4.61	5.99	7.38	7.82	9.21	10.60	11.98	13.82
3	4.11	4.64	5.32	6.25	7.81	9.35	9.84	11.34	12.84	14.32	16.27
4	5.39	5.99	6.74	7.78	9.49	11.14	11.67	13.28	14.86	16.42	18.47
5	6.63	7.29	8.12	9.24	11.07	12.83	13.39	15.09	16.75	18.39	20.51
6	7.84	8.56	9.45	10.64	12.59	14.45	15.03	16.81	18.55	20.25	22.46
7	9.04	9.80	10.75	12.02	14.07	16.01	16.62	18.48	20.28	22.04	24.32
8	10.22	11.03	12.03	13.36	15.51	17.53	18.17	20.09	21.95	23.77	26.12
9	11.39	12.24	13.29	14.68	16.92	19.02	19.68	21.67	23.59	25.46	27.88
10	12.55	13.44	14.53	15.99	18.31	20.48	21.16	23.21	25.19	27.11	29.59
11	13.70	14.63	15.77	17.28	19.68	21.92	22.62	24.72	26.76	28.73	31.26
12	14.85	15.81	16.99	18.55	21.03	23.34	24.05	26.22	28.30	30.32	32.91
13	15.98	16.98	18.20	19.81	22.36	24.74	25.47	27.69	29.82	31.88	34.53
14	17.12	18.15	19.41	21.06	23.68	26.12	26.87	29.14	31.32	33.43	36.12
15	18.25	19.31	20.60	22.31	25.00	27.49	28.26	30.58	32.80	34.95	37.70
16	19.37	20.47	21.79	23.54	26.30	28.85	29.63	32.00	34.27	36.46	39.25
17	20.49	21.61	22.98	24.77	27.59	30.19	31.00	33.41	35.72	37.95	40.79
18	21.60	22.76	24.16	25.99	28.87	31.53	32.35	34.81	37.16	39.42	42.31
19	22.72	23.90	25.33	27.20	30.14	32.85	33.69	36.19	38.58	40.88	43.82
20	23.83	25.04	26.50	28.41	31.41	34.17	35.02	37.57	40.00	42.34	45.31
21	24.93	26.17	27.66	29.62	32.67	35.48	36.34	38.93	41.40	43.78	46.80
22	26.04	27.30	28.82	30.81	33.92	36.78	37.66	40.29	42.80	45.20	48.27
23	27.14	28.43	29.98	32.01	35.17	38.08	38.97	41.64	44.18	46.62	49.73
24	28.24	29.55	31.13	33.20	36.42	39.36	40.27	42.98	45.56	48.03	51.18
25	29.34	30.68	32.28	34.38	37.65	40.65	41.57	44.31	46.93	49.44	52.62
26	30.43	31.79	33.43	35.56	38.89	41.92	42.86	45.64	48.29	50.83	54.05
27	31.53	32.91	34.57	36.74	40.11	43.19	44.14	46.96	49.64	52.22	55.48
28	32.62	34.03	35.71	37.92	41.34	44.46	45.42	48.28	50.99	53.59	56.89
29	33.71	35.14	36.85	39.09	42.56	45.72	46.69	49.59	52.34	54.97	58.30
30	34.80	36.25	37.99	40.26	43.77	46.98	47.96	50.89	53.67	56.33	59.70
40	45.62	47.27	49.24	51.81	55.76	59.34	60.44	63.69	66.77	69.70	73.40
50	56.33	58.16	60.35	63.17	67.50	71.42	72.61	76.15	79.49	82.66	86.66
60	66.98	68.97	71.34	74.40	79.08	83.30	84.58	88.38	91.95	95.34	99.61
80	88.13	90.41	93.11	96.58	101.9	106.6	108.1	112.3	116.3	120.1	124.8
100	109.1	111.7	114.7	118.5	124.3	129.6	131.1	135.8	140.2	144.3	149.4

Index